Progress in Nonlinear Differential Equations and Their Applications
Volume 63

Editor

Haim Brezis
Université Pierre et Marie Curie
Paris
and
Rutgers University
New Brunswick, N.J.

Editorial Board

Elliptic and Parabolic Problems

A Special Tribute to the Work of Haim Brezis

Catherine Bandle
Henri Berestycki
Bernhard Brighi
Alain Brillard
Michel Chipot
Jean-Michel Coron
Carlo Sbordone
Itai Shafrir
Vanda Valente
Giorgio Vergara Caffarelli
Editors

Birkhäuser
Basel · Boston · Berlin

Editors:

Catherine Bandle
Mathematisches Institut
Universität Basel
Rheinsprung 21
4051 Basel, Switzerland
bandle@math.unibas.ch

Henri Berestycki
Ecole des hautes études
en sciences sociales (EHESS)
CAMS
54, Boulevard Raspail
75006 Paris, France
Henri.Berestycki@ehess.fr

Bernard Brighi
Université de Haute-Alsace
Faculté des Sciences et Techniques
4 rue des frères Lumières
68093 Mulhouse Cedex, France
b.brighi@uha.fr

Alain Brillard
Université de Haute-Alsace
Laboratoire de Gestion
des Risques et Environnement
25, rue de Chemnitz
68200 Mulhouse, France
A.Brillard@uha.fr

Michel Chipot
Universität Zürich
Angewandte Mathematik
Winterthurerstr. 190
8057 Zürich, Switzerland
chipot@amath.unizh.ch

Jean-Michel Coron
Département de mathématiques
Bâtiment 425
Université de Paris-Sud
91405 Orsay, France
Jean-Michel.Coron@math.u-psud.fr

Carlo Sbordone
Dipartimento di Matematica e Applicazioni
Università di Napoli "Federico II"
Via Cintia
80126 Napoli, Italy
sbordone@unina.it

Itai Shafrir
Department of Mathematics
Technion – Israel Institute of Technology
32000 Haifa, Israel
shafrir@math.technion.ac.il

Vanda Valente
CNR-IAC
Viale del Policlinico, 137
00161 Roma, Italy
valente@iac.rm.cnr.it

Giorgio Vergara Caffarelli
Dipartimento di metodi e modelli
matematici per le Scienze Aplicate
Università di Roma "La Sapienza"
Via A. Scarpa 16
00161 Roma, Italy
vergara@dmmm.uniroma1.it

2000 Mathematics Subject Classification 35Bxx, 35Jxx, 35Kxx

A CIP catalogue record for this book is available from the Library of Congress,
Washington D.C., USA

Bibliographic information published by Die Deutsche Bibliothek
Die Deutsche Bibliothek lists this publication in the Deutsche Nationalbibliografie;
detailed bibliographic data is available in the Internet at <http://dnb.ddb.de>.

ISBN 3-7643-7249-4 Birkhäuser Verlag, Basel – Boston – Berlin

© 2005 Birkhäuser Verlag, P.O. Box 133, CH-4010 Basel, Switzerland
Part of Springer Science+Business Media
Printed on acid-free paper produced of chlorine-free pulp. TCF ∞
Printed in Germany
ISBN 10: 3-7643-7249-4
ISBN 13: 978-3-7643-7249-1

9 8 7 6 5 4 3 2 1   www.birkhauser.ch

Contents

Preface

The goal of these proceedings and of the meeting of Gaeta was to celebrate and honor the mathematical achievements of Haim Brezis. The prodigious influence of his talent and his personality in the domain of nonlinear analysis is unanimously acclaimed! This impact is visible in the huge number of his former students (dozens), students of former students (hundreds) and collaborators (hundreds). Thus the Gaeta meeting was, to some extent, the family reunion of part of this large community sharing a joint interest in the field of elliptic and parabolic equations and pushing it to a very high standard.

Italy has a long tradition and taste for analysis and we could not find a better place neither a more complete support for the realisation of our project. We have to thank here the university of Cassino, Napoli, Roma "la Sapienza", the GNAMPA-Istituto di Alta Matematica, CNR-IAC, MEMOMAT, RTN Fronts-Singularities, the commune of Gaeta. Additional founding came from the universities of Mulhouse and Zürich. Finally, we are grateful to Birkhäuser and Dr. Hempfling who allowed us to record the talks of this conference in a prestigious volume.

The organizers

Progress in Nonlinear Differential Equations
and Their Applications, Vol. 63, 1–12

One-Layer Free Boundary Problems with Two Free Boundaries

Andrew Acker

Abstract. We study the uniqueness and successive approximation of solutions of a class of two-dimensional steady-state fluid problems involving infinite periodic flows between two periodic free boundaries, each characterized by a flow-speed condition related to Bernoulli's law.

1. Introduction

We study a class of generic double-free-boundary problems involving ideal fluid-flows in two-dimensional, periodic, strip-like flow-domains. The periodic flow boundaries are both free, and each is characterized by a condition expressing the boundary flow-speed as a given function of position. Such conditions often arise from application of Bernoulli's law. We study the questions of existence and uniqueness of solutions of the above double-free-boundary problems, as well as the successive approximation of their solutions by a trial free boundary method called the Operator Method.

The existence, uniqueness, and convergence questions were all first studied in the context of the corresponding one-free-boundary problem, in which one boundary component is specified. The one-free-boundary existence results essentially follow (in 2 dimensions) from a theorem of Beurling [9], and the corresponding uniqueness results follow from the Lavrentiev Principle (called the Lindelof Principle by M.A. Lavrentiev [12]). The analytical trial-free boundary convergence proof for this case was obtained by the author in [3, 4]. The author has also previously studied some aspects of the double-free-boundary problem, including existence (see [1, 2]). He has more recently generalized the Operator Method to various related free boundary problems (see [5, 6], for example). Other generalizations and modified versions have been studied by Meyer [13], Acker and Meyer [8], Kadakal [11], and Acker, Kadakal and Miller [7]. The operator method has been implemented numerically with some striking results in [7, 11]. The present convergence study is the first one to apply to completely-free flows, referring to flow configurations in which no flow surface is specified.

We will study the periodic two-free-boundary problem in the following very general parametric form:

Problem 1. For given values $\sigma, N > 0$, let $a^\pm(p) = a^\pm(x,y)\colon \Re^2 \to \Re_+$ denote given strictly-positive continuous functions such that

$$a^\pm(x+\sigma, y) = a^\pm(x,y) \tag{1}$$

for all $(x,y) \in \Re^2$. Also assume

$$\pm(a^\pm(x,y) - a^\pm(x_0, y_0)) \geq 0 \text{ whenever } y - y_0 \geq N|x - x_0| \text{ in } \Re^2, \tag{2}$$

$$\lim_{y\to\pm\infty} a^\pm(x,y) = \infty; \ \lim_{y\to\mp\infty} a^\pm(x,y) = 0 \tag{3}$$

for any $x \in \Re$. Let $\phi(\lambda) : \Re_+ \to \Re_+$ denote a given positive, continuously-differentiable function such that $\phi'(\lambda) < 0$. For any $\lambda > 0$, we seek a pair of σ-periodic Lipschitz-continuous functions $f_\lambda^\pm(x)\colon \Re \to \Re$ such that $f_\lambda^- < f_\lambda^+$, and such that

$$|\nabla U_\lambda(p_0)| := \lim_{p\to p_0} |\nabla U_\lambda(p)| = \lambda a^-(p_0) \text{ for } p_0 \in f_\lambda^-, \tag{4}$$

$$|\nabla U_\lambda(p_0)| := \lim_{p\to p_0} |\nabla U_\lambda(p)| = \phi(\lambda) a^+(p_0) \text{ for } p_0 \in f_\lambda^+. \tag{5}$$

Here, $U_\lambda(p)\colon \mathrm{Cl}(\Omega_\lambda) \to \Re$ denotes the unique solution of the boundary value problem:

$$\Delta U_\lambda = 0 \text{ in } \Omega_\lambda, U_\lambda = \pm 1 \text{ on } f_\lambda^\pm, \ U_\lambda(x+\sigma, y) = U_\lambda(x,y), \tag{6}$$

where $\Omega_\lambda = \{p = (x,y) \in \Re^2 : f_\lambda^-(x) < y < f_\lambda^+(x)\}$.

Remarks. The author has shown in [2] that Problem 1 has at least one solution-pair $(f_\lambda^-, f_\lambda^+)$ for each $\lambda \in \Re_+$ (see [2; Theorem 3.1]). We will show that for any fixed $\lambda \in \Re_+$, the solution is unique under general conditions on the functions $a^\pm(p)$ (see Theorem 2, Part (a)). We define a trial free boundary method for the successive approximation of the solution of Problem 1 at $\lambda = \phi(\lambda) = 1$, based on a generalization of the Operator Method of [3, 4] to the two-free-boundary case (see Section 2). Our main purpose is to give an analytical proof of the global convergence of the operator iterates to the solution (see Theorem 3).

Remark. Problem 1 has a hard-barrier version, in which the flow is subjected to certain geometric constraints. This situation can be treated as a limiting case of the Problem 1 in which $a^\pm(x,y) = 1$ for $\pm(y - f_0^\pm(x)) < 0$ and $a^\pm(x,y) = \infty$ for $\pm(y - f_0^\pm(x)) > 0$, where the functions $f_0^- < f_0^+$ are given (see [1] and [10; Chapt. 6]).

2. Notation, definitions, preliminary results

Generalized distance. The Euclidean distance between two sets $\Sigma_1, \Sigma_2 \subset \Re^2$ is defined by $d(\Sigma_1, \Sigma_2) := \inf\{|p - q| : p \in \Sigma_1, q \in \Sigma_2\}$. In terms of the given functions $a^{\pm}(p) : \Re^2 \to \Re_+$, we define $d_{\pm}(\Sigma_1, \Sigma_2) := \inf\{||\gamma||_{\pm} : \gamma$ is a rectifiable arc joining Σ_1 to $\Sigma_2\}$, where, in terms of the arc-length parameter s, we define $||\gamma||_{\pm} = \int_{\gamma} a^{\pm}(p)ds$. For a point $p_0 \in \Re^2$ and set $\Sigma \subset \Re^2$, we define $d(p_0, \Sigma) = d(\{p_0\}, \Sigma\}$ and $d_{\pm}(p_0, \Sigma) = d_{\pm}(\{p_0\}, \Sigma)$.

Vector-space notation. Let X denote the family of all σ-periodic continuous functions $f(x) : \Re \to \Re$. We write $f = \{(x, f(x)) : x \in \Re\}$ for any $f \in X$, thus conforming to the standard practice of identifying the function f with its graph. Then X is a vector space, where, for any $f, g \in X$ and $\alpha, \beta \in \Re$, we define $\alpha f + \beta g = \{(x, \alpha f(x) + \beta g(x)) : x \in \Re\}$. Moreover, X is a Banach space in the norm: $M(f) = \max\{|f(x)| : x \in \Re\}$. In X, we define $f < (\leq)g$ if $f(x) < (\leq)g(x)$ for all $x \in \Re$. For $f \in X$ and $\alpha \in \Re$, we define $f + \alpha = \{(x, f(x) + \alpha) : x \in \Re\} \in X$. In X, we write $f_n \to f$ as $n \to \infty$ if $\lim_{n \to \infty} M(f_n - f) = 0$. For given $N \in \Re_+$, we let $X(N)$ denote the family of all functions $f \in X$ such that $|f(x) - f(x')| \leq N|x - x'|$ for all $x, x' \in \Re$. Let X^2 denote the family of all ordered pairs $\mathbf{f} := (f^-, f^+)$ such that $f^{\pm} \in X$. Then X^2 is a vector space under component-wise addition and scaler multiplication, and a Banach space in the norm $M(\mathbf{f}) = \max\{M(f^-), M(f^+)\}$. In X^2, we write $\mathbf{f} < (\leq)\mathbf{g} \Leftrightarrow f_i < (\leq)g_i$ for $i = 1, 2$, and $\mathbf{f}_n \to \mathbf{f}$ as $n \to \infty$ if $f_n^{\pm} \to f^{\pm}$ as $n \to \infty$. We also let $\mathbf{Y}$ (resp. $\mathbf{Y}(N)$) denote the family of all ordered pairs $\mathbf{f} := (f^-, f^+)$ in X^2 (resp. $(X(N))^2$ such that $f^- < f^+$.

Operator definitions. For any $\mathbf{f} := (f^-, f^+) \in \mathbf{Y}$, we use $U(p) = U(\mathbf{f}; p) : \mathrm{Cl}(\Omega(\mathbf{f})) \to \Re$ to denote the unique solution of the periodic boundary value problem:

$$\Delta U = 0 \text{ in } \Omega(\mathbf{f}), U = \pm 1 \text{ on } f^{\pm}, U(x, y) = U(x + \sigma, y),$$

where $\Omega(\mathbf{f}) := \{p = (x, y) \in \Re^2 : f^-(x) < y < f^+(x)\}$. For any $\varepsilon \in (0, 1)$, we define the mapping $\mathbf{R}_{\varepsilon}(\mathbf{f}) : \mathbf{Y} \to \mathbf{Y}$ such that $\mathbf{R}_{\varepsilon}(\mathbf{f}) = (R_{\varepsilon}^-(\mathbf{f}), R_{\varepsilon}^+(\mathbf{f}))$, where

$$R_{\varepsilon}^{\pm}(\mathbf{f}) := \{p \in \Omega(\mathbf{f}) : U(\mathbf{f}; p) = \pm(1 - \varepsilon)\}.$$

We also define the mappings $\mathbf{S}_{\varepsilon}(\mathbf{f}) : \mathbf{Y} \to \mathbf{Y}, \varepsilon \in (0, 1)$, such that $\mathbf{S}_{\varepsilon}(\mathbf{f}) = (S_{\varepsilon}^-(f^-), S_{\varepsilon}^+(f^+))$ for $\mathbf{f} := (f^-, f^+)$, where, for any $f \in X$, we define $\Omega^{\pm}(f) := \{p = (x, y) \in \Re^2 : \pm(y - f(x)) > 0\}$ and

$$S_{\varepsilon}^{\pm}(f) := \{p \in \Omega^{\pm}(f) : d_{\pm}(p, f) = \varepsilon\} \in X.$$

Finally, we define the mappings $\mathbf{T}(\mathbf{f}, \varepsilon) = \mathbf{T}_{\varepsilon}(\mathbf{f}) : \mathbf{Y} \to \mathbf{Y}, \varepsilon \in (0, 1)$, such that $\mathbf{T}_{\varepsilon} = \mathbf{S}_{\varepsilon} \circ \mathbf{R}_{\varepsilon}$. In other words, we define $\mathbf{T}_{\varepsilon}(\mathbf{f}) = (T_{\varepsilon}^-(\mathbf{f}), T_{\varepsilon}^+(\mathbf{f}))$, where

$$T_{\varepsilon}^{\pm}(\mathbf{f}) = S_{\varepsilon}^{\pm}(R_{\varepsilon}^{\pm}(\mathbf{f})) = \{p \in \Omega^{\pm}(R_{\varepsilon}^{\pm}(\mathbf{f})) : d_{\pm}(p, R_{\varepsilon}^{\pm}(\mathbf{f})) = \varepsilon\} \in X.$$

Lemma 1. *Under Assumptions (1) and (2), we have that $\mathbf{R}_\varepsilon, \mathbf{S}_\varepsilon, \mathbf{T}_\varepsilon : \mathbf{Y}(N) \to \mathbf{Y}(N)$ for any $\varepsilon \in (0,1)$. Also, we have that $\mathbf{R}_\varepsilon(\mathbf{f}) \leq \mathbf{R}_\varepsilon(\mathbf{g})$, $\mathbf{S}_\varepsilon(\mathbf{f}) \leq \mathbf{S}_\varepsilon(\mathbf{g})$, and $\mathbf{T}_\varepsilon(\mathbf{f}) \leq \mathbf{T}_\varepsilon(\mathbf{g})$ for any $\mathbf{f}, \mathbf{g} \in \mathbf{Y}$ such that $\mathbf{f} \leq \mathbf{g}$.*

Proof. See [3,6] for similar proofs based on the maximum principle and properties of generalized distance.

3. Existence, uniqueness, monotonicity, and continuous dependence of solutions

Theorem 1. *For any $\lambda > 0$, Problem 1 has at least one solution-pair $\mathbf{f}_\lambda := (f_\lambda^-, f_\lambda^+) \in \mathbf{Y}(N)$.*

Proof. See [2, Theorem 2(a)].

Assumption A. Assume that there do not exist any values $y_0^\pm \in \Re$ and $a_0^\pm, \lambda, \eta \in \Re_+$ such that $y_0^+ - y_0^- = (2/\lambda a_0^-) = (2/\phi(\lambda) a_0^+)$ and $a^\pm(x, y_0^\pm) = a_0^\pm = a^\pm(x, y_0^\pm + \eta)$ for all $x \in \Re$.

Theorem 2. *Assume in Problem 1 that Assumption A holds. Then (a) there is exactly one solution $\mathbf{f}_\lambda = (f_\lambda^-, f_\lambda^+) \in \mathbf{Y}(N)$ at each $\lambda \in \Re_+$. Moreover (b) $\mathbf{f}_\lambda \leq \mathbf{f}_\mu$ whenever $\lambda \leq \mu$ in $\Re_+$, (c) $M(\mathbf{f}_\mu - \mathbf{f}_\lambda) \to 0$ as $\mu \to \lambda \in \Re_+$, and (d) $\min\{\pm f_\lambda^\mp(x) : x \in \Re\} \uparrow \infty$ as $\lambda^{\pm 1} \uparrow \infty$.*

Lemma 2. *We have $\mathbf{f}_\lambda \leq \mathbf{f}_\mu$ for any solutions $\mathbf{f}_\lambda = (f_\lambda^-, f_\lambda^+), \mathbf{f}_\mu = (f_\mu^-, f_\mu^+) \in \mathbf{Y}$ of Problem 1 at values $\lambda, \mu \in \Re_+$ with $\lambda < \mu$. Moreover, we have*

$$\mathbf{f}_\mu \geq \mathbf{f}_\lambda + a_0^{-1}(\min\{((\mu/\lambda) - 1), (1 - (\phi(\lambda)/\phi(\mu)))\}), \tag{7}$$

where $a_0(t) : [0, \infty) \to [0, \infty)$ denotes any strictly increasing function such that $a_0(0) = 0$, and such that, for all $p, q \in \mathrm{Cl}(\Omega(f_\lambda^-, f_\mu^+))$, we have

$$(|a^\pm(p) - a^\pm(q)|/\min\{a^\pm(p), a^\pm(q)\}) \leq a_0(|p - q|). \tag{8}$$

Proof. For given $\lambda, \mu \in \Re_+$ with $\lambda < \mu$, choose η to be maximum subject to the requirement that $\mathbf{f}_\mu \geq \mathbf{f}_\lambda + \eta$. Then either $f_\mu^+ \cap (f_\lambda^+ + \eta) \neq \emptyset$ or $f_\mu^- \cap (f_\lambda^- + \eta) \neq \emptyset$. If $p_0 \in f_\mu^- \cap (f_\lambda^- + \eta)$, then

$$\mu a^-(p_0) = |\nabla U_\mu(p_0)| \leq |\nabla U(\mathbf{f}_\lambda + \eta; p_0)| = |\nabla U_\lambda(p_0 - \eta)| = \lambda a^-(p_0 - \eta). \tag{9}$$

Similarly, if $p_0 \in f_\mu^+ \cap (f_\lambda^+ + \eta)$, then for $\phi_1 = \phi(\lambda), \phi_2 = \phi(\mu)$, we have

$$\phi_2 a^+(p_0) = |\nabla U_\mu(p_0)| \geq |\nabla U(\mathbf{f}_\lambda + \eta; p_0)| = |\nabla U_\lambda(p_0 - \eta)| = \phi_1 a^+(p_0 + \eta). \tag{10}$$

Therefore $\eta > 0$, since if $\eta \leq 0$, then (9) and (10) both contradict the assumed monotonicity of the functions $a^\pm(p)$ (see (2)). For $\eta > 0$, it follows from (9) (resp. (10)) that $(\mu/\lambda) \leq a_0(\eta) + 1$ (resp. $(\phi(\lambda)/\phi(\mu)) \leq a_0(\eta) + 1$). The assertion (7) follows.

Lemma 3. *For each $n \in \mathbb{N}$, let $\mathbf{f}_n = (f_n^-, f_n^+)$ denote a solution of Problem 1 at $\lambda_n \in \Re_+$. If $\mathbf{f}_n \to \mathbf{f} = (f^-, f^+) \in \mathbf{Y}$ and $\lambda_n \to \lambda \in \Re_+$ for $n \to \infty$, then $\mathbf{f}$ is a solution of Problem 1 at λ.*

Proof. Let $U(p) = U(\mathbf{f}; p)$ in the closure of $\Omega := \Omega(\mathbf{f})$, and let $W(p)$ denote the σ-periodic solution of the boundary value problem:

$$\Delta W = 0 \text{ in } \Omega, W = \ln(\mu^{\pm} a^{\pm}(p)) \text{ on } f^{\pm}, \tag{11}$$

where $\mu^- = \lambda$ and $\mu^+ = \phi(\lambda)$. Then $U_n(p) \to U(p)$ and $W_n(p) \to W(p)$, both uniformly in any compact subset of Ω, where $U_n(p) := U(\mathbf{f}_n; p)$ and $W_n(p) := \ln(|\nabla U_n(p)|)$ in $\Omega_n := \Omega(\mathbf{f}_n)$ for each $n \in \mathbb{N}$. This follows from (11), in view of the fact that (by (4), (5), (6)) $W_n(p)$ is a σ-periodic solution of the boundary value problem:

$$\Delta W_n = 0 \text{ in } \Omega_n, W_n = \ln(\mu_n^{\pm} a^{\pm}(p)) \text{ on } f_n^{\pm},$$

where $\mu_n^- = \lambda_n$ and $\mu_n^+ = \phi(\lambda_n)$. Since $\nabla U_n(p) \to \nabla U(p)$ as $n \to \infty$ for any $p \in \Omega$, it follows that $W(p) = \ln(|\nabla U|)$ throughout Ω. The assertion follows.

Lemma 4. *For $i = 0, 1$, let $U_i(p) = U(\mathbf{f}_i; p)$ in the closure of $\Omega_i := \Omega(\mathbf{f}_i)$, where $\mathbf{f}_i = (f_i^-, f_i^+) \in \mathbf{Y}$. Assume that the functions $|\nabla U_i(p)| : \Omega_i \to \Re, i = 0, 1$, have continuous, strictly-positive extensions to $\mathrm{Cl}(\Omega_i)$, $i = 0, 1$. If $\mathbf{f}_0 \leq \mathbf{f}_1$ and $\mathbf{f}_0 \neq \mathbf{f}_1$, then*

$$\pm(|\nabla U_1(p^{\pm})| - |\nabla U_0(p^{\pm})|) > 0 \tag{12}$$

at any point $p^{\pm} \in f_0^{\pm} \cap f_1^{\pm}$.

Proof. We will prove (12) only in the "$-$" case. Define $\widehat{f}_1^{\pm} \in X$ such that $\widehat{f}_1^+ = f_1^+$ and $\widehat{f}_1^-(x) = \min(f_0^+(x) - \varepsilon, f_1^-(x))$, where we choose $0 < \varepsilon \leq (1/2)d(f_0^-, f_0^+)$. Then $\widehat{U}_1(p) := U(\widehat{f}_1^-, \widehat{f}_1^+; p) \geq U_1(p)$ in Ω_1 by the maximum principle, and $U_0(p) > \widehat{U}_1(p)$ in $\Omega_0 \cap \widehat{\Omega}_1$ by the strict maximum principle. By the choice of $\widehat{f}_1^-$, there exists a function $\tilde{f} \in X$ such that $\widehat{f}_1^- < \tilde{f} < f_0^+$. Since

$$U_0(p) - \widehat{U}_1(p) \geq \alpha(\widehat{U}_1(p) + 1) \tag{13}$$

on $\widehat{f}_1^- \cup \tilde{f}$ for some $\alpha > 0$, the same holds for all $p \in \Omega(\widehat{f}_1^-, \tilde{f})$. It follows from (13) that

$$U_0(p) + 1 \geq \beta(\widehat{U}_1(p) + 1) \geq \beta(U_1(p) + 1) \geq 0 \tag{14}$$

in $\Omega(\widehat{f}_1^-, \tilde{f})$, where $\beta = \alpha + 1$. We also have $U_0(p_0) + 1 = U_1(p_0) + 1 = 0$ for any $p_0 \in f_0^- \cap f_1^-$. Thus, for $p_0 \in f_0^- \cap f_1^-$, we have

$$\beta \int_{\gamma} |\nabla U_1| ds = \beta(U_1(p_1) + 1) \leq (U_0(p_1) + 1) = \int_{\gamma} (\partial U_0 / \partial \nu) ds \leq \int_{\gamma} |\nabla U_0| ds, \tag{15}$$

where γ denotes any arc of steepest ascent of U_1 which originates at p_0 and is short enough so that $\gamma \subset \mathrm{Cl}(\Omega(\widehat{f}_1^-, \tilde{f})$. Here, p_1 denotes the terminal endpoint of γ, ν denotes the forward tangent vector to γ, and s denotes the arc-length parameter. By (15), there exists a sequence of points $(p_n)_{n=1}^{\infty}$ in $\Omega(\widehat{f}_1^-, \tilde{f})$ such that $p_n \to p_0$ as $n \to \infty$ and $\beta|\nabla U_1(p_n)| \leq |\nabla U_0(p_n)|$ for each $n \in \mathbb{N}$. The assertion follows.

Proof of Theorem 2, *Part* (a). Assume for given $\lambda \in \Re_+$ that Problem 1 has two solutions $\widehat{\mathbf{f}}_\lambda = (\widehat{f}_\lambda^-, \widehat{f}_\lambda^+)$, $\widetilde{\mathbf{f}}_\lambda = (\widetilde{f}_\lambda^-, \widetilde{f}_\lambda^+) \in \mathbf{Y}$. We will show that $\widehat{\mathbf{f}}_\lambda = \widetilde{\mathbf{f}}_\lambda$. Choose the value $\eta \in \Re$ to be minimum subject to the requirement that $\widetilde{\mathbf{f}}_\lambda \le \widehat{\mathbf{f}}_\lambda + \eta$. If $\widetilde{\mathbf{f}}_\lambda \ne \widehat{\mathbf{f}}_\lambda + \eta$, then, by (4), (5), and Lemma 4, we have

$$\lambda a^-(p_0) = |\nabla \widetilde{U}_\lambda(p_0)| > |\nabla U(\widehat{\mathbf{f}}_\lambda + \eta, p_0)| \tag{16}$$

$$= |\nabla \widehat{U}_\lambda(p_0 - \eta)| = \lambda a^-(p_0 - \eta) \ge \lambda a^-(p_0)$$

for any point $p_0 \in \widetilde{f}_\lambda^- \cap (\widehat{f}_\lambda^- + \eta)$, and

$$\phi(\lambda) a^+(p_0) = |\nabla \widetilde{U}_\lambda(p_0)| < |\nabla U(\widehat{\mathbf{f}}_\lambda + \eta, p_0)| \tag{17}$$

$$= |\nabla \widehat{U}_\lambda(p_0 - \eta)| = \lambda a^+(p_0 - \eta) \le \phi(\lambda) a^+(p_0)$$

for any point $p_0 \in \widetilde{f}_\lambda^+ \cap (\widehat{f}_\lambda^+ + \eta)$, where $\widetilde{U}_\lambda(p) = U(\widetilde{\mathbf{f}}_\lambda; p)$ (resp. $\widehat{U}_\lambda(p) = U(\widehat{\mathbf{f}}_\lambda; p)$) in the closure of $\widetilde{\Omega}_\lambda = \Omega(\widetilde{\mathbf{f}}_\lambda)$ (resp. $\widehat{\Omega}_\lambda = \Omega(\widehat{\mathbf{f}}_\lambda)$). This contradiction shows that

$$\widetilde{\mathbf{f}}_\lambda = \widehat{\mathbf{f}}_\lambda + \eta. \tag{18}$$

Therefore, both $\widehat{\mathbf{f}}_\lambda$ and $\widehat{\mathbf{f}}_\lambda + \eta$ are solutions of Problem 1 at λ. It follows that

$$a^\pm(p) = a^\pm(p + \eta) \tag{19}$$

for all $p \in \widehat{f}_\lambda^\pm$. For $\eta \ne 0$, it follows from (2) and (19) that

$$a^\pm(p) = a_0^\pm > 0 \tag{20}$$

on $\widehat{f}_\lambda^\pm$, where $a_0^\pm$ denote positive constants. It follows from (20) that

$$\widehat{W}_\lambda(p) := \ln(|\nabla \widehat{U}_\lambda(p)|) = (1/2)(\ln(b_0^- b_0^+) + \ln(b_0^+/b_0^-)\widehat{U}_\lambda(p)), \tag{21}$$

where $b_0^- = \lambda a_0^-$, $b_0^+ = \phi(\lambda) a_0^+$, since both sides of (21) are σ-periodic harmonic functions in $\widehat{\Omega}_\lambda$ satisfying the same boundary conditions on the boundary components $\widehat{f}_\lambda^\pm$. Therefore

$$\kappa^\pm(p) = \partial \widehat{W}_\lambda(p)/\partial \nu = \ln(b_0^+/b_0^-)\partial \widehat{U}_\lambda(p)/\partial \nu = \ln(b_0^+/b_0^-)b_0^\pm \tag{22}$$

for all $p \in \widehat{f}_\lambda^\pm$, where $\kappa^\pm(p)$ denotes signed curvature and ν denotes the "upward" unit normal to the curve $\widehat{f}_\lambda^\pm$ at the point $p \in \widehat{f}_\lambda^\pm$. By (22), the curves $\widehat{f}_\lambda^\pm$ have constant curvature. Since they are also σ-periodic, it easily follows that $\kappa^\pm(p) = 0$ on $\widehat{f}_\lambda^\pm$, and therefore that $\widehat{f}_\lambda^\pm(x) = y_0^\pm$ for $x \in \Re$, where $(y_0^+ - y_0^-) = (2/b_0^+) = (2/b_0^-) > 0$. Therefore

$$a^\pm(x, y_0^\pm) = a^\pm(x, y_0^\pm + \eta) = a_0^\pm \tag{23}$$

for all $x \in \Re$. This contradiction of Assumption A shows that actually $\eta = 0$. Therefore $\widetilde{\mathbf{f}}_\lambda = \widehat{\mathbf{f}}_\lambda$, and the assertions follow.

Proof of Theorem 2, *Parts* (b), (c). Part (b) follows from Part (a) and Lemma 2. Concerning Part (c), for given $\lambda \in \Re$, there exist (by Lemma 2 and the theorem of Ascoli-Arzela) pairs $\widehat{\mathbf{f}}_\lambda = (\widehat{f}_\lambda^-, \widehat{f}_\lambda^+)$, $\widetilde{\mathbf{f}}_\lambda = (\widetilde{f}_\lambda^-, \widetilde{f}_\lambda^+) \in \mathbf{Y}$ such that $\mathbf{f}_\mu \uparrow \widehat{\mathbf{f}}_\lambda$ as $\mu \uparrow \lambda$

and $\mathbf{f}_\mu \downarrow \tilde{\mathbf{f}}_\lambda$ as $\mu \downarrow \lambda$. By Lemma 3, $\widehat{\mathbf{f}}_\lambda$ and $\tilde{\mathbf{f}}_\lambda$ are both solutions of Problem 1 at λ. Therefore $\widehat{\mathbf{f}}_\lambda = \mathbf{f}_\lambda = \tilde{\mathbf{f}}_\lambda$ by Part (a), completing the proof.

Proof of Theorem 2, Part (d). We will prove only that $\min\{f_\lambda^-(x) : x \in \Re\} \uparrow \infty$ as $\lambda \uparrow \infty$. Assuming this assertion to be false, we have (in view of Lemma 2) that $\min\{f_\lambda^-(x) : x \in \Re\} \leq B_0 < \infty$ for all $\lambda \in \Re_+$. It follows that

$$\max\{f_\lambda^-(x) : x \in \Re\} \leq B_0 + N\sigma, \tag{24}$$

since $\mathrm{Var}(f_\lambda^\pm) \leq N\sigma$ for all $\lambda \in \Re_+$, due to the σ-periodicity and Lipschitz continuity of $f_\lambda^\pm$. It follows from (4) and (24) that

$$K_\lambda \geq \lambda\sigma A_0^-, \tag{25}$$

for all $\lambda \in \Re_+$, where $A_0^- := \min\{a^-(x,y) : y \leq B_0 + N\sigma\} > 0$, and where K_λ denotes the capacity of one period of Ω_λ (which is the arc-length integral of $|\nabla U_\lambda(p)|$ along one period of any level curve of U_λ). It follows by well-known properties of capacity that

$$\min\{f_\lambda^+(x) : x \in \Re\} - \max\{f_\lambda^-(x) : x \in \Re\} \leq (\sigma/K_\lambda) \leq (1/\lambda A_0^-). \tag{26}$$

Since $\mathrm{Var}(f_\lambda^\pm) \leq N\sigma$, it follows from (26) that $f_\lambda^+(x) \leq B_0 + 2N\sigma + (1/A_0^-)$ for $\lambda \geq 1$, from which it follows by (5) that

$$K_\lambda \leq \sigma\phi(\lambda)(1 + N^2)^{1/2} A_0^+, \tag{27}$$

for $\lambda \geq 1$, where $A_0^+ = \max\{a^+(x,y) : y \leq B_0 + 2N\sigma + (1/A_0^-)\}$. Clearly (25) and (27) cannot both be true for sufficiently large $\lambda \geq 1$.

Remarks. (a) The author stated (essentially) Theorem 2, Part (a) earlier, in [2, Thm. 2(b)]. However, the proof given in [2] is defective. (b) Assumption A is satisfied if, for example, either one of the functions $A^\pm(y) : \Re \to \Re_+$ is strictly increasing, where $A^\pm(y) := \int_0^\sigma a^\pm(x, \pm y)dx$.

4. Successive approximation of solutions

Theorem 3. *Assume in Problem 1 that the given functions $a^\pm(p) : \Re^2 \to \Re_+$ satisfy Assumption A. Let $\mathbf{f} \in \mathbf{Y}(N)$ denote the unique solution of Problem 1 at $\lambda = 1$, where we assume that $\phi(1) = 1$. For any given initial guess $\mathbf{f}_0 \in \mathbf{Y}(N)$, let the sequence of successive approximations $(\mathbf{f}_n)_{n=0}^\infty$ in $\mathbf{Y}(N)$ be defined recursively such that $\mathbf{f}_{n+1} = \mathbf{T}(\mathbf{f}_n; \varepsilon_n)$ for $n = 0, 1, 2, \ldots$, where $(\varepsilon_n)_{n=0}^\infty$ denotes a sequence of values in the interval $(0, 1)$. Then there exists a strictly increasing function $p_0(t) : [0, \infty) \to \Re$ with $p_0(0) = 0$ such that $\mathbf{f}_n \to \mathbf{f}$ as $n \to \infty$ provided that $\varepsilon_n \downarrow 0$ as $n \to \infty$ and the sum of the values $p_0(\mu\varepsilon_n), n = 1, 2, \ldots \infty$ diverges for any (arbitrarily small) $\mu > 0$. Moreover, for any constant $E > 0$, convergence (relative to a particular admissible sequence $(\varepsilon_n)_{n=0}^\infty$) is uniform over all initial guesses $\mathbf{f}_0 \in \mathbf{Y}(N)$ such that $M(\mathbf{f}_0 - \mathbf{f}) \leq E$.*

Lemma 5. *Assume in Problem* 1 *that Assumption A holds. Then for any closed interval* $I \subset \Re_+$, *there exists a continuous function* $z_0(t) : [0, \infty) \to \Re$ *with* $z_0(0) = 0$ *such that*

$$|\psi_\lambda^\pm(p)| \leq z_0(d(p, f_\lambda^\pm)) \tag{28}$$

uniformly for all $p \in \mathrm{Cl}(\Omega_\lambda)$ *and* $\lambda \in I$, *where* $\psi_\lambda^\pm(p) := \ln(|\nabla U_\lambda(p)|/\mu^\pm a^\pm(p))$, $\mu^- = \lambda$, *and* $\mu^+ = \phi(\lambda)$.

Proof. The function $\psi_\lambda^\pm(p)$ is continuous, and therefore uniformly continuous function of p and λ in the compact set $S := \{(p, \lambda) : p \in \mathrm{Cl}(\Omega_\lambda), \lambda \in I\}$. Thus the result follows from the fact that $\psi_\lambda^\pm(p) = 0$ on the boundary-portion $S^\pm := \{(p, \lambda) : p \in f_\lambda^\pm, \lambda \in I\}$ of S.

Lemma 6. *Given the constants* $0 < \alpha \leq \beta$ *and the functions* $F^- < F^+$ *in* X, *we have*

$$g^+ + (\beta - \alpha)/\tilde{a}_+ \leq h^+; h^- + (\beta - \alpha)/\tilde{a}_- \leq g^- \tag{29}$$

for any functions $f^\pm, g^\pm, h^\pm \in X$ *such that*

$$F^- \leq h^- < g^- < f^- \leq F^+; F^- \leq f^+ < g^+ < h^+ \leq F^+,$$

$d_\pm(p, f^\pm) \leq \alpha$ *for all* $p \in g^\pm$, *and* $d_\pm(p, f^\pm) \geq \beta$ *for all* $p \in h^\pm$. *Here, we define* $\tilde{a}_\pm := \sup\{a^\pm(p) : p \in \Omega(F^-, F^+)\}$.

Proof. We study only the "+" case of (29). For any fixed $x_0 \in \Re$, let $\gamma_0 := (x_0) \times [g^+(x_0), h^+(x_0)]$. For given $\varepsilon > 0$, let γ_1 denote an arc joining the point $p_0 := (x_0, g^+(x_0)) \in g^+$ to the set f^+ such that $||\gamma_1||_+ \leq d_+(p_0, f^+) + \varepsilon \leq \alpha + \varepsilon$. For the arc $\gamma := \gamma_0 \cup \gamma_1$, we have $||\gamma||_+ = ||\gamma_0||_+ + ||\gamma_1||_+$. Since γ joins the point $p_1 := (x_0, h^+(x_0))$ to the set f^+, we have $||\gamma||_+ \geq \beta$. It follows that $||\gamma_0||_+ \geq (\beta - \alpha - \varepsilon)$. Since $||\gamma_0||_+ \leq \tilde{a}_+(h^+(x_0) - g^+(x_0))$, we have $h^+(x_0) - g^+(x_0) \geq (\beta - \alpha - \varepsilon)/\tilde{a}_+$. The "+"-case of the assertion (29) follows, since $x_0 \in \Re$ is arbitrary and $\varepsilon > 0$ can be chosen arbitrarily small.

Lemma 7. *Assume in Problem* 1 *that Assumption A holds. Let* $\phi(\lambda) = 1/\lambda$, *so that conditions* (4), (5) *reduce to the form:* $|\nabla U_\lambda(p)| = \lambda^{\mp 1} a^\pm(p)$ *on* $f_\lambda^\pm$, *where* $\mathbf{f}_\lambda := (f_\lambda^-, f_\lambda^+) \in \mathbf{Y}(N)$ *denotes a solution at* $\lambda \in \Re_+$. *Then, for given* $\kappa > 0$, *there exist a constant* $C_0 > 0$ *and a continuous, increasing function* $\eta_0(\varepsilon) : [0, 1) \to \Re$ *such that* $\eta_0(0) = 1$, *and such that*

$$T_\varepsilon^\pm(\mathbf{f}_\lambda) \geq f_\lambda^\pm + C_0(\theta_0(\varepsilon)) - \lambda)\varepsilon \tag{30}$$

for all $\lambda \in (1/\kappa, 1]$ *and* $\varepsilon \in (0, 1)$ *satisfying* $\lambda \leq \theta_0(\varepsilon) := (1/\eta_0(\varepsilon))$, *and such that*

$$T_\varepsilon^\pm(\mathbf{f}_\lambda) \leq f_\lambda^\pm - C_0(\lambda - \eta_0(\varepsilon))\varepsilon \tag{31}$$

for all $\lambda \in (1, \kappa]$ *and* $\varepsilon \in (0, 1)$ *satisfying* $\eta_0(\varepsilon) \leq \lambda$.

Proof. By Lemma 5, we have

$$|\ln(\lambda^{\pm 1}|\nabla U_\lambda(p)|/a^\pm(p))| \leq z_0(d(p, f_\lambda^\pm))$$

in Ω_λ for all $\lambda \in [(1/\kappa), \kappa]$. Therefore, there exists a constant $C > 0$ such that

$$|\ln(\lambda^{\pm 1}|\nabla U_\lambda(p)|/a^\pm(p))| \leq z_0(C\varepsilon), \tag{32}$$

uniformly for all $\lambda \in [(1/\kappa), \kappa]$ and $p \in \Omega_\lambda$ such that $|U_\lambda(p) \mp 1| \leq \varepsilon$, where the constant C is chosen such that $d(p; f_\lambda^\pm) \leq C|U_\lambda(p) \mp 1| \leq C\varepsilon$ for all $\varepsilon \in (0,1)$. By exponentiation of (32), we have

$$\theta_0(\varepsilon)a^\pm(p) \leq \lambda^{\pm 1}|\nabla U_\lambda| \leq \eta_0(\varepsilon)a^\pm(p) \tag{33}$$

for all $p \in \Omega_\lambda$ such that $|U_\lambda(p) \mp 1| \leq \varepsilon$, where the strictly increasing function $\eta_0(\varepsilon) := \exp(z_0(C\varepsilon))$ is such that $\eta_0(0) = 1$. Therefore, if $\gamma_\pm$ denotes any arc of steepest ascent of U_λ joining $f_\lambda^\pm$ to $R_\varepsilon^\pm(\mathbf{f}_\lambda) := \{p \in \Omega_\lambda : U_\lambda(p) = \pm(1 - \varepsilon)\}$, then $\varepsilon = \int_{\gamma_\pm} |\nabla U_\lambda| ds \geq \int_{\gamma_\pm} \lambda^{\mp 1}\theta_0(\varepsilon)a^\pm(p)ds = \lambda^{\mp 1}\theta_0(\varepsilon)||\gamma_\pm||_\pm$ by (33). Therefore,

$$d_\pm(p, R_\varepsilon^\pm(\mathbf{f}_\lambda)) \leq \varepsilon\lambda^{\pm 1}\eta_0(\varepsilon) \tag{34}$$

for all $p \in f_\lambda^\pm$. On the other hand, by (33) we have $||\gamma_\pm||_\pm = \int_{\gamma_\pm} a^\pm(q)ds \geq \int_{\gamma_\pm} \lambda^{\pm 1}\theta_0(\varepsilon)|\nabla U_\lambda| ds \geq \varepsilon\lambda^{\pm 1}\theta_0(\varepsilon)$ for any rectifiable arc $\gamma_\pm$ joining $f_\lambda^\pm$ to $R_\varepsilon^\pm(\mathbf{f}_\lambda)$. Therefore

$$d_\pm(f_\lambda^\pm, R_\varepsilon^\pm(\mathbf{f}_\lambda)) \geq \varepsilon\lambda^{\pm 1}\theta_0(\varepsilon) \tag{35}$$

Now, the "+" cases of (30) and (31) follow directly from (29), Part (a), in conjunction with (34) and (35) (in that order), and the "−" cases of (30) and (31) follow directly from (29), Part (b), in conjunction with (34) and (35) in that order.

Lemma 8. *Let $\mathbf{f}(\lambda)$ denote the solution of Problem 1 at $\lambda \in \Re_+$, where Assumption A holds and we assume that $\phi(\lambda) = 1/\lambda$. Then, for any given $\kappa > 1$, there exists a strictly increasing function $p_0(\delta) : [0, \infty) \to \Re$, with $p_0(0) = 0$ and $p_0(\delta) \leq (1/2\kappa)$, such that (in notation of Lemma 7), we have*

$$\mathbf{T}_\varepsilon(\mathbf{f}(\lambda)) \geq \mathbf{f}(\lambda + p_0([\theta_0(\varepsilon) - \lambda]\varepsilon))$$

whenever $\varepsilon \in (0,1)$ and $1/\kappa \leq \lambda \leq \theta_0(\varepsilon) < 1$, and

$$\mathbf{T}_\varepsilon(\mathbf{f}(\lambda)) \leq \mathbf{f}(\lambda - p_0([\lambda - \eta_0(\varepsilon)]\varepsilon))$$

whenever $\varepsilon \in (0,1)$ and $1 < \eta_0(\varepsilon) \leq \lambda \leq \kappa$.

Proof. By Theorem 2, Part (c), there exists a continuous, strictly increasing function $p(t) : [0, \infty) \to \Re$, with $p(0) = 0$ and $p(t) \leq (1/2\kappa)$, such that

$$\mathbf{f}(\lambda) - t \leq \mathbf{f}(\lambda - p(t)) \leq \mathbf{f}(\lambda + p(t)) \leq \mathbf{f}(\lambda) + t$$

for all $\lambda \in [1/\kappa, \kappa]$ and $\delta \geq 0$. In view of this, the assertion follows directly from Lemma 7, where we define $p_0(t) = p(C_0 t)$ for all $t \geq 0$.

Lemma 9. *In the context of Problem 1 and Lemma 8, for given $\kappa > 1$, there exists a null function $z_0(\varepsilon)$ such that*

$$\mathbf{f}(\lambda - z_0(\varepsilon)) \leq \mathbf{T}_\varepsilon(\mathbf{f}(\lambda)) \leq \mathbf{f}(\lambda + z_0(\varepsilon))$$

uniformly for all $\lambda \in [1/\kappa, \kappa]$ and $\varepsilon \in (0,1)$, where $\mathbf{f}(\lambda)$ solves Problem 1 at λ.

Proof sketch. It is easily seen that

$$M(\mathbf{T}_\varepsilon(\mathbf{f}(\lambda)) - \mathbf{f}(\lambda)) \to 0 \text{ as } \varepsilon \to 0+$$

for any $\lambda \in \Re_+$, where the convergence is uniform over all $\lambda \in [1/\kappa, \kappa]$. The assertion follows from this in view of Lemma 2.

Proof of Theorem 3. Choose $\kappa > 1$ so large that

$$\mathbf{f}(1/\kappa) \leq \mathbf{T}_\varepsilon(\mathbf{f}(1/\kappa)) \leq \mathbf{T}_\varepsilon(\mathbf{f}(\kappa)) \leq \mathbf{f}(\kappa)$$

for all $\varepsilon \in (0,1)$, and such that $\mathbf{f}_0 \in \mathbf{Y}(N, \kappa)$ for all $\mathbf{f}_0 \in \mathbf{Y}(N)$ such that $M(\mathbf{f}_0 - \mathbf{f}) \leq E$, where $\mathbf{f}(\lambda)$ denotes the unique solution of Problem 1 at $\lambda \in \Re_+$, and where we define $\mathbf{Y}(N, \kappa) := \{\widehat{\mathbf{f}} \in \mathbf{Y}(N) : \mathbf{f}(1/\kappa) \leq \widehat{\mathbf{f}} \leq \mathbf{f}(\kappa)\}$. Thus $\mathbf{T}_\varepsilon : \mathbf{Y}(N, \kappa) \to \mathbf{Y}(N, \kappa)$ for any $\varepsilon \in (0,1)$, and we have $\mathbf{f}_n \in \mathbf{Y}(N, \kappa)$ for all $n = 0, 1, \ldots$ and for all $\mathbf{f}_0 \in \mathbf{Y}(N)$ such that $M(\mathbf{f}_0 - \mathbf{f}) \leq E$. For each $n = 0, 1, 2, \ldots$, choose $\alpha_n \in [(1/\kappa), 1]$ and $\beta_n \in [1, \kappa]$ such that α_n is maximum and β_n is minimum subject to the requirements that $\alpha_n \leq 1 \leq \beta_n$ and

$$\mathbf{f}(\alpha_n) \leq \mathbf{f}_n \leq \mathbf{f}(\beta_n).$$

Then, by Lemma 8, we have

$$\mathbf{f}(\alpha_n + p_0([\theta_0(\varepsilon_n) - \alpha_n]\varepsilon_n)) \leq \mathbf{T}(\mathbf{f}(\alpha_n), \varepsilon_n) \leq \mathbf{f}_{n+1}$$

$$\leq \mathbf{T}(\mathbf{f}(\beta_n), \varepsilon_n) \leq \mathbf{f}(\beta_n - p_0([\beta_n - \eta_0(\varepsilon_n)]\varepsilon_n)),$$

provided that $\alpha_n \leq \theta_0(\varepsilon_n)$ and $\beta_n \geq \eta_0(\varepsilon_n)$. Thus

$$\min\{1, \alpha_n + p_0([(\theta_0(\varepsilon_n)) - \alpha_n]\varepsilon_n)\} \leq \alpha_{n+1}$$

$$\leq \beta_{n+1} \leq \max\{1, \beta_n - p_0([\beta_n - \eta_0(\varepsilon_n)]\varepsilon_n)\}$$

for any $n \in \{0\} \cup \mathbb{N}$ such that $\alpha_n \leq \theta_0(\varepsilon_n)$ and $\beta_n \geq \eta_0(\varepsilon_n)$. It follows that

$$0 \leq E^\pm_{n+1} \leq \max\{0, E^\pm_n - p_0([E^\pm_n - \rho^\pm_0(\varepsilon_n)]\varepsilon_n)\} \leq \pm(\kappa^{\pm 1} - 1), \tag{36}$$

for all $n \in \{0\} \cup \mathbb{N}$ such that $\rho^\pm_0(\varepsilon_n) \leq E^\pm_n$, where $E^+_n := \beta_n - 1$, $E^-_n := 1 - \alpha_n$, $\rho^+_0(\varepsilon) := \eta_0(\varepsilon) - 1$, and $\rho^-_0(\varepsilon) := 1 - \theta_0(\varepsilon)$. For any $\mu > 0$, there exists an integer $n_0(\mu)$ such that

$$E^\pm_{n+1} \leq E^\pm_n + z_0(\varepsilon_n) \leq E^\pm_n + \mu \tag{37}$$

and $\rho^\pm_0(\varepsilon_n) \leq \mu$, both for all $n \geq n_0(\mu)$, where $z_0(\cdot)$ denotes the null function in Lemma 9. We have $(E^\pm_n - \rho^\pm_0(\varepsilon_n))\varepsilon_n \geq \mu\varepsilon_n$ for any $n \geq n_0(\mu)$ such that $E^\pm_n \geq 2\mu$. In view of this, it follows from (36) that

$$E^\pm_{n+1} \leq \max\{0, E^\pm_n - p_0(\mu\varepsilon_n)\} \tag{38}$$

for any $n \geq n_0(\mu)$ such that $E^\pm_n \geq 2\mu$. In view of the fact that the sum of $p_0(\mu\varepsilon_n)$ over all $n = 1, 2, \ldots$ diverges, it easily follows from (38) that there exist smallest integers $n^\pm_1(\mu) \geq n_0(\mu)$ such that $E^\pm_n \leq 2\mu$ for $n = n^\pm_1(\mu)$. It then follows easily from (37) and (38) that $E^\pm_n \leq 3\mu$ for all $n \geq n^\pm_1(\mu)$. The assertion follows, since $\mu > 0$ is arbitrary in the above argument.

To prove uniform convergence of the iterates corresponding to all admissible initial guesses $\mathbf{f}_0 \in \mathbf{Y}(N, \kappa)$, it suffices to apply the above estimates for the "+"

(resp. "$-$") case to the initial guess $\mathbf{f}_0 = \mathbf{f}(\kappa)$ (resp. $\mathbf{f}_0 = \mathbf{f}(1/\kappa)$), and then apply Lemma 1 to sandwich the iterates corresponding to any other admissible initial guess between these iterates.

5. Concluding remarks

Problem 1, together with all our results concerning it, can be easily modified in a two ways: For $\sigma = 2\pi$, Problem 1 easily converts under the conformal mapping $w = \log(iz)$ to an equivalent annular double-free-boundary problem for which all results in this paper apply (under the same equivalence). This is perhaps the strongest direct incentive for studying the periodic case. In another equivalent version, the σ-periodicity assumption can be everywhere omitted, while restricting the functions $a^{\pm}(p)$, $U_\lambda(p)$, and $f_\lambda^{\pm}(x)$ to $0 \leq x \leq b$ for some $b > 0$, and adding (in (6)) the requirement that $(\partial U_\lambda(p)/\partial x) = 0$ in $((0, b) \times \Re) \cap \Omega_\lambda$. This problem reduces to the $\sigma = 2b$ case of Problem 1 by even continuation (in x) of all functions to $|x| \leq b$, followed by the 2b-periodic (in x) continuation of all functions.

References

[1] A. Acker, Some free boundary optimization problems and their solutions. In *Numerische Behandlung von Differentialgleichungen mit besonderer Berücksichtigung freier Randwertsaufgaben*. Hrsg. von J. Albrecht, L. Collatz, G. Haemmerlin. Birkhäuser Verlag, Basel, 1978.

[2] A. Acker, A free boundary optimization problem involving weighted areas. J. Appl. Math and Phys (ZAMP) **29** (1978), 395–408.

[3] A. Acker, How to approximate the solutions of certain free boundary problems for the Laplace equation by using the contraction principle. J. Appl. Math. Phys. (ZAMP) **32** (1981), 22–33.

[4] A. Acker, Convergence results for the trial free-boundary method. IMA Journal on Numerical Analysis **8** (1988), 357–364.

[5] A. Acker, On the convexity and successive approximation of solutions in a free boundary problem with two fluid phases. Comm. in Part. Diff. Eqs. **14** (1989), 1635–1652.

[6] A. Acker, On 2-layer free boundary problems with generalized joining conditions: convexity and successive approximation of solutions. In: *Comparison Methods and Stability Theory* (X. Liu and D. Siegel, editors), Lecture notes in Pure and Applied Mathematics, Vol. 162. New York: Marcel Dekker, Inc., 1994.

[7] A. Acker, E. Kadakal, K. Miller, A trial free boundary method for computing Batchelor flows. J. Comput. Appl. Math. **80** (1997), 31–48.

[8] A. Acker, R. Meyer, A free boundary problem for the p-Laplacian: uniqueness, convexity, and successive approximation of solutions. Electronic J. Diff. Eqs. **1995**, no. 8 (June 21, 1995).

[9] A. Beurling, On free boundary problems for the Laplace equation. In *Seminars on Analytic Functions*, Vol. I. Institute for Advanced Study, Princeton, N.J. (1957), 248–263.

[10] A. Beurling, *The Collected Works of Arne Beurling, Vol. 1* (edited by Carleson, L., Malliavan, P., Neuberger, J., and Werner, J.), Birkhäuser Verlag, Boston.

[11] E. Kadakal, On the successive approximation of solutions of some elliptic free boundary problems. Doctoral Dissertation. Department of Mathematics and Statistics, Wichita State University, June, 1996.

[12] M.A. Lavrentiev, *Variational Methods.* P. Noordhoff, Groningen, the Netherlands (1963).

[13] R. Meyer, Approximation of the solution of free boundary problems for the p-Laplace equation. Doctoral dissertation. Department of Mathematics and Statistics, Wichita State University, 1993.

Andrew Acker
Department of Mathematics and Statistics
Wichita State University
Wichita KS 67260-0033
USA
e-mail: `acker@math.wichita.edu`

Progress in Nonlinear Differential Equations
and Their Applications, Vol. 63, 13–21
© 2005 Birkhäuser Verlag Basel/Switzerland

New Numerical Solutions for the Brezis-Nirenberg Problem on $\mathbb{S}^n$

Catherine Bandle and Simon Stingelin

*Dedicated to Haim Brezis in appreciation for having initiated
the study of this fascinating problem.*

1. Introduction

Let $\mathbb{S}^n = \{x \in \mathbb{R}^{n+1} : |x| = 1\}$ be the n-dimensional unit sphere. Denote by $B(\theta_1) \subset \mathbb{S}^n$ the geodesic ball centered at the North pole with geodesic radius θ_1, and by $\triangle_{\mathbb{S}^n}$ the Laplace-Beltrami operator on $\mathbb{S}^n$. We consider the Brezis-Nirenberg problem [4]

$$\triangle_{\mathbb{S}^n} u + \lambda u + u^{2^*-1} = 0, \quad u > 0 \quad \text{in} \quad B(\theta_1),$$
$$u = 0 \quad \text{on} \quad \partial B(\theta_1), \tag{1.1}$$

where $2^* = \frac{2n}{n-2}$ is the critical Sobolev exponent and $n \geq 3$.

It is not difficult to see that a necessary condition for the existence of positive solutions is $\lambda < \lambda_1$ where λ_1 is the lowest eigenvalue of $\triangle_{\mathbb{S}^n}$ with Dirichlet boundary conditions. It was shown in [1] for $n = 3$ and in [7] for $n \geq 4$ that for $\lambda \in (\lambda^*, \lambda_1)$ where

$$\lambda^* = \begin{cases} (\pi^2 - 4\theta_1^2)/4\theta_1^2 & \text{if} \quad n = 3 \\ -n(n-2)/4 & \text{if} \quad n \geq 4 \end{cases}$$

Problem (1.1) has a unique solution. Furthermore this solution is a minimizer of the energy associated to (1.1), namely

$$I_\lambda = \min \int_{B(\theta_1)} \left(\frac{\|\nabla v\|}{2} - \frac{\lambda v^2}{2} - \frac{|v|^{2^*}}{2^*} \right) d\mu, \quad v \in H^1(d\mu),$$

where $\| \cdot \|$ denotes the length, and $d\mu$ the volume element with respect to the metric on $\mathbb{S}^n$. It was also shown by means of a Pohozaev type identity that for balls contained in the upper hemisphere ($\theta_1 < \pi/2$) no other solution exists.

In this paper we present numerical solutions for large balls $B(\theta_1)$ with $\theta_1 > \pi/2$ and small $\lambda < -n(n-2)/4$. The observation that in contrast to the problem in $\mathbb{R}^n$, additional solutions may appear goes back to [1]. Recently Brezis and Peletier [3] announced the following result for (1.1) in S^3:

> *If $\theta_1 > \pi/2$ there exists $A = A(k, \theta_1) > 0$ such that for $\lambda < -A$ there exist at least $2k$ solutions such $u(0) \in (0, |\lambda|^{1/4})$.*

2. Numerical strategy

If the solution depends only on the geodesic distance θ to the north pole (1.1) assumes the form

$$\frac{1}{\sin^{n-1}(\theta)} \left(\sin^{n-1}(\theta)u'\right)' + \lambda u + u^{2^*-1} = 0 \quad \text{in} \quad (0, \theta_1),$$

$$u'(0) = 0 \quad \text{and} \quad u(\theta_1) = 0. \tag{2.1}$$

We often find it more convenient to project $\mathbb{S}^n$ stereographically onto $\mathbb{R}^n$ such that the upper hemisphere is mapped into $\{r < 1\}$. A geodesic ball $B(\theta_1)$ is then mapped into a ball $B_R \in \mathbb{R}^n$ centered at the origin. The relation between the geodesic distance θ and the distance of the image to the origin r is given by $r = \tan(\theta/2)$. In terms of r, (2.1) becomes

$$\rho^{-n}r^{1-n} \left(\rho^{n-2}r^{n-1}u'\right)' + \lambda u + u^{2^*-1} = 0 \quad \text{in} \quad (0, R)$$

$$u'(0) = 0 \quad \text{and} \quad u(R) = 0, \quad \rho(r) = \frac{2}{1+r^2}. \tag{2.2}$$

Observe that $R = 1$ corresponds to the equator.

First we compute $R(u_0) = \min\{r \in \mathbb{R}^+ : u(r; u_0) = 0\}$ where $u(r; u_0)$ is the solution of (2.2) with $u(0) = u_0$ (Fig. 1). We use a 4th-order Runge-Kutta method with an error estimator for the step size control. We stop the computation if $u(r) < 0$ or if r is larger than some prescribed value $r_{\max}$.

From this figure we can determine the number of solutions for a given radius R. In order to calculate $u(r)$ for given R, we use a Newton iteration

$$u_0^{i+1} = u_0^i - \left[\frac{\partial}{\partial u_0} u(R; u_0^i)\right]^{-1} u(R; u_0^i).$$

By substituting

$$x := \rho^{n-2}r^{n-1}u'$$
$$y := \rho^{n-2}r^{n-1}v'(r) \tag{2.3}$$

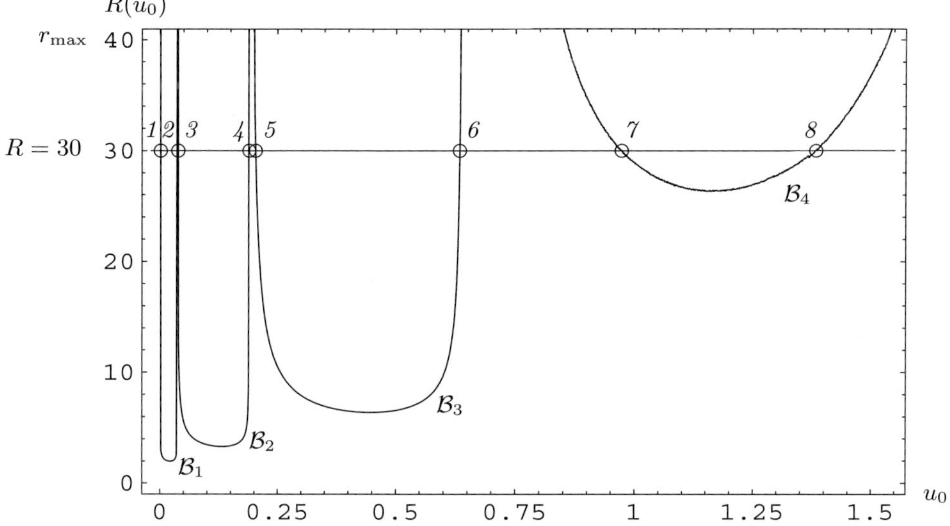

FIGURE 1. $R(u_0)$ for $n = 3$, $\lambda = -20$. The number of solutions for $u_0 \in (0, |\lambda|^{1/4})$ with $R(u_0) = 30$ is 8. The example solutions 1–8 are shown in Fig. 2.

into the equation we get

$$
\begin{aligned}
u'(r) &= \operatorname{sign}(x)\left(\rho^{n-2}r^{n-1}\right)^{-1}|x| & u(0) &= u_0 \\
x'(r) &= -\rho^n r^{n-1}(\lambda u + |u|^{2^*-2}u) & x(0) &= 0 \\
v'(r) &= \left(\rho^{n-2}r^{n-1}\right)^{-1} y & v(0) &= 1 \\
y'(r) &= -\rho^n r^{n-1}(\lambda + (p^* - 1)|u|^{2^*-2})v & y(0) &= 0.
\end{aligned}
\tag{2.4}
$$

In every Newton step we need $u(R; u_0)$ and $\frac{\partial}{\partial u_0}u(R; u_0^i)$. Because of the flexible step size of our Runge-Kutta method we have to solve (2.4) up to $r > R$ and then interpolate between $u(r_{k+1}; u_0^i)$ and $u(r_k; u_0^i)$ to get $u(R; u_0^i)$. The value of $\frac{\partial}{\partial u_0}u(R; u_0^i)$ is obtained in a similar way. In Fig. 2 the solutions corresponding to the different branches in Fig. 1 are depicted in terms of the θ-variable. It is interesting to note that the solutions corresponding to $\mathcal{B}_k$ have k maxima.

Next we fix the radius R or θ_1, resp. and compute the solution branches $(u(r; \lambda), \lambda)$. For this purpose we use linear finite elements and the Newton iteration. We are looking for weak solutions $u \in \{W^{1,2}([0, R]), u(R) = 0\} =: W^{1,2}$ satisfying

$$
\begin{aligned}
0 &= \int_0^R \rho^{n-2}r^{n-1}u'\,\varphi' - \lambda\rho^n r^{n-1}u\,\varphi dr - \int_0^R \rho^n r^{n-1}|u|^{2^*-2}u\varphi dr \\
&=: G(u, \lambda)[\varphi] \quad \forall \varphi \in C^\infty([0, R]), \varphi(R) = 0.
\end{aligned}
\tag{2.5}
$$

C. Bandle and S. Stingelin

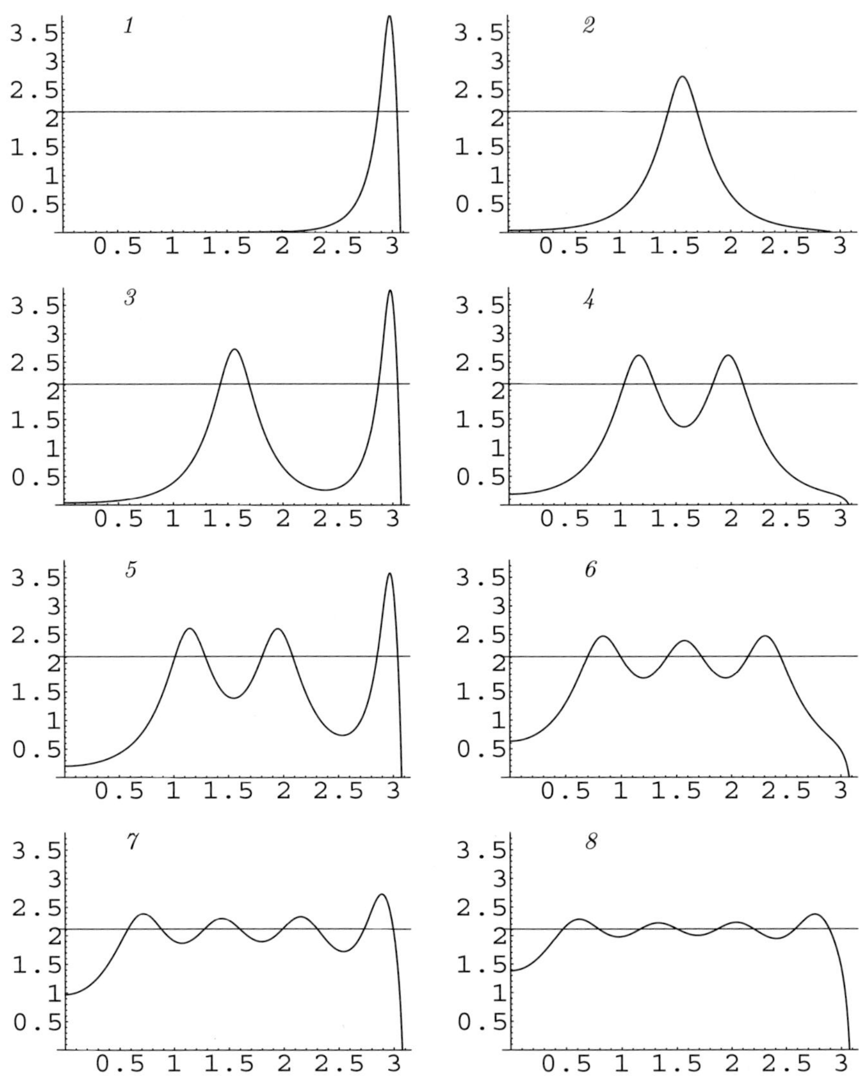

FIGURE 2. Solutions depicted in terms of the θ-variable for $n = 3$, $\lambda = -20$ with $R = 30$.

If the solution branch $\Gamma = \{(u, \lambda) : G(u, \lambda) = 0\}$ has a singular point where the Fréchet derivative G_u vanishes, we use Keller's method [5] which consists in introducing an appropriate parametrization s of the solution branch $x(s) = (u(s), \lambda(s))$. It is determined by the equation

$$P(x(s), s) = \left(\begin{array}{c} G(u, \lambda) \\ N(u, \lambda, s) \end{array} \right) = 0,$$

where $N(u, \lambda, s)$ is chosen such that s is the pseudo arclength introduced by Keller. In our case we set

$$N(u, \lambda, s) = \frac{1}{2} \int_0^R \rho^{n-2} r^{n-1} \dot{u}'(s_0)(u'(s) - u'(s_0)) dr$$
$$+ \frac{1}{2} \dot{\lambda}(s_0)(\lambda(s) - \lambda(s_0)) - (s - s_0), \quad \text{where} \quad \dot{} = \frac{d}{ds}.$$

On a smooth branch $x(s)$ satisfies

$$\mathcal{A}(s) \cdot \dot{x}(s) = - \left(\begin{array}{c} 0 \\ N_s(u, \lambda, s) \end{array} \right) \tag{2.6}$$

with

$$\mathcal{A}(s) := \left(\begin{array}{cc} G_u(u, \lambda) & G_\lambda(u, \lambda) \\ N_u(u, \lambda, s) & N_\lambda(u, \lambda, s) \end{array} \right).$$

Equation (2.6) will be solved numerically in a finite-dimensional subspace of $W^{1,2}$. For more details we refer to [6] and [7].

3. Numerical results

The structure of the solution branches becomes most transparent if we choose as parameter axis (λ, u_0) and $(\lambda, -\partial_\lambda I_\lambda)$. In our case we have

$$-\partial_\lambda I_\lambda = \frac{1}{2} \int_{B_R} u^2 \rho^n dx.$$

We denote by $\mathcal{B}_k$, resp. $\mathcal{B}_{-k}$ the solution branches with $u_0 < |\lambda|^{\frac{1}{2^*-2}}$ and k maxima or with $u_0 > |\lambda|^{\frac{1}{2^*-2}}$ and k minima, resp. The branch $\mathcal{B}_0$ is the branch of minimizers emanating from λ_1. Notice that the branches $\mathcal{B}_k$ have a different meaning than in Fig. 1 where λ is kept fixed and the radius varies.

Observations and open problems

1. It is interesting to note that in the (λ, u_0) diagram the branches $\mathcal{B}_k$ and $\mathcal{B}_{-k}$ are separated by the constant solution $|\lambda|^{\frac{n-2}{4}}$. From the $(\lambda, -\partial I_\lambda)$ diagram we infer that $-\partial I_\lambda(u_\lambda)$ is bounded above by some function which behaves like $-\partial_\lambda I_\lambda(|\lambda|^{\frac{n-2}{4}}) = \frac{|\lambda|^{\frac{n-2}{2}}}{2} \int_{B_R} \rho^n dx$ for large $|\lambda|$.

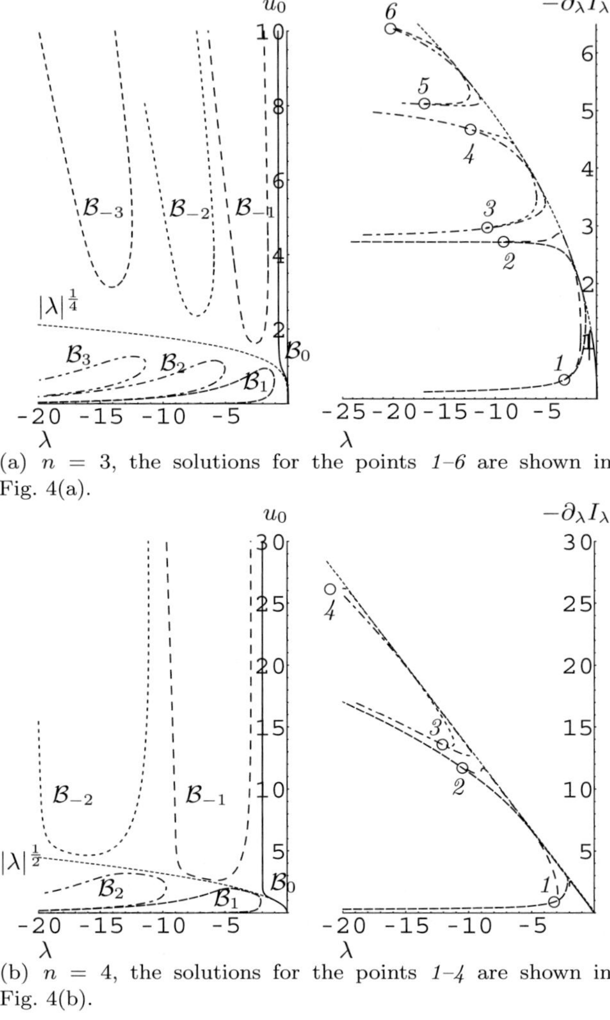

(a) $n = 3$, the solutions for the points *1–6* are shown in Fig. 4(a).

(b) $n = 4$, the solutions for the points *1–4* are shown in Fig. 4(b).

FIGURE 3. Solution paths for $R = 25$.

2. In $\mathbb{S}^3$ the existence of $\mathcal{B}_k$ was established in [3]. The asymptotic analysis for $-\lambda \to \infty$ will be carried out in a forthcoming paper [2]. There the phenomenon of clustering will be discussed in a more general setting. For small λ the solutions of $\mathcal{B}_k$ have a clustered layer at $\theta = \pi/2$ and some have also an additional boundary layer.

3. In contrast to $\mathcal{B}_k$ the branches $\mathcal{B}_{-k}$ seem to be linked to the critical Sobolev exponent. No analytical existence proof is available yet. In the diagram

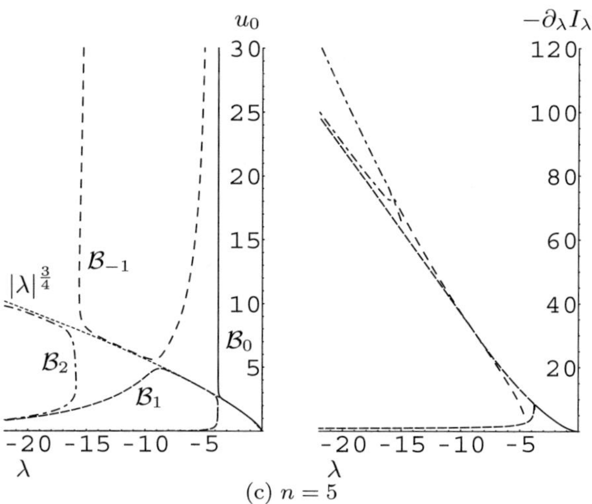

FIGURE 3. (Cont.) Solution paths for $R = 25$.

$(\lambda, -\partial_\lambda I_\lambda)$, cf. Fig. 3 at the points $1, 2, \ldots$, we can make the following observation. Let $u_{\lambda^{1,2}}$ denote the solutions on $\mathcal{B}_{-k}$. Then there exist λ_k such that $-\partial I_{\lambda_k}(u_{\lambda_k^{1,2}}) = -\partial I_{\lambda_k}(v_{\lambda^{1,2}})$, cf. Fig. 4, where $v_{\lambda^{1,2}}$ are solutions of $\mathcal{B}_k$ with a singularity at the origin.

4. The behavior of the solution branches $\mathcal{B}_k$ can be described in the following way: on the first branch $\mathcal{B}_1$, the solution on the lower part (denoted by $u_{1,1}$ in Fig. 4), with a maximum nearby the boundary, is stable and goes over into an unstable solution (denoted by $u_{1,2}$) with a maximum at the equator. The lower part of the second branch starts from a linear combination of the stable solution $u_{1,1}$ and the unstable solution $u_{1,2}$ and goes also over into a solution with two maxima symmetric around the equator. This behavior can be generalized in the following way: the solutions on $\mathcal{B}_k$ starts on the lower part of the branch given by a linear combination of $u_{1,1}$ and $u_{k,2}$ and goes over into a solution with k maxima symmetric around the equator.

$$u_{1,1} \qquad\qquad \rightarrow u_{1,2}$$
$$u_{2,1} \approx u_{1,2} + u_{1,1} \quad \rightarrow u_{2,2}$$
$$\vdots$$
$$u_{k,1} \approx u_{k-1,2} + u_{1,1} \rightarrow u_{k,2}$$
$$u_{k+1,1} \approx u_{k,2} + u_{1,1} \rightarrow u_{k+1,2}$$
$$\vdots$$

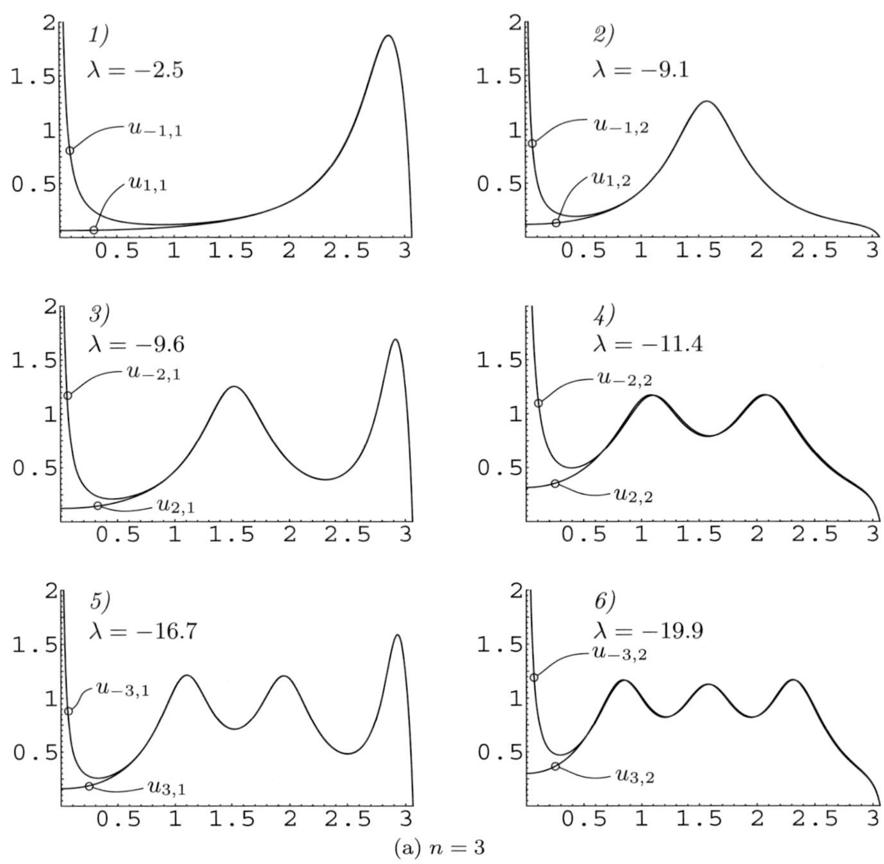

FIGURE 4. Solutions $u_{\pm k,j} \in \mathcal{B}_{\pm k}$ with $j = 1, 2$ depicted in terms of the θ-variable for $R = 25$, scaled with $|\lambda|^{-1/(2^*-2)}$.

References

[1] C. Bandle and R. Benguria, *The Brezis-Nirenberg Problem on* $\mathbb{S}^n$, J. Diff. Equ. **178** (2002), 264–279.

[2] C. Bandle, S. Stingelin and Juncheng Wei, *Multiple clustered layer solutions for semilinear elliptic problems on* $\mathbb{S}^n$, in preparation.

[3] H. Brezis and L.A. Peletier, *Elliptic equations with critical exponent on* $\mathbb{S}^3$*: new non-minimising solutions*, C. R. A. S. Paris Ser., **339** (2004), 391–394.

[4] H. Brezis and L. Nirenberg, *Positive solutions of nonlinear elliptic equations involving critical Sobolev exponents*, Comm. Pure Appl. Math. **36** (1983), 437–477.

[5] H.B. Keller, *Numerical solution of bifurcation and nonlinear eigenvalue problems*, in Applications of bifurcation theory (Proc. Advanced Sem. Univ. Wisconsin, Madison Wis. 1976), Academic Press, New York (1977), 359–384.

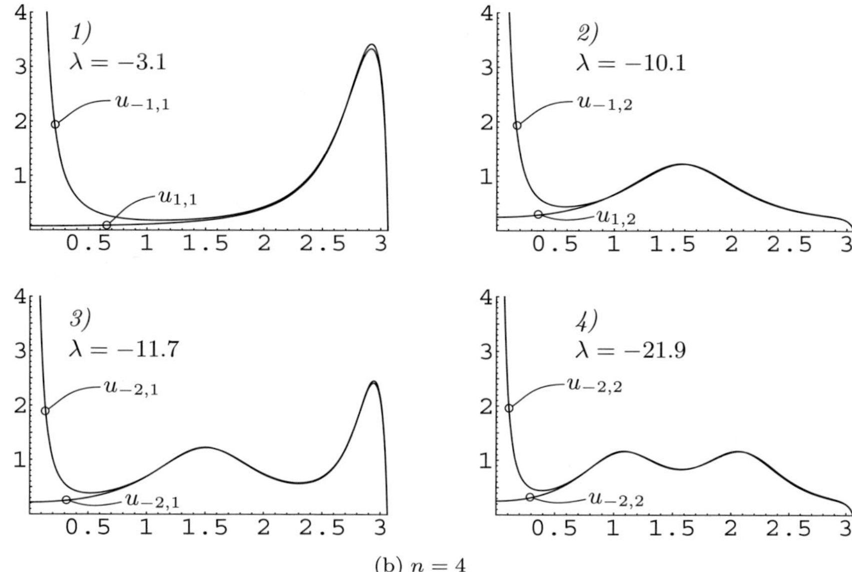

(b) $n = 4$

FIGURE 4. (Cont.)
Solutions $u_{\pm k,j} \in \mathcal{B}_{\pm k}$ with $j = 1,2$ depicted in terms of the θ-variable for $R = 25$, scaled with $|\lambda|^{-1/(2^*-2)}$.

[6] S. Stingelin, *New numerical solutions for the Brezis-Nirenberg problem on* $\mathbb{S}^n$, Preprint 2003-15, Department of Mathematics, University of Basel, `www.math.unibas.ch`, (2003)

[7] S. Stingelin, *Das Brezis-Nirenberg Problem auf der Sphäre* $\mathbb{S}^n$, Inauguraldissertation, Universität Basel, 2004.

Catherine Bandle
Mathematisches Institut
Universität Basel
Rheinsprung 21
CH-4051 Basel, Switzerland
e-mail: `catherine.bandle@unibas.ch`

Simon Stingelin
Endress+Hauser Flowtec AG
Kägenstr.7
CH-4153 Reinach, Switzerland
e-mail: `simon.stingelin@flowtec.endress.com`

Progress in Nonlinear Differential Equations
and Their Applications, Vol. 63, 23–32

On some Boundary Value Problems for Incompressible Viscous Flows with Shear Dependent Viscosity

H. Beirão da Veiga

*Dedicated to Haim Brezis in the occasion of his sixtieth birthday
with my warmest wishes and admiration*

Abstract. In the sequel we discuss some regularity results *up to the boundary* for solutions to the Navier-Stokes equations with shear dependent viscosity, under slip and non-slip boundary conditions, proved in references [3] and [4]. In this talk we show the main lines of the proofs.

1. Introduction

In the following we consider the well-known Navier-Stokes system of equations with shear dependent viscosity

$$\begin{cases} \dfrac{\partial u}{\partial t} + u \cdot \nabla u - \nabla \cdot T(u, \pi) = f, \\[2mm] \nabla \cdot u = 0, \end{cases} \tag{1.1}$$

where

$$T = -\pi I + \nu_T(u)\, \mathcal{D} u, \quad \mathcal{D} u = \nabla u + \nabla u^T, \quad \nu_T(u) = \nu_0 + \nu_1 |\mathcal{D} u|^{p-2}$$

and ν_0, ν_1 are strictly positive constants. The above stress tensor T satisfies the Stokes Principle, see [11] page 231. For $p = n = 3$, this is the classical Smagorinsky turbulence model, see [12].

The first mathematical studies on the above kind of equations go back to [5], [6] and [7]. For some references see, for instance, [8] and [3]. Essential existence, uniqueness and regularity results for these type of models under the non-slip boundary condition (1.3) are proved in [8], where new ideas and techniques are developed. In particular it is proved that

$$u \in L^{\frac{2}{p-1}}(0, T; W^{2, \frac{6}{p+1}}) \tag{1.2}$$

for each $p \in\,]2 + \frac{1}{4}, 3[$; see [8] Theorem 1.17.

Theoretical contributions to the system (1.1) mostly concern the nonslip boundary condition

$$u_{|\Gamma} = 0. \tag{1.3}$$

However, the nonhomogeneous slip type boundary condition

$$\begin{cases} (u \cdot \underline{n})_{|\Gamma} = 0, \\ \beta\, u_\tau + \underline{\tau}(u)_{|\Gamma} = b(x), \end{cases} \tag{1.4}$$

appears to be quite important in many fields. Here $\underline{n}$ is the unit outward normal to the domain's boundary Γ, $\beta \geq 0$ is a given constant and $b(x)$ is a given tangential vector field. We denote by $\underline{t} = T \cdot \underline{n}$ the normal component of the tensor T, by $u_\tau = u - (u \cdot \underline{n})\,\underline{n}$ the tangential component of u and by $\underline{\tau}$ the tangential component of $\underline{t}$

$$\underline{\tau}(u) = \underline{t} - (\underline{t} \cdot \underline{n})\underline{n}. \tag{1.5}$$

In the pioneering paper [13] the authors give a fundamental contribution to mathematical study of this type of boundary conditions. A very complete mathematical study is presented in references [1] and [2] (to which we also refer for some references).

In reference [4], by appealing to the results proved in reference [3], we prove regularity results for solutions to the full Navier-Stokes evolution system (1.1) under the boundary conditions (1.3) or (1.4) and given initial data. However, to simplify exposition and notation we consider in the sequel only the non-slip boundary condition (1.3) and assume $n = 3$. We denote by Ω a bounded, open, regular set.

We set

$$V = \left\{ v \in W_0^{1,p}(\Omega) : \nabla \cdot v = 0 \right\}$$

endowed with the canonical $W^{1,p}(\Omega)$-norm and recall the following Korn's type inequality. For each $v \in V$

$$\|v\|_{1,p} \leq c(p, \Omega)\|\mathcal{D}\, v\|_p \,.$$

Let us write (1.1) in the more explicit form

$$\begin{cases} \frac{\partial u}{\partial t} + (u \cdot \nabla)\, u - \\ \nu_0 \nabla \cdot \mathcal{D}\, u - \nu_1 \nabla \cdot \left(|\mathcal{D}\, u|^{p-2}\, \mathcal{D}\, u \right) + \nabla \pi = f, \\ \nabla \cdot u = 0, \\ u(0) = u_0(x). \end{cases} \tag{1.6}$$

In [4] we prove the following results (which holds, in a similar form, under the boundary condition (1.4)). The exponents $\overline{p}$, l and m are defined as follows.

$$\overline{p} = \frac{2p}{3p - 4}\,; \quad l = \frac{3\,(4 - p)}{5 - p}\,; \quad m = \frac{6\,(4 - p)}{8 - p}\,. \tag{1.7}$$

Theorem 1.1. *Let be $2 + \frac{2}{5} \le p < 4$ and let u be a weak solution to problem (1.6), (1.3). Assume, moreover, that $u_0 \in V$ and $f \in L^2(0, T; L^2(\Omega))$. Then*

$$\begin{cases} u \in L^2(0, T; W^{2,p'}) \cap L^\infty(0, T; W^{1,p}) \,, \\ \nabla \pi \in L^2(0, T; L^{\bar{p}}) \,, \\ \frac{\partial u}{\partial t} \in L^2(0, T; L^2) \,. \end{cases} \tag{1.8}$$

Theorem 1.2. *Let be $2 + \frac{2}{5} \le p < 3$. Moreover, let u be a weak solution to problem (1.6), (1.3) where $u_0 \in V$ and $f \in L^2(0, T; L^2(\Omega))$. Then*

$$\begin{cases} u \in L^{4-p}(0, T; W^{2,l}) \cap L^\infty(0, T; W^{1,p}) \,, \\ \nabla \pi \in L^{\frac{2(4-p)}{p}}(0, T; L^m) \,, \\ \frac{\partial u}{\partial t} \in L^2(0, T; L^2) \,. \end{cases} \tag{1.9}$$

Theorem 1.3. *The regularity results stated in the above two theorems hold as well for $2 + \frac{1}{4} \le p \le 2 + \frac{2}{5}$.*

This last result follows by appealing to the regularity result (1.2) proved in reference [8]. Hence $n = 3$ and the boundary value problem (1.3) are formally required here.

We point out that if we drop the term $(u \cdot \nabla) u$, then all the above results hold for $p \ge 2$. It is significant that all the exponents that appear in equations (1.2), (1.8) and (1.9) are equal to 2 when $p = 2$.

It is worth noting that the aim of these notes is to show the structure of the proofs developed in references [3] and [4]. In proving regularity results *up to the boundary* for weak solutions of the system (1.6) the interaction between the nonlinear terms containing $\nabla u + \nabla u^T$ and the boundary conditions are the really *new obstacles* to face here. The presence of the convection term, a non flat boundary, and the time dependence can be tackled by appealing to more or less involved but well-known techniques. This leads us in reference [3] to concentrate our attention on the following stationary problem:

$$\begin{cases} -\nu_0 \nabla \cdot \left(\nabla u + \nabla u^T\right) - \\ \nu_1 \nabla \cdot \left(|\nabla u + \nabla u^T|^{p-2} \left(\nabla u + \nabla u^T\right)\right) + \nabla \pi = f, \\ \nabla \cdot u = 0 \,. \end{cases} \tag{1.10}$$

2. The Stokes stationary problem

Theorem 2.1. *Assume that $2 < p$ and $f \in L^2(\Omega)$. Let u, π be a weak solution in Ω to problem (1.10) under the boundary condition (1.3). Then $u \in W^{2,p'}(\Omega)$ and*

$$\|u\|_{2,p'} \le c \left(\|\mathcal{D} u\|_p^{p-1} + \|f\| + \|\mathcal{D} u\|_p^{\frac{p-2}{2}} \|f\|\right) \,. \tag{2.1}$$

Moreover, if $p < 4$, $\nabla\pi \in L^{\bar{p}}(\Omega)$ and

$$\|\nabla\pi\|_{\bar{p}} \leq c\left(\|\mathcal{D}u\|_p^{p-1} + \|f\| + \|\mathcal{D}u\|_p^{p-2}\|f\|\right). \tag{2.2}$$

Note that $\|\mathcal{D}u\|_p$ is bounded. See equation (2.6) below.

Theorem 2.2. *Assume that $n = 3$ and $2 < p < 3$. Let f, u and π be as in Theorem 2.1. Then $u \in W^{2,l}(\Omega)$ and $\nabla\pi \in L^m(\Omega)$. Moreover,*

$$\|u\|_{2,l} \leq c\left(\|\mathcal{D}u\|_p^{p-1} + \|f\| + \|\mathcal{D}u\|_p + \|f\|^{\frac{2}{4-p}}\right), \tag{2.3}$$

and

$$\|\nabla\pi\|_m \leq c\left(|\mathcal{D}u\|_p + \|\mathcal{D}u\|_p^{\frac{p(p-1)}{2}} + \|f\| + \|f\|^{\frac{p}{4-p}}\right). \tag{2.4}$$

Theorems 2.1 and 2.2 with Ω replaced by the half-space $\mathbb{R}_+^n$ are proved in reference [3]. The extension of the proofs to Ω can be essentially done by appealing to the results proved for the half-space case in [3] together with techniques of localization and flatten the boundary. In particular, the technique introduced in reference [2] applies here. For an alternative technique see [8].

As in [4], we illustrate the core of the proof by considering the problem in the half-space. In this case the set of points for which $x_3 = 0$ correspond to the boundary of Ω.

In the half-space case the fact that the canonical inclusions $L^{p_1} \hookrightarrow L^{p_0}$, $p_0 < p_1$, fails leads to additional technicalities and heavier notation. To avoid here this situation (regularity has a local character) assume for convenience that our solution u has compact support in a half-sphere $B_R^+ = \{x : |x| < R, x_3 \geq 0\}$, for some $R > 0$.

A vector field $u \in V$ is a *weak solution* to problem (1.10), (1.3) if

$$\frac{1}{2}\int_\Omega \nu_T(u)\,\mathcal{D}u \cdot \mathcal{D}v\,dx = \int_\Omega f \cdot v\,dx, \tag{2.5}$$

for all $v \in V$. For some details see the equations (2.2) and (2.5) in reference [3]. By replacing v by u in equation (2.5) one easily proves the basic estimate

$$\frac{\nu_0^2}{2}\|\nabla u\|^2 + \nu_0\,\nu_1\,\|\mathcal{D}u\|_p^p \leq c\,[f]_{-1}^2. \tag{2.6}$$

Note that $\|\nabla u\|_p \leq c\|\mathcal{D}u\|_p$.

In the sequel we denote by $D^2 u$ the set of all the second derivatives of u. The meaning of expressions like $\|D^2 u\|$ is clear. The symbol $D_*^2 u$ may denote any of the second order derivatives $\partial^2 u_j/\partial x_i\,\partial x_k$ except for the derivatives $\partial^2 u_j/\partial x_n^2$, if $j < n$. Moreover,

$$|D_*^2 u|^2 := \left|\frac{\partial^2 u_n}{\partial x_n^2}\right|^2 + \sum_{\substack{i,j,k=1 \\ (i,k)\neq(n,n)}}^{n} \left|\frac{\partial^2 u_j}{\partial x_i\,\partial x_k}\right|^2.$$

Similarly, ∇^* may denote any first order partial derivative, except for $\partial\,/\partial x_n$.

The proof of theorems 2.1 and 2.2 in the "half-space version" given in reference [3] consists on a sequence of steps denoted below by (a), (b), (c), (b1) and (c1). Steps (a), (b) and (c) correspond to Theorem 2.1. Steps (b1) and (c1) correspond to Theorem 2.2. In each step we prove the results shown in the following box.

(a) $\left\{ \begin{array}{l} D_*^2 \, u \\ |\mathcal{D}u|^{p-2} \, \nabla^* \, \mathcal{D}u \end{array} \right\} \in L^2(\Omega).$

(b) $\left\{ \begin{array}{l} D^2 \, u \\ |\mathcal{D}u|^{\frac{p-2}{2}} \, \nabla^* \, \mathcal{D}u \\ \nabla^* \, \pi \end{array} \right\} \in L^{p'}(\Omega).$

(c) $\nabla \, \pi \in L^{\bar{p}}(\Omega).$

(b1)   in step (b) replace p' by $l.$

(c1)   in step (c) replace $\bar{p}$ by $m.$

Note that there is a loss of regularity in going from tangential to normal derivatives and in going from u to π. In Theorems 2.1 and 2.2 (bounded set Ω) we do not take into account "additional regularity" in the tangential directions.

Next we illustrate the more significant points in the proofs of each of the above steps.

Step (a): This is the very basic step. The crucial estimate is

$$\nu_0 \int |\mathcal{D}\frac{\partial u}{\partial x_k}|^2 \, dx \tag{2.7}$$

$$+ \nu_1 \int \left\{ |\mathcal{D}u|^{p-2} \, |\mathcal{D}\frac{\partial u}{\partial x_k}|^2 + (p-2) \, |\mathcal{D}u|^{p-4} \left(\mathcal{D}u \cdot \mathcal{D}\frac{\partial u}{\partial x_k} \right)^2 \right\} \, dx \leq c \, \nu_0^{-1} \, \|f\|^2,$$

for each index k, $k \neq n$. This estimate yields

$$\nu_0 \, \|D_*^2 \, u\|^2 + \nu_1 \sum_{k=1}^{n-1} \left\| \, |\mathcal{D}u|^{\frac{p-2}{2}} \, \mathcal{D}\frac{\partial u}{\partial x_k} \right\|^2 \leq c \, \nu_0^{-1} \, \|f\|^2. \tag{2.8}$$

The main ingredients in the proof of steep (a) are Nirenberg's translation method, a Taylor expansion's lemma in a symmetric form and Fatou's Lemma.

Remark. It is worth noting that the application of the translation method to obtain the L^2 estimates for the D_*^2 derivatives is not obstructed by the presence of the non linear second-order term. This is a main point in proving regularity for second order derivatives of solutions in the presence of the shear viscosity. The convexity of the function $\psi(U) = |U|^p$ plays here a fundamental role. To show this point to the reader in a simple context we prove in the appendix the estimate (2.8) in a very particular one-dimensional case.

Step (b): Statement $(b)_2$ (i.e., the second statement (b)) follows immediately from $(a)_2$ by appealing to Hölder's inequality and to the boundedness of $\|\mathcal{D}u\|_p$.

Statement $(b)_3$: Differentiation of the first equation (1.10) with respect to x_k, $k \neq 3$, shows that

$$\nabla \frac{\partial \pi}{\partial x_k} = \nabla \cdot \left[-\nu_0 \mathcal{D} \frac{\partial u}{\partial x_k} \right] + \nabla \cdot \left[-\nu_1 \frac{\partial}{\partial x_k} \left(|\mathcal{D}u|^{p-2} \mathcal{D}u \right) \right] + \nabla \cdot G, \qquad (2.9)$$

where, for uniformity of notation we introduce $G_{ij} = \delta_{kj} f_i$ ($\nabla \cdot G = \frac{\partial f}{\partial x_k}$ and $\|G\| = \|f\|$).

By appealing to the estimates (a) and $(b)_2$ one shows that the term inside square brackets belongs to $L^{p'}$. A classical result due to J. Nečas shows that

$$\frac{\partial \pi}{\partial x_k} \in L^{p'},$$

for each $k \neq n$. Let us recall this result (see [9]) in the form needed here. Let $g(x)$ be a scalar field defined in B_R^+ such that

$$g = \nabla \cdot w_0, \text{ and } \nabla g = \nabla \cdot W,$$

where w_0 and W belong to $L^\alpha(B_R^+)$, for some $\alpha > 1$. Then

$$\|g\|_{L^\alpha(B_R^+)} \leq c \left(R \|w_0\|_{L^\alpha(B_R^+)} + \|W\|_{L^\alpha(B_R^+)} \right),$$

where c is independent of R (by a scaling argument).

Statement $(b)_1$: For almost all $x \in \mathbb{R}_+^3$ we consider the $(n-1) \times (n-1)$ linear system consisting on the $n-1$ first equations (1.10) in terms of the unknowns

$$\frac{\partial^2 u_j}{\partial x_n^2}, \quad \text{for} \quad j \neq n.$$

Straightforward calculations show that the matrix A of this system is symmetric positive definite and that its eigenvalues are larger or equal to $\nu_0 + \nu_1 |\mathcal{D}u|^{p-2}$. Hence,

$$det\, A \geq (\nu_0 + \nu_1 |\mathcal{D}u|^{p-2})^{n-1}.$$

By solving the system for the above unknowns and by appealing to the estimates proved in the previous steps one easily proves that that

$$\frac{\partial^2 u_j}{\partial x_n^2} \in L^{p'}, \quad j \neq n.$$

Step (c): The nth equation (1.10) gives $\frac{\partial \pi}{\partial x_n}$ in terms of quantities already estimated in the previous steps. Straightforward calculations lead to

$$\frac{\partial \pi}{\partial x_n} \in L^{\bar{p}}.$$

Statements (b_1) and (c_1): Actually, in the proof of Theorem 2.2 the statements (b1) and (c1) come out together. Let us just do some comment on this result. Let

$$u \in W^{1,p} \qquad\qquad (2.10)$$

be a weak solution. Then, by Theorem 2.1, u belongs to $W^{2,p'}$ and a Sobolev embedding theorem shows that

$$u \in W^{1,q}, \tag{2.11}$$

where $\frac{1}{q} = \frac{1}{p'} - \frac{1}{n}$. Hence $q > p$ if $p < \frac{2n}{n-1}$. In conclusion, if $p < 3$, we win some regularity for the first-order derivatives. Now the question is if this win allows to prove better regularity for the second-order derivatives by turning back to the beginning of the proof of Theorem 2.1 and by replacing in the proofs (2.10) by (2.11). In this regard we have to take into account that this second tour does not correspond to the previous one since the power p in the system (1.10) is not replaced by q.

Hence the real question is the following. Assume that for some $q \geq p$ the weak solution u of system (1.10) belongs to $W^{1,q}$. What kind of $W^{2,s}-$ regularity can we prove for u?

Since we know that $u \in W^{2,p'}$ we may replace the above question by the following one. If $u \in W^{2,s}$ for some $s \geq p'$, may we prove that $u \in W^{2,r}$ for some $r \geq s$? In [3] we prove the following result.

Proposition 2.1. *Let u be a weak solution to the problem (1.10) and assume that $D^2 u \in L^s$. Then*

$$D^2 u, \nabla^* \pi, |Du|^{p-2} \nabla^* Du \in L^r, \tag{2.12}$$

where

$$r = \phi_p(s) := \frac{6\,s}{(5 - p)\,s + 3\,(p - 2)}.$$

In particular,

$$u \in W^{2,s} \quad \Rightarrow \quad u \in W^{2,r}.$$

The above proposition allows us, by a bootstrap argument, to make any finite number of regularizing steps. The next question is wether we may "go to the limit" in order to arrive to the exponent r for which $r = s$ (i.e., r is the fixed point $l = \phi_p(l)$ of the map ϕ_p). This requires sharp estimates at each stage of the proof. We succeed in proving these estimates and the desired result. In fact, the above fixed point l is just the exponent l defined in (1.7).

3. The evolution problem

Multiplication side by side of (1.6) by u, integration in Ω and suitable integrations by parts show that

$$\frac{d}{dt}\|u(t)\|^2 + \nu_0\,\|u\|_1^2 + \nu_1\,\|u\|_{1,p}^p \leq \frac{c}{\nu_0}\,[f]_{-1}^2, \tag{3.1}$$

30   H. Beirão da Veiga

where $[f]_{-1}$ denotes the norm in H^{-1}. By integration of (3.1) with respect to time and by taking into account the boundary condition (1.3) it follows

$$\|u(t)\|^2_{L^\infty(0,T;L^2)} + \nu_0 \|u\|^2_{L^2(0,T;H^1)} + \nu_1 \|u\|^p_{L^p(0,T;W^{1,p})}$$
$$\leq c \|u(0)\|^2 + \tfrac{c}{\nu_0} \|f\|^2_{L^2(0,T;H_0^{-1})}. \tag{3.2}$$

Next we multiply side by side (1.6) by $\frac{\partial u}{\partial t}$ and integrate in Ω. Suitable integrations by parts lead to the following equation

$$\|\tfrac{\partial u}{\partial t}\|^2 + \tfrac{\nu_0}{2} \tfrac{d}{dt} \|\mathcal{D}u\|^2 + \tfrac{\nu_1}{2p} \tfrac{d}{dt} \|\mathcal{D}u\|^p_p + \int_\Omega (u \cdot \nabla) u \cdot \tfrac{\partial u}{\partial t}\, dx$$
$$= \int_\Omega f \cdot \tfrac{\partial u}{\partial t}\, dx. \tag{3.3}$$

On the other hand, by appealing in particular to a Sobolev embedding theorem one shows that

$$\|(u \cdot \nabla) u\| \leq \|\mathcal{D}u\|^2_p, \tag{3.4}$$

provided that $p \geq \frac{4n}{n+2}$. For $n = 3$ this leads to the value $2 + \frac{2}{5}$ referred in Theorems 1.1 and 1.2. From (3.3) and (3.4) one gets

$$\|\tfrac{\partial u}{\partial t}\|^2 + \nu_0 \tfrac{d}{dt} \|\mathcal{D}u\|^2 + \nu_1 \tfrac{d}{dt} \|\mathcal{D}u\|^p_p \leq$$
$$c \left(\|f\|^2 + \|\mathcal{D}u\|^{4-p}_p \|\mathcal{D}u\|^p_p \right). \tag{3.5}$$

Furthermore (3.2) shows that $\|\mathcal{D}u\|^{4-p}_p \in L^1(0, T)$. It readily follows that

$$\frac{\partial u}{\partial t} \in L^2(0,T; L^2(\Omega)) \quad \text{and} \quad \mathcal{D}u \in L^\infty(0,T; L^p(\Omega)).$$

Hence

$$F(t) \in L^2(0,T; L^2(\Omega)) \quad \text{and} \quad u \in L^\infty(0,T; W^{1,p}(\Omega)), \tag{3.6}$$

where

$$F(t) = f - (u \cdot \nabla) u - \frac{\partial u}{\partial t}.$$

Next we write the equation (1.6) in the form

$$-\nu_0 \nabla \cdot \mathcal{D}u - \nu_1 \nabla \cdot \left(|\mathcal{D}u|^{p-2} \mathcal{D}u\right) + \nabla \pi = F \tag{3.7}$$

and apply, for each $t \in (0,T)$, the estimate (2.1) with f replaced by F. By integrating side by side this last estimate in $(0,T)$ and by taking into account (3.6) one proves that

$$u \in L^2(0,T; W^{2,p'}),$$

as claimed in Theorem 2.1. Finally, by replacing in the above argument the estimate (2.1) by the estimate (2.3) one shows that

$$u \in L^{4-p}(0,T; W^{2,l}),$$

as claimed in Theorem 2.2.

4. Appendix

Here we prove the estimate (2.8) in the particular case $n = 1$. We start by the following Taylor expansion's formula for $n = 1$. By setting $\psi(U) = |U|^p$ one has

$$\psi(U) = \psi(V) + \psi'(V)(U - V) + \frac{1}{2}\psi''(\overline{V})(U - V)^2,$$

where $\overline{V}$ is a point between U and V. By exchanging U and V in the above equation and by adding side by side the two equations one easily gets

$$(|U|^{p-2}U - |V|^{p-2}V)(U - V) = \frac{p-1}{2}(|\overline{U}|^{p-2} + |\overline{V}|^{p-2})|U - V|^2, \qquad (4.1)$$

where

$$\overline{U} = \alpha U + (1 - \alpha)V, \quad \overline{V} = \beta U + (1 - \beta)V \quad \text{and} \quad 0 < \alpha, \beta < 1.$$

Clearly, in the case $n = 1$ we drop pressure and boundary conditions and, moreover, we do not subject the solution u to the divergence free constraint. The variational formulation (2.5) of our problem reads: $u \in V = H^1(\mathbb{R})$ satisfies

$$\nu_0 \int_{\mathbb{R}} u'(x)\,v'(x)\,dx + \nu_1 \int_{\mathbb{R}} |u'(x)|^{p-2}\,u'(x)\,v'(x)\,dx = \int_{\mathbb{R}} f\,v\,dx, \qquad (4.2)$$

for all $v \in V$. Application of the classical translation method in the very usual way leads to the estimate

$$\nu_0 \int_{\mathbb{R}} \left|\frac{u'(x) - u'(x - h)}{h}\right|^2 dx$$

$$+ \nu_1 \int_{\mathbb{R}} \frac{(|u'(x)|^{p-2}\,u'(x) - |u'(x - h)|^{p-2}\,u'(x - h))\,(u'(x) - u'(x - h))}{h} \cdot \frac{1}{h}\,dx$$

$$\leq \|f\| \left\|\frac{u'(x) - u'(x - h)}{h}\right\|, \qquad (4.3)$$

By setting $U = u'(x)$ and $V = u'(x - h)$ in (4.1) and by introducing these relations in the equation (4.3) one easily shows that

$$\frac{\nu_0}{2}\int_{\mathbb{R}}\left|\frac{u'(x) - u'(x - h)}{h}\right|^2 dx \qquad (4.4)$$

$$+ \nu_1\frac{p-1}{2}\int_{\mathbb{R}}\left|\frac{u'(x) - u'(x - h)}{h}\right|^2 (|\widehat{U}(x)|^{p-2} + |\widehat{V}(x)|^{p-2})\,dx \leq \frac{1}{2\nu_0}\|f\|^2,$$

where

$$\widehat{U}(x) = \alpha(x)\,u'(x) + (1 - \alpha(x))\,u'(x - h),$$

$$\widehat{V}(x) = \beta(x)\,u'(x) + (1 - \beta(x))\,u'(x - h)$$

and $0 < \alpha(x), \beta(x) < 1$ a.e. in $\mathbb{R}$. In particular, as $h \to 0$, $\widehat{U}(x) \to u'(x)$ and $\widehat{V}(x) \to u'(x)$ a.e. in $\mathbb{R}$. On the other hand, as a first consequence of (4.4) one gets $\nu_0 \, \|u''\|^2 \leq \nu_0^{-1} \, \|f\|^2$. In particular, as $h \to 0$,

$$\frac{u'(x) - u'(x - h)}{h} \to u''(x)$$

a.e. in $\mathbb{R}$.

The above picture allows us to pass to the limit in (4.4) by appealing to Fatou's lemma. This proves the estimate (2.8) in the particular case $n = 1$.

References

[1] Beirão da Veiga, H.; Regularity of solutions to a nonhomogeneous boundary value problem for general Stokes systems in $\mathbb{R}^n_+$. *Math. Annalen*, **328** (2004), 173–192.

[2] Beirão da Veiga, H.; Regularity for Stokes and generalized Stokes systems under nonhomogeneous slip type boundary conditions. *Advances Diff. Eq.*, **9**, no. 9–10, (2004), 1079–1114.

[3] Beirão da Veiga, H.; On the regularity of flows with Ladyzhenskaya shear dependent viscosity and slip and non-slip boundary conditions. *Comm. Pure Appl. Math.*, **58** (2005), 552–577.

[4] Beirão da Veiga, H.; On the regularity of flows with Ladyzhenskaya shear dependent viscosity and slip and non-slip boundary conditions. Part II, to appear.

[5] Ladyzhenskaya, O.A.; On nonlinear problems of continuum mechanics. *Proc. Int. Congr. Math.(Moscow, 1966)*, 560–573. Nauka, Moscow, 1968. English transl. in Amer.Math. Soc. Transl.(2) 70, 1968.

[6] Ladyzhenskaya, O.A.; Sur des modifications des équations de Navier-Stokes pour des grand gradients de vitesses. *Séminaire Inst. Steklov* **7** (1968), 126–154.

[7] Ladyzhenskaya, O.A.; *The Mathematical Theory of Viscous Incompressible Flow*. Second edition. Gordon and Breach, New-York, 1969.

[8] Malek, J., Nečas, J., Ružička, M.; On weak solutions to a class of non-Newtonian incompressilble fluids in bounded three-dimensional domains: the case $p \geq 2$. *Advances in Diff. Equations* **6** (2001), 257–302.

[9] Nečas, J.; *Équations aux Dérivées Partielles*. Presses de l'Université de Montréal, Montréal, 1965.

[10] Nirenberg, L.; On elliptic partial differential equations. *An. Sc. Norm. Sup. Pisa* **13** (1959), 116–162.

[11] Serrin, J.; Mathematical Principles of Classical Fluid Mechanics. *Encyclopedia of Physics* VIII, 125–263. Springer-Verlag, Berlin, 1959.

[12] Smagorinsky, J.S.; General circulation experiments with the primitive equations. I. The basic experiment. *Mon. Weather Rev.* **91** (1963), 99–164.

[13] Solonnikov, V.A., Ščadilov, V.E.; On a boundary value problem for a stationary system of Navier-Stokes equations. *Proc. Steklov Inst. Math.* **125** (1973), 186–199.

H. Beirão da Veiga
Department of Applied Mathematics
Pisa University, via Diotisalvi, 2
I-56126 Pisa, Italy
e-mail: `bveiga@dma.unipi.it`

Progress in Nonlinear Differential Equations
and Their Applications, Vol. 63, 33–42
© 2005 Birkhäuser Verlag Basel/Switzerland

Radiative Heat Transfer in Silicon Purification

A. Bermúdez, R. Leira, <u>M.C. Muñiz</u> and F. Pena

Abstract. We present a numerical model describing the thermal behavior of
a silicon purification process which takes place into a so-called *casting ladle*.
We consider, simultaneously, the phase change in the silicon and a nonlinear
non-local boundary condition arising from the Stefan-Boltzmann radiation
condition at the enclosure surfaces within the ladle. We also propose a numerical approximation using a finite element method. An iterative algorithm
and numerical results are presented.

1. Introduction

In many engineering applications involving high-temperature processes numerical
simulation provides an insight into the radiative analysis of these complex systems
and it promotes improvements of several process optimization (see [3, 5]).

The motivation of this work is to compute the numerical solution of the
problem addressed in [6] applied to a silicon purification process – see [1]. We consider, simultaneously, the phase change in the silicon and the non-local boundary
condition arising from the Stefan-Boltzmann radiation condition at the enclosure
surfaces within the ladle.

The outline of this paper is as follows. In Section 2 the physical problem
is introduced. In Section 3 using the axisymmetry of the domain, we formulate
the mathematical problem in a two-dimensional domain by means of cylindrical
coordinates. Section 4 is devoted to introduce space and time discretization of the
aforementioned problem and to present an iterative algorithm. Finally, in Section
5, several numerical results are shown.

2. The physical problem

Metallurgical grade silicon (MG-Si) is obtained from a silicon oxide in electrical
submerged arc furnaces. A technique of MG-Si purification is to melt it and to
induce its directional solidification. This method of removing impurities is based

This work has been supported by MCYT-FEDER DPI2003-01316 and FERROATLANTICA
I+D.

on the fact that most impurities tend to remain in a molten region rather than re-solidify.

This purification process is taking place into a casting ladle which consists of a finite axisymmetric cylinder containing a cylindrical enclosure. After the casting ladle being electrically heated, its lid is open and molten silicon is poured into its inner cavity keeping a gap between the top of the silicon and the upper part of the inner ladle surface where several heating elements are located. The objective is now to push upwards the metal impurities by means of inducing its one-directional solidification switching on the heating elements and then keeping molten the top of the silicon ingot. In doing so, the solid silicon grows gradually upwards into the liquid and the metallic impurities are segregated into the melt region during solidification; thus, at the end of the process most of impurities are concentrated at the top of the silicon ingot.

Radiation heat transfer is considered at the inner cavity and materials of the enclosure are assumed to be opaque (see [4]); therefore radiation may be treated as a surface phenomenon. Moreover, we assume both that the walls of the cylindrical enclosure behave as black surfaces and that the medium within the enclosure is radiatively nonparticipating so that it has no effect on the radiation transfer between inner surfaces. We also assume that radiative properties are independent of wavelength.

3. The mathematical model

In this section we present a thermal model for a transient conductive-radiative heat transfer problem with phase change taking place in the casting ladle. This container is axisymmetric with respect to the z-axis and it has an inner cavity. Using cylindrical coordinates the three-dimensional problem is transformed into a two-dimensional one written on a vertical section of the ladle.

We denote by $\Omega \subset \mathbb{R}^3$ the casting ladle with silicon, which consists of a domain generated by the rotation about the z-axis of a bounded polygonal connected set $D \subset \{(r, z) \in \mathbb{R}^2; r \geq 0\}$, called meridian section of Ω (see [1]). We assume that Ω has a Lipschitz boundary and that the intersection of the set ∂D with the z-axis does not contain isolated points. We assume that the boundary of Ω is the union of two disjoint sets: an outer part, denoted by Γ_c, and an inner boundary, called Γ. Moreover, we assume that the boundary of D is the union of $\overline{\gamma}_0$, $\overline{\gamma}_c$ and $\overline{\gamma}$ where γ_0, γ_c and γ are disjoint open sets and (see Figure 1),

- γ_0 is a subset of the z-axis,
- γ_c is the outer part of ∂D,
- γ is the boundary generating Γ.

We denote by γ_1, γ_2 and γ_3 the subsets of γ depicted in Figure 2; we denote by Γ_1 the subset of Γ which is generated by rotation of set γ_1 about the z-axis.

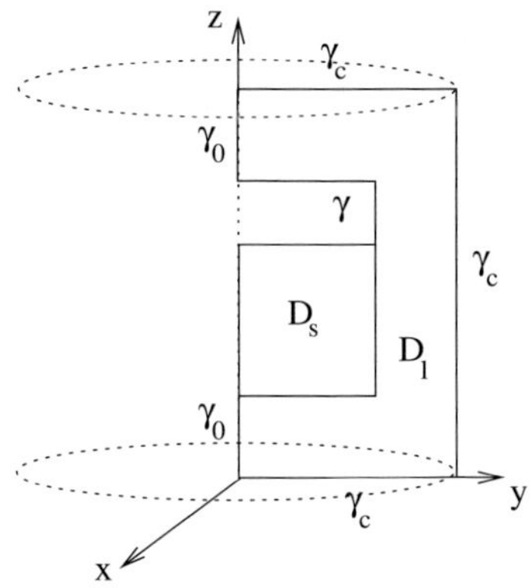

FIGURE 1. D: generating surface of Ω.

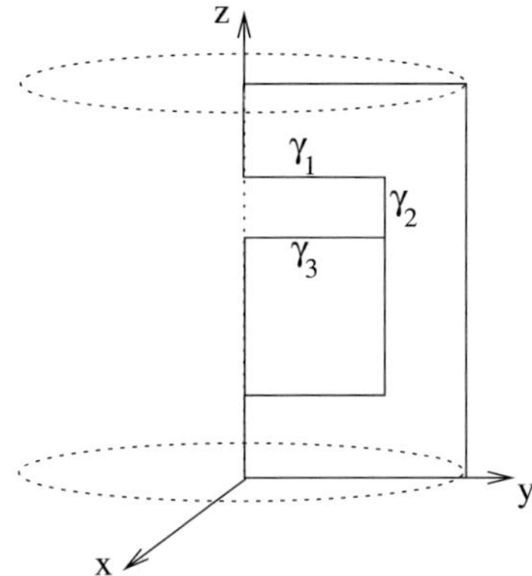

FIGURE 2. Subsets of γ.

Similarly, let Ω_s and Ω_l be the subsets of Ω corresponding to silicon and ladle which are generated by rotation of sets D$_s$ and D$_l$, respectively, about the z-axis (see Figure 1).

Let $[0, \mathcal{T}] \subset \mathbb{R}$ be the time interval with $\mathcal{T} > 0$. Moreover we set $Q_{\mathcal{T}} = \Omega \times (0, \mathcal{T})$. Throughout this paper we denote by $T(\mathbf{x}, t)$ the absolute temperature at each point $\mathbf{x} \in \Omega$ and each time $t \in [0, \mathcal{T}]$. Assuming both there is no convection and there is no internal heat source, transient heat conduction is described by

$$\rho(\mathbf{x}, T)c(\mathbf{x}, T)\frac{\partial T}{\partial t}(\mathbf{x}, t) - \nabla.(k(\mathbf{x}, T)\nabla T) = 0 \text{ in } Q_{\mathcal{T}}, \tag{3.1}$$

where ρ is density, c is specific heat and k is thermal conductivity, all depending on the spacial variable $\mathbf{x}$ and time t. We assume that the medium is piecewise homogeneous and nonlinear. Moreover, a change of phase is taking place in Ω_s at a single temperature T_s. Thus, introducing an enthalpy function $\mathrm{H}(\mathbf{x}, T)$ we have

$$\mathrm{H}(\mathbf{x}, T) = \chi_{\Omega_s}(\mathbf{x})\, \mathrm{H}_s(T) + \chi_{\Omega_l}(\mathbf{x})\, \mathrm{H}_l(T) \tag{3.2}$$

where

$$\mathrm{H}_s(T) = \begin{cases} \Phi_s(T) & T < T_s, \\ [\Phi_s(T_s), \Phi_s(T_s) + L\,\rho(T_s)] & T = T_s, \\ \Phi_s(T) + L\,\rho(T_s) & T > T_s, \end{cases} \tag{3.3}$$

$$\mathrm{H}_l(T) = \Phi_l(T), \tag{3.4}$$

and $\Phi_j(T)$ being the function

$$\Phi_j(T) = \int_0^T \rho_j(\zeta)c_j(\zeta)\, d\zeta. \tag{3.5}$$

The subscript j being both s or l represents silicon or ladle, respectively, and L denotes the latent heat per unit mass. Notice that multivalued function (3.3) is accounting for phase change at temperature T_s in the silicon, whereas enthalpy in (3.4) is considered in the ladle where there is no phase transition.

Hence the heat transfer equation reads

$$\frac{\partial e}{\partial t}(\mathbf{x}, t) - \nabla.(k(\mathbf{x}, T)\nabla T) = 0 \text{ in } Q_T, \tag{3.6}$$

where e denotes enthalpy density which is expressed, in terms of temperature, as follows

$$e(\mathbf{x}, t) \in \mathrm{H}(\mathbf{x}, T(\mathbf{x}, t)). \tag{3.7}$$

3.1. Boundary conditions

On Γ_c a convection boundary condition is considered,

$$k\frac{\partial T}{\partial \mathbf{n}} = \alpha(T_c - T) \text{ on } \Gamma_c, \tag{3.8}$$

$\mathbf{n}$ being the outward unit normal vector to the boundary, α the convection heat transfer coefficient and T_c the temperature of surroundings.

On Γ_1 heating elements are considered to heat the top of the silicon in order for inducing the silicon directional solidification. These thermal devices are switched off if the highest temperature at this boundary, denoted by $\tilde{T}$, is greater or equal than $\bar{\theta}$. Thus the boundary condition at this boundary is given by

$$k\frac{\partial T}{\partial \mathbf{n}} = g(\tilde{T}, t) \text{ on } \Gamma_1, \tag{3.9}$$

where

$$\tilde{T}(t) = \max_{\mathbf{x} \in \Gamma_1} T(\mathbf{x}, t), \tag{3.10}$$

and the function g represents the power given to the heating elements. Let us assume

$$g(\theta, t) = \begin{cases} p(t) & \text{if } \theta \leq \underline{\theta}, \\ p(t)(\bar{\theta} - \theta)/(\bar{\theta} - \underline{\theta}) & \text{if } \underline{\theta} \leq \theta \leq \bar{\theta}, \\ 0 & \text{if } \bar{\theta} \geq \theta, \end{cases} \tag{3.11}$$

where $\underline{\theta}$ is the threshold temperature at which the power begins to decrease and $p(t)$ is a time-dependent function related with the maximum power of the heating device.

On the inner boundary Γ we consider a non-local radiative boundary condition assuming that surfaces may be approximated as black surfaces (see [4]). Therefore energy only leaves the surface as a result of emission and all incident radiation is absorbed. Hence on Γ we set

$$k\frac{\partial T}{\partial \mathbf{n}}(\mathbf{x}) + \mathcal{G}(\sigma T^4)(\mathbf{x}) = 0, \tag{3.12}$$

where

$$\mathcal{G}(\sigma T^4)(\mathbf{x}) = (\mathcal{I} - \mathcal{K})(\sigma T^4)(\mathbf{x}), \tag{3.13}$$

$\mathcal{I}$ being the identity operator, σ the Stefan-Boltzmann radiation constant and the integral operator $\mathcal{K}$ defined by

$$\mathcal{K}(\zeta)(\mathbf{x}) = \int_{\Gamma} F(\mathbf{x}, \mathbf{y})\zeta(\mathbf{y})dS_{\mathbf{y}}, \quad \mathbf{x} \in \Gamma, \tag{3.14}$$

with $F(\mathbf{x}, \mathbf{y})$ denoting the view factor between points $\mathbf{x}$ and $\mathbf{y}$ of Γ, and $dS_{\mathbf{y}}$ a differential surface element.

Remark 3.1. The view factor between points $\mathbf{x}$ and $\mathbf{y}$ of Γ quantifies the visibility between these two points and it is given by (see [6] and references therein)

$$F(\mathbf{x}, \mathbf{y}) = \frac{\mathbf{n_x} \cdot (\mathbf{y} - \mathbf{x})\, \mathbf{n_y} \cdot (\mathbf{x} - \mathbf{y})}{\pi\, |\mathbf{x} - \mathbf{y}|^4}, \quad a.e.\ (\mathbf{x}, \mathbf{y}) \in \Gamma \times \Gamma, \ \mathbf{x} \neq \mathbf{y}, \tag{3.15}$$

$\mathbf{n_x}$ and $\mathbf{n_y}$ being the outward unit normal vectors to Γ at $\mathbf{x}$ and $\mathbf{y}$, respectively, directed outwards Ω.

On the boundary γ_0 we consider

$$k\frac{\partial T}{\partial \mathbf{n}} = 0 \text{ on } \gamma_0. \tag{3.16}$$

3.2. Cylindrical coordinates and weak formulation

Due to the axisymmetry of the domain and assuming that all the fields involved in this problem are independent of the angular variable θ, we transform the 3D problem into a 2D one using cylindrical coordinates – see [1].

Setting $R_{\mathcal{T}} = \mathrm{D} \times (0, \mathcal{T})$, a straightforward computation from (3.6) and (3.7) leads to

$$\frac{\partial e}{\partial t} - \frac{1}{r}[\frac{\partial}{\partial r}(rk\frac{\partial T}{\partial r}) + \frac{\partial}{\partial z}(rk\frac{\partial T}{\partial z})] = 0 \text{ in } R_{\mathcal{T}}, \tag{3.17}$$

$$e(r, z, t) \in H(r, z, T(r, z, t)). \tag{3.18}$$

Initial condition. We consider the initial condition $T(r, z, 0) = T_0(r, z)$ in D, T_0 being a temperature distribution in D.

3.3. Weak formulation

Multiplying (3.17) by a test function, integrating in the meridian section D, using the Green formula and taking into account boundary conditions (3.8), (3.9), (3.12) and (3.16) we get

$$\int_{\mathrm{D}} \frac{\partial e}{\partial t} v\, r\, drdz \ + \ \int_{\mathrm{D}} k\nabla T.\nabla v\, r\, drdz + \int_{\gamma} \mathcal{G}(\sigma T^4)vrd\gamma \tag{3.19}$$

$$= \int_{\gamma_c} \alpha(T_c - T)vrd\gamma + \int_{\gamma_1} g(\tilde{T}, t)vrd\gamma, \quad a.e. \text{ in } [0, \mathcal{T}].$$

4. Numerical solution: time and space discretization

In this section we use a one-step implicit scheme for time discretization of equation (3.19) and a finite element method for space discretization. Function T is approximated by piecewise linear finite elements on a triangular mesh.

We consider the time interval $[0, \mathcal{T}]$, $N \in \mathbb{N}$, $N > 0$ and we set $\Delta t = \mathcal{T}/N$. Now we introduce the mesh $\Pi = \{t^0, \ldots, t^N\}$ of the time interval given by

$$t^0 = 0,$$

$$t^{n+1} = t^n + \Delta t, \quad n = 0, 1, \ldots, N - 1.$$

We denote by $F^n(r, z)$ the value of field F at point $(r, z) \in D$ and time $t = t^n$. Now the value of $\dot{e}((r, z), t)$ at $(r, z) \in D$ and $t = t^{n+1}$ is approximated by Euler scheme:

$$\dot{e}((r, z), t^{n+1}) \approx \frac{e^{n+1}(r, z) - e^n(r, z)}{\Delta t}.$$

Moreover, associated with a family of triangular meshes τ_h of the domain D, we consider the finite element spaces V_h given by

$$V_h = \{v_h \in \mathcal{C}(\bar{D}), v_h|_K \in P_1(K), \forall K \in \tau_h\}, \tag{4.1}$$

$P_1(K)$ being the space of polynomials of degree ≤ 1 defined on an element K.

Therefore from (3.19) we obtain **the discretized problem**:

For each $n = 0, 1, \ldots, N - 1$ find the functions T_h^{n+1} and e_h^{n+1} in V_h such that

$$\frac{1}{\Delta t} \int_D e_h^{n+1} v_h \, r \, dr dz + \int_D k(T_h^{n+1}) \nabla T_h^{n+1} . \nabla v_h \, r \, dr dz + \int_\gamma \mathcal{G}(\sigma(T_h^{n+1})^4) v_h \, r \, d\gamma$$

$$= \int_{\gamma_c} \alpha(T_c - T_h^{n+1}) v_h \, r \, d\gamma + \int_{\gamma_1} g(\tilde{T}_h^{n+1}, t^{n+1}) v_h \, r \, d\gamma + \frac{1}{\Delta t} \int_D e_h^n v_h \, r \, dr dz, \forall v_h \in V_h,$$

$$\tag{4.2}$$

where

$$\tilde{T}_h^{n+1} = \max_{\mathbf{x} \in \gamma_1} T_h^{n+1}(\mathbf{x}), \tag{4.3}$$

$$e_h^{n+1}(q) \in H(q, T_h^{n+1}(q)), \tag{4.4}$$

for all vertices q in $\bar{D}$.

Notice that the multivalued function H is relating temperature to enthalpy; to deal with this nonlinearity, we use an iterative algorithm introduced in Bermúdez & Moreno (see [2]), defining the function $p_h^{n+1} = e_h^{n+1}(q) - \omega T_h^{n+1}(q)$, $\omega > 0$, at each vertex q and using the equivalence

$$p_h^{n+1}(q) \in H(q, T_h^{n+1}(q)) - \omega T_h^{n+1}(q) \iff p_h^{n+1}(q) = H_\lambda^\omega(q, T_h^{n+1}(q) + \lambda p_h^{n+1}(q)), \tag{4.5}$$

with $0 < \lambda \leq 1/(2\omega)$ and H_λ^ω being the Yosida approximation of the operator $H - \omega I$.

In order to numerically solve (4.2) we replace e_h^{n+1} by $p_h^{n+1} + \omega T_h^{n+1}$.

4.1. An iterative algorithm

In this section we present the iterative algorithm consisting of three embedded loops as the flow chart of the algorithm shows in Figure 3, (see [1]).Hereafter we omit subscript h associated with the space discretization for the sake of simplicity.

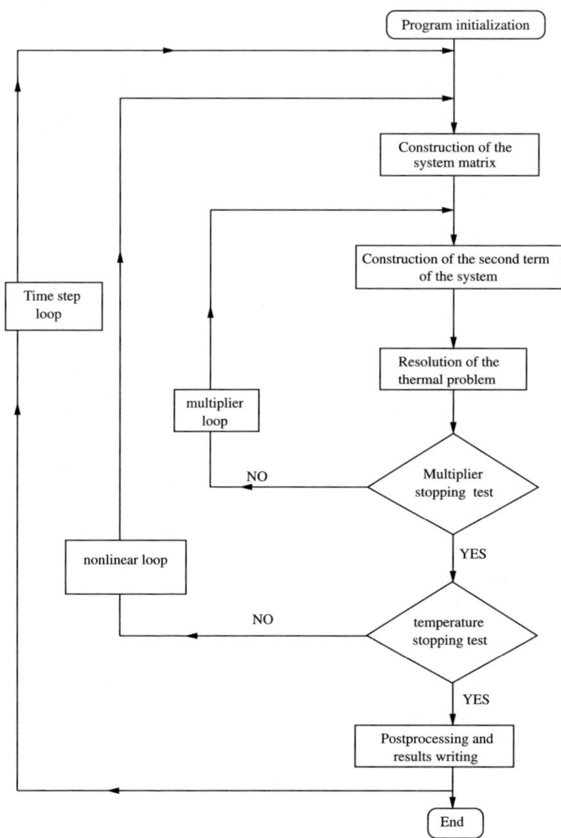

FIGURE 3. Flow chart of the algorithm.

Time loop: Let us suppose that T^n and p^n are known. At time t^{n+1}, function T^{n+1} is obtained as the limit of the sequence T_s^{n+1} coming from the following iterative algorithm,

(Nonlinear loop) **(a) Initialization:** $T_0^1 = T_0$ and $T_0^{n+1} = T^n$ for $n \geq 1$,

$$p_0^{n+1} = \begin{cases} p^n & \text{if } n \geq 1, \\ e^0 - \omega T_0 & \text{if } n = 0, \end{cases}$$

where

$$e^0(\mathbf{x}) = \int_0^{T_0(\mathbf{x})} \rho(\mathbf{x}, \tau)c(\mathbf{x}, \tau)d\tau$$

and T_0 being the initial condition.

    (b) $(s+1)$**th step:** T_s^{n+1} and p_s^{n+1} are known. Function T_{s+1}^{n+1} is obtained as the limit of the sequence $T_{s+1,k}^{n+1}$ computed by

(Multiplier loop) (i) Initial step: $T_{s+1,0}^{n+1} = T_s^{n+1}$ and $p_{s+1,0}^{n+1} = p_s^{n+1}$.

(ii) $(k+1)$**th step:** $p_{s+1,k}^{n+1}$ is known and in order to compute $T_{s+1,k+1}^{n+1}$ and $p_{s+1,k+1}^{n+1}$ we proceed as follows:

    (a) $T_{s+1,k+1}^{n+1}$ is the solution of the linear problem

$$\frac{\omega}{\Delta t} \int_D T_{s+1,k+1}^{n+1} v\, r\, drdz + \int_D k(T_s^{n+1})\nabla T_{s+1,k+1}^{n+1} \cdot \nabla v\, r\, drdz$$

$$+ \int_\gamma \mathcal{G}(\sigma(T_s^{n+1})^4)v\, r\, d\gamma$$

$$= \int_{\gamma_c} \alpha(T_c - T_{s+1,k+1}^{n+1})v\, r\, d\gamma + \int_{\gamma_1} g(\tilde{T}_s^{n+1}, t^{n+1})v\, r\, d\gamma$$

$$+ \frac{1}{\Delta t} \int_D (e^n - p_{s+1,k}^{n+1})v\, r\, drdz, \tag{4.6}$$

    for all $v \in V_h$ where $\tilde{T}_s^{n+1}(\mathbf{x}) = \max_{\mathbf{x} \in \gamma_1} T_s^{n+1}(\mathbf{x})$

    (b) Then $p_{s+1,k+1}^{n+1}$ is calculated at every vertex of the mesh using the formula

$$p_{s+1,k+1}^{n+1}(q) = \mathrm{H}_\lambda^\omega(q, T_{s+1,k+1}^{n+1}(q) + \lambda p_{s+1,k}^{n+1}(q)). \tag{4.7}$$

5. Numerical results

The aforementioned algorithm has been implemented in a computer code written in Fortran and in this section we present several numerical results concerning this code. The computation was performed on a personal computer working under Windows. Figure 4 shows the geometry and the mesh used for finite element discretization. It was made with Modulef library (see [7]). Figure 5 presents the configuration of the materials forming the ladle.

    The ladle is preheated using the inner heating elements; isotherms of the preheated ladle are shown in Figure 6. Afterwards silicon at $1480\,^{\circ}$C is poured into the inner cavity. The power of the heating elements is 10 kW during the first $40000\,s$; then the power is decreasing to zero as time increases, i.e., the function $p(t)$ of (3.11) is given by

$$p(t) = \begin{cases} 10 \text{ kW} & \text{if } t \leq 40000\,s, \\ \dfrac{10}{(t - 39999)^{1/5}} \text{ kW} & \text{if } t > 40000\,s. \end{cases}$$

Moreover, $\underline{\theta} = 1450\,^{\circ}$C and $\bar{\theta} = 1550\,^{\circ}$C.

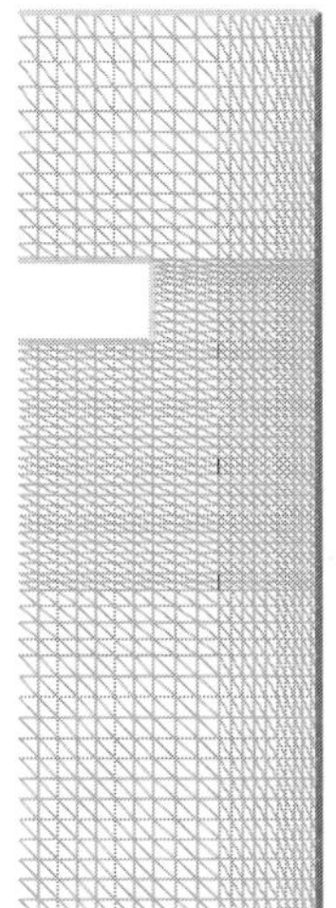

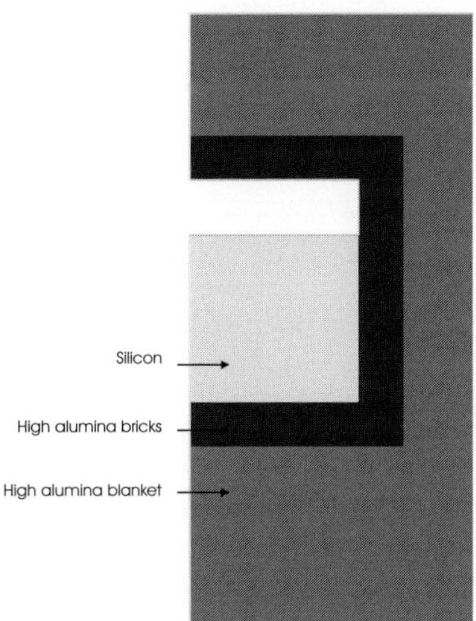

FIGURE 4. Geometry and mesh of the ladle with silicon.

FIGURE 5. Materials of the ladle with silicon.

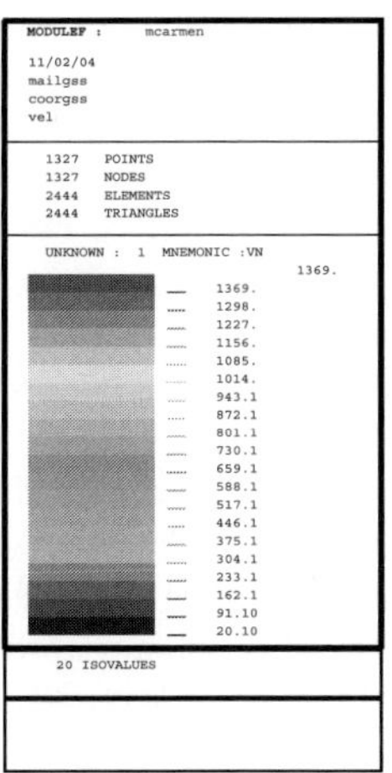

FIGURE 6. Isotherms (°C) of the preheated ladle.

Figures 7 a) and 7 b) show the solidification front corresponding to times $t = 1000\,s$ and $t = 36000\,s$, respectively. We remark that silicon solidifies around the walls of the inner enclosure at the beginning of the process, and then the solidification front is progressively getting flatter. With Figure 7 c), showing the solidification front at time $t = 128000\,s$, we emphasize the fact that the solidification front grows upwards as time increases; thus the top of the silicon ingot is the last part to solidify.

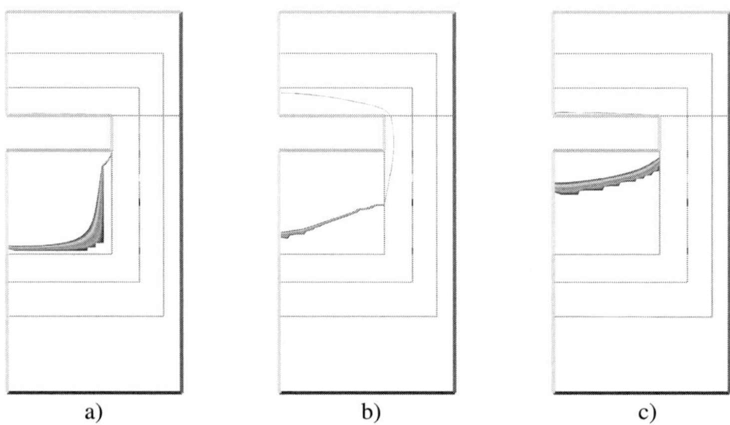

FIGURE 7. Solidification front at a) $t = 1000\,s$, b) $t = 36000\,s$, c) $t = 128000\,s$.

References

[1] Bermúdez, A., Leira, R., Muñiz, M.C. & Pena, F., Numerical modelling of a transient conductive-radiative thermal problem arising in silicon purification, submitted paper.

[2] Bermúdez, A. & Moreno, C., Duality methods for solving variational inequalities. Comput. Math. Appl., **7** (1981), 43–58.

[3] Hsu, P.F. & Ku, J.C., Radiative heat transfer in finite cylindrical enclosures with nonhomogeneous participating media. Journal of Thermophysics and Heat Transfer, **8** (3) (1994), 434–440.

[4] Incropera, F.P. & De Witt, D.P., *Fundamentals of heat and mass transfer*, Wiley, (New York, 1990).

[5] Nunes, E.M., Modi, V. & Naraghi, M.H.N., Radiative transfer in arbitrarily-shaped axisymmetric enclosures with anisotropic scattering media. International Journal of Heat and Mass Transfer, **43** (2000), 3275–3285.

[6] Tiihonen, T., Stefan-Boltzmann radiation on non-convex surfaces, Mathematical Methods in the Applied Sciences, **20** (1997), 47–57.

[7] Toit, H. D., *Introduction à MODULEF. Guide #1*, INRIA, (France, 1991).

A. Bermúdez, R. Leira, <u>M.C. Muñiz</u> and F. Pena
Departamento de Matemática Aplicada
Universidade de Santiago de Compostela
15706 Santiago, Spain
e-mail: `mcarmen@usc.es`

Progress in Nonlinear Differential Equations
and Their Applications, Vol. 63, 43–50

A Decay Result for a Quasilinear Parabolic System

Said Berrimi and Salim A. Messaoudi

Dedicated to Pr. Haim. Brezis on the occasion of his 60th birthday

Abstract. In this paper we consider a quasilinear parabolic system of the form
$$A(t)\,|u_t|^{m-2}\,u_t - \Delta u = u\,|u|^{p-2}\,,$$
$m \geq 2$, $p > 2$, in a bounded domain associated with initial and Dirichlet boundary conditions.We show that, for suitable initial datum, the energy of the solution decays " in time" exponentially if $m = 2$ whereas the decay is of a polynomial order if $m > 2$.

Mathematics Subject Classification (2000). 35K05–35K65.

Keywords. Quasilinear, Parabolic, Nonlinear source, Decay.

1. Introduction

Research of global existence and finite time blow-up of solutions for the initial boundary value problem
$$
\begin{aligned}
u_t - div(|\nabla u|^{\alpha-2}\nabla u) + f(u) &= 0, & x &\in \Omega, & t &> 0 \\
u(x,t) &= 0, & x &\in \partial\Omega, & t &\geq 0 \\
u(x,0) &= u_0(x), & x &\in \Omega,
\end{aligned}
\tag{1}
$$

where $\alpha \geq 2$ and Ω is a bounded domain of $\mathbb{R}^n$ $(n \geq 1)$, with a smooth boundary $\partial\Omega$, has attracted a great deal of people. The obtained results show that global existence and nonexistence depend roughly on α, the degree of nonlinearity in f, the dimension n, and the size of the initial datum. In the early 70's, Levine [8] introduced the concavity method and showed that solutions with negative energy blow-up in finite time. Later, this method had been improved by Kalantarov and Ladyzhenskaya [7] to accommodate more situations. Ball [2] also studied (1) with f depending on u as well as on ∇u and established a nonglobal existence result in bounded domains. This result was generalized to unbounded domains by Alfonsi and Weissler [1].

For the case $\alpha > 2$, Junning [6] studied (1) with f depending also on u and ∇u. He proved a nonglobal existence result under the condition

$$\frac{1}{\alpha} \int_\Omega |\nabla u_0(x)|^\alpha \, dx - \int_\Omega F(u_0(x))dx$$
$$\leq -\frac{4(\alpha - 1)}{\alpha T (\alpha - 2)^2} \int_\Omega u_0^2(x)dx, \tag{2}$$

where $F(u) = \int_0^u f(s)ds$. This type of results have been extensively generalized and improved by Levine, Park, and Serrin in a paper [9], where the authors proved some global, as well as nonglobal, existence theorems. Their result, when applied to problem (1), requires that

$$\frac{1}{\alpha} \int_\Omega |\nabla u_0(x)|^\alpha dx - \int_\Omega F(u_0(x))dx < 0. \tag{3}$$

We note that the inequality (3) implies (2). In 1999, Erdem [4] discussed the initial Dirichlet-type boundary problem for

$$u_t - \sum_{i=1}^n \frac{\partial}{\partial x_i}\left((d + |\nabla u|^{m-2})\frac{\partial u}{\partial x_i}\right) + g(u, \nabla u) = f(u), \quad x \in \Omega, \quad t > 0$$

and established a blow-up result. Messaoudi [10] showed that the blow-up result can also be obtained for solutions satisfying

$$\frac{1}{m} \int_\Omega |\nabla u_0(x)|^m dx - \int_\Omega F(u_0(x))dx \leq 0.$$

On the other hand if f has at most a linear growth then we can find global solutions (see [5]).

Concerning the asymptotic behavior, Engler, Kawohl, and Luckhaus [3] considered problem (1) with $\alpha = 2$ and showed that, for, $f(0) = 0$, $f'(u) \geq a > 0$, and sufficiently small initial datum u_0, the solution satisfies a gradient estimate of the type

$$\|\nabla u\|_p \leq C e^{-\delta t} \|\nabla u_0\|_p.$$

For initial boundary problems to the quasilinear equation

$$u_t - div(\sigma(|\nabla u|^2)\nabla u) + f(u, \nabla u) = 0,$$

results concerning global existence and gradient estimates have been established, under certain geometric conditions on $\partial\Omega$, by Nakao and Ohara [12], [13] and Nakao and Chen [14].

Pucci and Serrin [15] discussed the following quasilinear parabolic system

$$A(t)|u_t|^{m-2}u_t = \Delta u - f(x, u),$$

for $m > 1$ and f satisfying $(f(x, u), u) \geq 0$. They established a global result of solutions and showed that these solutions tend to the rest state as $t \to \infty$, however no rate of decay has been given.

In this work we consider a similar problem of the form

$$A(t)\left|u_t\right|^{m-2}u_t - \Delta u = |u|^{p-2}u, \quad x \in \partial\Omega, \quad t \in J$$
$$u(x,t) = 0, \qquad\qquad x \in \partial\Omega, \quad t \in J \tag{4}$$
$$u(x,0) = u_0, \qquad\qquad x \in \Omega,$$

where $J = [0, \infty)$ and Ω is a bounded open subset of $\mathbb{R}^n$. The values of u are taken in $\mathbb{R}^N$, $N \geq 1$ and $A \in C(J; \mathbb{R}^{N\times N})$. We assume that A is bounded and satisfies the condition

$$(A(t)v, v) \geq c_0 |v|^2, \qquad \forall t \in J, \quad v \in \mathbb{R}^N,$$

where $(.,.)$ is the inner product in $\mathbb{R}^N$ and $c_0 > 0$. We will show that, for small initial energy, the solution of (4) decays exponentially if $m = 2$ whereas the decay is of a polynomial order if $m > 2$. Our method of proof relies on the use of a lemma by Nakao [11].

2. Preliminaries

In order to state and prove our result, we introduce the following notation:

$$I(u(t)) = I(t) = \|\nabla u(t)\|_2^2 - \|u(t)\|_p^p$$
$$E(u(t)) = E(t) = \tfrac{1}{2}\|\nabla u(t)\|_2^2 - \tfrac{1}{p}\|u(t)\|_p^p \tag{5}$$
$$H = \left\{v \in \left(H_0^1\right)^N : I(v) > 0\right\} \cup \{0\}.$$

By multiplying the equation in (4) by u_t and integrating over Ω, using the boundary conditions, we get

$$\frac{d}{dt}E(t) = -\int_\Omega A(t)\left|u_t\right|^{m-2}u_t.u_t dx \leq 0, \tag{6}$$

for regular solutions. The same result is obtained for weak solutions by a simple density argument.

Next, we prove the invariance of the set H. For this aim we note that, by the embedding $H_0^1 \hookrightarrow L^q$, we have

$$\|u\|_q \leq C\|\nabla u\|_2, \tag{7}$$

for $2 \leq q \leq \frac{2n}{n-2}$ if $n \geq 3$, $q > 2$ if $n = 1, 2$ where $C = C(n, q, \Omega)$ is the best constant.

Lemma 2.1. (Nakao[11]) *Let $\varphi(t)$ be a nonincreasing and nonnegative function defined on $[0, T]$, $T > 1$, satisfying*

$$\varphi^{1+r}(t) \leq k_0(\varphi(t) - \varphi(t+1)), \qquad t \in [0, T],$$

for $k_0 > 1$ and $r \geq 0$. Then we have , for each $t \in [0, T]$,

$$\varphi(t) \quad \leq \quad \varphi(0)e^{-k[t-1]^+}, \qquad r = 0$$
$$\varphi(t) \quad \leq \quad \left\{\varphi(0)^{-r} + k_0 r\,[t-1]^+\right\}^{\frac{-1}{r}} \qquad r > 0$$

where $[t-1]^+ = \max\{t-1, 0\}$ and $k = \ln(\frac{k_0}{k_0-1})$.

Lemma 2.2. *Suppose that*

$$2 < p \leq \tfrac{2n}{n-2}, \qquad n \geq 3$$
$$p > 2, \qquad n = 1, 2. \tag{8}$$

If $u_0 \in H$, and satisfying

$$C^p \left(\frac{2p}{p-2} E(0) \right)^{\frac{p-2}{2}} < 1 \tag{9}$$

then the solution $u(t) \in H$ for each $t \in [0, T)$.

Proof. Since $I(u_0) > 0$, then there exists (by continuity) $T_m < T$ such that

$$I(u(t)) \geq 0, \ \forall t \in [0, T_m] ;$$

this gives

$$E(t) = \left(\frac{p-2}{2p} \right) \|\nabla u(t)\|_2^2 + \frac{1}{p} I(t) \geq \left(\frac{p-2}{2p} \right) \|\nabla u(t)\|_2^2. \tag{10}$$

So,

$$\|\nabla u(t)\|_2^2 \leq \left(\frac{2p}{p-2} \right) E(t) \leq \left(\frac{2p}{p-2} \right) E(0), \ \forall t \in [0, T_m]. \tag{11}$$

We then use (7)–(9) and (11) to obtain, for each $t \in [0, T_m]$,

$$\|u(t)\|_p^p \leq C^p \|\nabla u(t)\|_2^p = C^p \|\nabla u(t)\|_2^{p-2} \|\nabla u(t)\|_2^2$$
$$\leq C^p \left(\tfrac{2p}{p-2} E(0) \right)^{\frac{p-2}{2}} \|\nabla u(t)\|_2^2 < \|\nabla u(t)\|_2^2. \tag{12}$$

Therefore, by virtue of (5) and (12), we obtain

$$I(t) = \|\nabla u(t)\|_2^2 - \|u(t)\|_p^p > 0. \tag{13}$$

This shows that $u(t) \in H$, for all $t \in [0, T_m]$. By repeating this procedure, and using the fact that

$$\lim_{t \to T_m} C^p \left(\frac{2p}{p-2} E(t) \right)^{\frac{p-2}{2}} \leq \beta < 1,$$

T_m is extended to T.

Lemma 2.3. *Suppose that* (8) *and* (9) *hold, then*

$$\eta \|\nabla u(t)\|_2^2 \leq I(t). \tag{14}$$

Proof. It suffices to rewrite (12) as:

$$\|u(t)\|_p^p \ \leq \ C^p \left(\frac{2p}{p-2} E(0) \right)^{\frac{p-2}{2}} \|\nabla u(t)\|_2^2 = (1 - \eta) \|\nabla u(t)\|_2^2$$
$$\leq \ \|\nabla u(t)\|_2^2 - \eta \|\nabla u(t)\|_2^2. \tag{15}$$

Thus (14) follows for

$$\eta = 1 - C^p \left(\frac{2p}{p-2} E(0) \right)^{\frac{p-2}{2}} > 0. \tag{16}$$

Theorem. *Suppose that (8) holds. Assume further that $u_0 \in H$ and satisfies (9), then the solution satisfies the following decay estimations:*

$$E(t) \le E(0)e^{-[t-1]^+}, \qquad m = 2 \tag{17}$$

$$E(t) \le \left\{ (E(0))^{-\left(\frac{m-2}{2}\right)} + \frac{C_5}{c_0} \frac{m-2}{2} [t-1]^+ \right\}^{-\left(\frac{2}{m-2}\right)}, \qquad m > 2. \tag{18}$$

Proof. We integrate (6) over $[t, t+1]$ to obtain

$$E(t) - E(t+1) = \int_t^{t+1} \int_\Omega |u_t(s)|^{m-2} A(s)u_t.u_t dxds$$

$$\ge c_0 \int_t^{t+1} \int_\Omega |u_t(s)|^m dxds = c_0(F(t))^m, \tag{19}$$

where

$$(F(t))^m = \int_t^{t+1} \|u_t(s)\|_m^m ds. \tag{20}$$

Now we multiply the equation in (4) by u and integrate over $\Omega \times [t, t+1]$ to arrive at

$$\int_t^{t+1} I(s)ds \le \int_t^{t+1} \|A(s)\| \int_\Omega |u_t(s)|^{m-1} |u(s)| \, dxds.$$

By the Cauchy-Schwarz inequality, we have the following

$$\int_t^{t+1} I(s)ds \le \int_t^{t+1} \|A(s)\| \|u_t(s)\|_m^{m-1} \|u(s)\|_m \, ds$$

$$\le A \int_t^{t+1} \|u_t(s)\|_m^{m-1} \|u(s)\|_m \, ds, \tag{21}$$

where

$$A = \sup_J \|A(s)\| < \infty.$$

Exploiting (7) and (10), we obtain

$$\int_t^{t+1} I(s)ds \le CA \left(\frac{2p}{p-2} \right)^{\frac{1}{2}} \left(\sup_{t \le s \le t+1} E^{\frac{1}{2}}(s) \right) \left(\int_t^{t+1} \|u_t(s)\|_m^{m-1} \, ds \right). \tag{22}$$

Now we use the fact that

$$\int_t^{t+1} \left(\int_\Omega |u_t(s)|^m \, dx \right)^{\frac{m-1}{m}} ds \le \left(\int_t^{t+1} \int_\Omega |u_t(s)|^m \, dxds \right)^{\frac{m-1}{m}} = (F(t))^{m-1} \tag{23}$$

to get

$$\int_t^{t+1} I(s)ds \leq CA \left(\frac{2p}{p-2}\right)^{\frac{1}{2}} \left(E^{\frac{1}{2}}(t)\right)(F(t))^{m-1}. \tag{24}$$

From (5) we have

$$E(t) = \left(\frac{p-2}{2p}\right)\|\nabla u(t)\|_2^2 + \frac{1}{p}I(t). \tag{25}$$

Integrating both sides of (25) over $[t, t+1]$ and using (14), one can write

$$\int_t^{t+1} E(s)ds \leq \left(\frac{1}{p} + \frac{p-2}{2p\eta}\right)\int_t^{t+1} I(s)ds. \tag{26}$$

A combination of (24) and (26) leads to

$$\int_t^{t+1} E(s)ds \leq CA \left(\frac{2p}{p-2}\right)^{\frac{1}{2}} \left(\frac{1}{p} + \frac{p-2}{2p\eta}\right) \left(E^{\frac{1}{2}}(t)\right)(F(t))^{m-1}. \tag{27}$$

By using (6) again, we have

$$E(s) \geq E(t+1), \quad \forall s \leq t+1;$$

hence

$$\int_t^{t+1} E(s)ds \geq E(t+1). \tag{28}$$

Inserting (28) in (19) and using (27), we easily have

$$\begin{aligned}
E(t) &\leq \int_t^{t+1} E(s)ds + \int_t^{t+1}\int_\Omega A(s)\,|u_t(s)|^{m-2}\,u_t(s).u_t(s)dxds \\
&\leq CA\left(\frac{2p}{p-2}\right)^{\frac{1}{2}}\left(\frac{1}{p} + \frac{p-2}{2p\eta}\right)E^{\frac{1}{2}}(t)(F(t))^{m-1} \\
&\quad + \int_t^{t+1}\int_\Omega A(s)\,|u_t(s)|^m\,dxds \\
&\leq C_1\left[E^{\frac{1}{2}}(t)(F(t))^{m-1} + (F(t))^m\right],
\end{aligned} \tag{29}$$

for C_1 a constant depending on A, C, p and η only. We then use Young's inequality to get, from (29),

$$E(t) \leq C_2\left((F(t))^{2(m-1)} + (F(t))^m\right). \tag{30}$$

At this end, we distinguish two cases:

1) $m = 2$. In this case, we have from (30)

$$E(t) \leq 2C_2 F^2(t) \leq C_3 F^2(t) \leq \frac{C_3}{c_0}\left(E(t) - E(t+1)\right). \tag{31}$$

Lemma 2.1 then yields

$$E(t) \leq E(0)e^{-k[t-1]^+}, \quad k = \ln\left(\frac{C_3}{C_3 - c_0}\right). \tag{32}$$

2) $m > 2$. In this case, we note that, by (19), we have

$$F^m(t) \leq \frac{E(t)}{c_0} \leq \frac{E(0)}{c_0}.$$

Therefore (30) gives

$$
\begin{aligned}
E(t) \;\; &\leq \;\; C_2 \left((F(t))^{2(m-2)} + (F(t))^{m-2} \right) F^2(t) \\
&\leq \;\; C_3 \left((\frac{E(0)}{c_0})^{\frac{2(m-2)}{m}} + (\frac{E(0)}{c_0})^{\frac{m-2}{m}} \right) F^2(t) \\
&\leq \;\; C_4 F^2(t);
\end{aligned}
\tag{33}
$$

hence

$$E^{\frac{m}{2}}(t) \leq C_5 F^m(t) \leq \frac{C_5}{c_0} \left(E(t) - E(t+1) \right). \tag{34}$$

Again Lemma 2.1 for

$$r = \frac{m-2}{2} > 0, \tag{35}$$

gives

$$E(t) \leq \left\{ E(0)^{-\left(\frac{m-2}{2}\right)} + \frac{C_5}{c_0} \frac{m-2}{2} [t-1]^+ \right\}^{-\frac{2}{m-2}}$$

This completes the proof.

Acknowledgment

This work was completed while the first author was in a visit to KFUPM. Both authors would like to express their sincere thanks to KFUPM for its support. This work has been funded by KFUMP under Project# MS/VISCO ELASTIC 270.

References

[1] Alfonsi L. and Weissler F., Blow-up in $\mathbb{R}^n$ for a parabolic equation with a damping nonlinear gradient term, *Progress in nonlinear differential equations and their applications* **7** (1992), 1–20.

[2] Ball J., Remarks on blow-up and nonexistence theorems for nonlinear evolution equations, *Quart. J. Math. Oxford Ser.* **28** (1977), 473–486.

[3] Englern H., Kawohl B. and Luckhaus S., Gradient estimates for solutions of parabolic equations and systems, *J. Math. Anal. Appl.* **147** (1990), 309–329.

[4] Erdem D., Blow-Up of solutions to quasilinear parabolic equations, *Applied Math. Letters* **12** (1999), 65–69.

[5] Friedman A., Partial differential equations of parabolic type, *Prentice-Hall* Englewood NJ 1964.

[6] Junning Z., Existence and nonexistence of solutions for $u_t = \text{div}(|\nabla u|^{p-2}\nabla u) + f(\nabla u, u, x, t)$, *J. Math. Anal. Appl.* **172** (1993), 130–146.

[7] Kalantarov V.K. and Ladyzhenskaya O.A., Formation of collapses in quasilinear equations of parabolic and hyperbolic types. (Russian) Boundary value problems of mathematical physics and related questions in the theory of functions, *10. Zap. Naučn. Sem. Leningrad. Otdel. Mat. Inst. Steklov (LOMI)* **69** (1977), 77–102.

[8] Levine H., Some nonexistence and instability theorems for solutions of formally parabolic equations of the form $Pu_t = -Au + F(u)$, *Archive Rat. Mech. Anal.* **51** (1973), 371–386.

[9] Levine H., Park S., and Serrin J., Global existence and nonexistence theorems for quasilinear evolution equations of formally parabolic typ e. *J. Diff. Eqns.* **142** (1998), 212–229.

[10] Messaoudi S.A., A note on blow-up of solutions of a quasilinear heat equation with vanishing initial energy, *J. Math. Anal. Appl.* **273** (2002), 243–247.

[11] Nakao M., Asymptotic stability of the bounded or almost periodic solutions of the wave equations with nonlinear damping terms, *J. Math. Anal. Applications* **58** (1977), 336–343.

[12] Nakao M. and Ohara Y., Gradient estimates of periodic solutions for quasilinear parabolic equations, *J. Math. Anal. Appl.* **204** (1996), 868–883.

[13] Nakao M. and Ohara Y., Gradient estimates for a quasilinear parabolic equation of the mean curvature type, *J. Math. Soc. Japan* **48 # 3** (1996), 455–466.

[14] Nakao M. and Chen C., Global existence and gradient estimates for the quasilinear parabolic equations of m-Laplacian type with a nonlinear convection term, *J. Diff. Eqns.* **162** (2000), 224–250.

[15] Pucci P. and Serrin J., Asymptotic stability for nonlinear parabolic systems, Energy methods in continuum mechanics, (*Oviedo*, 1994), 66–74, *Kluwer Acad. Publ., Dordrecht*, 1996.

Said Berrimi
Math. Department
University of Setif
Setif, Algeria
e-mail: `berrimi@yahoo.fr`

Salim A. Messaoudi
Mathematical Sciences Department
KFUPM, Dhahran 31261
Saudi Arabia
e-mail: `messaoud@kfupm.edu.sa`

Progress in Nonlinear Differential Equations
and Their Applications, Vol. 63, 51–58

The Shape of Charged Drops: Symmetry-breaking Bifurcations and Numerical Results

S.I. Betelú, M.A. Fontelos and U. Kindelán

Abstract. We prove the existence of both stable and unstable stationary non-spherical shapes for charged, isolated liquid drops of a conducting Newtonian fluid. These shapes are spheroids whose eccentricity is an increasing function of the total charge. We also develop a numerical method based on the Boundary Integral Method in order to compute the stable shapes.

1. Introduction

When a drop of a conducting Newtonian fluid is electrically charged, charges tend to migrate to the surface because of their mutual electrostatic repulsion. Once in the surface, they produce stresses opposing those due to surface tension. If the total electric charge is small enough, then surface tension is the dominant force at the surface and isolated levitating drops remain spherical. On the other hand if the electric charge is larger than some critical value then, as Lord Rayleigh shows (see [12]), a spherical drop becomes unstable and should disintegrate by emitting jets of fluid from arbitrary points of the surface. For a drop with total charge Q, surface tension coefficient γ and radius R_0 suspended in a medium of dielectric constant ε_0, this instability occurs when the dimensionless parameter

$$X \equiv \frac{Q^2}{32\gamma\pi^2\varepsilon_0 R_0^3} \tag{1}$$

exceeds unity.

However, both numerical simulation and experiments show that Rayleigh's breakup mechanism is not completely correct. Firstly, it was shown numerically in [2] the existence of nonspherical drop configurations for values of X larger than 1. In [2] the authors argue that some of these configurations should be linearly stable. Second, as it has recently been reported in [6], the way in which a drop disintegrates is very different of the one Rayleigh proposed. In fact the drop becomes first a prolate spheroid and after it reaches a certain eccentricity emits two extremely thin jets from the poles.

Our goal in this paper is to review the result obtained in collaboration with A. Friedman (see [7]) in which it is proved the existence of those spheroids found in [2] by using Crandall and Rabinowitz's bifurcation theory (cf.[4], [5], [10]) suitably adapted to the study of free boundary problems (see also [3] and the references contained therein, where these ideas have been applied to other free boundary problems) and develop a numerical method based on the boundary integral formulation for Stokes fluids in order to test the degree of stability of the spheroids found theoretically.

Outside of the drop (the domain that we will denote by Ω) there is an electric field with its corresponding potential V, which satisfies Laplace equation and decays at infinity (see Eqs. (2–4) below). At the drop's surface, the electric potential is constant because the liquid is a conductor. The pressure difference between the inside and outside of the drop, δp, balances with the capillary forces (proportional to the mean curvature κ) and the electrostatic repulsion. The electrostatic forces are proportional to the surface charge density and the normal component of the electric field (which is equal to the normal derivative of V). It can be shown that in the surface of a conductor the charge distributes proportionally to the normal derivative of the potential (cf. for instance [8]). Putting all these facts together we obtain the system:

$$\Delta V = 0 \quad \text{in } \mathbb{R}^3 \backslash \Omega \,, \tag{2}$$

$$V = C \quad \text{on } \partial\Omega \,, \tag{3}$$

$$V(\mathbf{r}) \rightarrow 0 \quad \text{as } |\mathbf{r}| \rightarrow \infty \,, \tag{4}$$

$$\delta p = \gamma\kappa - \frac{\varepsilon_0}{2}\left(\frac{\partial V}{\partial n}\right)^2 \quad \text{on } \partial\Omega \tag{5}$$

where $\partial\Omega$ is the surface of the drop, and κ the mean curvature ($\kappa > 0$ if Ω is a sphere). The constant C in (3) has to be chosen such that condition (4) is satisfied.

The system (2)–(5) has a simple explicit solution when Ω is a sphere of radius R_0,

$$V(\mathbf{r}) = \frac{Q}{4\pi\varepsilon_0 r} \,, \tag{6}$$

$$\delta p = \frac{\gamma}{R_0} - \frac{\varepsilon_0}{2}\left(\frac{Q}{4\pi\varepsilon_0 R_0^2}\right)^2 \,, \tag{7}$$

$$\text{and} \quad C = \frac{Q}{4\pi\varepsilon_0 R_0} \,. \tag{8}$$

In Section 2 we first establish the existence of bifurcation branches of solutions with non-spherical domains Ω. We choose as the bifurcation parameter the total charge Q. (In [7], γ was chosen as the bifurcation parameter, but here we prefer to use Q, because it is more convenient for the comparison with the numerical results later in this work). We assume that γ, ε_0 and R_0 as fixed; then the constant C in (8) is also fixed, and the pressure difference δp is taken to be the same as for the spherical drop, that is,

$$\delta p = \delta p_0(Q) \equiv \frac{\gamma}{R_0} - \frac{\varepsilon_0}{2} \left(\frac{Q}{4\pi\varepsilon_0 R_0^2} \right)^2 .$$

Later we prove that there exists a sequence of bifurcation branches with

$$Q = Q_l + \varepsilon Q_{l1} + \varepsilon^2 Q_{l2} + \cdots \quad (l = 2, 3, \dots)$$

and free boundary

$$r = R_0 + x(\theta) = R_0 + \varepsilon Y_{l,0}(\theta) + \varepsilon^2 \Lambda_{l2}(\theta) + \cdots$$

where $Y_{l,0}(\theta)$ is the spherical harmonics $Y_{l,m}$ with $m = 0$, and $Q_2 < Q_3 < \cdots$.

The first bifurcation point $Q = Q_2$ is the one most physically relevant (all the other branches should be unstable). We finally discuss the stability along the bifurcation branch emanating from Q_2 and obtain the result represented in Figure 1. Finally, in Section 3 we solve numerically the evolution problem in a Stokes flow.

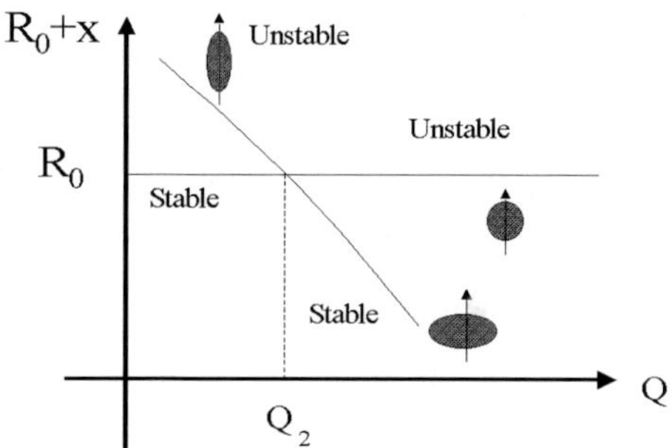

FIGURE 1. Graphical representation of the bifurcation branches. The drop's shape is axisymmetric in all cases and we represent the axis of symmetry with an arrow.

2. The bifurcation result

Consider the family of bounded domains Ω with boundary $\partial\Omega$ defined, in spherical coordinates, by

$$\partial\Omega : r = r(\theta, \varphi) = R_0 + x(\theta, \varphi)$$

where $x(\theta, \varphi)$ belongs to the set

$$X^{m+2+\alpha} = \left\{ x \in C^{m+2+\alpha}(\Sigma), \ \pi \text{ -periodic in } \ \theta, 2\pi \text{ -periodic in } \varphi \right\}$$

with $\Sigma = [0, 2\pi] \times [0, \pi]$ and $m \geq 1$. Let us define also the Banach spaces

$$X_1^{m+2+\alpha} = \text{closure of the linear space}$$
$$\text{spanned by } \{Y_{j,0}(\theta), \ j = 0, 1, 2, \dots\} \text{ in } X^{m+2+\alpha}.$$

Let

$$F(x,Q) \equiv \frac{\gamma}{R_0} - \frac{\varepsilon_0}{2}\left(\frac{Q}{4\pi\varepsilon_0 R_0^2}\right)^2 - \gamma\kappa + \frac{\varepsilon_0}{2}\left(\frac{\partial V}{\partial n}\bigg|_{\partial\Omega}\right)^2. \tag{1}$$

In order to solve the bifurcation problem $F(x,Q) = 0$ we use the following Crandall-Rabinowitz theorem [4]:

Theorem 1. *Let X, Y be real Banach spaces and $F(x, Q)$ a C^p map, $p \geq 3$, of a neighborhood of $(0, Q_0)$ in $X \times \mathbb{R}$ into Y with $F(0, Q) = 0$. Suppose*

(i) *$F_Q(0, Q_0) = 0$,*
(ii) *$KerF_x(0, Q_0)$ is one-dimensional, spanned by x_0,*
(iii) *$ImF_x(0, Q_0) = Y_1$ has codimension 1,*
(iv) *$F_{QQ}(0, Q_0) \in Y_1$ and $F_{Qx}(0, Q_0)x_0 \notin Y_1$.*

Then $(0, Q_0)$ is a bifurcation point of the equation $F(x, Q) = 0$ in the following sense: In a neighborhood of $(0, Q_0)$ the set of solutions of $F(x, Q) = 0$ consists of two C^{p-2} smooth curves Γ_1 and Γ_2 which intersect only at the point $(0, Q_0)$; Γ_1 is the curve $(0, Q)$ and Γ_2 can be parametrized as follows:

$$\Gamma_2 \quad : \quad (x(\varepsilon), Q(\varepsilon)) \ , \ |\varepsilon| \text{ small,}$$
$$(x(0), Q(0)) \ = \ (0, Q_0) \ , \ x'(0) = x_0 \ .$$

This theorem, when applied to our problem with $X = X_1^{m+2+\alpha}$ and $Y = X_1^{m+\alpha}$ yields the existence of bifurcation branches of solutions at

$$Q_l = \sqrt{8\gamma\varepsilon_0\,(l+2)}R_0^{\frac{3}{2}}\pi \ , \quad l = 2, 3, 4, \ldots \tag{2}$$

and since $KerF_x(0, Q_l) = Y_{l,0}(\theta)$ as one can compute (see [7]), free boundaries along these bifurcation branches of the form $r = R_0 + \varepsilon Y_{l,0}(\theta) + O(\varepsilon^2)$. More precisely:

Theorem 2. *For each Q_l given by (2) there exists a C^∞ bifurcation branch of solutions to the free boundary problem (2)–(5) with free boundary in $C^{m+2+\alpha}(\Sigma)$ of the form*

$$r = R_0 + \varepsilon Y_{l,0}(\theta) + \sum_{k=2}^{m+1} \varepsilon^k \Lambda_{lk}(\theta) + O(\varepsilon^{m+2}) \tag{3}$$

and

$$Q = Q_l + \sum_{k=1}^{m} \varepsilon^k Q_{lk} + O(\varepsilon^{m+1}) \tag{4}$$

for any integer $m > 0$.

Remark 1. In the case $l = 2$, (3) represents a prolate spheroid if $\varepsilon > 0$ or an oblate spheroid if $\varepsilon < 0$.

Remark 2. Although (2) is well defined and positive for $l = 0$ and $l = 1$ we do not consider the existence of bifurcation branches at Q_0 and Q_1 since the corresponding free boundaries of the form (3) would represent at order ε just a dilated or translated sphere.

The analysis of the stability along the bifurcation branches is done for Stokes flow. Thus, the fluid velocity $\vec{u}$ and the fluid pressure p inside the drop satisfy the equations:

$$-\nabla p + \mu_1 \Delta \vec{u} \ = \ 0 \ \text{in} \ \Omega(t) \,, \tag{5}$$

$$\nabla \cdot \vec{u} \ = \ 0 \ \text{in} \ \Omega(t) \tag{6}$$

where $\Omega(t)$ is the volume occupied by the drop's fluid. Equations identical to (5), (6) must be satisfied by the velocity and pressure of the fluid in $\mathbb{R}^3 \backslash \Omega(t)$ with μ_1 replaced by μ_2. In addition we have the boundary conditions

$$(T^{(2)} - T^{(1)})\vec{n} = \left[\gamma\kappa - \frac{\varepsilon_0}{2}\left(\frac{\partial V}{\partial n}\right)^2 \right] \vec{n} \quad \text{on} \ \partial\Omega(t) \,, \tag{7}$$

$$v_N = \vec{u} \cdot \vec{n} \quad \text{on} \ \partial\Omega(t) \tag{8}$$

where

$$T_{ij}^{(k)} = -p\delta_{ij} + \mu_k \left(\frac{\partial u_i}{\partial x_j} + \frac{\partial u_j}{\partial x_i} \right) \,, \ k = 1, 2 \,, \tag{9}$$

$\vec{n}$ is the outward normal to $\partial\Omega(t)$, and v_N is the velocity of the free boundary $\partial\Omega(t)$ in the direction $\vec{n}$.

As it is explained in [7], the stability depends critically on the sign of the first eigenvalue of F_x along the two bifurcation branches. Let us denote these two bifurcation branches by $\Gamma_1 : (0, Q(\delta))$ and $\Gamma_2 : (x(\varepsilon), \widetilde{Q}(\varepsilon))$ where $\widetilde{Q}(\varepsilon)$ is the function given by (4) with $l = 2$ so that $\widetilde{Q}(0) = Q_2$ and $\widetilde{Q}'(0) = Q_{21}$, and $Q(\delta) = Q_2 + \delta$ so that $Q(0) = Q_2$ and $Q'(0) = 1$. It can easily be shown that along the branch Γ_1 the Fréchet derivative of F has $Y_{l,0}$ as eigenfunction, i.e.,

$$[F_x(0, Q(\delta))] \, Y_{l,0} = \lambda_l(\delta) Y_{l,0} \tag{10}$$

where

$$\lambda_l(\delta) = -\frac{\gamma}{R_0^2}\left(\frac{1}{2}l(l+1) - 1 \right) + \frac{\varepsilon_0 Q^2(\delta)}{(4\pi\varepsilon_0)^2 R_0^5}(l-1) \tag{11}$$

so that, in particular $\lambda_2(0) = 0$ if $Q(\delta) = Q_2$. Also, if $Q(\delta) > Q_2$ then $\lambda_l(\delta) > 0$ for all $l \geq 2$, while $\lambda_2(\delta) < 0$ if $Q(\delta) < Q_2$. In order to study stability along Γ_2 it will be important to know if there exist eigenvalues of the operator $F_x(x(\varepsilon), \widetilde{Q}(\varepsilon))$ with positive real part. So we need to study the equation

$$F_x(x(\varepsilon), \widetilde{Q}(\varepsilon))w(\varepsilon) = \mu(\varepsilon)w(\varepsilon) \tag{12}$$

for $\widetilde{Q}$ close to Q_2. The main tool is the following lemma whose proof is just an application of Theorem 1.16 of Crandall and Rabinowitz [5] to our particular problem (see also Theorem 3.6.3 in [10]):

Lemma. *For ε such that $|\varepsilon| < \varepsilon^*$ (ε^* sufficiently small) there exists a unique $\mu(\varepsilon)$ and a unique $w(\varepsilon)$ such that (12) is satisfied. Moreover,*

$$w(\varepsilon) = Y_{2,0} + \varepsilon\widetilde{x} \quad \text{with} \ \|\widetilde{x}\|_{X_1^{m+2+\alpha}} \leq C \tag{13}$$

$$\mu(\varepsilon) = -\varepsilon\widetilde{Q}'(\varepsilon)\lambda_2'(0) + O(\varepsilon^2) \tag{14}$$

After a tedious computation analogous to the one in Section 5 of [7] one can easily find that $Q_{21} < 0$ and since $\widetilde{Q}'(\varepsilon) = Q_{21} + O(\varepsilon)$ we have then $\mu(\varepsilon) > 0$ if $\varepsilon > 0$ and small enough and $\mu(\varepsilon) < 0$ if $\varepsilon < 0$ and small enough. Arguing as in Section 7 of [7] this fact yields stability of oblate spheroids ($\varepsilon < 0$) and instability of prolate spheroids ($\varepsilon > 0$).

3. Numerical method

Our numerical method to track the evolution of the drop in time is based on the boundary integral method for the Stokes system (see [11], [1] for a comprehensive explanation of the method). It yields the following integral equation for the velocity at $\partial\Omega(t)$:

$$
\begin{aligned}
u_j(\mathbf{r}_0) \;=\; & -\frac{1}{4\pi}\frac{1}{\mu_1 + \mu_2}\int_{\partial\Omega(t)} f_i(\mathbf{r})G_{ij}(\mathbf{r},\mathbf{r}_0)dS(\mathbf{r}) \\
& -\frac{1}{4\pi}\frac{\mu_2 - \mu_1}{\mu_2 + \mu_1}\int_{\partial\Omega(t)} u_i(\mathbf{r})T_{ijk}(\mathbf{r},\mathbf{r}_0)n_k(\mathbf{r})dS(\mathbf{r}) \;.
\end{aligned}
\tag{15}
$$

where

$$
G_{ij}(\mathbf{r},\mathbf{r}_0) \;=\; \frac{\delta_{ij}}{|\mathbf{r}-\mathbf{r}_0|} + \frac{(r_i - r_{0,i})(r_j - r_{0,j})}{|\mathbf{r}-\mathbf{r}_0|^3}
$$

$$
T_{ijk}(\mathbf{r},\mathbf{r}_0) \;=\; -6\frac{(r_i - r_{0,i})(r_j - r_{0,j})(r_k - r_{0,k})}{|\mathbf{r}-\mathbf{r}_0|^5}
$$

$$
f_i(\mathbf{r}) \;=\; \left[\gamma\kappa(\mathbf{r}) - \frac{\varepsilon_0}{2}\left(\frac{\partial V}{\partial n}\right)^2(\mathbf{r})\right] n_i(\mathbf{r}).
$$

This equation simplifies considerably if $\mu_1 = \mu_2$, since in this case the velocity is directly given by the first surface integral at the right-hand side of (15) which depends uniquely on the geometry of $\partial\Omega(t)$ and the normal component of the electric field.

The solution for the electric potential can also be written in integral form as

$$
V(\mathbf{r}_0) = \frac{1}{2\pi}\int_{\partial\Omega(t)} V(\mathbf{r})\frac{\partial\left(1/|\mathbf{r}-\mathbf{r}_0|\right)}{\partial n} - \frac{\partial V}{\partial n}(\mathbf{r})\frac{1}{|\mathbf{r}-\mathbf{r}_0|}dS(\mathbf{r}).
\tag{16}
$$

Eq. (16) is an integral equation for the unknown values of $\partial V/\partial n$ at the surface, where V is a constant that is determined by the total charge Q by integrating the charge density over the surface,

$$
Q = \varepsilon_0 \int_{\partial\Omega(t)} \frac{\partial V}{\partial n}(\mathbf{r})dS(\mathbf{r}).
\tag{17}
$$

Then we discretize the axisymmetric surface with N rings where the velocity and the potential are approximated by constants. This leads to a system of linear equations that is solved using LU decomposition. Our method increases its stability by applying a singularity removal procedure as explained in Section 6.4

(formula 6.4.3) of [11] and is computationally much more efficient when restricted to axisymmetric configurations since the integrals in the polar coordinate θ in expression (15) can be computed analytically (see Section 2.4 in [11]). The surface integrals in (15) transform then into line integrals with kernels given in terms of elliptic functions. In order to evaluate these integrals we perform an approximation of the drop in N axisymmetric slices and hence the surface will be divided in N rings with center in the axis of symmetry.

4. Numerical results and discussion

We have performed experiments in axially symmetric geometry by keeping the volume of the drops constant and changing the total charge. The resulting dimensionless parameter X (see Eq. (1)) are $\simeq$ 0.81, 1.03, 1.27, 1.53, 1.82, 2.14, 2.48, 2.85, 3.24, 3.66. We have taken equal viscosities inside and outside the drop in order to simplify the boundary integral calculation and started our dynamical simulation with spherical and nearly spherical initial data. In all cases, the profiles converge to those depicted in Figure 2.

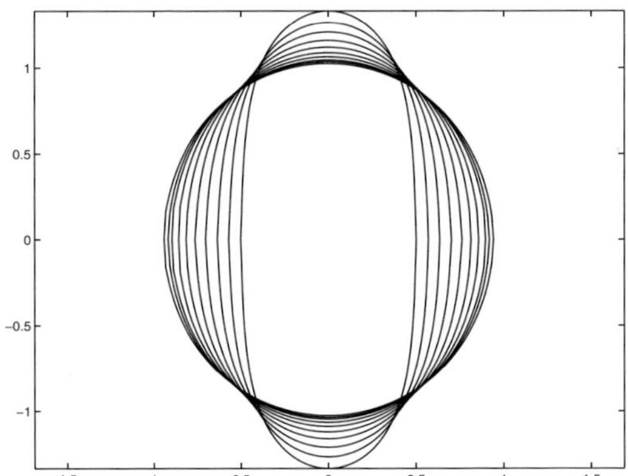

FIGURE 2. The limit axisymmetric profiles at large times for increasing values of the electric charge. The axis of symmetry is horizontal.

Notice the increasing eccentricity and how the minimum radius decreases monotonically. Our code does not work properly for values of X larger than 3.66 indicating the possible presence of a new instability for X larger than this value.

It is simple to prove that in the limit in which the drop becomes a flat disk, one has $X = \frac{16}{3}$ so that the maximum amount of charge that a drop can possibly store is bounded. We are currently developing a code to implement the Boundary integral method without assuming axial symmetry. We discretize the surface by means of triangles and compute the charge density by formulating linear systems as above. Our goal is to test stability of spheroids under non-axisymmetric perturbations.

References

[1] J.M. Rallison, A. Acrivos, A numerical study of the deformation and burst of a viscous drop in an external flow, *J. Fluid Mech.* **89** (1978), 191–200.

[2] O.A. Basaran, L.E. Scriven, Axisymmetric shapes and stability of isolated charged drops, *Phys. Fluids A* **1** − 5 (1989), 795–798.

[3] A. Borisovich and A. Friedman, Symmetry-breaking bifurcations for free boundary problems, to appear.

[4] M.G. Crandall and L.H. Rabinowitz, Bifurcation from simple eigenvalues, *J. Functional Analysis*, **8** (1971), 321–340.

[5] M.G. Crandall and L.H. Rabinowitz, Bifurcation, perturbation of simple eigenvalues and linearized stability, *Arch. Rat. Mech. Anal.*, **52** (1973), 161–180.

[6] D. Duft, T. Achtzehn, R. Müller, B. A. Huber and T. Leisner, Rayleigh jets from levitated microdroplets, *Nature*, vol. **421**, 9 January 2003, pg. 128.

[7] M.A. Fontelos, A. Friedman, Symmetry-breaking bifurcations of charged drops, *Arch. Ration. Mech. Anal.* **172**, 2 (2004), 267–294.

[8] J.D. Jackson, Classical Electrodynamics, John Wiley & Sons, New York; 3rd edition,1999.

[9] M.J. Miksis, Shape of a drop in an electric field, *Phys. of Fluids*, **24**-11 (1981), 1967–1972.

[10] L. Nirenberg, Topics in Nonlinear Functional Analysis, Courant Inst. of Math. Sci., New York, 1974.

[11] C. Pozrikidis, Boundary integral methods for linearized viscous flow, Cambridge texts in Applied Mathematics, Cambridge University Press, 1992.

[12] Lord Rayleigh, On the equilibrium of liquid conducting masses charged with electricity, *Phil. Mag.* **14** (1882), 184–186.

S.I. Betelú
Department of Mathematics
University of North Texas
P.O. Box 311430
Denton, TX 76203-1430, USA

M.A. Fontelos
Departamento de Matemática Aplicada
Universidad Rey Juan Carlos
28933, Móstoles, Spain

U. Kindelán
Departamento de Matemática Aplicada
y Métodos Informáticos
Universidad Politécnica de Madrid
28003, Madrid, Spain

Progress in Nonlinear Differential Equations
and Their Applications, Vol. 63, 59–66
© 2005 Birkhäuser Verlag Basel/Switzerland

On the One-dimensional Parabolic Obstacle Problem with Variable Coefficients

A. Blanchet, J. Dolbeault and R. Monneau

Abstract. This note is devoted to continuity results of the time derivative of the solution to the one-dimensional parabolic obstacle problem with variable coefficients. It applies to the smooth fit principle in numerical analysis and in financial mathematics. It relies on various tools for the study of free boundary problems: blow-up method, monotonicity formulae, Liouville's results.

Mathematics Subject Classification (2000). 35R35.

Keywords. parabolic obstacle problem, free boundary, blow-up, Liouville's result, monotonicity formula, smooth fit.

1. Introduction

Consider a parabolic obstacle problem in an open set. We look for local properties, which do not depend on the boundary conditions and the initial conditions, but only depend on the equation in the interior of the domain. Consider a function u with a one-dimensional space variable x in $Q_1(0)$ where by $Q_r(P_0)$ we denote the parabolic box of radius r and of centre $P_0 = (x_0, t_0)$:

$$Q_r(P_0) = \left\{ (x,t) \in \mathbb{R}^2, \quad |x - x_0| < r, \quad |t - t_0| < r^2 \right\} .$$

Assume that u is a solution of the one-dimensional parabolic obstacle problem with variable coefficients:

$$\begin{cases} a(x,t)u_{xx} + b(x,t)u_x + c(x,t)u - u_t = f(x,t) \cdot \mathbb{1}_{\{u>0\}} & \text{a.e. in } Q_1(0) \\ u \geq 0 \quad \text{a.e. in } Q_1(0) \end{cases} \tag{1}$$

where u_t, u_x, u_{xx} respectively stand for $\frac{\partial u}{\partial t}, \frac{\partial u}{\partial x}, \frac{\partial^2 u}{\partial x^2}$, and $\mathbb{1}_{\{u>0\}}$ is the characteristic function of the positive set of u. Here the free boundary Γ is defined by

$$\Gamma = (\partial \{u = 0\}) \cap Q_1(0) .$$

To simplify the presentation, we assume that the coefficients

$$a, \ b, \ c \text{ and } f \text{ are } C^1 \text{ in } (x,t) \,, \tag{2}$$

but Hölder continuous would be sufficient in what follows.

A natural assumption is that the differential operator is uniformly elliptic, *i.e.*, the coefficient a is bounded from below by zero. If we do not make further assumptions on a and on f, we cannot expect any good property of the free boundary Γ. Suppose that:

$$\exists \delta > 0, \quad a(x,t) \geq \delta, \ f(x,t) \geq \delta \ \text{ a.e. in } Q_1(0) \,. \tag{3}$$

Up to a reduction of the size of the box (see [4]), any weak solution u of (1) has a bounded first derivative in time and bounded first and second derivatives in space. Assume therefore that this property holds on the initial box:

$$|u(x,t)|, \ |u_t(x,t)| \,, \ |u_x(x,t)| \text{ and } |u_{xx}(x,t)| \text{ are bounded in } Q_1(0) \,. \tag{4}$$

This problem is a generalisation to the case of an operator with variable coefficients of Stefan's problem (case where the parabolic operator is $\partial^2/\partial x^2 - \partial/\partial t$). Stefan's problem describes the interface of ice and water (see [10, 13, 8]). The problem with variable coefficients arises in the pricing of American options in financial mathematics (see [3, 2, 14, 11, 9, 15, 1]).

If P is a point such that $u(P) > 0$, by standard parabolic estimates u_t is continuous in a neighborhood of P. On the other hand if P is in the interior of the region $\{u = 0\}$, u_t is obviously continuous. The only difficulty is therefore the regularity on the free boundary Γ. By assumption u_t is bounded but may be *discontinuous* on Γ. The regularity of u_t is a crucial question to apply the "smooth-fit principle" which amounts to the C^1 continuity of the solution at the free boundary. This principle is often assumed, especially in the papers dealing with numerical analysis (see P. Dupuis and H. Wang [6] for example).

In a recent work L. Caffarelli, A. Petrosyan and H. Shagholian [5] prove the C^∞ regularity of the free boundary locally around some points which are energetically characterised, without any sign assumption neither on u nor on its time derivative. This result holds in higher dimension but in the case of constant coefficients. We use tools similar to the ones of [5] and the ones the last author developed previously for the elliptic obstacle problem in [12]. Our main result is the following:

Theorem 1.1 (Continuity of u_t for almost every time). *Under assumptions* (1)–(2)–(3)–(4), *for almost every time t, the function u_t is continuous on $Q_1(0)$.*

This result is new, even in the case of constant coefficients. The continuity of u_t cannot be obtained everywhere in t, as shown by the following example. Let $u(x,t) = \max\{0, -t\}$. It satisfies $u_{xx} - u_t = \mathbb{1}_{\{u>0\}}$ and its time derivative is obviously discontinuous at $t = 0$.

If additionally we assume that $u_t \geq 0$ we achieve a more precise result:

Theorem 1.2 (Continuity of u_t for all t when $u_t \geq 0$). *Under assumptions* (1)–(2)–(3)–(4), *if* $u_t \geq 0$ *in* $Q_1(0)$ *then* u_t *is continuous everywhere in* $Q_1(0)$.

The assumption that $u_t \geq 0$ can be established in some special cases (special initial conditions, boundary conditions, and time independent coefficients). See for example the results of Friedman [7], for further results on the one-dimensional parabolic obstacle problem with particular initial conditions.

In Section 2 we introduce blow-up sequences, which are a kind of zoom at a point of the free boundary. They converge, up to a sub-sequence, to a solution on the whole space of the obstacle problem with constant coefficients. Thanks to a monotonicity formula for an energy we prove in Section 3 that the blow-up limit is scale-invariant. This allows us to classify in Section 4 all possible blow-up limits in a Liouville's theorem. Then we sketch the proof of Theorem 1.1. We even classify energetically the points of the free boundary into the set of regular and singular points. In Section 5 we prove the uniqueness of the blow-up limit at singular points. Then we give the sketch of the proof of the Theorem 1.2. For further details we refer to [4].

2. The notion of blow-up

Given a point $P_0 = (x_0, t_0)$ on the free boundary Γ, we can define the blow-up sequence by

$$u^\varepsilon_{P_0}(x, t) = \frac{u(x_0 + \varepsilon x, t_0 + \varepsilon^2 t)}{\varepsilon^2}, \qquad \varepsilon > 0 . \tag{5}$$

Roughly speaking the action of this rescaling is to zoom on the free boundary at scale ε (see figure 1).

By assumption, $u(P_0) = 0$. Because u is non-negative, we also have $u_x(P_0) = 0$. Moreover $u^\varepsilon_{P_0}$ has a bounded first derivative in time and bounded second derivatives in space. For this reason, using Ascoli-Arzelà's theorem, we can find a sequence $(\varepsilon_n)_n$ which converges to zero such that $\left(u^{\varepsilon_n}_{P_0}\right)_n$ converges on every compact set of $\mathbb{R}^2 = \mathbb{R}_x \times \mathbb{R}_t$ to a function u^0 (called the blow-up limit) and which *a priori* depends on the choice of the sequence $(\varepsilon_n)_n$.

The limit function u^0 satisfies the parabolic obstacle problem with constant coefficients on the whole space-time:

$$a(P_0)u^0_{xx} - u^0_t = f(P_0) \cdot \mathbb{1}_{\{u>0\}} \quad \text{in} \quad \mathbb{R}^2 .$$

By the non-degeneracy assumption (3), it is possible to prove that $0 \in \partial \{u^0 = 0\}$.

To characterise the blow-up limit u^0, we need to come back to the original equation satisfied by u and to obtain additional estimates. In order to simplify the presentation we make a much stronger assumption on u: assume that u is a

    A. Blanchet, J. Dolbeault and R. Monneau

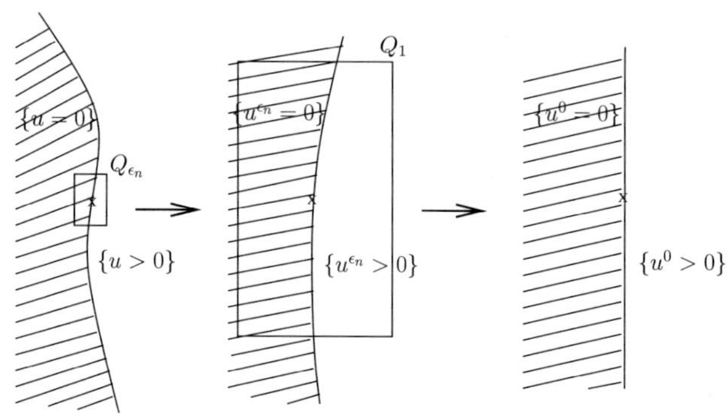

FIGURE 1. Blow-up

solution on the whole space-time of the equation with constant coefficients $a \equiv 1$, $f \equiv 1$, $b \equiv 0$ and $c \equiv 0$:

$$u_{xx} - u_t = 1\!\!1_{\{u>0\}} \quad \text{in} \quad \mathbb{R}^2 . \tag{6}$$

Without this assumption, all tools have to be localised. See [4] for more details.

3. A monotonicity formula for energy

For every time $t < 0$, we define the quantity

$$\mathcal{E}(t; u) = \int_{\mathbb{R}} \left\{ \frac{1}{-t} \left(|u_x(x,t)|^2 + 2\,u(x,t) \right) - \frac{1}{t^2}\,u^2(x,t) \right\} G(x,t)\,dx$$

where G satisfies the backward heat equation $G_{xx} + G_t = 0$ in $\{t < 0\}$ and is given by

$$G(x,t) = \frac{1}{\sqrt{2\pi(-t)}} \exp\left(\frac{-x^2}{4(-t)} \right) .$$

Theorem 3.1 (Monotonicity formula for energy). *Assume that u is a solution of (6). The function $\mathcal{E}$ is non-increasing in time for $t < 0$, and satisfies*

$$\frac{d}{dt}\mathcal{E}(t; u) = -\frac{1}{2(-t)^3} \int_{\mathbb{R}} |\mathcal{L}u(x,t)|^2 G(x,t)\,dx \tag{7}$$

where

$$\mathcal{L}u(x,t) = -2\,u(x,t) + x \cdot u_x(x,t) + 2\,t \cdot u_t(x,t) .$$

A similar but different energy is introduced in [5, 16].

Corollary 3.2 (Homogeneity of the blow-up limit). *Any blow-up limit u^0 of $(u_{P_0}^{\varepsilon_n})_n$ defined in (5), satisfies*

$$u^0(\lambda x, \lambda^2 t) = \lambda^2 u^0(x,t) \quad \text{for every} \quad x \in \mathbb{R},\ t < 0,\ \lambda > 0 . \tag{8}$$

Proof. We prove it in the case $P_0 = 0$. The crucial property is the scale-invariance of $\mathcal{E}$:

$$\mathcal{E}(\varepsilon_n^2 t; u) = \mathcal{E}(t; u_0^{\varepsilon_n}) \, .$$

Taking the limit $\varepsilon_n \to 0$, we get

$$\mathcal{E}(0^-; u) := \lim_{\substack{\tau \to 0 \\ \tau < 0}} \mathcal{E}(\tau; u) = \mathcal{E}(t; u^0) \quad \text{for every} \quad t < 0 \, . \tag{9}$$

From the monotonicity formula (7), we get $\mathcal{L}u^0(x,t) = 0$ for $t < 0$. This implies the homogeneity (8) of u^0 for $t < 0$. $\qquad\square$

4. A Liouville's theorem and consequences

Let

$$v_+(x,t) = \frac{1}{2}(\max\{0, x\})^2 \, ,$$

$$v_-(x,t) = \frac{1}{2}(\max\{0, -x\})^2 \, ,$$

and for $m \in [-1, 0]$

$$v_m(x,t) = \begin{cases} m\,t + \dfrac{1+m}{2}\, x^2 & \text{if} \quad t \le 0 \, , \\[2ex] t\, V_m\left(\dfrac{|x|}{t}\right) > 0 & \text{if} \quad 0 < t < C_m \cdot x^2 \, , \\[2ex] 0 & \text{if} \quad t \ge C_m \cdot x^2 \, , \end{cases}$$

where the coefficient C_m is an increasing function of m, satisfying $C_m = 0$ if $m = -1$, and $C_m = +\infty$, if $m = 0$. The precise expression of V_m is given in [4]. In particular we get $v_{-1}(x,t) = \max\{0, -t\}$ and $v_0(x,t) = \frac{1}{2}x^2$.

Theorem 4.1 (Classification of global homogeneous solutions in $\mathbb{R}^2$). *Let $u^0 \neq 0$ be a non-negative solution of (6) satisfying the homogeneity condition (8). Then u^0 is one of v_+, v_- or v_m for some $m \in [-1, 0]$.*

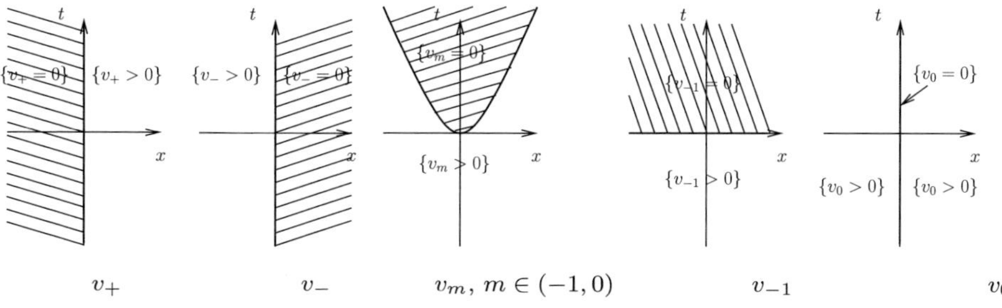

FIGURE 2. Solutions of Theorem 4.1

Similar versions of this theorem are also proved in [5].

Theorem 1.2 is a consequence of Theorem 4.1. Every blow-up limit satisfies $u_t^0 \leq 0$. A more detailed analysis leads to

$$\liminf_{P \to P_0} u_t(P) \leq 0 \, . \tag{10}$$

From the assumption $u_t \geq 0$ we so infer that $u_t = 0$.

We also have an energy criterion to characterise points of the free boundary

Theorem 4.2 (Regular and singular points). *Let u be a solution of* (6). *Then either* $\mathcal{E}(0^-; u) = \sqrt{2}$ *or* $\mathcal{E}(0^-; u) = \sqrt{2}/2$.

In the first case (*i.e.*, $\mathcal{E}(0^-; u) = \sqrt{2}$) P_0 is called a *singular point*. Otherwise (*i.e.*, $\mathcal{E}(0^-; u) = \sqrt{2}/2$), P_0 is a *regular point*.

Proof. By (9) we have $\mathcal{E}(0^-; u) = \mathcal{E}(-1; u^0)$. Blow-up limits have been classified in Theorem 4.1. A simple calculation gives $\mathcal{E}(-1; v_+) = \mathcal{E}(-1; v_+) = \sqrt{2}/2$, and $\mathcal{E}(-1; v_m) = \sqrt{2}$ for every $m \in [-1, 0]$. $\square$

5. A monotonicity formula for singular points

One of the crucial idea of [12] can be adapted to the parabolic framework.

Theorem 5.1 (Monotonicity formula for singular points). *Let u be a solution of* (6) *and assume that $P_0 = 0$ is a singular point. For any $m \in [-1, 0]$ the function*

$$t \mapsto \Phi^{v_m}(t; u) = \int_{\mathbb{R}} \frac{1}{t^2} \left(u(x, t) - v_m(x, t)\right)^2 G(x, t) \, dx \, , \quad t < 0 \tag{11}$$

is non-increasing.

As a consequence $\lim_{\tau \to 0, \, \tau < 0} \Phi^{v_m}(t; u) := \Phi^{v_m}(0^-; u)$ is well defined.

Corollary 5.2 (Blow-up limit at singular points). *Under the assumption of Theorem 5.1, there exists a real $m \in [-1, 0]$ such that any blow-up limit of u at 0 is equal to v_m.*

Proof. Consider two sequences $(\varepsilon_n)_n$ and $(\tilde{\varepsilon}_n)_n$ converging to 0 such that $(u_0^{\varepsilon_n})_n$ and $(u_0^{\tilde{\varepsilon}_n})_n$ respectively converge to two blow-up limits $u^0 = v_m$, for some $m \in [0, 1]$, and $\tilde{u}^0$. Using the scale-invariance we get

$$\Phi^{v_m}(-1; u_0^{\varepsilon_n}) = \Phi^{v_m}(-\varepsilon_n^2 t; u) \quad \text{and} \quad \Phi^{v_m}(-1; u_0^{\tilde{\varepsilon}_n}) = \Phi^{v_m}(-\tilde{\varepsilon}_n^2 t; u) \, .$$

Passing to the limit in the scale-invariance we obtain

$$0 = \Phi^{v_m}(-1; u^0) = \Phi^{v_m}(0^-; u) = \Phi^{v_m}(-1; \tilde{u}^0) \, .$$

This proves that $\tilde{u}^0 = v_m = u^0$, *i.e.*, the uniqueness of the blow-up limit. $\square$

Heuristically the singular points whose blow-up limit is v_m with $m \in [-1, 0)$ have a free boundary with horizontal tangent in the (x, t)-plane: this almost never occurs. On the other hand the singular points with v_0 as blow-up limit and the regular points have a blow-up limit which satisfies $u_t^0 = 0$. This last argument can be refined to show that the time derivative of u is continuous at such points, and consequently for almost every time. This proves Theorem 1.1.

Acknowledgements

We thank D. Lamberton and B. Lapeyre for their comments and suggestions. This study has been partially supported by the ACI NIM "EDP et finance, # 2003-83".

References

[1] Y. ACHDOU, *An inverse problem for parabolic variational inequalities in the calibration of American options.* to appear in SIAM J. Control Optim.

[2] A. BENSOUSSAN AND J.-L. LIONS, *Applications des inéquations variationnelles en contrôle stochastique*, Dunod, Paris, 1978. Méthodes Mathématiques de l'Informatique, No. 6.

[3] F. BLACK AND M. SCHOLES, *The pricing of options and corporate liabilities*, J. Polit. Econ., 81 (1973), pp. 637–659.

[4] A. BLANCHET, J. DOLBEAULT, AND R. MONNEAU, *On the continuity of the time derivative of the solution to the one-dimensional parabolic obstacle problem with variable coefficients.* In preparation.

[5] L. CAFFARELLI, A. PETROSYAN, AND H. SHAHGHOLIAN, *Regularity of a free boundary in parabolic potential theory*, J. Amer. Math. Soc., 17 (2004), pp. 827–869 (electronic).

[6] P. DUPUIS AND H. WANG, *Optimal stopping with random intervention times*, Adv. in Appl. Probab., 34 (2002), pp. 141–157.

[7] A. FRIEDMAN, *Parabolic variational inequalities in one space dimension and smoothness of the free boundary*, J. Functional Analysis, 18 (1975), pp. 151–176.

[8] ———, *Variational principles and free-boundary problems*, Pure and Applied Mathematics, John Wiley & Sons Inc., New York, 1982. A Wiley-Interscience Publication.

[9] P. JAILLET, D. LAMBERTON, AND B. LAPEYRE, *Variational inequalities and the pricing of American options*, Acta Appl. Math., 21 (1990), pp. 263–289.

[10] D. KINDERLEHRER AND G. STAMPACCHIA, *An introduction to variational inequalities and their applications*, vol. 88 of Pure and Applied Mathematics, Academic Press Inc. [Harcourt Brace Jovanovich Publishers], New York, 1980.

[11] D. LAMBERTON AND B. LAPEYRE, *Introduction au calcul stochastique appliqué à la finance*, Ellipses Édition Marketing, Paris, second ed., 1997.

[12] R. MONNEAU, *On the number of singularities for the obstacle problem in two dimensions*, J. Geom. Anal., 13 (2003), pp. 359–389.

[13] J.-F. RODRIGUES, *Obstacle problems in mathematical physics*, vol. 134 of North-Holland Mathematics Studies, North-Holland Publishing Co., Amsterdam, 1987. Notas de Matemática [Mathematical Notes], 114.

[14] P. VAN MOERBEKE, *An optimal stopping problem with linear reward*, Acta Math., 132 (1974), pp. 111–151.

[15] S. VILLENEUVE, *Options américaines dans un modèle de Black-Scholes multidimensionnel*, PhD thesis, Université De Marne la Vallée, 1999.

[16] G.S. WEISS, *Self-similar blow-up and Hausdorff dimension estimates for a class of parabolic free boundary problems*, SIAM J. Math. Anal., 30 (1999), pp. 623–644 (electronic).

A. Blanchet
CEREMADE
Université Paris Dauphine
place de Lattre de Tassigny
F-75775 Paris Cédex 16, France

and

CERMICS
Ecole Nationale des Ponts et Chaussées
6 et 8 avenue Blaise Pascal
Cité Descartes Champs-sur-Marne
F-77455 Marne-la-Vallée Cédex 2, France

J. Dolbeault
CEREMADE
Université Paris Dauphine
place de Lattre de Tassigny
F-75775 Paris Cédex 16, France

R. Monneau
CERMICS
Ecole Nationale des Ponts et Chaussées
6 et 8 avenue Blaise Pascal
Cité Descartes Champs-sur-Marne
F-77455 Marne-la-Vallée Cédex 2, France

Progress in Nonlinear Differential Equations
and Their Applications, Vol. 63, 67–82

Hardy Potentials and Quasi-linear Elliptic Problems Having Natural Growth Terms

Lucio Boccardo

"...seder tra filosofica famiglia.
Tutti lo miran, tutti onor li fanno."
(Inferno, IV)

Abstract. In this paper we consider nonlinear boundary value problems whose simplest model is the following:

$$\begin{cases} -\Delta u = \gamma |\nabla u|^2 + \frac{A}{|x|^2} & \text{in } \Omega \qquad (\gamma,\, A \in \mathbb{R}) \\ \quad\;\; u = 0 & \text{on } \partial\Omega. \end{cases} \tag{0.1}$$

where Ω is a bounded open set in $\mathbb{R}^N$, $N > 2$.

Keywords. quasi-linear elliptic equations, natural growth terms, quadratic growth with respect to the gradient, Hardy inequality.

1. Introduction

It is well known that the minimization in $W_0^{1,2}(\Omega)$ (Ω is a bounded domain in $\mathbb{R}^N$, $N > 2$) of simple functionals like

$$I(v) = \frac{1}{2} \int_\Omega a(x,v)|\nabla v|^2 - \int_\Omega f(x)v(x),$$

where a is a bounded, smooth function and $f \in L^2(\Omega)$, leads to the following Euler-Lagrange equation

$$\begin{cases} -\text{div}(a(x,u)\nabla u) + \frac{1}{2}a'(x,u)|\nabla u|^2 = f & \text{in } \Omega \\ \qquad\qquad\qquad u = 0 & \text{on } \partial\Omega. \end{cases} \tag{1.1}$$

Thus Calculus of Variations (and also Stochastic Control) is a motivation to the study of quasi-linear Dirichlet problems having lower order terms with quadratic growth with respect to the gradient, even if the equation is not the Euler-Lagrange equation of integral functionals.

In this paper we give some contributions to the existence results for quasi-linear elliptic problems having natural growth terms. We discuss multiplicity of solutions only in Remark 2.1.

We are interested in existence and nonexistence of weak solutions of

$$\begin{cases} -\operatorname{div}(M(x,u)\nabla u) = b(x,u,\nabla u) + f(x) & \text{in } \Omega, \\ \qquad\qquad u = 0 & \text{on } \partial\Omega \end{cases} \tag{1.2}$$

where Ω is a bounded open subset of $\mathbb{R}^N$, $0 \in \Omega$, mainly if $|f| \leq \frac{A}{|x|^2}$ (observe that $\frac{A}{|x|^2}$ does not belong to $L^{\frac{N}{2}}(\Omega)$).

We assume that $M(x,s)$ is a Carathéodory matrix and $b(x,s,\xi)$ is a Carathéodory function (that is, measurable with respect to x for every $(s,\xi) \in \mathbb{R} \times \mathbb{R}^N$, and continuous with respect to (s,ξ) for almost every $x \in \Omega$) which satisfy, for some positive constants α, β, γ, a.e. in $x \in \Omega$, $\forall s \in \mathbb{R}$, $\forall \xi \in \mathbb{R}^N$

$$M(x,s)\xi \cdot \xi \geq \alpha|\xi|^2 \tag{1.3}$$

$$|M(x,s)| \leq \beta \tag{1.4}$$

$$|b(x,s,\xi)| \leq \gamma|\xi|^2. \tag{1.5}$$

In papers ([7], [8], [9], [10]), we proved existence of bounded weak solutions in $W_0^{1,p}(\Omega)$, $p > 1$, for (1.2) under suitable assumptions on the data (in particular on the summability of f). We have developed a method which essentially allows to prove the existence of a solution once one can provide an L^∞ estimate for the solutions of a family of approximate equations.

The main goal of the present paper is to prove existence results if the data are not regular enough in order to have bounded solutions, but no sign condition is assumed on the function $b(x,s,\xi)$; in [6], [1], [15], [2], the assumptions (1.5) and $b(x,s,\xi)s \leq 0$ easily allow us to prove a priori estimates in $W_0^{1,p}(\Omega)$ and also in $L^\infty(\Omega)$, if $f \in L^m(\Omega)$, $m > \frac{N}{2}$, so that, thanks to the above remark, $f \in L^m(\Omega)$, with $m > \frac{N}{2}$, implies easily existence of solutions. Moreover, with the assumptions used in [3], [4], [5] (mainly $b(x,s,\xi)s \leq 0$) it is possible to prove a priori estimates in $W_0^{1,p}(\Omega)$ even if f belongs only to $L^1(\Omega)$ (but not if f is a measure: see [15], [5]).

For the sake of simplicity, our framework is the Sobolev space $W_0^{1,2}(\Omega)$, instead of $W_0^{1,p}(\Omega)$, and we have $-\operatorname{div}(M(x,v)\nabla v))$ as principal part of the differential operator, instead of $-\operatorname{div}(a(x,v,\nabla v))$ and positive right-hand side. The proof of the general case will follow using the methods of the present paper together with the ideas of [8].

2. Right-hand side like $\frac{1}{|x|^2}$

Under our hypotheses, the operator $Q(v) = -\operatorname{div}(M(x,v)\nabla v))$ is a bounded, continuous, coercive and pseudomonotone operator from $W_0^{1,2}(\Omega)$ into $W^{-1,2}(\Omega)$ and therefore it is surjective; but we miss the properties of Q, if we add the term $b(x,v,\nabla v)$ and we consider our operator $-\operatorname{div}(M(x,v)\nabla v)) + b(x,v,\nabla v)$.

Remark 2.1. *Here we discuss the following case:* $\Omega = B(0,1)$, $f(x) = \frac{A}{|x|^2}$. *If we look for radial solutions:* $u(x) = w(|x|) = w(r)$, *then* $u_{x_i} = w'(r)\frac{x_i}{|x|}$, $-\Delta u = -w''(r) - w'(r)\frac{N-1}{|x|}$ *and* $|\nabla u|^2 = w'(r)^2$. *Thus the model problem* (0.1) *becomes*

$$+w'' + \frac{N-1}{r}w' + \gamma w'(r)^2 = -\frac{A}{r^2}, \qquad w(1) = 0.$$

If we look for solutions of the type $w(r) = B\log_e(r)$, $B \in \mathbb{R}$, *we have* $-\frac{B}{r^2} + \frac{B(N-1)}{r^2} + \frac{B^2\gamma}{r^2} = -\frac{A}{r^2}$, $\gamma B^2 + (N-2)B + A = 0$. *So we have solutions of the type* $w(r) = B\log(r)$, *if* $(N-2)^2 - 4A\gamma \geq 0$. *If* $(N-2)^2 - 4A\gamma > 0$, *that is* $A < \left(\frac{N-2}{2}\right)^2 \frac{1}{\gamma}$, *we have the couple of solutions* $w_i(r) = B_i\log(r)$, *with*

$$\begin{cases} B_1 = \frac{-(N-2)-\sqrt{(N-2)^2-4A\gamma}}{2\gamma} \\ B_2 = \frac{-(N-2)+\sqrt{(N-2)^2-4A\gamma}}{2\gamma}. \end{cases}$$

We point out that we have found solutions if $A < \left(\frac{N-2}{2}\right)^2 \frac{1}{\gamma}$. *Moreover, under this condition, there exist two solutions.*

We shall use the following inequality.

Proposition 2.2. [Hardy-Sobolev inequality] *The Hardy inequality states that*

$$\mathcal{H}\int_\Omega \frac{|v|^2}{|x|^2} \leq \int_\Omega |\nabla v|^2, \quad \forall v \in W_0^{1,2}(\Omega). \tag{2.1}$$

Moreover $\mathcal{H} = \left(\frac{N-2}{2}\right)^2$ *is optimal.*

In this section we study existence and nonexistence of solutions of (1.2). The main point of our framework is the assumption on the right-hand side f: $|f(x)| \leq \frac{A}{|x|^2}$; so that, in general, f does not belong to $L^{\frac{N}{2}}(\Omega)$: f belongs to the Marcinkiewicz space $M^{\frac{N}{2}}(\Omega)$. Existence and nonexistence for right-hand side in the Marcinkiewicz space $M^{\frac{N}{2}}(\Omega)$ is studied in [20]. Our proofs hinge on the Hardy-Sobolev inequality and strongly use the approach of [7], [8], [9].

2.1. Existence

Now we state the following theorem which contains the main result of this paper.

Theorem 2.3. *Let us assume that* (1.3), (1.4), (1.5) *hold true and that*

$$0 \leq f(x) \leq \frac{A}{|x|^2}, \quad A < \frac{\alpha^2 \mathcal{H}}{\gamma}. \tag{2.2}$$

70   L. Boccardo

Then there exists a solution u of the Dirichlet problem (1.2) in the following (weak) sense

$$\begin{cases} u \in W_0^{1,2}(\Omega) : \\ \displaystyle\int_\Omega M(x,u)\nabla u \nabla \varphi = \int_\Omega b(x,u,\nabla u)\varphi + \int_\Omega f\,\varphi, \\ \forall \varphi \in W_0^{1,2}(\Omega) \cap L^\infty(\Omega). \end{cases} \tag{2.3}$$

Remark 2.4. *Notice that the limitation on A in Remark 2.1 is very close to that in (2.2).*

Remark 2.5. *It is possible to prove existence results concerning equations with right hands dominated by $\frac{A}{dist(x,\Sigma)^2}$, $\Sigma \subset \Omega$, (instead of $\frac{A}{|x|^2}$) thanks to the inequalities proved in [14], [18], [25].*

In order to prove the existence result, let us define for $n \in \mathbb{N}$, the approximations

$$b_n(x,s,\xi) = \frac{b(x,s,\xi)}{1 + \frac{1}{n}|b(x,s,\xi)|}, \qquad f_n(x) = \frac{f(x)}{1 + \frac{1}{n}|f(x)|}. \tag{2.4}$$

We notice that

$$|b_n(x,s,\xi)| \le |b(x,s,\xi)|,\ 0 \le f_n(x)| \le f(x)$$

and

$$|b_n(x,s,\xi)| \le n,\ 0 \le f_n(x) \le n.$$

Consider the approximate boundary value problems

$$\begin{cases} -\operatorname{div}(M(x,u_n)\nabla u_n)) = b_n(x,u_n,\nabla u_n) + f_n(x) & \text{in } \Omega, \\ u_n = 0 & \text{on } \partial\Omega. \end{cases} \tag{2.5}$$

The existence of weak, bounded (see [26]) solution $u_n \in W_0^{1,2}(\Omega)$ of (2.5) follows by the classical results of [24] (see also [12], [22]).

Lemma 2.6. *Assume (2.2). Let be $M > 1$, such that $A < \frac{\alpha^2 \mathcal{H}}{M\gamma} < \frac{\alpha^2 \mathcal{H}}{\gamma}$ and*

$$\lambda \in \left(\frac{\alpha\mathcal{H} - \sqrt{\alpha^2\mathcal{H}^2 - \gamma AM\mathcal{H}}}{AM}, \frac{\alpha\mathcal{H} + \sqrt{\alpha^2\mathcal{H}^2 - \gamma AM\mathcal{H}}}{AM} \right). \tag{2.6}$$

Then there exists a positive constant $C_{\lambda,M} = C(\lambda, M, A, N, \alpha, \gamma)$ such that

$$\int_\Omega e^{2\lambda u_n}|\nabla u_n|^2 \le C_{\lambda,M}. \tag{2.7}$$

Proof. Let

$$\phi(t) = (e^{2\lambda t} - 1).$$

As u_n belongs to $W_0^{1,2}(\Omega) \cap L^\infty(\Omega)$, so does $v = \phi(u_n)$ and it is easy to prove that

$$v = \phi(u_n), \quad \nabla v = \phi'(u_n)\nabla u_n.$$

First of all, the use of $-\phi(u_n^-)$ as test function in (2.5) implies that u_n is positive. Then, following [9], we use $\phi(u_n)$ as test function in (2.5):

$$\int_\Omega |\nabla u_n|^2 [\alpha \phi'(u_n) - \gamma \phi(u_n)] \le \int_\Omega f(x)\phi(u_n).$$

Notice that

$$\frac{\gamma}{2\alpha} \le \frac{\alpha \mathcal{H} - \sqrt{\alpha^2 \mathcal{H}^2 - \gamma A M \mathcal{H}}}{AM}$$

so that $\lambda > \frac{\gamma}{2\alpha}$. Then

$$(2\lambda\alpha - \gamma)\int_\Omega |\nabla u_n|^2 e^{2\lambda u_n} \le A \int_\Omega \frac{(e^{2\lambda u_n} - 1)}{|x|^2}.$$

With M as in the statement, and since, if $t \ge \log_e \frac{M+1}{M-1}$, we have $(e^{2t} - 1) \le M(e^t - 1)^2$, then

$$\frac{2\lambda\alpha - \gamma}{\lambda^2} \int_\Omega |\nabla(e^{\lambda u_n} - 1)|^2$$

$$\le A \int_{u_n \le \frac{1}{\lambda} \log_e \frac{M+1}{M-1}} \left(\frac{(e^{2\lambda u_n} - 1)}{|x|^2} - \frac{M(e^{\lambda u_n} - 1)^2}{|x|^2}\right) + AM \int_\Omega \frac{(e^{\lambda u_n} - 1)^2}{|x|^2}.$$

Note that on the subset $\{x \in \Omega : 0 \le u_n(x) \le \frac{1}{\lambda}\log_e \frac{M+1}{M-1}\}$ we have $((e^{2\lambda u_n} - 1) - M(e^{\lambda u_n} - 1)^2) \le \frac{1}{M-1}$, so that the use of the Hardy-Sobolev inequality (2.1) implies

$$\left(\frac{2\lambda\alpha - \gamma}{\lambda^2} - \frac{AM}{\mathcal{H}}\right)\int_\Omega |\nabla(e^{\lambda u_n} - 1)|^2 \le \frac{A}{M-1}\int_\Omega \frac{1}{|x|^2}. \qquad (2.8)$$

In the previous estimate, we need $(2\lambda\alpha - \gamma)\mathcal{H} - \lambda^2 AM > 0$, that is $AM\lambda^2 - 2\alpha\mathcal{H}\lambda + \gamma\mathcal{H} < 0$. There exists λ such that the previous inequality is satisfied, since $(\alpha\mathcal{H})^2 - AM\gamma\mathcal{H} > 0$, because of the choice $A < \frac{\alpha^2\mathcal{H}}{M\gamma}$. The roots of $AM\lambda^2 - 2\alpha\mathcal{H}\lambda + \gamma\mathcal{H} = 0$ are

$$\frac{\alpha\mathcal{H} - \sqrt{\alpha^2\mathcal{H}^2 - \gamma AM\mathcal{H}}}{AM} \qquad \text{and} \qquad \frac{\alpha\mathcal{H} + \sqrt{\alpha^2\mathcal{H}^2 - \gamma AM\mathcal{H}}}{AM}$$

so that any

$$\lambda \in \left(\frac{\alpha\mathcal{H} - \sqrt{\alpha^2\mathcal{H}^2 - \gamma AM\mathcal{H}}}{AM}, \frac{\alpha\mathcal{H} + \sqrt{\alpha^2\mathcal{H}^2 - \gamma AM\mathcal{H}}}{AM}\right)$$

is suitable in the previous inequalities. Then (2.8) implies (2.7). $\qquad\square$

Corollary 2.7. *From inequality* (2.7) *we deduce*

$$\int_\Omega |\nabla u_n|^2 \le C_{\lambda,M}, \qquad \mathcal{H}\int_\Omega \frac{(e^{\lambda u_n} - 1)^2}{|x|^2} \le C_{\lambda,M}$$

and

$$\int_{\{x\in\Omega:u_n(x)>k\}} |\nabla u_n|^2 \leq \frac{C_{\lambda,M}}{e^{2\lambda k}}. \tag{2.9}$$

Remark 2.8. *The use and the proof of the last inequality is the most important difference with the techniques of* [7], [8], [9].

By Corollary (2.7), the sequence $\{u_n\}$ is bounded in $W_0^{1,2}(\Omega)$. Then there exist a function u in $W_0^{1,2}(\Omega)$ and a subsequence, still denoted $\{u_n\}$, such that u_n converges to u weakly in $W_0^{1,2}(\Omega)$ and a.e. (so that also u is positive).

Lemma 2.9. *The sequence $\{u_n\}$ converges strongly to u in $W_0^{1,2}(\Omega)$.*

Proof. Recall the definitions

$$T_k(s) = \begin{cases} s & \text{if } |s| \leq k, \\ k\frac{s}{|s|} & \text{if } |s| > k \end{cases} \quad \text{and} \quad G_k(s) = s - T_k(s).$$

Following [4], we use $\varphi[T_k(u_n) - T_k(u)]$ as test function in (2.5) (even if in the present paper the inequality $b(x,s,\xi)s \geq 0$ is not assumed), where

$$\varphi(t) = (e^{\mu|t|} - 1)\mathrm{sgn}(t), \quad \frac{2\gamma}{\alpha} < \mu < 2\lambda$$

and λ as in (2.6). We obtain

$$\int_\Omega M(x,u_n)\nabla[T_k(u_n) - T_k(u)]\nabla[T_k(u_n) - T_k(u)]\varphi'[T_k(u_n) - T_k(u)]$$

$$\leq \gamma \int_\Omega |\nabla u_n|^2|\varphi[T_k(u_n) - T_k(u)] + \epsilon_k(n)$$

$$= \gamma \int_\Omega |\nabla T_k(u_n)|^2|\varphi[T_k(u_n) - T_k(u)] + \gamma \int_\Omega |\nabla G_k(u_n)|^2|\varphi[T_k(u_n) - T_k(u)] + \epsilon_k(n)$$

$$\leq 2\gamma \int_\Omega |\nabla[T_k(u_n) - T_k(u)]|^2|\varphi[T_k(u_n) - T_k(u)]| + \omega_k(n)$$

$$+\gamma \int_\Omega |\nabla G_k(u_n)|^2|\varphi[T_k(u_n) - T_k(u)]| + \epsilon_k(n)$$

where

$$\epsilon_k(n) = \int_\Omega f_n\varphi[T_k(u_n) - T_k(u)]$$

$$-\int_\Omega M(x,u_n)\nabla T_k(u)\nabla[T_k(u_n) - T_k(u)]\varphi'[T_k(u_n) - T_k(u)]$$

$$+\int_\Omega M(x,u_n)\nabla G_k(u_n)\nabla T_k(u)\varphi'[T_k(u_n) - T_k(u)]$$

and

$$\omega_k(n) = 2\gamma \int_\Omega |\nabla T_k(u)|^2 |\varphi[T_k(u_n) - T_k(u)].$$

Observe that both $\epsilon_k(n)$ and $\omega_k(n)$ converge to zero, as $n \to +\infty$ (with k fixed) and that $0 \leq T_k(u) \leq k$, so that

$$\gamma \int_\Omega |\nabla G_k(u_n)|^2 |\varphi[T_k(u_n) - T_k(u)]|$$

$$= \gamma \int_{\{x \in \Omega: u_n(x) > k\}} |\nabla u_n|^2 |\varphi[k - T_k(u)]| \leq \gamma\varphi[k] \int_{\{x \in \Omega: u_n(x) > k\}} |\nabla u_n|^2.$$

Now (use (2.9) and previous inequality)

$$(\mu\alpha - 2\gamma) \int_\Omega |\nabla[T_k(u_n) - T_k(u)]|^2 \tag{2.10}$$

$$\leq \int_\Omega |\nabla[T_k(u_n) - T_k(u)]|^2 \{\alpha\varphi'[T_k(u_n) - T_k(u)] - 2\gamma\varphi[T_k(u_n) - T_k(u)]\}$$

$$\leq \gamma \frac{C_{\lambda,M}}{e^{(2\lambda-\mu)k}} + \epsilon_k(n) + \omega_k(n).$$

Here, fix $\epsilon > 0$ and choose k_0 such that $\frac{\gamma C_{\lambda,M}}{e^{2(\lambda-\mu)k_0}} < \epsilon$; then $n_\epsilon \in \mathbb{N}$ such that, $\epsilon_{k_0}(n)$ and $\omega_{k_0}(n) < \epsilon$, if $n > n_\epsilon$; with these choices (2.10) gives

$$\frac{(\mu\alpha - 2\gamma)}{3} \int_\Omega |\nabla[T_{k_0}(u_n) - T_{k_0}(u)]|^2 \leq \epsilon, \quad \forall n > n_\epsilon.$$

Then, for $n > n_\epsilon$, using again (2.9) and the choice of k_0, we have

$$\frac{(\mu\alpha - 2\gamma)}{6} \int_\Omega |\nabla(u_n - u)]|^2 \leq \frac{(\mu\alpha - 2\gamma)}{3} \int_\Omega |\nabla[T_{k_0}(u_n) - T_{k_0}(u)]|^2$$

$$+ \frac{2(\mu\alpha - 2\gamma)}{3} \int_\Omega |\nabla G_{k_0}(u_n)|^2 + \frac{2(\mu\alpha - 2\gamma)}{3} \int_\Omega |\nabla G_{k_0}(u)|^2$$

$$\leq \epsilon + \frac{4(\mu\alpha - 2\gamma)}{3} \frac{C_{\lambda,M}}{e^{(2\lambda-\mu)k_0}} \leq \epsilon[1 + \frac{4(\mu\alpha - 2\gamma)}{3\gamma}],$$

i.e., the strong convergence in $W_0^{1,2}(\Omega)$ of the sequence $\{u_n\}$. $\qquad\square$

Proof of Theorem 2.3. As a consequence of the previous lemma, up to a subsequence still denoted by u_n, $\nabla u_n(x)$ is almost everywhere convergent to $\nabla u(x)$. In order to pass to the limit in the approximate equation, we are going to prove that

$$b_n(x, u_n, \nabla u_n) \to b(x, u, \nabla u) \qquad \text{strongly in } L^1(\Omega). \tag{2.11}$$

Since $b_n(x, u_n, \nabla u_n)$ converges almost everywhere to

$$b(x, u, \nabla u) \quad \text{and} \quad |b_n(x, u_n, \nabla u_n)| \leq \gamma |\nabla u_n(x)|^2,$$

we have also $|b(x, u, \nabla u)| \leq \gamma |\nabla u(x)|^2$. We adapt the classical proof of the Lebesgue theorem. We can use the Fatou Lemma, since

$$\gamma |\nabla u_n(x)|^2 + \gamma |\nabla u(x)|^2 - |b_n(x, u_n, \nabla u_n) - b(x, u, \nabla u)| \geq 0.$$

Then

$$2\gamma \int_\Omega |\nabla u(x)|^2 \leq \liminf \int_\Omega [\gamma |\nabla u_n(x)|^2 + \gamma |\nabla u(x)|^2 - |b_n(x, u_n, \nabla u_n) - b(x, u, \nabla u)|]$$

$$= \lim \gamma \int_\Omega |\nabla u_n(x)|^2 + \gamma \int_\Omega |\nabla u(x)|^2 + \liminf \int_\Omega [-|b_n(x, u_n, \nabla u_n) - b(x, u, \nabla u)|]$$

$$= 2\gamma \int_\Omega |\nabla u(x)|^2 + \liminf \int_\Omega [-|b_n(x, u_n, \nabla u_n) - b(x, u, \nabla u)|]$$

which implies

$$\limsup \int_\Omega |b_n(x, u_n, \nabla u_n) - b(x, u, \nabla u)| \leq 0,$$

that is

$$\int_\Omega |b_n(x, u_n, \nabla u_n) - b(x, u, \nabla u)| \to 0.$$

The convergences of Lemma 2.9 and (2.11) allow us to pass to the limit in the weak formulation of (2.5), in order to obtain that u is a weak solution of (2.3). $\quad \square$

Remark 2.10. *Remark that the assumptions (2.6) and $M > 1$ imply that*

$$\lambda < \frac{\alpha \mathcal{H} + \sqrt{\alpha^2 \mathcal{H}^2 - \gamma A \mathcal{H}}}{A}.$$

That is

$$\int_\Omega |\nabla (e^{\lambda u} - 1)|^2 < \infty, \quad \forall \lambda < \frac{\alpha \mathcal{H} + \sqrt{\alpha^2 \mathcal{H}^2 - \gamma A \mathcal{H}}}{A}. \tag{2.12}$$

In particular

$$\int_\Omega |\nabla (e^{\frac{\gamma}{\alpha} u} - 1)|^2 < \infty.$$

Note that, in Remark 2.1 (where $\alpha = 1$), only the solution w_2 satisfies (2.12).

2.2. Nonexistence

We prove that the linear form of the boundary problem (1.2) does not have solutions, if we assume $A > \frac{\alpha^2 \mathcal{H}}{\gamma}$.

Theorem 2.11. *Suppose that $M(x,s) = M(x)$, where M is a symmetric matrix, which satisfies (1.3) and (1.4), and consider the model problem of (2.3), that is*

$$\begin{cases} u \in W_0^{1,2}(\Omega) : \\ \int_\Omega M(x)\nabla u \nabla \varphi = \gamma \int_\Omega |\nabla u|^2 \varphi + \int_\Omega \frac{A}{|x|^2}\,\varphi, \\ \forall \varphi \in W_0^{1,2}(\Omega) \cap L^\infty(\Omega). \end{cases} \qquad (2.13)$$

If we assume that $A > \frac{\alpha^2 \mathcal{H}}{\gamma}$, there are no solutions of (2.13).

Proof. By contradiction we assume the existence of a weak solution u of (2.11).

We use a result of [21]. Let ψ_n be the first eigenfunction (positive) and H_n the first eigenvalue of

$$\psi_n \in W_0^{1,2}(\Omega) : -\mathrm{div}(M(x)\nabla\psi_n) = H_n \frac{\psi_n}{\frac{1}{n} + |x|^2}. \qquad (2.14)$$

The sequence $\{H_n\}$ converges to $\mathcal{H}$. So that, for some $\nu \in \mathbb{N}$, we have also $A\gamma - \alpha^2 H_\nu > 0$. Remark that a solution u of (2.13) is positive. Thanks to Remark 2.10, we can use $(e^{\frac{\gamma}{\alpha}u} - 1)$ as test function in (2.14), written for ψ_ν, and $e^{\frac{\gamma}{\alpha}u}\psi_\nu$ as test function in (2.13). Then (after simplifications)

$$\begin{cases} \dfrac{\gamma}{\alpha} \int_\Omega M(x)\nabla\psi_\nu \nabla u\, e^{\frac{\gamma}{\alpha}u} = H_\nu \int_\Omega \dfrac{\psi_\nu(e^{\frac{\gamma}{\alpha}u} - 1)}{\frac{1}{\nu} + |x|^2} \\ \int_\Omega M(x)\nabla u \nabla\psi_\nu\, e^{\frac{\gamma}{\alpha}u} = A \int_\Omega \dfrac{e^{\frac{\gamma}{\alpha}u}\psi_\nu}{|x|^2}. \end{cases}$$

Thus

$$A\int_\Omega \frac{\psi_\nu}{|x|^2} + \frac{A}{\nu} \int_\Omega \frac{(e^{\frac{\gamma}{\alpha}u} - 1)\psi_\nu}{|x|^2(\frac{1}{\nu} + |x|^2)} + (A\gamma - \alpha^2 H_\nu)\int_\Omega \frac{(e^{\frac{\gamma}{\alpha}u} - 1)\psi_\nu}{\gamma(\frac{1}{\nu} + |x|^2)} = 0$$

which is impossible, since every term is nonnegative, being $u \geq 0$. $\qquad\square$

Remark 2.12. *Theorem 2.3 proves existence under the assumption $A < \frac{\alpha^2 \mathcal{H}}{\gamma}$; Theorem 2.11 shows that there are no solutions of (2.13) if $A > \frac{\alpha^2 \mathcal{H}}{\gamma}$. We are not able to study the case $A = \frac{\alpha^2 \mathcal{H}}{\gamma}$ in the general case (not even with the use of Improved Hardy-Sobolev inequality [1] of [16]).*

In Remark 2.1 we have indeed shown that $w(r) = -\frac{N-2}{2}\log(r)$ is a solution in the radial case.

---

[1] **Improved Hardy-Sobolev inequality:** For $1 < q < \frac{2N}{N-2}$, there exists $h_q \in \mathbb{R}^+$ such that

$$h_q\|v\|^2_{L^q(\Omega)} \leq \int_\Omega |\nabla v|^2 - \left(\frac{N-2}{2}\right)^2 \int_\Omega \frac{|v|^2}{|x|^2}, \qquad \forall v \in W_0^{1,2}(\Omega). \qquad (2.15)$$

3. $f \in L^{\frac{N}{2}}(\Omega)$

In this section we assume that f belongs to $L^{\frac{N}{2}}(\Omega)$.

Lemma 3.1. *Assume that f is a positive function which belongs to $L^{\frac{N}{2}}(\Omega)$ and*

$$\|f\|_{\frac{N}{2}} < \frac{\alpha^2 S}{\gamma},$$

where S is the Sobolev constant. Let be $M > 1$, such that $\|f\|_{\frac{N}{2}} < \frac{\alpha^2 S}{M\gamma} < \frac{\alpha^2 S}{\gamma}$ and

$$\lambda \in \left(\frac{\alpha S - \sqrt{\alpha^2 S^2 - \gamma\|f\|_{\frac{N}{2}} MS}}{\|f\|_{\frac{N}{2}} M} , \frac{\alpha S + \sqrt{\alpha^2 S^2 - \gamma\|f\|_{\frac{N}{2}} MS}}{\|f\|_{\frac{N}{2}} M} \right).$$

Then there exists a positive constant $C_{\lambda,M} = C(\lambda, M, f, N, \alpha, \gamma)$ such that the estimate (2.7) still holds.

Proof. We repeat the proof of Lemma 2.6 till (2.1), but we use Hölder and Sobolev inequalities instead of Hardy-Sobolev inequality:

$$\frac{2\lambda\alpha - \gamma}{\lambda^2} \int_\Omega |\nabla(e^{\lambda u_n} - 1)|^2 \leq \int_\Omega |f|((\frac{M+1}{M-1})^2 - 1)$$

$$+ M\|f\|_{\frac{N}{2}} \left(\int_\Omega (e^{\lambda u_n} - 1)^{\frac{2N}{N-2}} \right)^{\frac{N-2}{N}}$$

$$\leq C_M \|f\|_1 + \frac{M\|f\|_{\frac{N}{2}}}{S} \int_\Omega |\nabla(e^{\lambda u_n} - 1)|^2.$$

Thus

$$\left(\frac{2\lambda\alpha - \gamma}{\lambda^2} - \frac{M\|f\|_{\frac{N}{2}}}{S} \right) \int_\Omega |\nabla(e^{\lambda u_n} - 1)|^2 \leq C_M \|f\|_1. \qquad \square$$

Theorem 3.2. *Let us assume that (1.3), (1.4), (1.5) hold true and that $\|f\|_{\frac{N}{2}} < \alpha^2 S \gamma^{-1}$. Then there exists a weak solution u of the Dirichlet problem (2.3).*

Proof. Under the same assumption we proved Lemma 3.1 which says that the estimate (2.7) still holds. Thus also the proof of Theorem 3.2 still holds. $\qquad \square$

Remark 3.3. *Also in the proof of previous theorem the starting point is the approach of [9]. The theorem is proved also in [19] (see also [17]).*

4. Lower order terms and weak summability of the data

In this section we will show how the presence of a lower order term $g(u)$ allows us to weaken the summability assumption on $f(x)$. In particular, in (2.3), we may not assume $A < \alpha^2 \mathcal{H} \gamma^{-1}$, if we add a suitable lower order term $g(u)$.

Consider the following boundary value problems

$$\begin{cases} -\mathrm{div}(M(x,u)\nabla u) + g(u) = b(x,u,\nabla u) + f(x) & \text{in } \Omega, \\ \qquad\qquad\qquad u = 0 & \text{on } \partial\Omega \end{cases} \tag{4.1}$$

and

$$\begin{cases} -\mathrm{div}(M(x,u_n)\nabla u_n)) + g(u_n) = b_n(x,u_n,\nabla u_n) + f_n(x) & \text{in } \Omega, \\ \qquad\qquad\qquad u_n = 0 & \text{on } \partial\Omega \end{cases} \tag{4.2}$$

under the assumptions (1.3)–(1.5), where b_n and f_n are as in (2.4) and

$$\begin{cases} g(t) = (e^{Bt} - 1), & B > \dfrac{\gamma}{\alpha} \cdot \dfrac{2}{N-2}, \\[2mm] 0 \le f \in L^p(\Omega), & 1 + \dfrac{\gamma}{B\alpha} < p \le \dfrac{N}{2}. \end{cases} \tag{4.3}$$

The existence of weak solution $u_n \in W_0^{1,2}(\Omega)$ of (4.2) follows by the results of [13]. We assume $p \le \frac{N}{2}$, since the case $p > \frac{N}{2}$ is studied in [9] (see Introduction).

Lemma 4.1. *There exists a positive constant $C_{p,B} = C(p, B, N, \alpha, \gamma)$ such that*

$$\int_\Omega e^{B(p-1)u_n}|\nabla u_n|^2 \le C_{p,B}$$

Moreover the sequence $\{g(u_n)\}$ converges in $L^1(\Omega)$.

Proof. As in Lemma 2.6, the solutions u_n are positive. Then use (again) $\phi(u_n)$ as test function in (4.2)

$$\phi(t) = (e^{2\lambda t} - 1), \quad \lambda = \frac{(p-1)B}{2}.$$

Note that $\frac{\gamma}{2\alpha} < \lambda = \frac{(p-1)B}{2}$.

$$(2\lambda\alpha - \gamma)\int_\Omega |\nabla u_n|^2 e^{2\lambda u_n} + \int_{\{x\in\Omega:\, g(u_n)\le 2f(x)\}} g(u_n)\phi(u_n)$$

$$+\frac{1}{2}\int_{\{x\in\Omega:\, g(u_n)\ge 2f(x)\}} g(u_n)\phi(u_n) + \frac{1}{2}\int_{\{x\in\Omega:\, g(u_n)\ge 2f(x)\}} f\phi(u_n)$$

$$\le \int_{\{x\in\Omega:\, g(u_n)\ge 2f(x)\}} f(x)\phi(u_n) + \int_{\{x\in\Omega:\, g(u_n)< 2f(x)\}} f(x)\phi(u_n).$$

78 L. Boccardo

Thus

$$(2\lambda\alpha - \gamma) \int_\Omega |\nabla u_n|^2 e^{2\lambda u_n} + \frac{1}{2} \int_\Omega g(u_n)(e^{2\lambda u_n} - 1)$$

$$\leq \int_{\{x\in\Omega:u_n<g^{-1}(2f(x))\}} f(x)(e^{2\lambda u_n} - 1) \leq \int_\Omega f(x)(e^{2\lambda g^{-1}(2f(x))} - 1) \leq C_1 + C_1 \int_\Omega f^p.$$

That is

$$\begin{cases} (2\lambda\alpha - \gamma) \int_\Omega |\nabla u_n|^2 e^{2\lambda u_n} \leq C_1 + C_1 \int_\Omega f^p \\ \int_\Omega g(u_n)(e^{2\lambda u_n} - 1) \leq 2C_1 + 2C_1 \int_\Omega f^p = C_0. \end{cases} \tag{4.4}$$

Thus (again) the estimate (2.7) still holds. So the sequence $\{u_n\}$ is bounded in $W_0^{1,2}(\Omega)$ and there exist a function u in $W_0^{1,2}(\Omega)$ and a subsequence, still denoted $\{u_n\}$, such that u_n converges weakly to u in $W_0^{1,2}(\Omega)$ and almost everywhere. The use of the Fatou Lemma in the second inequality (4.4) implies that

$$\int_\Omega g(u)(e^{2\lambda u} - 1) \leq C_0.$$

Thus $g(u) \in L^1(\Omega)$, since

$$\int_\Omega g(u) \leq \int_\Omega g(u)\chi_{\{x\in\Omega:0\leq 2\lambda u(x)\leq \log 2\}} + C_0.$$

Moreover, for any $k > 0$,

$$\int_\Omega g(u_n(x)) \leq \int_\Omega g(u_n(x))\chi_{\{x\in\Omega:0\leq u_n(x)\leq k\}} + \frac{C_0}{e^{2\lambda k}}$$

which implies

$$\limsup \int_\Omega g(u_n(x)) \leq \int_\Omega g(u(x))\chi_{\{x\in\Omega:0\leq u(x)\leq k\}} + \frac{C_0}{e^{2\lambda k}}$$

that is

$$\limsup \int_\Omega g(u_n(x)) \leq \int_\Omega g(u(x)).$$

On the other hand, the Fatou Lemma implies that

$$\liminf \int_\Omega g(u_n(x)) \geq \int_\Omega g(u(x)).$$

So we proved that

$$\int_\Omega g(u_n(x)) \to \int_\Omega g(u(x)). \tag{4.5}$$

Theorem 4.2. *Let us assume that (1.3), (1.4), (1.5), (4.3) hold true. Then there exists a weak solution u of the Dirichlet problem (4.1). Moreover*

$$\int_\Omega e^{(2\lambda+B)u} \le C_2 + C_2 \int_\Omega f^p.$$

Proof. It is possible to repeat the proof of Lemma 2.9, in order to show (again) the strong convergence of the sequence $\{u_n\}$ to u in $W_0^{1,2}(\Omega)$, (2.11), and then the proof of Theorem 2.3. Notice that the new term

$$\int_\Omega g(u_n(x))\varphi[T_k(u_n) - T_k(u)]$$

goes to zero, since, as consequence of (4.5), $g(u) \in L^1(\Omega)$ and Lebesgue's Theorem, we have

$$\int_\Omega |g(u_n(x)) - g(u(x))|$$

$$= \int_\Omega [g(u_n(x)) - g(u(x))] + 2 \int_{\{x\in\Omega:0\le u_n(x)\le u(x)\}} [-g(u_n(x)) + g(u(x))] \to 0.$$

These steps allow us to pass to the limit in the weak formulation of (4.2), in order to obtain that u is a weak solution of (4.1). $\square$

Remark 4.3. *Existence of bounded solutions of quasi-linear equations having lower order terms with quadratic growth with respect to the gradient, even if the data belong to $L^1(\Omega)$, is proved in [23], thanks to the presence of a "blow-up term": roughly speaking $\lim_{s\to\sigma} g(s) = \infty$, for some σ.*

5. Degenerate coercivity of the principal part

In this section a problem with degenerate coercivity in the principal part is studied:

$$\begin{cases} -\mathrm{div}(a(x,u)\nabla u) = b(x,u,\nabla u) + f(x) & \text{in } \Omega, \\ u = 0 & \text{on } \partial\Omega, \end{cases} \tag{5.1}$$

where $a(x,s)$ and $b(x,s,\xi)$ are Carathéodory functions satisfying the following conditions:

$$\frac{\alpha}{(1+|s|)^\theta} \le a(x,s) \le \beta, \quad \alpha > 0, \tag{5.2}$$

for some real number θ such that

$$0 \le \theta < 1, \tag{5.3}$$

$$\begin{cases} |b(x,s,\xi)| \le \gamma \dfrac{|\xi|^2}{(1+|s|)^{1+\theta}}, \\ \gamma < \alpha(1-\theta). \end{cases} \tag{5.4}$$

Remark 5.1. *If $b(x, s, \xi) \equiv 0$, the existence of bounded solutions in studied in [11] (see also the references quoted therein).*

The previous assumptions are motivated by the behavior of the two terms of the Euler-Lagrange equation of the functional

$$\frac{1}{2} \int_\Omega \frac{|\nabla v|^2}{(1 + |v|)^\theta}.$$

Theorem 5.2. *Let us assume that $0 \le f \in L^m(\Omega)$, $m > \frac{N}{2}$, and (5.2), (5.3), (5.4), hold true. Then there exists a bounded weak solution u of the Dirichlet problem (5.1).*

Proof. We can repeat the technique of [11]. Define $Q(s) = \int_0^s \frac{1}{(1+t)^\theta} \, dt$. For $k > 0$, if we take $G_k(Q(u_n))$ as test function in the approximate boundary value problems

$$\begin{cases} -\operatorname{div}(a_n(x, u_n)\nabla u_n) = b_n(x, u_n, \nabla u_n) + f(x) & \text{in } \Omega, \\ u_n = 0 & \text{on } \partial\Omega, \end{cases}$$

where

$$a_n(x, s) = a(x, T_n(s)), \quad b_n(x, s, \xi) = \frac{b(x, s, \xi)}{1 + \frac{1}{n}|b(x, s, \xi)|}$$

and use the assumptions, we obtain

$$\alpha \int_{\{Q(u_n(x)) > k\}} \frac{|\nabla u_n|^2}{(1 + u_n)^{2\theta}}$$

$$\le \int_{\{Q(u_n(x)) > k\}} \gamma \frac{|\nabla u_n|^2}{(1 + u_n)^{1+\theta}} |G_k(Q(u_n))| + \int_{\{Q(u_n(x)) > k\}} f \, G_k(Q(u_n)) \quad (5.5)$$

that is

$$\left(\alpha - \frac{\gamma}{1 - \theta}\right) \int_{\{Q(u_n(x)) > k\}} |\nabla G_k(Q(u_n))|^2 \le \int_{\{Q(u_n(x)) > k\}} f \, G_k(Q(u_n)). \quad (5.6)$$

Inequality (5.6) is exactly the starting point of Stampacchia's L^∞-regularity proof (see [26], [22]), so that there exists a positive constant L such that (recall also the assumption $f \in L^m(\Omega)$, $m > \frac{N}{2}$)

$$\|Q(u_n)\|_{L^\infty(\Omega)} \le L. \quad (5.7)$$

The properties of the function Q (in particular the fact that $\lim_{s \to +\infty} Q(s) = +\infty$, $\lim_{s \to -\infty} Q(s) = -\infty$) yield a bound for u_n in $L^\infty(\Omega)$ from (5.7):

$$\|u_n\|_{L^\infty(\Omega)} \le Q^{-1}(L).$$

It is proved in [8] that the L^∞ estimate for the solutions of the approximate equations implies also the compactness of the sequence $\{u_n\}$ in $W_0^{1,2}(\Omega)$. Note that in our case the terms

$$\frac{1}{(1+u_n)^\theta}, \quad \frac{1}{(1+u_n)^{1+\theta}}$$

do not give any problems, due to the L^∞-bound. Then the existence of a solution follows easily.

Acknowledgments

The author would like to thank Michaela Porzio, Luigi Orsina and Ireneo Peral for several useful discussions on the subject of this paper.

This paper was presented at the Conference Fifth E.C.E.P.P.
– A special tribute to the work of Haim Brezis (Gaeta, May 31, 2004).

References

[1] A. Bensoussan, L. Boccardo, F.Murat: On a nonlinear partial differential equation having natural growth terms and unbounded solution; Ann. Inst. H. Poincaré Anal. non lin. 5 (1988), 347–364.

[2] L. Boccardo: Positive solutions for some quasi-linear elliptic equations with natural growths; Atti Accad. Naz. Lincei 11 (2000), 31–39.

[3] L. Boccardo, T. Gallouet: Strongly nonlinear elliptic equations having natural growth terms and L^1 data; Nonlinear Anal. TMA 19 (1992), 573–579.

[4] L. Boccardo, T. Gallouet, F. Murat: A unified presentation of two existence results for problems with natural growth; in Progress in PDE, the Metz surveys 2, M. Chipot editor, in Research Notes in Mathematics 296, (1993) 127–137, Longman.

[5] L. Boccardo, T. Gallouet, L. Orsina: Existence and nonexistence of solutions for some nonlinear elliptic equations; J. Anal. Math. 73 (1997), 203–223.

[6] L. Boccardo, F.Murat, J.P. Puel: Existence de solutions non bornèes pour certaines equations quasi linèaires; Portugaliae Math. 41 (1982), 507–534.

[7] L. Boccardo, F.Murat, J.P. Puel: Résultats d'existence pour certains problèmes elliptiques quasi linéaires; Ann. Sc. Norm. Sup. Pisa 11 (1984), 213–235.

[8] L. Boccardo, F.Murat, J.P. Puel: Existence of bounded solutions for nonlinear elliptic unilateral problems; Ann. Mat. Pura Appl. 152 (1988), 183–196.

[9] L. Boccardo, F. Murat, J.P. Puel: L^∞-estimate for nonlinear elliptic partial differential equations and application to an existence result; SIAM J. Math. Anal. 23 (1992), 326–333.

[10] L. Boccardo, S. Segura, C. Trombetti: Existence of bounded and unbounded solutions for a class of quasi-linear elliptic problems with a quadratic gradient term; J. Math. Pures et Appl. 80 (2001), 919–940.

[11] L. Boccardo, H. Brezis: Some remarks on a class of elliptic equations with degenerate coercivity; Boll. Unione Mat. Ital. 6 (2003), 521–530.

[12] H. Brezis: Equations et inéquations non linéaires dans les espaces vectoriels en dualité; Ann. Inst. Fourier (Grenoble) 18 (1968), 115–175.

[13] H. Brezis, F.E. Browder: Some properties of higher order Sobolev spaces; J. Math. Pures Appl. 61 (1982), 245–259.

[14] H. Brezis, M. Marcus: Hardy's inequalities revisited. Dedicated to Ennio De Giorgi; Ann. Scuola Norm. Sup. Pisa Cl. Sci. 25 (1997), 217–237 (1998).

[15] H. Brezis, L. Nirenberg: Removable singularities for nonlinear elliptic equations; Topol. Methods Nonlinear Anal. 9 (1997), 201–219.

[16] H. Brezis, J.L. Vazquez: Blow-up solutions of some nonlinear elliptic problems; Rev. Mat. Univ. Complut. Madrid 10 (1997), 443–469.

[17] A. Dall'Aglio, D. Giachetti, J.P. Puel: Nonlinear elliptic equations with natural growth in general domains; Ann. Mat. Pura Appl. 181 (2002), 407–426.

[18] J. Davila, L. Dupaigne: Hardy-type inequalities; J. Eur. Math. Soc. (JEMS) 6 (2004), 335–365.

[19] V. Ferone, F. Murat: Nonlinear problems having natural growth in the gradient: an existence result when the source terms are small; Nonlinear Anal. TMA 42 (2000), 1309–1326.

[20] V. Ferone, F. Murat: Nonlinear elliptic equations with natural growth in the gradient and source terms in Lorentz spaces; to appear.

[21] J.P. Garcia Azorero, I. Peral: Hardy inequalities and some critical elliptic and parabolic problems; J. Differential Equations, 144 (1998), 441–476.

[22] P. Hartman, G. Stampacchia: On some nonlinear elliptic differential-functional equations ; Acta Math. 115 (1966), 271–310.

[23] T. Leonori: An existence result for some nonlinear elliptic equations having natural growth terms and strongly increasing lower order terms; preprint

[24] J. Leray, J.L. Lions: Quelques résultats de Višik sur les problèmes elliptiques semi-linéaires par les méthodes de Minty et Browder; Bull. Soc. Math. France, 93 (1965), 97–107.

[25] M. Marcus, V. Mizel, Y. Pinchover: On the best constant for Hardy's inequality in $\mathbb{R}^n$; Trans. Amer. Math. Soc. 350 (1998), 3237–3255.

[26] G. Stampacchia: Le problème de Dirichlet pour les équations elliptiques du second ordre à coefficients discontinus; Ann. Inst. Fourier (Grenoble), 15 n. 1 (1965), 189–258.

Lucio Boccardo
Dipartimento di Matematica
Università di Roma 1,
Piazza A. Moro 2
I-00185 Roma, Italia
e-mail: `boccardo@mat.uniroma1.it`

Progress in Nonlinear Differential Equations
and Their Applications, Vol. 63, 83–92
© 2005 Birkhäuser Verlag Basel/Switzerland

Recent Advances on Similarity Solutions Arising During Free Convection

Bernard Brighi and Jean-David Hoernel

Abstract. This paper reviews results about free convection near a vertical flat plate embedded in some saturated porous medium. We focus on a third order autonomous differential equation that gives a special class of solutions called similarity solutions. Two cases are under consideration: in the first one we prescribe the temperature on the plate and in the second one we prescribe the heat flux on it. We will also see that the same equation appears in other industrial processes.

Mathematics Subject Classification (2000). 34B15, 34C11, 76D10.

Keywords. Boundary layer, similarity solution, third order nonlinear differential equation, boundary value problem.

1. Introduction

Free convection boundary layer flows near a vertical flat plate embedded in some porous medium are studied for many years and a natural way to describe the convective flow is to look for similarity solutions. We consider two different sets of boundary conditions for the temperature on the plate: either we prescribe the temperature or we prescribe the heat flux. Both cases are leading to the same following third order non-linear autonomous differential equation

$$f''' + \alpha f f'' - \beta f'^2 = 0 \tag{1.1}$$

with the boundary conditions

$$f(0) = -\gamma, \ f'(\infty) = 0 \text{ and } f'(0) = 1, \tag{1.2}$$

or

$$f(0) = -\gamma, \ f'(\infty) = 0 \text{ and } f''(0) = -1. \tag{1.3}$$

The first set of boundary conditions (1.2) with $\alpha = \frac{m+1}{2}$ and $\beta = m$ for $m \in \mathbb{R}$ corresponds to prescribed heat on the plate as in [4], [5], [12], [14], [16], [21] and [24]. The second set of boundary conditions (1.3) with $\alpha = m + 2$ and $\beta = 2m + 1$

for $m \in \mathbb{R}$ is for the prescribed surface heat flux as done in [10] and [13]. In both cases the solutions depend on two parameters: m, the power-law exponent and γ, the mass transfer parameter. For $\gamma = 0$ we have an impermeable wall, $\gamma < 0$ corresponds to a fluid suction, and $\gamma > 0$ to a fluid injection.

Equation (1.1) with suitable boundary conditions also arises in other industrial processes such as boundary layer flow adjacent to stretching walls (see [1], [2], [15], [20], [22]) or excitation of liquid metals in a high-frequency magnetic field (see [25]).

2. The case of prescribed heat

2.1. Derivation of the model

We consider a vertical permeable flat plate embedded in a porous medium at the ambient temperature T_∞ and a rectangular Cartesian co-ordinate system with the origin fixed at the leading edge of the vertical plate, the x-axis directed upward along the plate and the y-axis normal to it. If we suppose that the porous medium is homogeneous and isotropic, that all the properties of the fluid and the porous medium are constants, that the fluid is incompressible and follows the Darcy-Boussinesq law and that the temperature along the plate is varying as x^m the governing equations are

$$\frac{\partial u}{\partial x} + \frac{\partial v}{\partial y} = 0,$$

$$u = -\frac{k}{\mu}\left(\frac{\partial p}{\partial x} + \rho g\right),$$

$$v = -\frac{k}{\mu}\frac{\partial p}{\partial y},$$

$$u\frac{\partial T}{\partial x} + v\frac{\partial T}{\partial y} = \lambda\left(\frac{\partial^2 T}{\partial x^2} + \frac{\partial^2 T}{\partial y^2}\right),$$

$$\rho = \rho_\infty(1 - \beta(T - T_\infty))$$

where u and v are the Darcy velocities in the x and y directions, ρ, μ and β are the density, viscosity and thermal expansion coefficient of the fluid, k is the permeability of the saturated porous medium and λ its thermal diffusivity, p is the pressure, T the temperature and g the acceleration of the gravity. The subscript ∞ is used for a value taken far from the plate. In our system of co-ordinates the boundary conditions along the plate are

$$v(x,0) = \omega x^{\frac{m-1}{2}}, \quad T(x,0) = T_w(x) = T_\infty + Ax^m, \quad m \in \mathbb{R},$$

with $A > 0$ and $\omega \in \mathbb{R}$ ($\omega < 0$ corresponds to a fluid suction, $\omega = 0$ is for an impermeable wall and $\omega > 0$ corresponds to a fluid injection). The boundary conditions far from the plate are

$$u(x,\infty) = 0, \quad T(x,\infty) = T_\infty.$$

If we introduce the stream function ψ such that

$$u = \frac{\partial \psi}{\partial y}, \quad v = -\frac{\partial \psi}{\partial x}$$

and assuming that convection takes place in a thin layer around the heating plate, we obtain the boundary layer approximation

$$\frac{\partial^2 \psi}{\partial y^2} = \frac{\rho_\infty \beta g k}{\mu} \frac{\partial T}{\partial y}, \tag{2.1}$$

$$\frac{\partial^2 T}{\partial y^2} = \frac{1}{\lambda} \left(\frac{\partial T}{\partial x} \frac{\partial \psi}{\partial y} - \frac{\partial T}{\partial y} \frac{\partial \psi}{\partial x} \right) \tag{2.2}$$

with

$$\frac{\partial \psi}{\partial x}(x,0) = -\omega x^{\frac{m-1}{2}} \text{ and } \frac{\partial \psi}{\partial y}(x,\infty) = 0.$$

Let us introduce the new dimensionless similarity variables

$$t = (Ra_x)^{\frac{1}{2}} \frac{y}{x}, \quad \psi(x,y) = \lambda (Ra_x)^{\frac{1}{2}} f(t), \quad \theta(t) = \frac{T(x,y) - T_\infty}{T_w(x) - T_\infty}$$

with $Ra_x = (\rho_\infty \beta g k (T_w(x) - T_\infty) x)/(\mu \lambda)$ the local Rayleigh number. In terms of these variables equations (2.1) and (2.2) become

$$f'' - \theta' = 0, \tag{2.3}$$

and

$$\theta'' + \frac{m+1}{2} f\theta' - m f'\theta = 0,$$

with the boundary conditions

$$f(0) = -\gamma, \quad \theta(0) = 1,$$

and

$$f'(\infty) = 0, \quad \theta(\infty) = 0, \tag{2.4}$$

where the prime denotes differentiation with respect to t and

$$\gamma = \frac{2\omega}{m+1} \sqrt{\frac{\mu}{\rho_\infty \beta g k A \lambda}}.$$

Integrating (2.3) and taking into account the boundary conditions (2.4) leads to

$$f' = \theta$$

and the problem (1.1)–(1.2) with $\alpha = \frac{m+1}{2}$ and $\beta = m$ for $m \in \mathbb{R}$ follows.

2.2. Useful tools

2.2.1. The initial value problem. Let $P_{m,\gamma,\mu}$ be the following initial value problem

$$\begin{cases} f''' + \frac{m+1}{2} f f'' - m f'^2 = 0 \\ f(0) = -\gamma, \\ f'(0) = 1, \\ f''(0) = \mu. \end{cases} \tag{2.5}$$

This first approach used is a shooting method that consists in finding values of $f''(0) = \mu$ for which f exists on $[0, \infty)$ and such that $f'(\infty) = 0$. This direct method allows us to consider vanishing solutions but does fail in some cases (see [8]).

2.2.2. The blowing-up co-ordinates. Let us notice that if f is a solution of (1.1) then for all $\kappa > 0$ the function $t \longmapsto \kappa f(\kappa t)$ is a solution too. Then, considering an interval I on which a solution f of (1.1) does not vanish, for $\tau \in I$ we can introduce the following blowing-up co-ordinates

$$\forall t \in I, \ s = \int_{\tau}^{t} f(\xi)d\xi, \ u(s) = \frac{f'(t)}{f(t)^2} \ \text{and} \ v(s) = \frac{f''(t)}{f(t)^3}. \tag{2.6}$$

Then, we easily get

$$\begin{cases} \dot{u} = P(u, v) := v - 2u^2, \\ \dot{v} = Q_m(u, v) := -\frac{m+1}{2}v + mu^2 - 3uv, \end{cases} \tag{2.7}$$

where the dot is for differentiating with respect to the variable s. To come back to the original problem it is sufficient to consider the initial value problem $P_{m,\gamma,\mu}$ with $\gamma \neq 0$ and look at the trajectories of the corresponding plane dynamical system (2.7). For details, see [11].

2.3. Main results

The problem (1.1)–(1.2) appears in engineering and physical literature, in very different context, in the middle of the previous century.

Rigorous mathematical results arise around the sixties. In [26] (Appendix 2) it is mentioned that for $\gamma = 0$ a simple explicit solution can be obtained in both cases $m = 1$ and $m = -\frac{1}{3}$ (see also [15], [21], [4] and [5]). On the other hand, the author notes that Mr J. Watson has given a simple proof that (1.1)–(1.2) has no solution for $\gamma = 0$ and $m \leq -1$.

An explicit solution is also given for $m = 1$ and any γ, first in [20], and later in [22] and [8]. In these latter papers one can also find the explicit solution for $m = -\frac{1}{3}$ and any γ.

Nonexistence for $\gamma = 0$ and $m = -\frac{1}{2}$ was noted in [1]. In [21], it is shown that for $\gamma = 0$ and $m < -\frac{1}{2}$ there are no solutions satisfying $f' f^2 \to 0$ at infinity.

Recently, further mathematical results concerning existence, nonexistence, uniqueness, nonuniqueness and asymptotic behavior, are obtained in [4], [5] for $\gamma = 0$, and in [18], [8], [11], [19] and [9] for the general case.

Numerical investigations can be found in [1], [6], [12], [14], [21], [22] and [27].

In view of all these papers, the following conclusions can be drawn.

- For $m < -1$, there exists $\gamma_* > 0$ such that problem (1.1)–(1.2) has infinitely many solutions if $\gamma > \gamma_*$, one and only one solution if $\gamma = \gamma_*$, and no solution if $\gamma < \gamma_*$. For $\gamma = \gamma_*$ we have that $f(t) \to \lambda < 0$ as $t \to \infty$, and for every $\gamma > \gamma_*$ there are two solutions f such that $f(t) \to \lambda < 0$ as $t \to \infty$ and all the other solutions verify $f(t) \to 0$ as $t \to \infty$ Moreover, if f is a solution to (1.1)–(1.2), then f is negative, strictly increasing and either concave or convex-concave. (See Fig. 1 for the two solutions such that $f(t) \to \lambda < 0$ as $t \to \infty$ and three other solutions in the case $m = -2$ and $\gamma = 5$.)

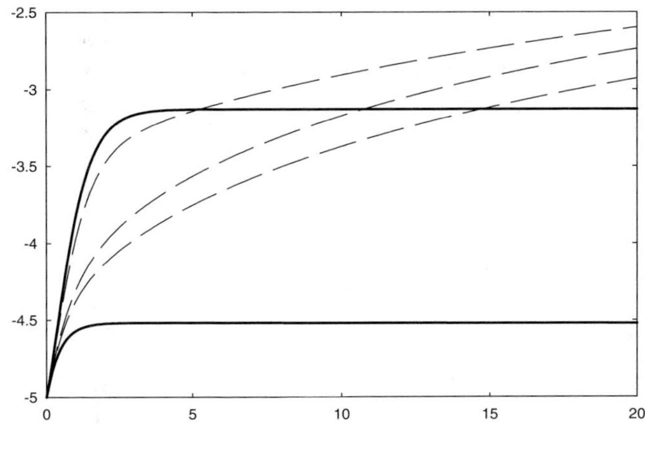

FIGURE 1

- For $m = -1$ and for every $\gamma \in \mathbb{R}$, the problem (1.1)–(1.2) has no solution.
- For $-1 < m \le -\frac{1}{2}$ and for every $\gamma \ge 0$, the problem (1.1)–(1.2) has no solution.
- For $-1 < m < -\frac{1}{2}$, there exists $\gamma_* < 0$ such that problem (1.1)–(1.2) has no solution for $\gamma_* < \gamma < 0$, one and only one solution which is bounded for $\gamma = \gamma_*$, and two bounded solutions and infinitely many unbounded solutions for $\gamma < \gamma_*$. These solutions are strictly increasing and either concave or convex-concave. (See Fig. 2 for the two bounded solutions and four unbouded solutions in the case $m = -0.75$ and $\gamma = -10$.)
- For $-\frac{1}{2} \le m < -\frac{1}{3}$ and for every $\gamma < 0$, the problem (1.1)–(1.2) has one bounded solution and infinitely many unbounded solutions. All these solutions are strictly increasing and either concave or convex-concave.
- For $-\frac{1}{3} \le m < 0$ and for every $\gamma \in \mathbb{R}$, the problem (1.1)–(1.2) has an infinite number of solutions. Moreover, if $\gamma \le 0$ one and only one solution is bounded, and if $\gamma > 0$ at least one is bounded, many infinitely are unbounded. All solutions are strictly increasing and either concave or convex-concave. If $\gamma > 0$, the solutions becomes positive for large t.

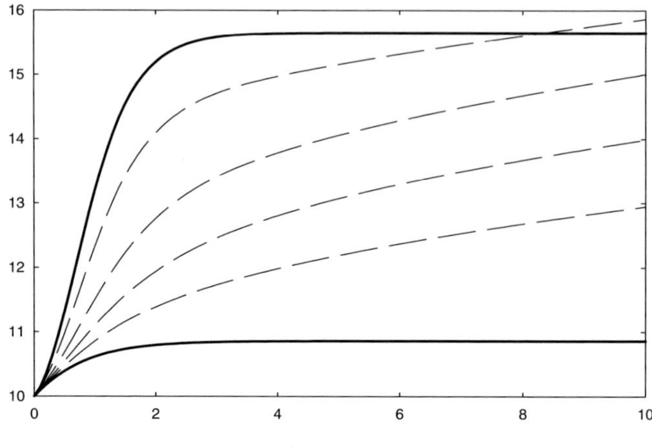

FIGURE 2

- For $m \in [0, 1]$ and for every $\gamma \in \mathbb{R}$, the problem (1.1)–(1.2) has one and only one solution, moreover this solution is concave and bounded. (See Fig. 3 for the unique solution in the case $m = 0.5$ and $\gamma = 0$.)

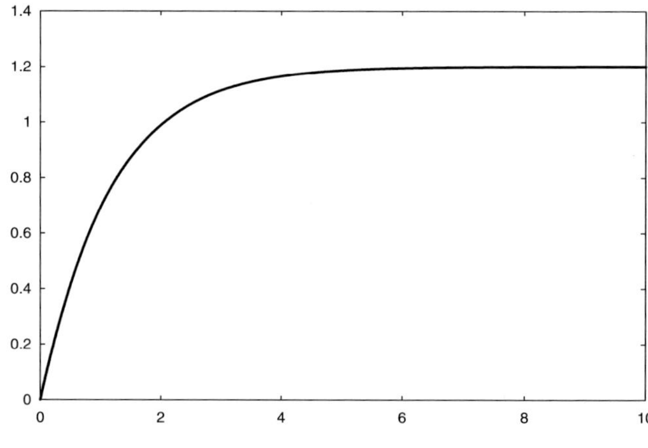

FIGURE 3

- For $m > 1$ and for every $\gamma \in \mathbb{R}$, the problem (1.1)–(1.2) has one and only one concave solution and an infinite number of concave-convex solutions. All these solutions are bounded. Moreover, there is an unique concave-convex solution that verifies $f(t) \to \lambda > 0$ as $t \to \infty$ and all the other concave-convex solutions are such that $f(t) \to 0$ as $t \to \infty$. (See Fig. 4 for the unique concave solution and three concave-convex solutions in the case $m = 1.1$ and $\gamma = 0$.)

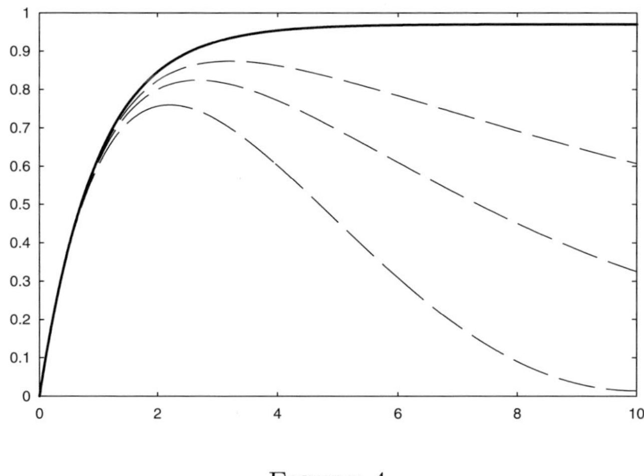

$$\textsc{Figure } 4$$

Remark 2.1. *The case $m = 0$ leads to the well-know Blasius equation (see [3], [7]) that also is a special case of the Falkner-Skan equation (see [17]).*

Remark 2.2. *In [23] the authors gives results about a slightly different problem for $m = -1$ that involves pseudo-similarity.*

We see from these results, that the unsolved questions concern the case $\gamma \geq 0$. More precisely, it should be interesting to try to answer to the following points

- For $-\frac{1}{2} < m < -\frac{1}{3}$, what happens for $\gamma \geq 0$?
- For $-\frac{1}{3} < m < 0$ and $\gamma > 0$, is there one or more bounded solutions?

Another purpose is to compute the critical values γ_* appearing in the results above.

3. The case of prescribed heat flux

We now suppose that the plate is subjected to a variable heat flux varying as x^m and a mass transfer rate varying as $x^{\frac{m-1}{3}}$ following [13] to obtain the problem (1.1)–(1.3) with $\alpha = m + 2$ and $\beta = 2m + 1$. The mathematical study is made in [10] and leads to the following results

- For $m < -2$ there exists $\gamma_* > \sqrt[3]{\frac{2}{(m+2)^2}}$ such that the problem (1.1)–(1.3) has no solution for $\gamma < \gamma_*$, one and only one solution for $\gamma = \gamma_*$ and infinitely many solutions for $\gamma > \gamma_*$. For $\gamma = \gamma_*$ we have that $f(t) \to \lambda < 0$ as $t \to \infty$, and for every $\gamma > \gamma_*$ there are two solutions f such that $f(t) \to \lambda < 0$ as $t \to \infty$ and all the other solutions verify $f(t) \to 0$ as $t \to \infty$ Moreover, if f is a solution of (1.1)–(1.3), then f is negative, strictly concave and increasing.
- For $m = -2$ and for every $\gamma \in \mathbb{R}$, the problem (1.1)–(1.3) has no solution.
- For $-2 < m < -1$, there exists $\gamma_* < 0$ such that the problem (1.1)–(1.3) has no solution for $\gamma > \gamma_*$, one and only one solution which is bounded for

$\gamma = \gamma_*$ and two bounded solutions and infinitely many unbounded solutions for $\gamma < \gamma_*$. Moreover, if f is a solution of (1.1)–(1.3), then f is positive, strictly concave, increasing and $f'(0) \geq -\frac{1}{(m+2)\gamma}$.

- For $m = -1$ the problem (1.1)–(1.3) only admits solutions for $\gamma < 0$. In this case there is an unique bounded solution with $f'(0) = -\frac{1}{\gamma}$ and an infinite number of unbounded solutions with $f'(0) > -\frac{1}{\gamma}$. Moreover all the solutions are positive, strictly concave and increasing.
- For $-1 < m < -\frac{1}{2}$ the problem (1.1)–(1.3) admits at least one bounded solution for $\gamma \in \mathbb{R}$ and many infinitely unbounded solutions for $\gamma < 0$. All these solutions are increasing and strictly concave and uniqueness of the bounded solution holds for $\gamma \leq 0$.
- For $m \geq -\frac{1}{2}$ all the solutions are bounded.
- For $-\frac{1}{2} \leq m \leq 1$ and for every $\gamma \in \mathbb{R}$ the problem (1.1)–(1.3) has one and only one solution. This solution is strictly concave and increasing. (Let us notice that for $m = -\frac{1}{2}$ we have the Blasius equation.)
- For $m > 1$ and $\gamma \in \mathbb{R}$ the problem (1.1)–(1.3) has one and only one concave solution and infinitely many concave-convex solutions. Moreover, there is an unique concave-convex solution that verifies $f(t) \to \lambda > 0$ as $t \to \infty$ and all the other concave-convex solutions are such that $f(t) \to 0$ as $t \to \infty$.

In this case it remains only two open questions

- For $-1 < m < -\frac{1}{2}$ and $\gamma > 0$, is the bounded solution unique?
- For $-1 < m < -\frac{1}{2}$ and $\gamma \geq 0$, is there unbounded solution?

4. Asymptotic behavior of the unbounded solutions

For the equation (1.1) we have the following asymptotic equivalent found in [9] and [19] that holds for unbounded solutions

Theorem 4.1. *Let f be an unbounded solution of* (1.1)–(1.2) *or* (1.1)–(1.3)*. There exists a constant $c > 0$ such that*

$$|f(t)| \sim ct^{\frac{\alpha}{\alpha-\beta}} \qquad as \quad t \to \infty.$$

References

[1] W.H.H. Banks, Similarity solutions of the boundary layer equations for a stretching wall, J. de Méchan. Théor. et Appl. 2 (1983), pp.375–392.

[2] W.H.H. Banks, M.B. Zaturska, Eigensolutions in boundary layer flow adjacent to a stretching wall, IMA J. Appl. Math. 36 (1986), pp. 263–273.

[3] Z. Belhachmi, B. Brighi & K. Taous, On the concave solutions of the Blasius equation, Acta Math. Univ. Comenianae, Vol. LXIX, 2 (2000), pp. 199–214.

[4] Z. Belhachmi, B. Brighi & K. Taous, Solutions similaires pour un problème de couche limite en milieux poreux, C. R. Mécanique 328 (2000), pp. 407–410.

[5] Z. Belhachmi, B. Brighi & K. Taous, On a family of differential equations for boundary layer approximations in porous media, Euro. Jnl of Applied Mathematics, Vol. 12, 4, Cambridge University Press (2001), pp. 513–528.

[6] Z. Belhachmi, B. Brighi, J.M. Sac-Epee & K. Taous, Numerical simulations of free convection about a vertical flat plate embedded in a porous medium, Computational Geosciences, vol. 7 (2003), pp. 137–166.

[7] H. Blasius, Grenzschichten in Flüssigkeiten mit kleiner Reibung, Z. Math. Phys. 56 (1908), pp. 1–37.

[8] B. Brighi, On a similarity boundary layer equation, Zeitschrift für Analysis und ihre Anwendungen, vol. 21 (2002) 4, pp. 931–948.

[9] B. Brighi, J.-D. Hoernel, Asymptotic behavior of the unbounded solutions of some boundary layer equation. To appear in Archiv der Mathematik.

[10] B. Brighi, J.-D. Hoernel, On similarity solutions for boundary layer flows with prescribed heat flux. Mathematical Methods in the Applied Sciences, vol. 28, 4 (2005), pp. 479–503.

[11] B. Brighi, T. Sari, Blowing-up coordinates for a similarity boundary layer equation. Discrete and Continuous Dynamical Systems (Serie A), Vol. 12 (2005) 5, pp. 929–948.

[12] M.A. Chaudhary, J.H. Merkin & I. Pop, Similarity solutions in free convection boundary-layer flows adjacent to vertical permeable surfaces in porous media: I prescribed surface temperature, Eur. J. Mech. B-Fluids, 14 (1995), pp. 217–237.

[13] M.A. Chaudhary, J.H. Merkin & I. Pop, Similarity solutions in free convection boundary-layer flows adjacent to vertical permeable surfaces in porous media: II prescribed surface heat flux, Heat and Mass Transfer 30, Springer-Verlag (1995), pp. 341–347.

[14] P. Cheng, W.J. Minkowycz, Free-convection about a vertical flat plate embedded in a porous medium with application to heat transfer from a dike, J. Geophys. Res. 82 (14) (1977), pp. 2040–2044.

[15] L.E. Crane, Flow past a stretching plane, Z. Angew. Math. Phys. 21 (1970), pp. 645–647.

[16] E.I. Ene, D. Poliševski, Thermal flow in porous media, D. Reidel Publishing Company, Dordrecht, 1987.

[17] V.M. Falkner, S.W. Skan, Solutions of the boundary layer equations, Phil. Mag., 7/12 (1931), pp. 865–896.

[18] M. Guedda, Nonuniqueness of solutions to differential equations for boundary layer approximations in porous media, C. R. Mécanique, 330 (2002), pp. 279–283.

[19] M. Guedda, Similarity solutions of differential equations for boundary layer approximations in porous media. To appear in ZAMP.

[20] P.S. Gupta, A.S. Gupta, Heat an mass transfer on a stretching sheet with suction or blowing, Can. J. Chem. Eng. 55 (1977), pp. 744–746.

[21] D.B. Ingham, S.N. Brown, Flow past a suddenly heated vertical plate in a porous medium, J. Proc. R. Soc. Lond. A 403 (1986), pp. 51–80.

[22] E. Magyari, B. Keller, Exact solutions for self-similar boundary-layer flows induced by permeable stretching wall. Eur. J. Mech. B-Fluids 19 (2000), pp. 109–122.

[23] E. Magyari, I. Pop, B. Keller, The "missing" self-similar free convection boundary-layer flow over a vertical permeable surface in a porous medium, Transport in Porous Media 46 (2002), pp. 91–102.

[24] J.H. Merkin, G. Zhang, On the similarity solutions for free convection in a saturated porous medium adjacent to impermeable horizontal surfaces, Wärme und Stoffübertr., 25 (1990), pp. 179–184.

[25] H.K. Moffatt, High-frequency excitation of liquid metal systems, IUTAM Symposium: Metallurgical Application of Magnetohydrodynamics, (1982) Cambridge.

[26] J.T. Stuart, Double boundary layers in oscillatory viscous flow, J. Fluid. Mech. 24 (1966), pp. 673–687.

[27] R.A. Wooding, Convection in a saturated porous medium at large Rayleigh number or Peclet number, J. Fluid. Mech., 15 (1963), pp. 527–544.

Bernard Brighi and Jean-David Hoernel
Université de Haute-Alsace
Laboratoire de Mathématiques et Applications
4, rue des frères Lumière
F-68093 Mulhouse (France)
e-mail: `bernard.brighi@uha.fr`
e-mail: `j-d.hoernel@wanadoo.fr`

Progress in Nonlinear Differential Equations
and Their Applications, Vol. 63, 93–102
© 2005 Birkhäuser Verlag Basel/Switzerland

Rellich Relations for
Mixed Boundary Elliptic Problems

R. Brossard, J.-P. Lohéac and M. Moussaoui

Abstract. For elliptic partial differential equations, mixed boundary conditions generate singularities in the solution, mainly when the boundary of the domain is connected. Following previous works concerning the Laplace equation, we here give Rellich relations involving singularities for the Lamé system.

These relations are useful in the problem of boundary stabilization of the waves equation and the elastodynamic system, respectively, when using the multiplier method.

Mathematics Subject Classification (2000). 35B30, 35J25, 35J55, 93D15.

Keywords. mixed elliptic problems, singularities, stabilization.

Introduction

Let Ω be a regular bounded open set of $\mathbb{R}^n$ and consider the following wave problem,

$$\begin{cases} u'' - \Delta u = 0\,, & \text{in} \quad \Omega \times (0, +\infty)\,, \\ u = 0\,, & \text{on} \quad \partial\Omega_D \times (0, +\infty)\,, \\ \partial_\nu u = F(u')\,, & \text{on} \quad \partial\Omega_N \times (0, +\infty)\,, \\ u(0) = u^0\,, & \text{in} \quad \Omega\,, \\ u'(0) = u^1\,, & \text{in} \quad \Omega\,. \end{cases}$$

Here the problem of boundary stabilization is to build some partition $(\partial\Omega_D, \partial\Omega_N)$ of the boundary $\partial\Omega$ and some feedback function F such that the energy of the solution u is (exponentially) decreasing with respect to time.

Many authors have studied this problem by using the multiplier method (see [8] and the references therein). This leads to the choice

$$\begin{aligned} \partial\Omega_N &= \{\mathbf{x} \in \partial\Omega \,/\, \mathbf{m}(\mathbf{x}).\boldsymbol{\nu}(\mathbf{x}) > 0\}\,, \\ \partial\Omega_D &= \partial\Omega \setminus \partial\Omega_N = \{\mathbf{x} \in \partial\Omega \,/\, \mathbf{m}(\mathbf{x}).\boldsymbol{\nu}(\mathbf{x}) \le 0\}\,, \\ F(u') &= -(\mathbf{m}.\boldsymbol{\nu})u'\,, \end{aligned}$$

where $\boldsymbol{\nu}(\mathbf{x})$ is the normal unit vector pointing outwards of Ω at some point $\mathbf{x} \in \partial\Omega$ and $\mathbf{m}$ is a function depending on a fixed point $\mathbf{x}_0 \in \mathbb{R}^n : \mathbf{m}(\mathbf{x}) = \mathbf{x} - \mathbf{x}_0$.

The main step of this method is to prove some Gronwall-type inequality concerning the energy of strong solutions (see Theorem 8.1 in [8]). This leads to define $H_D^1(\Omega) = \{v \in H^1(\Omega) \,/\, v = 0\,,\text{ on } \partial\Omega_D\,\}$ and to consider the operator $\mathcal{A}_w$

$$\mathcal{D}(\mathcal{A}_w) = \{(u, \hat{u}) \in H_D^1(\Omega) \times H_D^1(\Omega) \,/\, \Delta u \in L^2(\Omega)\,;\; \partial_\nu u = -(\mathbf{m}.\boldsymbol{\nu})\hat{u}\,,\text{ on } \partial\Omega_N\}\,,$$
$$\mathcal{A}_w(u, \hat{u}) = (-\hat{u}, -\Delta u)\,,\quad \forall(u, \hat{u}) \in \mathcal{D}(\mathcal{A}_w)\,.$$

The crucial point in the proof of above Gronwall-type inequality is to verify that if $(u, \hat{u})$ belongs to $\mathcal{D}(\mathcal{A}_w)$, then u satisfies a Rellich relation [11] in the following form

$$2\int_\Omega \Delta u\, \mathbf{m}.\nabla u\, d\mathbf{x} = (n-2)\int_\Omega |\nabla u|^2\, d\mathbf{x} + \int_{\partial\Omega} \left(2\,\partial_\nu u\, \mathbf{m}.\nabla u - \mathbf{m}.\boldsymbol{\nu}\, |\nabla u|^2\right) ds\,. \tag{1}$$

One can easily observe that this relation is satisfied if u is regular enough. A sufficient condition is that u is locally H^2 in $\overline{\Omega}$. For the above problem of boundary stabilization, this holds in the particular case when the interface $\Gamma = \overline{\partial\Omega_N} \cap \overline{\partial\Omega_D}$ is empty (this can be proved by using the method of difference quotients).

On the other hand, when the interface is not empty, some singular part can appear in u and the above "hidden regularity result" is generally false.

Anyway, in all cases, using a classical trace result, we can prove that if $(u, \hat{u})$ belongs to $\mathcal{D}(\mathcal{A}_w)$, then there exists $u_R \in H^2(\Omega)$ such that $U = u - u_R$ satisfies the following mixed boundary problem for the Laplace equation.

$$\begin{cases} -\Delta U = F\,, & \text{in } \Omega\,, \\ U = 0\,, & \text{on } \partial\Omega_D\,, \\ \partial_\nu U = 0\,, & \text{on } \partial\Omega_N\,, \end{cases} \tag{2}$$

where the right-hand side F belongs to $L^2(\Omega)$.

Under reasonable geometrical assumptions about Ω, when $\Gamma = \emptyset$, U is locally H^2 in some neighborhood of any point of $\overline{\Omega}$ and Rellich relation (1) is true.

When $\Gamma \neq \emptyset$, U can be singular even if F is very regular. In this case, formula (1) must be modified. A further term, which takes into account singularities, appears. This will be presented in the first part of this paper.

In the second part, we consider the case of the Lamé system which is related to the problem of the boundary stabilization of the elastodynamic system.

In the third part, we give a sketch of the proof of the Rellich relation for the Lamé system (detailed proofs can be found in [2, 3]).

1. Rellich relation for the Laplace equation

We first introduce the main geometrical assumptions.

Let Ω be a bounded open set of $\mathbb{R}^n$ ($n \geq 2$) such that its boundary $\partial\Omega$ satisfies, in the sense of Nečas [12],

$$\partial\Omega \text{ is of class } \mathcal{C}^2\,. \tag{3}$$

Given $\mathbf{x}$ a point of $\partial\Omega$, we denote by $\boldsymbol{\nu}(\mathbf{x})$ the normal unit vector pointing outwards of Ω.

We assume that there exists a partition $(\partial\Omega_N, \partial\Omega_D)$ of $\partial\Omega$ such that

$$\begin{aligned}
&\text{meas}(\partial\Omega_D) \neq 0\,,\ \text{meas}(\partial\Omega_N) \neq 0\,,\\
&\Gamma = \overline{\partial\Omega_D} \cap \overline{\partial\Omega_N} \text{ is a non-empty } \mathcal{C}^3\text{-manifold of dimension } n-2\,,\\
&\text{there exists a neighborhood } \omega \text{ of } \Gamma \text{ such that } \partial\Omega \cap \omega\\
&\qquad\qquad\qquad\qquad\qquad \text{is a } \mathcal{C}^3\text{-manifold of dimension } n-1\,.
\end{aligned} \qquad (4)$$

Furthermore, we suppose that there exists $\mathbf{x}_0 \in \mathbb{R}^n$ such that, setting $\mathbf{m}(\mathbf{x}) = \mathbf{x} - \mathbf{x}_0$, Γ satisfies

$$\mathbf{m}.\boldsymbol{\nu} = 0\,, \quad \text{on } \Gamma\,. \qquad (5)$$

We can consider $\partial\Omega_N$ as a submanifold of $\partial\Omega$, so that at each point $\mathbf{x}$ of its boundary Γ, we can define a normal unit vector $\boldsymbol{\tau}(\mathbf{x})$ pointing outwards of $\partial\Omega_N$. Observe that this vector is tangential with respect to $\partial\Omega$ (see Figure 1).

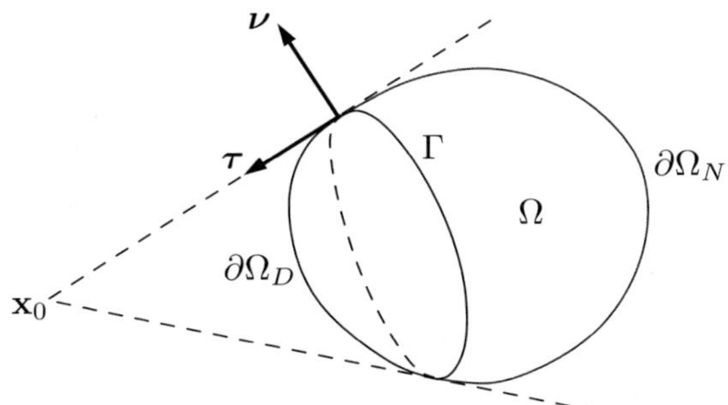

FIGURE 1. An example of domain Ω with a non-empty interface Γ.

Let us now give the extension of Rellich identity.

Theorem 1. *Under assumptions* (3)–(5), *let* $u \in \mathrm{H}^1(\Omega)$ *be such that*

$$\Delta u \in \mathrm{L}^2(\Omega)\,, \qquad u_{/\partial\Omega_D} \in \mathrm{H}^{3/2}(\partial\Omega_D)\,, \qquad \partial_\nu u_{/\partial\Omega_N} \in \mathrm{H}^{1/2}(\partial\Omega_N)\,.$$

Then, $2\,\partial_\nu u\,\mathbf{m}.\nabla u - \mathbf{m}.\boldsymbol{\nu}\,|\nabla u|^2$ *belongs to* $\mathrm{L}^1(\partial\Omega)$ *and there exists* $\zeta \in \mathrm{H}^{1/2}(\Gamma)$ *such that*

$$2\int_\Omega \Delta u\,\mathbf{m}.\nabla u\,dx = (n-2)\int_\Omega |\nabla u|^2\,dx + \int_{\partial\Omega} \left(2\,\partial_\nu u\,\mathbf{m}.\nabla u - \mathbf{m}.\boldsymbol{\nu}\,|\nabla u|^2\right) ds$$

$$+ \int_\Gamma |\zeta|^2\,\mathbf{m}.\boldsymbol{\tau}\,d\gamma\,.$$

The detailed proof of this result can be found in [1].

The first extension has been proved by P. Grisvard [6, 7] who has taken in account singularities generated by vortices of a polygonal domain. Observe in this case, that, if the polygonal domain is convex (with angles lower than π), formula (1) holds without any further term.

A further term appears when at some point of the interface, the angle is π. Indeed, this geometrical configuration generates a singularity which behaves locally like the Shamir function [13] given in polar coordinates by

$$U_S(r,\theta) = \varrho(r)\, \sqrt{r}\, \sin\frac{\theta}{2},$$

where ϱ is some cut-off function.

This function satisfies a problem in the form (2) and is not locally H^2 in any neighborhood of the origin (see Figure 2): observe only that near the origin, the normal derivative satisfies $\partial_r U_S(r,\pi) = O(r^{-1/2})$ (with Landau notations) and is not locally L^2 along the boundary.

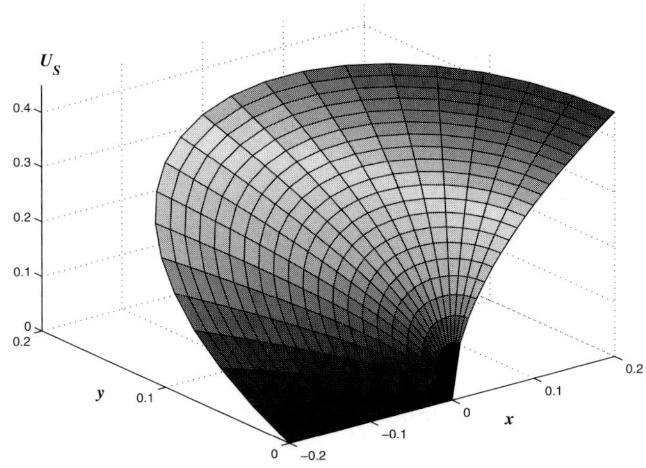

FIGURE 2. Local behavior of the Shamir function.

The second extension of Rellich formula has been proved by M. Moussaoui [9] when Ω is an infinite semi-cylinder (see Figure 5 at the end of this paper).

This has been extended for general n-dimensional smooth domains in [1] by using local coordinates at each point of the interface Γ.

2. Rellich relation for the Lamé system

We first introduce notations and motivate our work by the study of the boundary stabilization of the elastodynamic system. We end this section by giving main results.

2.1. Notations

We will use the following Lamé notations. Assume that $\mathbf{v} = (v_1, v_2, v_3)$ is a regular vector field, we define the strain tensor

$$\varepsilon_{ij}(\mathbf{v}) = \frac{1}{2}\left(\partial_j v_i + \partial_i v_j\right), \quad (i, j) \in \{1, 2, 3\}^2,$$

and the stress tensor

$$\sigma(\mathbf{v}) = 2\mu\,\varepsilon(\mathbf{v}) + \lambda\operatorname{div}(\mathbf{v})I_3,$$

where $\lambda > 0$ and $\mu > 0$ are the Lamé coefficients and I_3 is the identity matrix of $\mathbb{R}^3$.

We denote the classical inner product by:

$$\sigma(\mathbf{u}):\varepsilon(\mathbf{v}) = \operatorname{tr}(\sigma(\mathbf{u})\varepsilon(\mathbf{v})) = \sum_i \sum_j \sigma_{ij}(\mathbf{u})\varepsilon_{ij}(\mathbf{v}).$$

We write that $\mathbf{v}$ belongs to $\mathbb{L}^2(\Omega)$ (resp. $\mathbb{H}^s(\Omega)$), if every component of $\mathbf{v}$ belongs to $L^2(\Omega)$ (resp. $H^s(\Omega)$).

We also need to define: $\mathbb{H}^1_D(\Omega) = \{\mathbf{v} \in \mathbb{H}^1(\Omega) \,/\, \mathbf{v} = 0\,,\ \text{on}\ \partial\Omega_D\,\}$.

2.2. Boundary stabilization of the elastodynamic system

In [2, 4], we have considered the problem of the boundary stabilization of the elastodynamic system

$$\begin{cases} \mathbf{u}'' - \operatorname{div}(\sigma(\mathbf{u})) = 0\,, & \text{in}\ \ \Omega \times (0, +\infty)\,, \\ \mathbf{u} = 0\,, & \text{on}\ \ \partial\Omega_D \times (0, +\infty)\,, \\ \sigma(\mathbf{u})\boldsymbol{\nu} = -(\mathbf{m}.\boldsymbol{\nu})\mathbf{u}'\,, & \text{on}\ \ \partial\Omega_N \times (0, +\infty)\,, \\ \mathbf{u}(0) = \mathbf{u}^0\,, & \text{in}\ \ \Omega\,, \\ \mathbf{u}'(0) = \mathbf{u}^1\,, & \text{in}\ \ \Omega\,. \end{cases}$$

As well as for the wave equation, we can obtain a stabilization result by using multiplier method, provided that some Rellich relation is satisfied.

We follow a similar approach. Especially we have to consider the following operator

$$\mathcal{D}(\mathcal{A}_e) = \{(\mathbf{u}, \hat{\mathbf{u}}) \in \mathbb{H}^1_D(\Omega) \times \mathbb{H}^1_D(\Omega) \,/\, \operatorname{div}(\sigma(\mathbf{u})) \in \mathbb{L}^2(\Omega)\,;\ \sigma(\mathbf{u})\boldsymbol{\nu}$$

$$= -(\mathbf{m}.\boldsymbol{\nu})\hat{\mathbf{u}}\,,\ \text{on}\ \partial\Omega_N\}\,,$$

$$\mathcal{A}_e(\mathbf{u}, \hat{\mathbf{u}}) = (-\hat{\mathbf{u}}, -\operatorname{div}(\sigma(\mathbf{u})))\,, \quad \forall(\mathbf{u}, \hat{\mathbf{u}}) \in \mathcal{D}(\mathcal{A}_e)\,.$$

Again, under reasonable geometrical assumptions, we can use a trace result: if $(\mathbf{u}, \hat{\mathbf{u}})$ belongs to $\mathcal{D}(\mathcal{A}_e)$, one can build $\mathbf{u}_R \in \mathbb{H}^2(\Omega)$ and $\mathbf{F} \in \mathbb{L}^2(\Omega)$ such that $\mathbf{U} = \mathbf{u} - \mathbf{u}_R$ satisfies the following elasticity system

$$\begin{cases} -\operatorname{div}(\sigma(\mathbf{U})) = \mathbf{F}\,, & \text{in}\ \ \Omega\,, \\ \mathbf{U} = 0\,, & \text{on}\ \ \partial\Omega_D\,, \\ \sigma(\mathbf{U})\boldsymbol{\nu} = 0\,, & \text{on}\ \ \partial\Omega_N\,. \end{cases} \tag{6}$$

2.3. Main results

2.3.1. The regular case. We first give a result which is similar to (1) when $\mathbf{u}$ is regular enough.

Proposition 2. *Assume that the open bounded set Ω satisfies (3). If $\mathbf{u}$ belongs to $\mathbb{H}^2(\Omega)$, then $2\,(\sigma(\mathbf{u})\boldsymbol{\nu}).(\mathbf{m}.\nabla)\mathbf{u} - \mathbf{m}.\boldsymbol{\nu}\,\sigma(\mathbf{u}):\varepsilon(\mathbf{u})$ belongs to $\mathrm{L}^1(\partial\Omega)$ and*

$$2\int_\Omega \mathrm{div}(\sigma(\mathbf{u})).(\mathbf{m}.\nabla)\mathbf{u}\,dx = (n-2)\int_\Omega \sigma(\mathbf{u}):\varepsilon(\mathbf{u})\,dx$$
$$+ \int_{\partial\Omega} \left(2\,(\sigma(\mathbf{u})\boldsymbol{\nu}).(\mathbf{m}.\nabla)\mathbf{u} - \mathbf{m}.\boldsymbol{\nu}\,\sigma(\mathbf{u}):\varepsilon(\mathbf{u})\right) ds\,.$$

One can easily prove this result by applying two Green formulas.

Observe that the above relation holds when $\mathbf{u}$ is the solution of problem (6) if $\overline{\partial\Omega}_D \cap \overline{\partial\Omega}_N$ is empty.

Let us introduce the useful following notation:

$$\Theta(\mathbf{u},\mathbf{v}) = 2\,(\sigma(\mathbf{u})\boldsymbol{\nu}).(\mathbf{m}.\nabla)\mathbf{v} - \mathbf{m}.\boldsymbol{\nu}\,\sigma(\mathbf{u}):\varepsilon(\mathbf{v})\,.$$

In order to extend this result, we proceed as well as in the case of Laplace equation.

2.3.2. The case of a plane polygonal domain. Following works of P. Grisvard, we first consider the case of a plane polygonal domain.

We here suppose that Ω is a bounded convex polygonal open subset of $\mathbb{R}^2$ and its boundary is made of two broken lines $\partial\Omega_N$ and $\partial\Omega_D$ defined thanks to some point $\mathbf{x}_0$ belonging to $\mathbb{R}^2 \setminus \Omega$:

$$\partial\Omega_D = cl(\{\mathbf{x} \in \partial\Omega\,/\,\mathbf{m}(\mathbf{x}).\boldsymbol{\nu}(\mathbf{x}) \leq 0\})\,, \quad \partial\Omega_N = \partial\Omega \setminus \partial\Omega_D\,, \tag{7}$$

so that $\Gamma = \overline{\partial\Omega}_N \cap \overline{\partial\Omega}_D = \{\mathbf{s}_1, \mathbf{s}_2\}$ (see Figure 3). At $\mathbf{s}_1$ (resp. $\mathbf{s}_2$), let us define angle $\varpi_1 \in (0,\pi]$ (resp. $\varpi_2 \in (0,\pi]$) between $\partial\Omega_N$ and $\partial\Omega_D$. Let us define

$$J(\Omega) = \{\jmath\,/\,\varpi_\jmath = \pi\}\,.$$

For every $\jmath \in J(\Omega)$, we can define as well as above unit vectors $\boldsymbol{\nu}(\mathbf{s}_\jmath)$ and $\boldsymbol{\tau}(\mathbf{s}_\jmath)$ (see an example in Figure 3).

Theorem 3. *Let $\Omega \subset \mathbb{R}^2$ be a bounded convex polygonal open set such that its boundary $\partial\Omega$ satisfies (7). If $\mathbf{u} \in \mathbb{H}^1(\Omega)$ is such that*

$$\mathrm{div}(\sigma(\mathbf{u})) \in \mathbb{L}^2(\Omega)\,, \quad \mathbf{u}_{/\partial\Omega_D} \in \mathbb{H}^{3/2}(\partial\Omega_D)\,, \quad \sigma(\mathbf{u})\boldsymbol{\nu} \in \mathbb{H}^{1/2}(\partial\Omega_N)\,.$$

then $\Theta(\mathbf{u},\mathbf{u})$ belongs to $\mathrm{L}^1(\partial\Omega)$ and there exist at most two real coefficients $\Upsilon_\jmath$ such that

$$2\int_\Omega \mathrm{div}(\sigma(\mathbf{u})).(\mathbf{m}.\nabla)\mathbf{u}\,dx = \int_{\partial\Omega} \Theta(\mathbf{u},\mathbf{u})\,ds + \sum_{\jmath \in J(\Omega)} \Upsilon_\jmath^2\,\mathbf{m}(\mathbf{s}_\jmath).\boldsymbol{\tau}(\mathbf{s}_\jmath)\,.$$

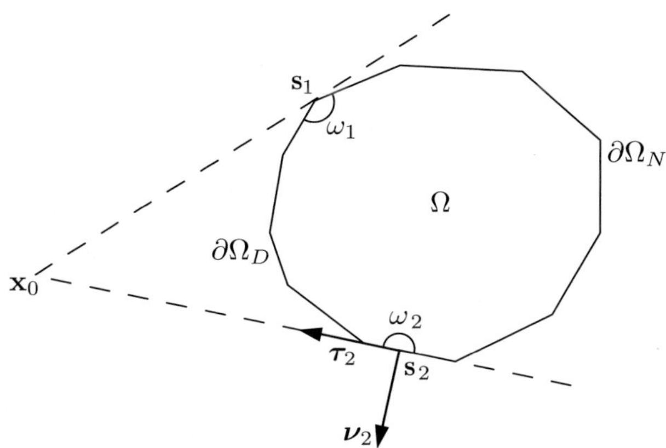

FIGURE 3. $\Omega \subset \mathbb{R}^2$ is open, bounded, polygonal and convex, $\Gamma = \{s_1, s_2\}$ and $J(\Omega) = \{2\}$.

This result has been announced in [2]. The reader will find a detailed proof in [3] and we only give main ideas of it.

We first observe that under above assumptions, $\mathbf{u}$ is locally $\mathbb{H}^2$ at every point of $\overline{\Omega}$ which is not a vertex. Hence we can apply Proposition 2 in a subdomain of Ω which does not contain small disks centered at vertices.

The main idea is to compute the limit of each integral term when the rays of these disks tend to 0. To this end, using some localization process, we can write $\mathbf{u}$ as the sum of a $\mathbb{H}^2$-part and a singular part which is given in [10] (see also [5]).

At each vertex which does not belong to Γ, the degree of the singularity is convenient and we get the limit without any further term.

A similar process holds at $\mathbf{s}_\imath \in \Gamma$ if $\varpi_\imath < \pi$.

If $\jmath \in J(\Omega)$, then in some neighborhood of $\mathbf{s}_\jmath$, we can write $\mathbf{u}$ as the sum of a $\mathbb{H}^2$-part $\mathbf{u}_R$ and a singular part $\mathbf{u}_S$ which locally behaves like the following function $\mathbf{U}_S$ given in [10]:

$$\mathbf{U}_S(r, \theta) = \Re\left(r^\alpha \mathbf{w}(\theta)\right),$$

where $\alpha \in \mathbb{C}$, $\Re\alpha = 1/2$, $\mathbf{w}$ is a complex-valued $\mathcal{C}^\infty$-function (see a detailed formula in [3]). Components of $\mathbf{U}_S$ are represented in Figure 4 for a particular choice of Lamé coefficients.

Hence, we write $\Theta(\mathbf{u}, \mathbf{u}) = \Theta(\mathbf{u}_R, \mathbf{u}_R) + \Theta(\mathbf{u}_R, \mathbf{u}_S) + \Theta(\mathbf{u}_S, \mathbf{u}_R) + \Theta(\mathbf{u}_S, \mathbf{u}_S)$ and we carefully compute the limits of corresponding integral terms. The fourth one gives $\Upsilon_\jmath^2\, \mathbf{m}(\mathbf{s}_\jmath).\boldsymbol{\tau}(\mathbf{s}_\jmath)$ ($\Upsilon_\jmath^2$ depends on the singularity coefficient of $\mathbf{u}$ at $\mathbf{s}_\jmath$), other ones give no further term.

Remark. Assumptions of Theorem 3 can be easily weakened. For a polygonal bounded domain, sufficient conditions are: $\varpi_\imath \in (0, \pi]$ for every $\imath$ and, if $\jmath \in J(\Omega)$, $\mathbf{m}(\mathbf{s}_\jmath).\boldsymbol{\nu}(\mathbf{s}_\jmath) = 0$.

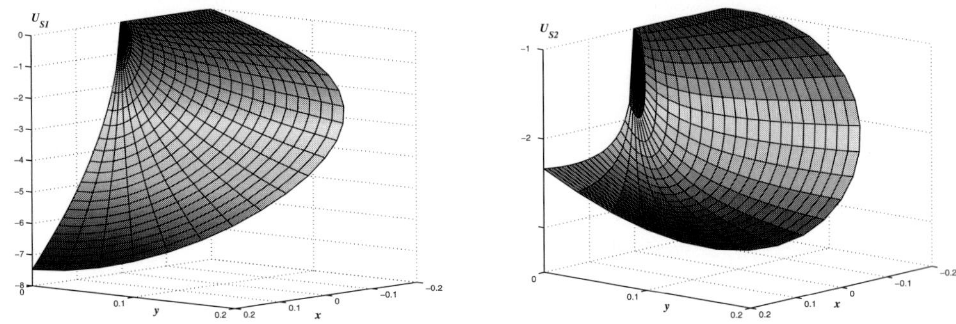

FIGURE 4. Local behavior of components of $\mathbf{U}_S$ (Lamé coefficients: $\lambda = \mu = 1$).

2.3.3. The general case. We here give an extension of Proposition 2 for the geometrical case described in Section 1.

Theorem 4. *Under assumptions (3)-(5), let* $\mathbf{u} \in \mathbb{H}^1(\Omega)$ *be such that*

$$\operatorname{div}(\sigma(\mathbf{u})) \in \mathbb{L}^2(\Omega), \quad \mathbf{u}_{/\partial\Omega_D} \in \mathbb{H}^{3/2}(\partial\Omega_D), \quad \sigma(\mathbf{u})\boldsymbol{\nu} \in \mathbb{H}^{1/2}(\partial\Omega_N).$$

Then $\Theta(\mathbf{u}, \mathbf{u})$ *belongs to* $\mathrm{L}^1(\partial\Omega)$ *and there exists* $\Upsilon \in \mathbb{L}^2(\Gamma)$ *such that*

$$2\int_{\Omega} \operatorname{div}(\sigma(\mathbf{u})).(\mathbf{m}.\nabla)\mathbf{u}\,d\mathbf{x}$$

$$= (n-2)\int_{\Omega} \sigma(\mathbf{u})\!:\!\varepsilon(\mathbf{u})\,d\mathbf{x} + \int_{\partial\Omega} \Theta(\mathbf{u}, \mathbf{u})\,ds + \int_{\Gamma} |\Upsilon|^2\,\mathbf{m}.\boldsymbol{\tau}\,d\gamma.$$

The following Section is devoted to a sketch of the proof of this result.

3. Sketch of the proof of Theorem 4

One can find a detailed proof in [3]. This proof is made of three main steps.

First step. We extend theorem 3 for a two-dimensional open set Ω which satisfies (3)-(5) such that, with notations of Subsection 2.3.2, $\Gamma = \{\mathbf{s}_1, \mathbf{s}_2\}$ and $J(\Omega) = \{1, 2\}$.

At each point $\mathbf{s}_i$, we introduce local coordinates and we get a local mixed boundary problem which involves an elasticity operator with non-constant coefficients. We prove that this operator is a small perturbation of Lamé operator and this leads to a similar structure of its solution.

Second step. We study the case considered in [9]: Ω is a semi-cylinder and Γ is its axis.

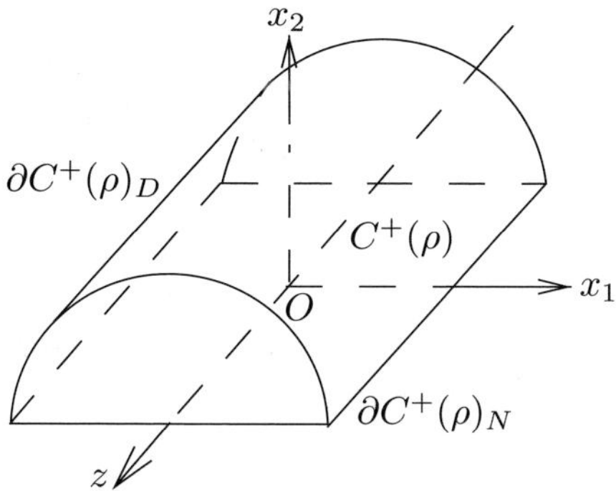

FIGURE 5. Case of a semi-cylinder.

We prove that involved singularities along Γ can be written $\begin{pmatrix} \mathbf{u}_S \\ U_S \end{pmatrix}$ where $\mathbf{u}_S$ is the singular function considered in Subsection 2.3.2 and U_S is the Shamir function (see Section 1).

We then proceed as well as for Theorem 3: we apply Proposition 2 in a convenient subdomain and we use this particular structure of singularities to get the result.

Third step. We consider a general three-dimensional open set Ω. As well as in the first step, we use a localization process in a neighborhood of each point of Γ. Similarly, in local coordinates (see Figure 6), we get a mixed boundary problem which is a small perturbation of previous one.

Thus we can use the above structure of singularities.

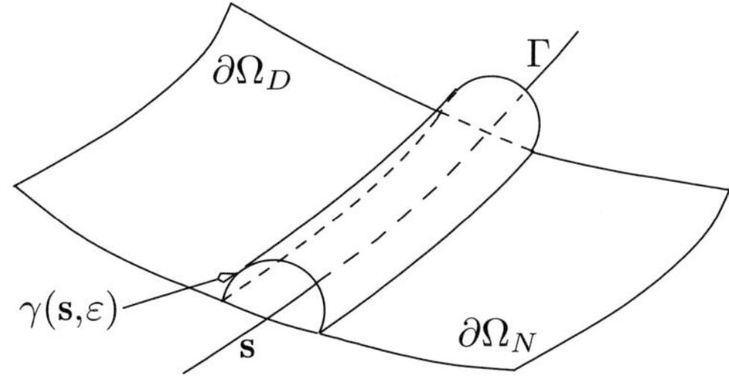

FIGURE 6. The general case.

We finally build Υ by using a compactness argument.

References

[1] BEY, R., LOHÉAC, J.-P., MOUSSAOUI, M., *Singularities of the solution of a mixed problem for a general second order elliptic equation and boundary stabilization of the wave equation*, J. Math. Pures Appl., 78 (1999), pp. 1043–1067.

[2] BROSSARD, R., LOHÉAC, J.-P., *Stabilisation frontière du système élastodynamique dans un polygone plan*, C.R. Math. Acad. Sci. Paris, 338 (2004), pp. 213–218.

[3] BROSSARD, R., LOHÉAC, J.P., *Boundary stabilization of elastodynamic systems. Part I: Rellich-type relations for a mixed boundary problem in elasticity*, CNRS UMR 5585 MAPLY, preprint 383 (2004), submitted.
http://maply.univ-lyon1.fr/publis/publiv/2004/publis.html

[4] BROSSARD, R., LOHÉAC, J.P., *Boundary stabilization of elastodynamic systems. Part II: Boundary stabilization of an elastodynamic system involving singularities*, CNRS UMR 5585 MAPLY, preprint 384 (2004), submitted.
http://maply.univ-lyon1.fr/publis/publiv/2004/publis.html

[5] GRISVARD, P., *Singularités en élasticité*, Arch. Ration. Mech. Anal., 107, no. 2 (1989), pp. 157–180.

[6] GRISVARD, P., *Contrôlabilité exacte des solutions de l'équation des ondes en présence de singularités*, J. Math. Pures Appl., 68 (1989), pp. 215–259.

[7] GRISVARD, P., *Elliptic problems in nonsmooth domains*. Monographs and Studies in Mathematics 24, Pitman, Boston, (1985).

[8] KOMORNIK, V., *Exact controllability and stabilization; The multiplier method*. Masson-John Wiley, Paris (1994).

[9] MOUSSAOUI, M., *Singularités des solutions du problème mêlé, contrôlabilité exacte et stabilisation frontière*. ESAIM Proc., Élasticité, Viscoélasticité et Contrôle optimal, Huitièmes Entretiens du Centre Jacques Cartier (1996), pp 157–168.

[10] MÉROUANI, B., *Solutions singulières du système de l'élasticité dans un polygone pour différentes conditions aux limites*, Maghreb Math. Rev., 5, no 1–2 (1996), pp. 95–112.

[11] RELLICH, F., *Darstellung der Eigenwerte von $\Delta u + \lambda u = 0$ durch ein Randintegral*, Math. Z., 46 (1940), pp. 635–636.

[12] NEČAS, J., *Les méthodes directes en théorie des équations elliptiques*. Masson, Paris (1967).

[13] SHAMIR, E., *Regularity of mixed second order elliptic problems*. Israel J. Math., 6 (1968), pp 150–168.

R. Brossard, J.-P. Lohéac and M. Moussaoui
MAPLY, C.N.R.S. U.M.R. 5585
École Centrale de Lyon, département M.I, B.P. 163
F-69131 Écully Cedex, France
e-mail: Romain.Brossard@ec-lyon.fr
e-mail: Jean-Pierre.Loheac@ec-lyon.fr
e-mail: Mohand.Moussaoui@ec-lyon.fr
URL: http://maply.univ-lyon1.fr

Progress in Nonlinear Differential Equations
and Their Applications, Vol. 63, 103–110

Lyapunov-type Inequalities and Applications to PDE

A. Cañada, J.A. Montero and S. Villegas

Abstract. This work is devoted to the study of resonant nonlinear boundary problems with Neumann boundary conditions. First, we consider the linear case doing a careful analysis which involves Lyapunov-type inequalities with the L_p- norms of the coefficient function. After this end, combining these results with Schauder fixed point theorem, we obtain some new results about the existence and uniqueness of solutions for resonant nonlinear problems.

Mathematics Subject Classification (2000). 34B15, 34B05.

Keywords. Neumann boundary value problems, resonance, Lyapunov inequalities, existence and uniqueness.

1. Introduction

Let us consider the Neumann problem

$$u''(x) + f(x, u(x)) = 0, \ x \in (0, L), \ u'(0) = u'(L) = 0 \tag{1.1}$$

where $f : [0, L] \times \mathbb{R} \to \mathbb{R}$, $(x, u) \to f(x, u)$, satisfies the condition

(H) f, f_u are continuous on $[0, L] \times \mathbb{R}$ and $0 \leq f_u(x, u)$ on $[0, L] \times \mathbb{R}$.

The existence of a solution of (1.1) implies

$$\int_0^L f(x, z) \, dx = 0 \tag{1.2}$$

for some $z \in \mathbb{R}$. However, conditions (H) and (1.2) are not sufficient for the existence of solutions of (1.1). Indeed, consider the problem

$$u''(x) + \pi^2 u(x) + \cos(\pi x) = 0, \ x \in (0, 1), \ u'(0) = u'(1) = 0. \tag{1.3}$$

The function $f(x, u) = \pi^2 u + \cos(\pi x)$ satisfies (H) and (1.2), but the Fredholm alternative theorem shows that there is no solution of (1.3).

The authors have been supported by the Ministry of Science and Technology of Spain (BFM2002-02649).

If (H) and (1.2) are assumed, and for instance, $L = 1$ for simplicity, different supplementary assumptions have been given which imply the existence of a solution of (1.1). For example

(h1) $f_u(x, u) \leq \beta(x)$ on $[0, 1] \times \mathbb{R}$ with $\beta \in L^\infty(0, 1)$, $\beta(x) \leq \pi^2$ on $[0, 1]$ and $\beta(x) < \pi^2$ on a subset of $(0, 1)$ of positive measure.

Conditions of this type are referred to as non-uniform non-resonance conditions with respect to the first positive eigenvalue of the associated linear homogeneous problem. By using variational methods, it is proved in [8] that (H), (1.2) and (h1) imply the existence of solutions of (1.1).

On the other hand, in [6] it is supposed

(h2) $f_u(x, u) \leq \beta(x)$ on $[0, 1] \times \mathbb{R}$ with $\beta \in L^1(0, 1)$ and $\displaystyle\int_0^1 \beta(x) \, dx \leq 4$.

The authors use Optimal Control theory methods to prove that (H), (1.2) and (h2) imply the existence and uniqueness of solutions of (1.1). Restriction (h2) is related to Lyapunov-type inequalities for linear second order equations (see, for instance Corollary 5.1 in [5] for the case of Dirichlet boundary conditions and [1] for a survey paper on Lyapunov inequalities).

Let us observe that supplementary conditions (h1) and (h2) are given respectively in terms of $\|\beta\|_\infty$ and $\|\beta\|_1$, the usual norms in the spaces $L^\infty(0, 1)$ and $L^1(0, 1)$. Also, it is clear that under the hypotheses (H) and (1.2), (h1) and (h2) are not related.

In this paper we provide supplementary conditions in terms of $\|\beta\|_p$, $1 < p < \infty$. In fact, this was the original motivation of our work, but the proofs are based in a previous analysis of the linear case which involves Lyapunov-type inequalities with the L_p−norm of the coefficient function for $1 \leq p \leq \infty$. Really, this is the main contribution of this paper where we carry out a complete treatment of the linear problem for any $p \geq 1$. As a consequence, a natural relation between (h1) and (h2) arises if one studies the limits of $\|\beta\|_p$ for $p \to 1^+$ and $p \to \infty$.

We restrict ourselves only to the case of Neumann boundary conditions for the sake of simplicity, but it is clear from the proofs that one can deal with other boundary conditions and more general second order equations. Also, some results for PDE problems may be obtained. The details of the proof may be seen in [3].

2. Lyapunov-type inequalities for the linear problem

This section will be concerned with the existence of nontrivial weak solutions of a homogeneous linear problem of the form

$$u''(x) + a(x)u(x) = 0, \ x \in (0, L), \ u'(0) = u'(L) = 0 \qquad (2.1)$$

where $a \in \Lambda$ and Λ is defined by

$$\Lambda = \{a \in L^1(0,L) \setminus \{0\} : \int_0^L a(x)\,dx \geq 0 \text{ and } (2.1) \text{ has nontrivial solutions. }\} \tag{2.2}$$

Obviously, the positive eigenvalues of the eigenvalue problem

$$u''(x) + \lambda u(x) = 0, \ x \in (0,L), \ u'(0) = u'(L) = 0, \tag{2.3}$$

belong to Λ. Therefore Λ is not empty and

$$\beta_p \equiv \inf_{a \in \Lambda} \|a\|_p, \ 1 \leq p \leq \infty \tag{2.4}$$

is a well-defined real number. The main result of this section is the following:

Theorem 2.1. *The following statements hold:*

1. *β_p is attained if and only if $1 < p \leq \infty$. In this case, β_p is attained in a unique element $a_p \in \Lambda$ which is not constant if $1 < p < \infty$.*
2. *The quantity β_p is given by*

$$\beta_p = \begin{cases} \dfrac{4}{L}, & \text{if } p = 1, \\[2ex] \dfrac{4(p-1)^{1+\frac{1}{p}}}{L^{2-\frac{1}{p}}p(2p-1)^{1/p}} \left(\int_0^{\pi/2} (\sin x)^{-1/p}\,dx \right)^2, & \text{if } 1 < p < \infty, \\[2ex] \dfrac{\pi^2}{L^2}, & \text{if } p = \infty. \end{cases} \tag{2.5}$$

3. *$a_\infty(x) \equiv \dfrac{\pi^2}{L^2}$. If $1 < p < \infty$, the function a_p is given by $a_p = |u_p|^{\frac{2}{p-1}}$, where in the interval $(0, \frac{L}{2})$, u_p is the unique positive solution of the problem*

$$-u''(x) = u(x)^{\frac{p+1}{p-1}}, \ u'(0) = 0, \ u(L/2) = 0$$

and in the interval $(L/2, L)$ u_p is defined by the formula $u(x) = -u(L - x), \ \forall\, x \in (L/2, L)$ (see the proof of Lemma 2.4).
4. *The mapping $[1,\infty) \to \mathbb{R}, \ p \to \beta_p$, is continuous and $\lim_{p\to\infty} \beta_p = \beta_\infty$. Moreover, the mapping $[1,\infty) \to \mathbb{R}, \ p \to L^{-1/p}\beta_p$ is strictly increasing.*

Proof. It is based on some lemmas. For the sake of simplicity we will consider the case $1 < p < \infty$. The results related to the cases $p = \infty$ can be obtained in a similar way. The case $p = 1$ is slightly more complicated (see [3]).

Lemma 2.2. *Assume $1 < p < \infty$ and let $X_p = \left\{ u \in H^1(0,L) : \int_0^L |u|^{\frac{2}{p-1}}u = 0 \right\}$.*

If $J_p : X_p \setminus \{0\} \to \mathbb{R}$ is defined by

$$J_p(u) = \int_0^L u'^2 \Big/ \left(\int_0^L |u|^{\frac{2p}{p-1}} \right)^{\frac{p-1}{p}} \tag{2.6}$$

and $m_p \equiv \inf_{X_p \setminus \{0\}} J_p$, m_p is attained. Moreover, if $u_p \in X_p \setminus \{0\}$ is a minimizer, then u_p satisfies the problem

$$u_p''(x) + A_p(u_p)|u_p(x)|^{\frac{2}{p-1}} u_p(x) = 0, \quad x \in (0, L), \quad u_p'(0) = u_p'(L) = 0, \qquad (2.7)$$

where

$$A_p(u_p) = m_p \left(\int_0^L |u_p|^{\frac{2p}{p-1}} \right)^{\frac{-1}{p}} \qquad (2.8)$$

Main ideas of the proof: The compact inclusion of $C[0, L]$ in $H^1(0, L)$ (the usual Sobolev space) implies that m_p is attained. On the other hand, using the Lagrange multiplier Theorem, it is deduced that u_p (a minimizer function) satisfies (2.7).

Lemma 2.3. *If $1 < p < \infty$, we have $\beta_p = m_p$.*

Proof. If $a \in \Lambda$ and $u \in H^1(0, L)$ is a nontrivial solution of

$$-u''(x) = a(x)u(x), \quad x \in (0, L), \quad u'(0) = u'(L) = 0, \qquad (2.9)$$

then

$$\int_0^L u'v' = \int_0^L auv, \quad \forall \, v \in H^1(0, L).$$

In particular, we have

$$\int_0^L u'^2 = \int_0^L au^2, \quad \int_0^L au = 0. \qquad (2.10)$$

Therefore, for each $k \in \mathbb{R}$, we have

$$\int_0^L (u+k)'^2 = \int_0^L u'^2 = \int_0^L au^2 \leq \int_0^L au^2 + k^2 \int_0^L a$$
$$= \int_0^L au^2 + \int_0^L k^2 a + 2k \int_0^L au = \int_0^L a(u+k)^2.$$

It follows from the Hölder inequality

$$\int_0^L (u+k)'^2 \leq \|a\|_p \|(u+k)^2\|_{\frac{p}{p-1}}$$

Also, since u is a nonconstant solution of (2.9), $u + k$ is a nontrivial function. Consequently

$$\|a\|_p \geq \int_0^L (u+k)'^2 \bigg/ \|(u+k)^2\|_{\frac{p}{p-1}}.$$

Now, choose $k_0 \in \mathbb{R}$ satisfying $u + k_0 \in X_p$. Then, $\|a\|_p \geq m_p$, $\forall \, a \in \Lambda$ and consequently $\beta_p \geq m_p$. Reciprocally, if $u_p \in X_p \setminus \{0\}$ is any minimizer of J_p, then u_p satisfies (2.7). Therefore, $A_p(u_p)|u_p|^{\frac{2}{p-1}} \in \Lambda$. Also,

$$\| A_p(u_p)|u_p|^{\frac{2}{p-1}} \|_p^p = A_p(u_p)^p \int_0^L |u_p|^{\frac{2p}{p-1}} = m_p^p.$$

Then $\beta_p \leq m_p$. $\qquad \square$

Lemma 2.4. *If $1 < p < \infty$, m_p is given by*

$$m_p = \frac{4(p-1)^{1+\frac{1}{p}}}{L^{2-\frac{1}{p}}p(2p-1)^{1/p}}\left(\int_0^{\pi/2}(\sin x)^{-1/p}\,dx\right)^2. \qquad (2.11)$$

Proof. By Lemma 2.2, if $u_p \in X_p \setminus \{0\}$ is a minimizer of J_p, then u_p satisfies a problem of the type

$$v''(x) + B|v(x)|^{\frac{2}{p-1}}v(x) = 0, \ x \in (0,L), \ \ v'(0) = v'(L) = 0, \qquad (2.12)$$

where B is some positive real constant. Also, let us observe that any nontrivial solution of (2.12) belongs to $X_p \setminus \{0\}$. On the other hand, it is possible to prove that for every $B \in \mathbb{R}^+$ and for every $n \in \mathbb{N}$ there exist exactly two solutions of (2.12) with exactly n zeros: one opposite to the other. If we called $v_{B,n}$ the solution with $v_{B,n}(0) > 0$ we deduce that

$$\inf_{B \in \mathbb{R}^+} \inf_{n \in \mathbb{N}} J_p(v_{B,n}) = m_p.$$

It is possible to prove that the first zero of $v_{B,n}$ is $\frac{L}{2n}$ and that $v_{B,n}$ is antisymmetric respect to the line $x = \frac{L}{2n}$ and symmetric respect to the line $x = \frac{L}{n}$; i.e.,

$$v_{B,n}(x) = -v_{B,n}\left(\frac{L}{n} - x\right), \quad \forall x \in \left[\frac{L}{2n}, \frac{L}{n}\right]$$

and

$$v_{B,n}(x) = (-1)^i v_{B,n}\left(x - \frac{L}{n}i\right), \quad \forall x \in \left[\frac{L}{n}i, \frac{L}{n}(i+1)\right]; \quad (i = 1, 2, \ldots, n-1).$$

Therefore it remains to define the function $v_{B,n}$ in the interval $[0, \frac{L}{2n}]$. A delicate calculus gives us

$$v(x) = \left(\frac{2nI}{L}\left(\frac{p}{B(p-1)}\right)^{1/2}\right)^{p-1}\varphi^{-1}\left(I - \frac{2nI}{L}x\right), \ \forall\, x \in \left[0, \frac{L}{2n}\right] \qquad (2.13)$$

where $\varphi : [0,1] \to \mathbb{R}$ is the strictly increasing function defined by

$$\varphi(t) = \int_0^t \frac{ds}{\left(1 - s^{\frac{2p}{p-1}}\right)^{1/2}} \qquad \text{and} \qquad I = \varphi(1).$$

Finally, and explicit calculus shows that

$$J_p(v_{B,n}) = \frac{4n^2(p-1)^{1+\frac{1}{p}}}{L^{2-\frac{1}{p}}p(2p-1)^{1/p}}\left(\int_0^{\pi/2}(\sin x)^{-1/p}\,dx\right)^2. \qquad (2.14)$$

At this point, one may observe two things. First, $J_p(v_{B,n})$ does not depend on B. Second, the optimal value of n is 1 which establishes the formula for m_p. $\square$

Lemma 2.5. *If $1 < p < \infty$ and the functions $u_p, v_p \in X_p \backslash \{0\}$ are minimizers of J_p, then there exists a nonzero constant $c \in \mathbb{R}$ such that $u_p = cv_p$. As a consequence, there is a unique function $a_p \in \Lambda$ such that $\beta_p = I_p(a_p)$. Moreover, the function a_p is given by $a_p = A_p(u_p)|u_p|^{\frac{2}{p-1}}$, where u_p is any minimizer of J_p in $X_p \setminus \{0\}$.*

Proof. It follows from (2.14) that both functions, u_p and v_p are of the form $u_p = v_{B_1,1}$ and $v_p = v_{B_2,1}$ for possibly different positive constants B_1 and B_2. Moreover, from (2.13) we deduce the existence of the constant c.

Now, let $a \in \Lambda$ be such that $\beta_p = I_p(a)$. Let $u \in H^1(0,L)$ be a nontrivial solution of

$$-u''(x) = a(x)u(x), \ x \in (0,L), \ u'(0) = u'(L) = 0.$$

Choose $k_0 \in \mathbb{R}$ satisfying $u + k_0 \in X_p$. Then, as in Lemma 2.3, we have

$$\int_0^L (u + k_0)'^2 \le \int_0^L a(u + k_0)^2 \le \|a\|_p \|(u + k_0)^2\|_{p'}$$

$$= m_p \|(u + k_0)^2\|_{p'} = m_p \left(\int_0^L |(u + k_0)|^{\frac{2p}{p-1}} \right)^{\frac{p-1}{p}} \le \int_0^L (u + k_0)'^2.$$

Then, $u + k_0$ is a minimizer of J_p in $X_p \setminus \{0\}$ and we have an equality in Hölder inequality. Therefore there exists a nontrivial constant d such that $a = d|(u + k_0)|^{\frac{2}{p-1}}$.

Now, if $\tilde{a} \in \Lambda$ is such that $\beta_p = I_p(\tilde{a})$, then there exists a nontrivial constant $\tilde{d}$ such that $\tilde{a} = \tilde{d}|(\tilde{u} + \tilde{k}_0)|^{\frac{2}{p-1}}$, where $\tilde{u} \in H^1(0,L)$ is a nontrivial solution of

$$-z''(x) = \tilde{a}(x)z(x), \ x \in (0,L), \ z'(0) = z'(L) = 0,$$

and $\tilde{k}_0 \in \mathbb{R}$ satisfies $\tilde{u} + \tilde{k}_0 \in X_p$. Since both functions $u + k_0$ and $\tilde{u} + \tilde{k}_0$ are minimizers of J_p in $X_p \setminus \{0\}$, there exists a positive constant c such that $u + k_0 = c(\tilde{u} + \tilde{k}_0)$. Then $a = dc^{\frac{2}{p-1}}|\tilde{u} + \tilde{k}_0|^{\frac{2}{p-1}}$. Moreover, since $\|a\|_p = \|\tilde{a}\|_p = \beta_p$ we must have $dc^{\frac{2}{p-1}} = \tilde{d}$ and consequently $a = \tilde{a}$.

Finally, u_p is any minimizer of J_p in $X_p \setminus \{0\}$, then we obtain from (2.7) that the function $= A_p(u_p)|u_p|^{\frac{2}{p-1}}$ belongs to Λ. Also the L_p- norm of this function is β_p. This proves the lemma. $\qquad \square$

It is trivial to prove that β_p is a continuous function if $p \in (1,\infty)$ and that $\lim_{p \to \infty} \beta_p = \beta_\infty$. Also, it is possible to prove that $\lim_{p \to 1} \beta_p = \beta_1$. To finish the proof of the theorem it remains to show that the mapping $[1,\infty) \to \mathbb{R}, p \to L^{-1/p}\beta_p$ is strictly increasing, which follows from the general inequality

$$L^{-1/p}\|f\|_p \le L^{-1/q}\|f\|_q \quad \forall f \in L^q(0,L)$$

where $1 < p < q < \infty$.

As an application of Theorem 2.1 to the linear problem

$$u''(x) + a(x)u(x) = f(x), \ x \in (0, L), \ u'(0) = u'(L) = 0, \qquad (2.15)$$

we have the following corollary, which clearly generalizes Theorem 3 in [6].

Corollary 2.6. *Let $a \in L^\infty \setminus \{0\}$, $0 \le a(x)$, a.e. in $(0, L)$, satisfying one of the following conditions:*

1. $\|a\|_1 \le \beta_1$,
2. *There is some $p \in (1, \infty)$ such that $\|a\|_p < \beta_p$ or $\|a\|_p = \beta_p$ and $a \ne a_p$.*
3. $\|a\|_\infty < \beta_\infty$ or $\|a\|_\infty = \beta_\infty$ and $a \ne a_\infty$.

Then for each $f \in L^\infty(0, L)$, the boundary value problem (2.15) has a unique solution.

3. The nonlinear problem

In this section we give some new results on the existence and uniqueness of solutions of nonlinear b.v.p. (1.1). To get our purpose, we combine the results obtained in the previous section with the Schauder's fixed point theorem.

Theorem 3.1. *Let us consider (1.1) where the following requirements are fulfilled:*

1. *f and f_u are continuous on $[0, L] \times \mathbb{R}$.*
2. *For some function $\beta \in L^\infty(0, L)$, we have $f_u(x, u) \le \beta(x)$ on $[0, L] \times \mathbb{R}$ and β satisfies some of the conditions given in Corollary 2.6.*
3. *$0 \le f_u(x, u)$ in $[0, L] \times \mathbb{R}$. Moreover, for each $u \in C[0, L]$ one has $f_u(x, u(x)) \ne$*

$$0, \ \text{a.e. on } [0, L] \ \text{and} \ \int_0^L f(x, 0) \, dx = 0.$$

Then, problem (1.1) has a unique solution.

The procedure of proving existence and uniqueness is standard and may be seen, for example, in [3], [6].

Remark 1. Since the change of variables $u(x) = v(x) + z$, $z \in \mathbb{R}$, transforms (1.1) into the problem

$$v''(x) + f(x, v(x) + z) = 0, \ x \in (0, L), \ v'(0) = v'(L) = 0,$$

the condition $\displaystyle\int_0^L f(x, 0) \, dx = 0$ in the previous Theorem, may be substituted by (1.2).

Previous result generalizes Theorem B in [6]. Also, under the hypothesis (1) of previous Theorem, it is a generalization of Theorem 2 in [8] for the case of ordinary differential equations.

Remark 2. One can expect that some results hold true in the case of Neumann boundary value problem for partial differential equations

$$\Delta u(x) + a(x)u(x) = 0, \ x \in \Omega, \ \frac{\partial u(x)}{\partial n} = 0, \ x \in \partial\Omega \qquad (3.1)$$

where Ω is a bounded and regular domain in $\mathbb{R}^N$. But here the role played by the dimension N may be important. For instance, if $N \geq 3$ and

$$\Lambda = \{a \in L^\infty(\Omega) : \int_\Omega a(x)\ dx > 0 \text{ and } (3.1) \text{ has nontrivial solutions}\}$$

then it may be proved that $\inf_{a \in \Lambda} \|a\|_p > 0 \iff p \geq \dfrac{N}{2}$. Moreover, it is possible to prove that this infimum is attained if and only if $p > \dfrac{N}{2}$.

References

[1] R.C. Brown and D.B. Hinton, *Lyapunov inequalities and their applications, Survey on Classical Inequalities*, T.M. Rassias, ed. Kluwer, Dordrecht, 2000, 1–25.

[2] P. Drábek, *Nonlinear eigenvalue problems and Fredholm alternative*, In Nonlinear Differential Equations, P. Drábek, P. Krejči and P. Takáč, Editors. Research Notes in Mathematics Series, 404, Chapman and Hall/CRC, London, 1–46, 1999

[3] A. Cañada, J.A. Montero and S. Villegas, *Liapunov-type inequalities and Neumann boundary value problems at resonance*, to appear in Math. Inequal. Appl.

[4] B. Dacorogna, W. Gangbo and N. Subia, *On a generalization of the Wirtinger inequality*, Ann. Inst. H. Poincaré Anal. Non Linéare, 9, (1992), no. 1, 29–50.

[5] P. Hartman, *Ordinary Differential Equations*, John Wiley and Sons Inc., New York-London-Sydney, 1964.

[6] W. Huaizhong and L. Yong, *Neumann boundary value problems for second-order ordinary differential equations across resonance*, SIAM J. Control and Optimization, 33, (1995), 1312–1325.

[7] L. Kotin, *A generalization of Liapunov's inequality*, J. Math. Anal. Appl., 102, (1984), 585–598.

[8] J. Mawhin, J.R. Ward and M. Willem, *Variational methods and semilinear elliptic equations*, Arch. Rational Mech. Anal., 95, (1986), 269–277.

A. Cañada, J.A. Montero and S. Villegas
Departamento de Análisis Matemático
Universidad de Granada
E-18071 Granada, Spain
e-mail: `acanada@ugr.es`
e-mail: `jmontero@ugr.es`
e-mail: `svillega@ugr.es`

Progress in Nonlinear Differential Equations
and Their Applications, Vol. 63, 111–118

Gaeta 2004. Elliptic Resonant Problems with a Periodic Nonlinearity

A. Cañada and D. Ruiz

Abstract. We study the existence of solution for a nonlinear PDE problem at resonance under Dirichlet boundary conditions. The nonlinear term considered comes from a periodic function: in particular, the problem is strongly resonant at infinity. In our proofs we shall use variational methods together with some asymptotic analysis.

Mathematics Subject Classification (2000). 35J25, 35B34, 35J20.

Keywords. Dirichlet boundary value problems, resonance, periodic nonlinearities, variational methods, asymptotics, Palais-Smale condition.

1. Introduction

In this work we are concerned with the problem:

$$\begin{cases} \Delta u + \lambda_1 u + g(u) = h(x) & x \in \Omega \\ u(x) = 0 & x \in \partial\Omega \end{cases} \tag{1.1}$$

where g is a periodic continuous function with zero mean value and Ω is a smooth bounded domain. Moreover, $h \in C(\overline{\Omega})$ and λ_1 stands for the principal eigenvalue of $-\Delta$ subject to homogeneous Dirichlet boundary conditions. Our aim is to study the existence of solutions for (1.1) depending on the external force $h \in C(\overline{\Omega})$.

The corresponding ODE problem has been largely studied in the literature, see [8, 14, 17, 19, 23]. Systems of ODE's at resonance with periodic nonlinear terms have also attracted much attention, mostly under periodic boundary conditions ([3, 15, 18]). Several advances have also been made in the case of Dirichlet boundary conditions, see [7, 9, 10, 11].

Problem (1.1) has been well understood when the domain is convex, see [13, 20, 21]. A general information is given, for any domain, in [16, 22]. In this work we are able to deal with arbitrary smooth domains and terms h not necessarily small (under some hypotheses on g, though).

The authors have been supported by the Ministry of Science and Technology of Spain (BFM2002-02649), and by J. Andalucía (FQM 116).

Observe that (1.1) is resonant and the linear part has a one-dimensional kernel; actually, if $g = 0$, problem (1.1) has a solution if and only if $\int_\Omega h(x)\phi_1(x)\,dx = 0$, where ϕ_1 is the principal eigenfunction. This motivates the split $h = \tilde{h} + r\phi_1(x)$, where $r \in \mathbb{R}$ and $\int_\Omega \tilde{h}(x)\phi_1(x)\,dx = 0$. Our objective is to fix $\tilde{h}$ and to study the existence of solution for (1.1) depending on the parameter r. The existence of solution when $r = 0$ was proved by [22, 23]. We will say that problem (1.1) is nondegenerate if it admits solutions for some $r \neq 0$. In this work we are concerned with the existence of solution when $|r|$ is small.

We use variational reasonings. Variational methods in the study of strong resonant problems have been much used, see for instance [1, 2, 4, 6, 22, 23], and more recently [5]. In this work we shall deal with the possible degeneracy of the expressions we intend to study. Our generic result stands for very general problems at strong resonance, and we think it is interesting by itself.

For any fixed $\tilde{h}$ we prove the existence of solution for (1.1) when r is small enough under certain hypotheses, which include the cases:

1. g is an even function.
2. $g(x) = \sin(x + \delta)$ for any $\delta \in (0, 2\pi)$.
3. $\hat{G}$ has nonzero mean value, where $\hat{G}$ is a periodic second primitive of g vanishing at zero.

To the best of our knowledge, this in the only result in literature about nondegeneracy of problem (1.1) in non-convex domains without the extra assumption that $\tilde{h}$ is small.

2. The variational approach: nondegeneracy

In this communication we deal with the problem:

$$\Delta u(x) + \lambda_1 u(x) + g(u(x)) = \tilde{h}(x) + r\phi_1(x), \quad x \in \Omega$$
$$u(x) = 0, \qquad\qquad x \in \partial\Omega \tag{2.1}$$

where g is a continuous periodic function with zero mean value (not constantly equal to zero). Moreover, $\Omega \subset \mathbb{R}^N$ is a bounded C^2 domain, λ_1 is the first eigenvalue of $-\Delta$ in Ω and ϕ_1 is the positive eigenfunction associated to λ_1. Finally, $r \in \mathbb{R}$ and $\tilde{h} \in C(\overline{\Omega})$ is such that:

$$\int_\Omega \tilde{h}(x)\phi_1(x)\,dx = 0.$$

We point out that the reasonings in [13, 20, 21] do not work for arbitrary domains. In this section we give some nondegeneracy results for any C^2 domain, but under some restrictions on g.

The solutions of (2.1) are critical points for the functional:

$$I_r : H_0^1(\Omega) \to \mathbb{R}, \quad I_r = J_r + N$$

$$J_r(u) = \int_\Omega \frac{1}{2}\left(|\nabla u(x)|^2 - \lambda_1 u(x)^2\right) + (\tilde{h}(x) + r\phi_1(x))u(x)\,dx$$

$$N(u) = -\int_\Omega G(u(x))\,dx$$

where G is a periodic primitive of g which is also chosen to have zero mean value.

If we denote $X = H_0^1(\Omega)$, define $\tilde{X} = \{u \in X : \int_\Omega u(x)\phi_1(x)\,dx = 0$, and $\bar{X} = \{a\phi_1(x) : a \in \mathbb{R}\}$. Clearly, any $u \in X$ can be decomposed as: $u = \tilde{u} + \bar{u}$, with $\tilde{u} \in \tilde{X}$ and $\bar{u} \in \bar{X}$.

For the sake of clarity, denote $I = I_0$, $J = J_0$. We now enumerate some basic properties of J and N; these can be easily checked by the reader by using analogous reasonings to those in [1, 2, 14]. First of all, the functionals J and N are weakly lower semicontinuous. It is easy to verify that the functional J is bounded below and $J(\tilde{u} + \bar{u}) = J(\tilde{u})$. Because of that, J does not verify the Palais-Smale condition; however, $J_{|\tilde{X}}$ is coercive and therefore it attains its infimum m at a certain u_0. In fact, this is the only solution belonging to $\tilde{X}$ of the associated linear problem:

$$\begin{aligned}
\Delta u(x) + \lambda_1 u(x)) &= \tilde{h}(x), & x \in \Omega \\
u(x) &= 0, & x \in \partial\Omega.
\end{aligned} \tag{2.2}$$

The following lemma is a version of the Riemann-Lebesgue lemma, and is stated and proved (in a much more general form) in [22], for instance. We enounce it here because it is crucial in our arguments.

Lemma 2.1. *Let $\Omega \subset \mathbb{R}^N$ be a bounded smooth domain, $U \subset H^1(\Omega)$ be a bounded set, and let f be a continuous periodic function with zero mean value. Then, we have:*

$$\lim_{a \to \infty} f(u + a\phi_1(\cdot)) = 0 \ \text{strongly in } H^{-1}(\Omega), \tag{2.3}$$

and

$$\lim_{a \to \infty} \int_\Omega f(u(x) + a\phi_1(x))\,dx = 0, \tag{2.4}$$

both convergences being uniform for $u \in U$.

Observe that since N is a bounded functional, I is also bounded below. We shall study the existence of minimum for I, which will provide us with a solution for (2.1) when $r = 0$. Afterwards we will use a perturbation argument to find solutions when r is small, in the same spirit as in [5].

As we said before, we intend to prove that the infimum of I is in fact a minimum. In order to do so, it suffices to prove that the (PS) property holds at the infimum level. In the next lemma we study the Palais-Smale condition for I.

Lemma 2.2. *The functional I verifies the property $(PS)_c$ for any $c \neq m$, where $m = \inf J$. Moreover, if $r \neq 0$, I_r verifies the property $(PS)_c$ for every $c \in \mathbb{R}$.*

Proof. Take a sequence u_n in X such that $I_r(u_n) \to c \neq m$ and $I'_r(u_n) \to 0$ in X'. We claim that u_n must be bounded; from this we can obtain the existence of a convergent subsequence through standard compactness arguments. Let us decompose $u_n = \bar{u}_n + \tilde{u}_n$, $\bar{u}_n \in \bar{X}$, $\tilde{u}_n \in \tilde{X}$.

The convergence $I'_r(u_n) \to 0$ in X' can be written in the form:

$$J'(\tilde{u}_n) + N'(u_n) + r \cdot \phi_1 \to 0.$$

If we denote $\tilde{X}' = \{\xi \in X' : \xi_{|\bar{X}} = 0\}$, it is known that $J' : \tilde{X} \to \tilde{X}'$ is affine and one-to-one; therefore, it has a continuous inverse. In particular, this implies that $\tilde{u}_n$ is bounded.

Suppose, reasoning by contradiction, that $\bar{u}_n$ is unbounded; we can assume, up to a subsequence, that $\bar{u}_n$ diverges. Recall that

$$I_r(u_n) = J_r(\tilde{u}_n) + N(\tilde{u}_n + \bar{u}_n) \to c$$

$$I'_r(u_n) = J'_r(\tilde{u}_n) + N'(\tilde{u}_n + \bar{u}_n) \to 0 \ \text{ in } X'.$$

From Lemma 2.1 it follows that $N(\tilde{u}_n + \bar{u}_n) \to 0$, $N'(\tilde{u}_n + \bar{u}_n) \to 0$. So, $J'(\tilde{u}_n) \to -r \cdot \phi_1$. Using again the map $J' : \tilde{X} \to \tilde{X}'$ we obtain, first, that r must be equal to zero (this proves the second statement of the lemma). If $r = 0$, because of the continuity of $(J')^{-1}$, $\tilde{u}_n \to u_0$. But then

$$I(u_n) = J(\tilde{u}_n) + N(\tilde{u}_n + \bar{u}_n) \to J(u_0) = m \neq c$$

which gives the desired contradiction. $\qquad\square$

It is easy to check (Riemann-Lebesgue Lemma) that the inequality $c := \inf I \leq m$ is always verified. Hence, we are interested in conditions which imply that $c < m$. The following proposition gives a result in that direction and, in such case, gives the existence of relative minima for I_r when r is small enough.

Proposition 2.3. *The following three assertions hold:*

1. *If $c < m$, I achieves its infimum. Moreover, there exists $\varepsilon > 0$ such that the functional I_r has at least two critical points when $|r| < \varepsilon$, $r \neq 0$.*
2. *If there exists $a \in \mathbb{R}$ such that $N(u_0 + a \cdot \phi_1) < 0$, then $c < m$.*
3. *If $N(u_0 + a \cdot \phi_1) = 0$ and $a \in \mathbb{R}$ is such that the range of $u_0 + a\phi_1$ is larger than the period of g, then $c < m$.*

Proof. First of all, let us focus on proving (1). If $c < m$, then the compactness property $(PS)_c$ holds for I, and therefore there exists an absolute minimum of I. Denote $K = \{u \in X : I(x) = c\}$, which must be compact, and $U \supset K$ an open bounded set in X. Since I is weakly lower semicontinuous, standard compactness arguments may be used to prove that $\inf\{I(x); u \in \partial U\} > c$.

So, by taking r small enough, we have that

$$\sup\{|I(u) - I_r(u)| : u \in \overline{U}\} \tag{2.5}$$

can be assumed to be as small as we need. A compactness argument (see [11] for more details) imply that I_r has a relative minimum at some $u \in U$.

If $r \neq 0$ the functional I_r is not bounded below; so, a mountain-pass critical point is also obtained (recall that I_r verifies the $(PS)_c$ condition for every $c \in \mathbb{R}$).

To prove (2), we only need to compute:

$$I(u_0 + a \cdot \phi_1) = J(u_0) + N(u_0 + a \cdot \phi_1) = m + N(u_0 + a \cdot \phi_1) < m.$$

Therefore, the infimum $c = \inf I$ must be less than m.

The proof of (3) is not as easy as previously. Suppose, reasoning by contradiction, that $c = \inf I = m$. The same computations as above yield:

$$I(u_0 + a \cdot \phi_1) = m + N(u_0 + a \cdot \phi_1) = m = c.$$

Therefore, I has a minimum, which is achieved at the point $u_0 + a \cdot \phi_1$. So, this is a solution for problem (2.1) with $r = 0$ (in fact, it is a strong solution). Recall now that $u_0 + a \cdot \phi_1$ is also a solution for the linear problem (2.2). Combining both equations, we have:

$$g(u_0(x) + a\phi_1(x)) = 0 \ \forall \, x \in \Omega.$$

It would imply that $g \equiv 0$, contradicting our hypotheses. $\qquad\qquad\square$

We remark here that our aim is to prove nondegeneracy of problem (2.1), that is, existence of solutions with $|r|$ small. In order to do so, we will use Proposition 2.3; we are then led with the estimate of the expression:

$$\Gamma(a) = \int_\Omega G(u_0(x) + a\phi_1(x)) \, dx. \tag{2.6}$$

Our aim is to prove that $\Gamma(a)$ is either positive for certain a or zero for some a large. To do that, we will study the primitive of Γ, which is given by the expression:

$$\Upsilon(a) = \int_\Omega \frac{\hat{G}(u_0(x) + a\phi_1(x))}{\phi_1(x)} \, dx \tag{2.7}$$

where $\hat{G}$ is a primitive of G vanishing at zero (equation (2.7) is well defined thanks to the Hopf lemma). Our approach will be to study the limits $\Upsilon(+\infty)$, $\Upsilon(-\infty)$.

Theorem 2.4. *Consider problem (2.1), and define m_0 as the mean value of $\hat{G}$. Then, we have the asymptotic estimate:*

$$\lim_{a \to \pm\infty} \Upsilon(a) - \int_{\partial\Omega} \frac{1}{|\nabla\phi_1(x)|} \, dx \int_0^a \frac{\hat{G}(t)}{t} \, dt = Cm_0$$

where $C \in \mathbb{R}$ is a constant. As a consequence, if we assume that one of the following hypotheses is verified:

1. $m_0 \neq 0$

2. $m_0 = 0$ *and* $\displaystyle \int_0^{+\infty} \frac{\hat{G}(t)}{t} \, dt \geq \int_0^{-\infty} \frac{\hat{G}(t)}{t} \, dt.$

Then, the functional I attains its infimum. Moreover, there exists $\varepsilon > 0$ such that if $|r| < \varepsilon$, $r \neq 0$, I_r has at least two critical points.

Proof. For any $\varepsilon > 0$, define $\Omega_\varepsilon = \{x \in \Omega : \phi_1(x) < \varepsilon\}$. Obviously, for any $\delta > 0$ there exists $\varepsilon > 0$ such that $dist(x, \partial\Omega) < \delta$ for any $x \in \Omega_\varepsilon$. Therefore, by choosing ε small enough, we can assume that Ω_ε is contained in an interior C^2 tubular neighborhood of $\partial\Omega$.

Thanks to the Hopf lemma we can also assume that $|\nabla\phi_1(x)| > 0$ for any $x \in \overline{\Omega_\varepsilon}$. From now on we denote $\Omega' = \Omega_\varepsilon$ for ε small but fixed.

Observe that:

$$\lim_{a \to \pm\infty} \int_{\Omega\setminus\Omega'} \frac{\hat{G}(u_0(x) + a\phi_1(x))}{\phi_1(x)}\, dx = \int_{\Omega\setminus\Omega'} \frac{m_0}{\phi_1(x)}\, dx$$

thanks to a generalized version of the Riemann-Lebesgue lemma (see [22], Proposition 2.1).

The fact that we can restrict ourselves to Ω' is very important in the following. This is what allows us to study arbitrary domains Ω, since the behavior of ϕ_1 near the boundary is well understood due to the Hopf lemma.

So we are interested in the asymptotics of the expression:

$$\Psi(a) = \int_{\Omega'} \frac{\hat{G}(u_0(x) + a\phi_1(x))}{\phi_1(x)}\, dx.$$

We first claim that $\Psi(a) \sim \int_{\Omega'} \frac{\hat{G}(a\phi_1(x))}{\phi_1(x)}\, dx$. Actually,

$$\int_{\Omega'} \frac{\hat{G}(u_0(x) + a\phi_1(x)) - \hat{G}(a\phi_1(x))}{\phi_1(x)}\, dx$$

$$= \int_0^1 \int_{\Omega'} G(su_0(x) + a\phi_1(x)) \frac{u_0(x)}{\phi_1(x)}\, dx\, ds \to 0 \quad (a \to \pm\infty).$$

This last limit holds since the quotient $\frac{u_0}{\phi_1}$ is bounded (thanks to C^1 regularity of u_0 and to Hopf lemma for ϕ_1).

For any $t \in [0, \varepsilon)$, let us define $S_t = \{x \in \Omega' : \phi_1(x) = t\}$, which is a $N - 1$ manifold. Then we can use the coarea formula (see [12]) to obtain:

$$\int_{\Omega'} \frac{\hat{G}(a\phi_1(x))}{\phi_1(x)}\, dx = \int_0^\varepsilon \int_{S_t} \frac{\hat{G}(a\phi_1(x))}{\phi_1(x)|\nabla\phi_1(x)|}\, dx\, dt = \int_0^\varepsilon \frac{\hat{G}(at)}{t} \int_{S_t} \frac{1}{|\nabla\phi_1(x)|}\, dx\, dt.$$

By using again the Hopf lemma, we deduce that the function

$$\rho : [0, \varepsilon) \to \mathbb{R}, \qquad \rho(t) = \int_{S_t} \frac{1}{|\nabla\phi_1(x)|}\, dx$$

is a C^1 function (see [21]). Then, by using again the Riemann-Lebesgue lemma, we obtain:

$$\int_0^\varepsilon \frac{\hat{G}(at)}{t} \rho(t) \sim \rho(0) \int_0^\varepsilon \frac{\hat{G}(at)}{t}\, dt + m_0 \int_0^\varepsilon \frac{\rho(t) - \rho(0)}{t}\, dt$$

$$= \int_{\partial\Omega} \frac{1}{|\nabla\phi_1(x)|}\, dx \int_0^{a\varepsilon} \frac{\hat{G}(t)}{t}\, dt + m_0 \int_0^\varepsilon \frac{\rho(t) - \rho(0)}{t}\, dt.$$

For the rest of the proof, we make use of the following lemma (for a proof, see [11]).

Lemma 2.5. *Let p be a C^1 periodic function vanishing at zero, and define $\Phi(a) = \int_0^a \dfrac{p(t)}{t}\, dt$. Then, we have the following estimate:*

$$\lim_{|a|\to\infty} |\Phi(a) - m_0 \ln |a|| < +\infty$$

where m_0 is the mean value of p.

Then, in case (1), obviously Φ must be increasing at a certain point, and we are done in virtue of Proposition 2.3. Furthermore, in case (2), if

$$\int_0^{+\infty} \frac{\hat{G}(t)}{t}\, dt > \int_0^{-\infty} \frac{\hat{G}(t)}{t}\, dt$$

we can argue as before. However, if we had an inequality, then we would obtain that $\Upsilon(+\infty) = \Upsilon(-\infty)$. In such case, either Υ is strictly increasing at a certain point or Υ is constant. In both cases, we can use Proposition 2.3 to conclude. $\square$

In particular, we can give the following result:

Theorem 2.6. *Suppose that g verifies one of the following hypotheses:*

1. *The function g is even.*
2. *$g(x) = \sin(x + \delta)$, for any $\delta \in (0, 2\pi)$.*

Then, the statement of Theorem 2.4 holds.

References

[1] D. Arcoya, *Periodic solutions of Hamiltonian systems with strong resonance at infinity*, Diff. and Int. Eq., 3 (1990), 909–921.

[2] D. Arcoya and A. Cañada, *Critical point theorems and applications to nonlinear boundary value problems*, Nonl. Anal. TMA, 14 (1990), 393–411.

[3] E.J. Banning, J.P. van der Weele, J.C. Ross and M.M. Kettenis, *Mode competition in a system of two parametrically driven pendulums with nonlinear coupling*, Physica A, 245 (1997), 49–98.

[4] P. Bartolo, V. Benci and D. Fortunato, *Abstract critical point theorems and applications to some nonlinear problems with "strong" resonance at infinity*, Nonl. Anal. TMA, 7 (1983), 981–1012.

[5] D. Bonheure, C. Fabry and D. Ruiz, *Problems at resonance for equations with periodic nonlinearities*, Nonl. Anal. 55 (2003), 557–581.

[6] A. Cañada, *A note on the existence of global minimum for a noncoercive functional*, Nonl. Anal. 21 (1993) 161–166.

[7] A. Cañada, *Nonlinear ordinary boundary value problems under a combined effect of periodic and attractive nonlinearities*, J. Math. Anal. Appl., 243 (2000), 174–189.

[8] A. Cañada and F. Roca, *Existence and multiplicity of solutions of some conservative pendulum-type equations with homogeneous Dirichlet conditions* Diff. Int. Eqns. 10 (1997) 1113–1122

[9] A. Cañada and D. Ruiz, *Resonant problems with multi-dimensional kernel and periodic nonlinearities,* Diff. Int. Eq. vol 16, No. 4 (2003), 499–512.

[10] A. Cañada and D. Ruiz, *Asymptotic analysis of oscillating parametric integrals and ordinary boundary value problems at resonance,* preprint.

[11] A. Cañada and D. Ruiz, *Periodic perturbations of resonant problems,* to appear in Calc. Var PDE.

[12] I. Chavel, Eigenvalues in Riemannian Geometry, Academic Press, New York 1984.

[13] D. Costa, H. Jeggle, R. Schaaf and K. Schmitt, *Oscillatory perturbations of linear problems at resonance,* Results Math. 14 (1988), No. 3-4, 275–287.

[14] E. N. Dancer, *On the use of asymptotics in nonlinear boundary value problems* Ann. Mat. Pura Appl. 131 (1982) 167–185.

[15] P. Drábek and S. Invernizzi, *Periodic solutions for systems of forced coupled pendulum-like equations,* J. Diff. Eq., 70, (1987), 390–402.

[16] D. Lupo and S. Solimini, *A note on a resonance problem,* Proc. Royal Soc. Edinburgh 102 A (1986), 1–7.

[17] J. Mawhin, *Problèmes non linéaires,* Sémin. Math. Sup. No. 104, Presses Univ. Montréal, 1987.

[18] J. Mawhin, *Forced second order conservative systems with periodic nonlinearity,* Ann. Inst. H. Poincaré, special issue dedicated to J.J. Moreau, 6, (1989), Suppl., 415–434.

[19] R. Schaaf and K. Schmitt, *A class of nonlinear Sturm-Liouville problems with infinitely many solutions,* Trans. AMS 306 (1988), 853–859.

[20] R. Schaaf and K. Schmitt, *Periodic perturbations of linear problems at resonance in convex domains,* Rocky Mount. J. Math. 20 (1990), 541–559.

[21] R. Schaaf and K. Schmitt, *Asymptotic behavior of positive branches of elliptic problems with linear part at resonance,* Z. angew. Math. Phys. 43 (1992), 645–676.

[22] S. Solimini, *On the solvability of some elliptic partial differential equations with the linear part at resonance,* J. Math. Anal. Appl. 117 (1986), 138–152.

[23] J.R. Ward, *A boundary value problem with a periodic nonlinearity,* Nonlinear Analysis 10 (1986), 207–213.

A. Cañada and D. Ruiz
Departamento de Análisis Matemático
Universidad de Granada
E-18071 Granada, Spain
e-mail: acanada@ugr.es
e-mail: daruiz@ugr.es

Progress in Nonlinear Differential Equations
and Their Applications, Vol. 63, 119–126

Harnack Inequality for p-Laplacians on Metric Fractals

Raffaela Capitanelli

Abstract. By using the approach of the *metric fractals*, we prove a Harnack inequality for non-negative local supersolutions of p-Laplacians – associated to p-Lagrangians – on metric fractals whose homogeneous dimension is less than p.

Keywords. Nonlinear energy forms, fractals, Harnack inequality.

1. Introduction

The notion of (measure-valued) homogeneous p-Lagrangian has been introduced in the paper of Malý and Mosco [11] and developed by Biroli and Vernole in [2] and [3].

A first result on the local regularity has been obtained in [3] in the case of absolute continuity of the homogeneous p-Lagrangian with respect to the underlying volume measure: in particular, a Harnack type inequality has been proved for the relative positive harmonics functions (in the linear case, see [1]).

In the present paper, we consider *possibly singular* homogeneous p-Lagrangians, motivated by the study of the dynamics of intrinsically irregular structures as the most common fractals, where the Lagrangians are typically singular with respect to the underlying volume measure (for example, see [10]).

We use the approach of "metric fractals" introduced by Mosco in [13], [14] and [15]. More precisely, a "metric fractal" is a connected topological space X, equipped with a quasimetric d (for which X is complete), a doubling measure μ and a measure-valued p-Lagrangian $\mathcal{L}^{(p)}$; moreover, d, μ, $\mathcal{L}^{(p)}$ are related by scaled Poincaré inequalities and by scaled estimates of the p-capacity on d-balls.

Just by using this setting in the linear case, on metric fractals whose homogeneous dimension is less than 2, uniform Harnack inequalities for local solutions and related Green function estimates have been proved in [14].

Now, we consider the nonlinear case and we show the Harnack inequality holds for non negative local supersolutions of p-Laplacians – associated to p-Lagrangians – on metric fractals whose homogeneous dimension is less than p.

The plan of the paper is as follows. In Section 2 we introduce the definitions and the main notations. In Section 3 we recall some functional inequalities proved in [5] and we state the Harnack inequality for non-negative local supersolutions proved in [7]. In Section 4 we give some applications: in particular, we recall that self-similar structures with self-similar Lagrangians are metric fractals and we show some concrete examples where the Harnack inequality holds.

2. Metric fractals

Let X be a locally compact Hausdorff topological space and μ a nonnegative bounded Radon measure on X with support X. Let $\tilde{\mathcal{L}}^{(p)}$ a Radon measure valued nonnegative map defined on a dense subalgebra $\mathcal{C}^{(p)}$ of the space $C_b(X)$ of bounded continuous functions on X. We make the following assumptions on $\tilde{\mathcal{L}}^{(p)}$, $(p > 1)$:

L1) $\tilde{\mathcal{L}}^{(p)}$ is positive semidefinite and convex in the space $\mathcal{M}$ of Radon measure.

L2) $\tilde{\mathcal{L}}^{(p)}$ is homogeneous of degree p.

L3) $\tilde{\mathcal{L}}^{(p)}$ is such that

$$||u|| = \left(\int_X |u|^p d\mu + \int_X d\tilde{\mathcal{L}}^{(p)}(u) \right)^{\frac{1}{p}} \tag{2.1}$$

is a norm in $\mathcal{C}^{(p)}$.

L4) $\tilde{\mathcal{L}}^{(p)}$ is strongly local: if $u - v = $ constant on $supp\varphi$, then

$$\int_X \varphi(x) d\tilde{\mathcal{L}}^{(p)}(u) = \int_X \varphi(x) d\tilde{\mathcal{L}}^{(p)}(v)$$

for any $\varphi \in C(X)$, $u, v \in \mathcal{C}^{(p)}$.

L5) for every $u, v \in \mathcal{C}^{(p)}$ there exists in the weakly* topology of $\mathcal{M}$ the following limit:

$$\lim_{t \to 0} \frac{\tilde{\mathcal{L}}^{(p)}(u + tv) - \tilde{\mathcal{L}}^{(p)}(u)}{t} = \langle \partial \tilde{\mathcal{L}}^{(p)}(u), v \rangle.$$

We define $\mathcal{L}^{(p)} : \mathcal{C}^{(p)} \times \mathcal{C}^{(p)} \to \mathcal{M}$ as

$$\mathcal{L}^{(p)}(u, v) = \langle \partial \tilde{\mathcal{L}}^{(p)}(u), v \rangle. \tag{2.2}$$

L6) The chain rules hold: if $u, v \in \mathcal{C}^{(p)}$ and $g \in C^1(\mathbb{R})$, with g' bounded on $\mathbb{R}$, then

$$g(u) : x \to g(u(x))$$

belongs to $\mathcal{C}^{(p)}$ and

$$\mathcal{L}^{(p)}(g(u), v) = |g'(u)|^{p-2} g'(u) \mathcal{L}^{(p)}(u, v),$$
$$\mathcal{L}^{(p)}(v, g(u)) = g'(u) \mathcal{L}^{(p)}(v, u).$$

Definition 2.1. *In the assumptions L1,...,L6, the measure $\mathcal{L}^{(p)}(u, v)$ in (2.2) will be called homogeneous p-Lagrangian.*

We denote by $\mathcal{D}^{(p)}$ the (abstract) completion of $\mathcal{C}^{(p)}$ with respect to the norm (2.1). Moreover, we assume that $\tilde{\mathcal{L}}^{(p)}$ is closable in $L^p(X, \mu)$, that is, $\mathcal{D}^{(p)}$ is injected in $L^p(X, \mu)$. We extend the homogeneous p-Lagrangian to $\mathcal{D}^{(p)}$ and we still denote it by $\mathcal{L}^{(p)}(u, v)$.

We now recall the definition of *metric fractals* (see [13]). We remark that the term "fractal" refers here to invariance under metric scaling instead of the usual meaning of "fractal" as invariance under self-similarities.

Definition 2.2. *A **metric fractal** is a quadruple $X \equiv (X, d, \mu, \mathcal{L}^{(p)})$ on a connected topological space X, with the following properties MF1,...,MF4:*

MF1)   *d is a quasi-distance on X (that is, d is a function on $X \times X$ with the properties of a usual distance, except for the triangle inequality satisfied in the form $d(x, y) \leq c_T(d(x, z) + d(z, y))$ with a fixed constant $c_T \geq 1$) and $X = (X, d)$ is a complete quasi-metric space;*

MF2)   *μ is a doubling measure on X, with fixed constants $\nu > 0$, $c_G > 0$ and $R_0 \in (0, \infty]$ (that is, it is a positive Borel measure μ supported on X, such that there exist constants $\nu > 0$, $c_G > 0$ and $R_0 \in (0, \infty]$, with*

$$0 < \mu(B_R) \leq c_G \left(\frac{R}{r}\right)^\nu \mu(B_r) < \infty \qquad (2.3)$$

for every $x \in X$, $B_R \equiv B(x, R) := \{y \in X : d(x, y) < R\} \subset\subset X$, $0 < r \leq R < R_0$;

MF3)   *$\mathcal{L}^{(p)}$ is a measure-valued homogeneous p-Lagrangian according the Definition 2.1;*

MF4)   *d, μ and $\mathcal{L}^{(p)}$ are related by the inequalities*

$$\frac{1}{\mu(B_R)} \int_{B_R} |u - u_{B_R}| d\mu \leq c_P R \left(\frac{1}{\mu(B_{qR})} \int_{B_{qR}} d\mathcal{L}^{(p)}(u, u)\right)^{\frac{1}{p}}, \qquad (2.4)$$

(where, for every ball B, $u_B = \mu(B)^{-1} \int_B u\, d\mu$), and

$$p - \mathrm{cap}(B_R, B_{2qR}) \leq c_C \frac{\mu(B_R)}{R^p}, \qquad (2.5)$$

for every $0 < R < R_0$ and every $B_R \subset B_{2qR} \subset\subset X$, with constants c_P, c_C, $q \geq 1$ independent of u, x and R.

Above,

$$p - \mathrm{cap}(B_R, B_{2qR})$$

$$:= \inf\left\{\int_X d\mathcal{L}^{(p)}(\Phi, \Phi) : \Phi \in \mathcal{C}^{(p)}, \Phi \geq 1 \text{ on } B_R, \ \mathrm{supp}\Phi \subset B_{2qR}\right\}.$$

By $\mathcal{D}^{(p)}_{loc}$ we denote the space of the functions $u \in L^p(X, \mu)$, such that for every relatively compact open subset A there exists a function $\tilde{u} \in \mathcal{D}^{(p)}$ such that $u = \tilde{u}$ on A.

The constants c_T, ν, c_G, c_P, c_C and q will be referred to, in the following, as the *structural constants* of X. We point out that, up to changing the structural constants without the homogeneous dimension ν, the structure of a metric fractal is *stable* under changes to equivalent quasi-distance and equivalent Lagrangians.

3. Harnack inequality

We recall some functional inequalities that give us estimates of the oscillation of finite energy functions (see [5]; in the linear case, see [14]). More precisely, starting from scaled Poincaré inequalities only, we have proved the following Morrey type estimates by using Riesz potentials techniques.

Theorem 3.1. *Let $\nu < p$. Let σ be such that $0 < c_T \sigma < 1$. Then, there exists a constant c_1, depending only on the structural constants of X, such that*

$$\sup_{B_{\sigma R}} v - \inf_{B_{\sigma R}} v \le c_1 (1 - \sigma)^{-\frac{\nu}{p}} \left(\frac{1}{\mu(B_R)} \int_{B_R} d\mathcal{L}^{(p)}(v, v) \right)^{\frac{1}{p}} R$$

for every $v \in \mathcal{D}_{loc}^{(p)}$ and every ball $B_R \subset\subset X$, $0 < (1 + c_T^2)R < R_0$.

Definition 3.2. *A local supersolution u in X is a function $u \in \mathcal{D}_{loc}^{(p)}$, such that*

$$\int_X d\mathcal{L}^{(p)}(u, \phi) \ge 0$$

for every non-negative $\phi \in \mathcal{C}^{(p)}$ with compact support in X.

By using the scaled capacity estimates, we obtain the following estimates of the energy of the positive local supersolutions.

Theorem 3.3. *Let $u \in \mathcal{D}_{loc}^{(p)}$ be a positive local supersolution in X. Then, there exists a constant c_2, depending only on the structural constants of X, such that*

$$\frac{1}{\mu(B_R)} \int_{B_R} d\mathcal{L}^{(p)}(\log u, \log u) \le c_2 R^{-p},$$

for every ball $B_R \subset B_{2qR} \subset\subset X$, $0 < R < R_0$.

From Theorem 3.1 and Theorem 3.3, we obtain the following Harnack type inequality for non-negative local supersolutions (see [7]).

Theorem 3.4. *Let X be a metric fractal, with $\nu < p$. Let $u \in \mathcal{D}_{loc}^{(p)}$ be a non-negative local supersolution in X. Then, there exists a constant c_3, depending only on the structural constants of X, such that*

$$\sup_{B_R} u \le c_3 \inf_{B_R} u$$

for every ball $B_{R/\sigma} \subset B_{2qR/\sigma} \subset\subset X$, $0 < 4c_T(1 + c_T^2)R < R_0$.

4. Examples

There are various classic and semi-classical examples of metric fractals (see [13] and [15]). In this section, we give examples of metric fractals with possibly singular p-Lagrangians for $p > 1$ (in the linear case, see [14] and [16]). We begin by recalling some notations and definitions. In the D-dimensional Euclidean space $\mathbb{R}^D$, $D \geq 1$, we consider the Euclidean distance $d_e(x, y) \equiv |x - y|$; let $B_e(x, r) := \{y \in \mathbb{R}^D : |x - y| < r\}$, $x \in \mathbb{R}^D$, $r > 0$, be the Euclidean balls. We suppose that $\Psi = \{\psi_1, \ldots, \psi_N\}$ is a given set of contractive similitudes $\psi_i : \mathbb{R}^D \to \mathbb{R}^D$, with contraction factors $\alpha_i^{-1} < 1$, that is,

$$|\psi_i(x) - \psi_i(y)| = \alpha_i^{-1}|x - y|$$

for every $x, y \in \mathbb{R}^D$, $i = 1, \ldots, N$. In [9], it is proved that there exists a unique closed bounded set K – the so-called *self-similar fractal* – which is invariant under Ψ, that is,

$$K = \bigcup_{i=1}^{N} \psi_i(K) \,. \tag{4.1}$$

Moreover, there exists a unique Borel regular measure μ in $\mathbb{R}^D$, with $\operatorname{supp}\mu = K$ and unit total mass, which is invariant under Ψ, that is, μ satisfies

$$\mu = \sum_{i=1}^{N} \alpha_i^{-d_f} \psi_{i\#}\mu \tag{4.2}$$

where $\psi_{i\#}\mu(\cdot) := \mu(\psi_i^{-1}(\cdot))$, $i = 1, \ldots, N$ and the real number d_f is uniquely determined by the relation

$$\sum_{i=1}^{N} \alpha_i^{-d_f} = 1.$$

Moreover, we suppose that the family $\Psi = \{\psi_1, \ldots, \psi_N\}$ satisfies the following *open set condition*: there exists a bounded open set $U \subset \mathbb{R}^D$, such that

$$\bigcup_{i=1}^{N} \psi_i(U) \subset U \,, \quad \text{with} \quad \psi_i(U) \cap \psi_j(U) = \emptyset \quad \text{if } i \neq j \,. \tag{4.3}$$

Then, the invariant measure μ coincides with the restriction to K of the d_f-dimensional Hausdorff measure of $\mathbb{R}^D$, $H^{d_f}\lfloor K$, normalized:

$$\mu = (H^{d_f}(K))^{-1} H^{d_f}\lfloor K.$$

We will use the notations $\psi_{i_1 \ldots i_n} := \psi_{i_1} \circ \psi_{i_2} \circ \cdots \circ \psi_{i_n}$, $A_{i_1 \ldots i_n} := \psi_{i_1 \ldots i_n}(A)$ for arbitrary n-tuples of indices $i_1, \ldots, i_n \in \{1, \ldots, N\}$ and arbitrary $A \subset K$.

We define the boundary Γ of K as

$$\Gamma = \bigcup_{i \neq j} \psi_i^{-1}(K_i \cap K_j).$$

In the following, we shall assume that Γ is a finite set,

$$\#\Gamma < \infty \tag{4.4}$$

and, for every $n \geqslant 1$ and every $i_1, \ldots, i_n \neq j_1, \ldots, j_n$,

$$K_{i_1 \ldots i_n} \cap K_{j_1 \ldots j_n} = \Gamma_{i_1 \ldots i_n} \cap \Gamma_{j_1 \ldots j_n}. \tag{4.5}$$

We now recall the definition of *variational fractal* (see [12] for $p = 2$).

Definition 4.1. *A variational fractal is a triple* $K \equiv (K, \mu, \mathcal{L}^{(p)})$ *where*

VF1) K *is the invariant set of a given family* $\Psi = \{\psi_1, \ldots, \psi_N\}$ *satisfying* (4.1), (4.3), (4.4) *and* (4.5);

VF2) μ *is the invariant measure* (4.2) *on* K;

VF3) $\mathcal{L}^{(p)}$ *is a nonlinear p-homogeneous Lagrangian with domain* $\mathcal{D}^{(p)}$ *in* $L^p(K, \mu)$ *in the sense of Definition 2.1; moreover, we suppose that for every* $u \in \mathcal{D}^{(p)}$ *and for every* $\varphi \in C(K)$, *we have*

$$\int_K \varphi d\mathcal{L}^{(p)}(u, u) = \sum_{i=1}^{N} \rho_i^{(p)} \int_K \varphi \circ \psi_i d\mathcal{L}^{(p)}(u \circ \psi_i, u \circ \psi_i) \tag{4.6}$$

with the real constants $\rho_i^{(p)} > 0$, $i = 1, \ldots, N$, *satisfying* $\rho_i^{(p)} = \mu(K_i)^\tau$, $i = 1, \ldots, N$, *for some real constant* $\tau < 1$ *independent of* $i = 1, \ldots, N$.

Given a variational fractal $K \equiv (K, \mu, \mathcal{L}^{(p)})$, we consider quasi-distances d on K with Euclidean scaling

$$d(x, y) = |x - y|^\delta , \quad x, y \in K \tag{4.7}$$

indexed by a real parameter $\delta > 0$. We choose δ by requiring d^p to obey on K the same scaling as $\mathcal{L}^{(p)}$ itself: more precisely, in [6], it has been proved that there exists one and only one constant $\delta > 0$, such that, the following relation holds

$$d^p(x, y) = \sum_{i=1}^{N} \rho_i^{(p)} d^p(\psi_i(x), \psi_i(y)) , \tag{4.8}$$

for every $x, y \in K$ and such δ is uniquely determined by the identity

$$\sum_{i=1}^{N} \rho_i^{(p)} \alpha_i^{-p\delta} = 1 \tag{4.9}$$

and is given by

$$\delta = d_f(1 - \tau)/p. \tag{4.10}$$

We denote the quasi-balls associated with d by $B(x, r)$, that is, $B(x, r) := \{y \in K : d(x, y) < r\}$, $x \in K$, $r > 0$. For every $x \in K$ and every $r > 0$, we have $B(x, r) = B_e(x, r^{\frac{1}{\delta}}) \cap K$.

In [6], we have proved that if the structure enjoys self-similar invariance and a global Poincaré inequality holds, then the *scaled Poincaré inequalities* on the homogeneous balls in MF4 hold. Moreover, in [7], we have proved that if a global estimate of capacity holds, then also the capacity inequalities on the homogeneous balls in MF4 hold.

Theorem 4.2. *Let $K \equiv (K, \mu, \mathcal{L}^{(p)})$ be a variational fractal endowed with its intrinsic metric d. Moreover, we suppose that there exists two constants c_Π and c_Γ such that*

$$\int_K |u - u(z)|^p d\mu \leq c_\Pi \int_{K-\Gamma} d\mathcal{L}^{(p)}(u, u) \tag{4.11}$$

for every $u \in \mathcal{D}^{(p)}$ and every $z \in \Gamma$ and

$$p - cap(\Gamma_1, \Gamma_2) \leqslant c_\Gamma \tag{4.12}$$

for all $\Gamma_1 \neq \emptyset$ and $\Gamma_2 \neq \emptyset$ such that $\Gamma_1 \cup \Gamma_2 = \Gamma$.
    Then, $(K, d, \mu, \mathcal{L}^{(p)})$ is a metric fractal according to the Definition 2.3.

Examples of variational fractals for which the global estimates (4.11) and (4.12) hold are provided by the Koch curve type fractals (see [4], [6], [7]) and by the Sierpinski gasket type fractals (see [8], [17] and [7]). By applying Theorem 4.2, these structures are metric fractals; moreover, by using the estimates of the renormalization factors $\rho_i^{(p)}$ (see [4], [8]), their homogeneous dimension ν is less then p and, in particular, Theorem 3.4 holds.

References

[1] Biroli M., Mosco U., *A Saint-Venant principle for Dirichlet forms on discontinuous media*, Ann. Mat. Pura Appl. IV 169 (1992), 125–181.

[2] Biroli M., Vernole P., *Strongly Local Nonlinear Dirichlet functionals and forms*, Quad. 585-P Dip. Mat. Politecnico di Milano, (2004).

[3] Biroli M, Vernole P., *Harnack inequality for nonlinear Dirichlet forms*, Quad. 586-P Dip. Mat. Politecnico di Milano, (2004).

[4] Capitanelli R., *Nonlinear energy forms on certain fractal curves*, Journal Nonlinear Convex Anal. 3 (2002), no.1, 67–80.

[5] Capitanelli R., *Functional inequalities for measure-valued Lagrangians on homogeneous spaces*, Adv. Math. Sci. Appl. 13 (2003), no. 1, 301–313.

[6] Capitanelli R., *Homogeneous p-Lagrangians and self-similarity*, Rend. Accad. Naz. Sci. XL Mem. Mat. Appl. 27 (2003), 215–235.

[7] Capitanelli R., *Harnack inequality for p-Laplacians associated to homogeneous p-Lagrangians*, preprint.

[8] Herman P.E., Peirone R., Strichartz R.S., *p-Energy and p-harmonic functions on Sierpinski gasket type fractals*, Potential Anal. 20 (2004), no. 2, 125–148.

[9] Hutchinson J.E., *Fractal and Self Similarity*, Indiana Univ. Math. J. 30 (1981), no. 5, 713–747.

[10] Kusuoka S., *Dirichlet forms on fractals and products of random matrices*, Publ. RIMS Kyoto Univ., 25, 659–680 (1989).

[11] Malý J., Mosco U., *Remarks on measure-valued Lagrangians on homogeneous spaces*, Ricerche Mat. 48 (1999), suppl., 217–231.

[12] Mosco U., *Variational fractals*, Ann. Scuola Norm. Sup. Pisa Cl. Sci.(4) 25 (1997), no. 3-4, 683–712 (1998).

[13] Mosco U., *Distance, mass and energy in analysis*, 60th anniversary of the Instituto de Matimático "Beppo Levi" (Spanish) (Rosario, 2000), 51–73, Cuadern. Inst. Mat. Beppo Levi, 30, Univ. Nac. Rosario, Rosario, 2001.

[14] Mosco U., *Harnack inequalities on recurrent metric fractals*, Dedicated to the 80th anniversary of Academician EvgeniĭFrolovich Mishchenko (Russian) (Suzdal, 2000). Tr. Mat. Inst. Steklova 236 (2002), Differ. Uravn. i Din. Sist., 503–508.

[15] Mosco U., *Energy functionals on certain fractal structures*. Special issue on optimization (Montpellier, 2000). J. Convex Anal. 9 (2002), no. 2, 581–600.

[16] Mosco U., *Harnack inequalities on scale irregular Sierpinski gaskets*, Nonlinear problems in mathematical physics and related topics, II, 305–328, Int. Math. Ser. (N. Y.), 2, Kluwer/Plenum, New York, 2002.

[17] Strichartz R.S., Wong C., *The p-Laplacian on the Sierpinski gasket*, Nonlinearity 17 (2004), no. 2, 595–616.

Raffaela Capitanelli
Dipartimento di Metodi e Modelli
Matematici per le Scienze Applicate
Università degli Studi di Roma "La Sapienza"
Via A. Scarpa 16
I-00161 Roma, Italy
e-mail: `raffaela.capitanelli@uniroma1.it`

Progress in Nonlinear Differential Equations
and Their Applications, Vol. 63, 127–134

Wave Propagation in Discrete Media

Ana Carpio

Abstract. We study wave front propagation in spatially discrete reaction-diffusion equations with cubic sources. Depending on the symmetry of the source, such wave fronts appear to be pinned or to glide at a certain speed. We describe the transition of travelling waves to stationary solutions and give conditions for front pinning. The nature of these depinning transitions seems to be preserved in higher dimensions. Finally, we discuss the different behavior observed when inertial terms are included in the model.

1. Introduction

Many systems in Nature are formed by a large number of small items (cells, atoms, wells, layers...). Sometimes such systems may be ideally described as a continuum. Then, their behavior can be understood by analyzing partial differential equations. Often, continuum limits fail to account for physical reality. Examples abound: atoms adsorbed on a periodic substrate, motion of dislocations in crystals, propagation of cracks in a brittle material, microscopic theories of friction between solid bodies, propagation of nerve impulses along myelinated fibers, pulse propagation through cardiac cells, calcium release waves in living cells, sliding of charge density waves, superconductor Josephson array junctions, or weakly coupled semiconductor superlattices. In all these contexts, the appropriate models must take into account the spatially discrete structure of the system.

We focus here on spatially discrete reaction-diffusion equations of the form:

$$\frac{du_n}{dt}(t) = d(u_{n+1}(t) - 2u_n(t) + u_{n-1}(t)) + g(u_n(t)) + F, \quad n \in \mathbb{N}, t > 0. \qquad (1)$$

The nonlinearity $g(u)$ is a smooth cubic function and has three zeros, $U_1(0) < U_2(0) < U_3(0)$. We assume that g is odd about its middle zero. $F > 0$ is a control parameter that alters the symmetry of the source. For an interval $[0, F_{max})$ of values of F, $g(u) + F$ has still three zeros, $U_1(F) < U_2(F) < U_3(F)$, with $g'(U_i(F)) < 0$ for $i = 1, 3$ and $g'(U_2(F)) > 0$. The first and the third zeros are stable whereas the second one is unstable. Typical choices are $g(u) = -\sin(u)$

or $g(u) = u(1 - u^2)$. The first choice corresponds to the overdamped Frenkel-Kontorova model for dislocation motion in atomic chains. Then, $u_n(t)$ is the displacement of the nth atom of the chain. The second choice yields the Nagumo model for propagation of nerve impulses along a myelinated nerve fiber. In this case, $u_n(t)$ describes the time evolution of the membrane voltage at the nth node of the fiber. The parameter $d > 0$ measures the strength of the coupling between either atoms or nodes.

The solutions of physical interest are *wave front solutions* joining the two stable constant states. When the coupling is weak (d small) wave fronts may fail to propagate, depending on the degree of symmetry of the source. The first rigorous proof of this fact is probably due to Keener [10]. Propagation failure is a well known feature in neurology: nerve impulses fail to propagate when the myelin sheath of the nerve is damaged, which results in small values for d. This is a reason to respect the discrete structure of the nerve. In the continuum limit:

$$u_t(x, t) = d\, u_{xx}(x, t) + g(u(x, t)) + F, \quad x \in \mathbb{R}, t > 0 \tag{2}$$

and the parameter d can be removed by scaling the continuous variable x. Therefore, it has no effect on propagation failure, which occurs only at $F = 0$ for any $d > 0$ [1]. Equation (2) fails to account for propagation failure at small d.

In this paper we study pinning and propagation of wave fronts in the spatially discrete reaction-diffusion model (1). We are interested in two types of solutions:

- stationary wave fronts u_n which decrease monotonically from $U_3(F)$ to $U_1(F)$ as n increases from $-\infty$ to ∞, as shown in Figure 1(a).
- travelling wave fronts $u_n(t) = u(n - ct)$ with speed $c > 0$ and a smooth profile $u(x)$ decreasing from $U_3(F)$ to $U_1(F)$ as x increases from $-\infty$ to ∞, as illustrated in Figures 1(b)–(c).

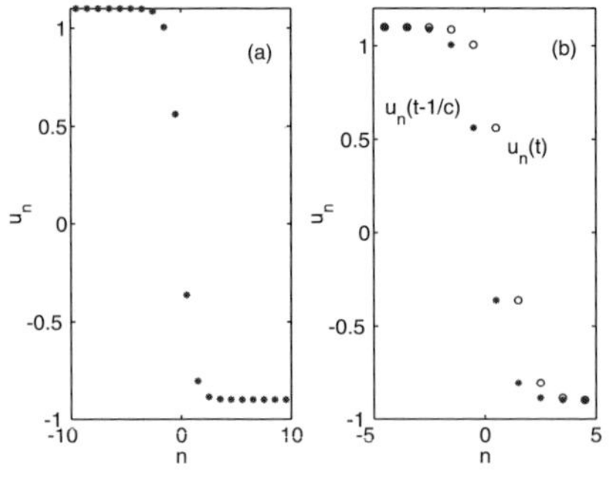
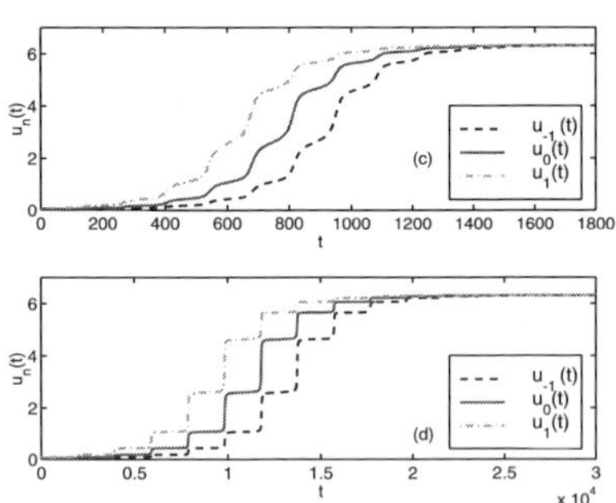

FIGURE 1. (a) Static wave front solution; (b) Travelling wave front solution: $u_n(t) = u_{n-1}(t - 1/c)$; (c)–(d) Generation of steps in the trajectories $u_n(t)$ as F approaches $F_s(D)$. The wave profiles are defined by $u(x) = u_0(\frac{x}{-c})$.

Section 2 deals with stationary wave fronts. Section 3 studies travelling wave fronts. In Section 4, we analyze the depinning transition that takes place in the system when fronts cease to be pinned and begin to move. Finally, Section 5 comments on extensions to models that include inertial terms.

2. Stationary wave fronts

In this section, we address the existence of stationary wave front solutions u_n of (1):

$$d(u_{n+1} - 2u_n + u_{n-1}) + g(u_n) + F = 0, \quad n \in \mathbb{N} \tag{3}$$

$$\lim_{n \to -\infty} u_n = U_3(F), \quad \lim_{n \to \infty} u_n = U_1(F) \tag{4}$$

under the hypotheses stated in the introduction. For the continuous counterpart, $d\, u_{xx} + g(u) + F = 0$, a simple phase plane argument shows the nonexistence of solutions joining $U_3(F)$ and $U_1(F)$ (heteroclinic orbits) unless $F = 0$, for any $d > 0$. In contrast, solutions of (3)–(4) can be found when the source is nearly symmetric [2], that is, for F small. The degree of symmetry required depends on d:

Theorem 1. *For any $d > 0$, there exists a threshold $F_s(d) \geq 0$ such that:*

1. *If $0 \leq F \leq F_s(d)$, there exists at least one monotone solution u_n of (3)–(4).*
2. *$F_s(d) > 0$ if d is small enough and $F_s(d) \to 0$ when $d \to \infty$.*
3. *Stationary wave fronts u_n may have the form $u_n = u(n)$ where $u(x)$ is a continuous solution of:*

$$d(u(x + 1) - 2u(x) + u(x - 1)) + g(u(x)) + F = 0, \quad x \in \mathbb{R} \tag{5}$$

$$\lim_{x \to -\infty} u(x) = U_3(F), \quad \lim_{x \to \infty} u(x) = U_1(F), \tag{6}$$

only when $F = 0$.

Existence of monotone wave fronts can be proved by a shooting technique when $F = 0$. In this case, at least two different profiles $u_n^{(1)}$ and $u_n^{(2)}$ are found, up to translations. One of them satisfies $u_1^{(1)} - U_2(0) = U_2(0) - u_0^{(1)}$ whereas the other one fulfills $u_0^{(2)} = U_2(0)$. When F and d are small enough, we can construct stationary upper and lower solutions for (1), which block propagation due to maximum principles. Stationary wave fronts are obtained as long time limits. Nonexistence of continuous profiles when $F \neq 0$ follows upon integration of (5).

Theorem 1 raises several questions: How many stationary solutions exist? Are they stable? Can $F_s(d) = 0$ for large d?

The structure of the set of sequences solving (3) becomes rather involved if we drop monotonicity [10]. With monotonicity assumptions we expect to find at least two different profiles (up to translations) for small F. Numerical experiments support this conjecture. Using the exponential convergence of the tails and a numerical continuation procedure in finite chains, we obtain two branches of wave fronts [13]. These branches start from the two monotone solutions found at

$F = 0$. A linear stability analysis shows that the branch generated by $u_n^{(1)}$ is stable whereas the branch generated by $u_n^{(2)}$ is unstable.

This picture changes dramatically when continuous solutions of (5)–(6) exist for $F = 0$. Then, we can construct a one parameter family of stationary monotone fronts: $u_n = u(n + a)$, $0 \leq a < 1$. Explicit examples are found by an inverse method. Let us choose an odd profile $u(x)$ decreasing from a constant value to a different one and tending to those constant values exponentially. For instance, we may select $u(x) = \tanh(-x)$. Then, $g(x) = -d(u(x + 1) - 2u(x) + u(x - 1))$ is a cubic odd source and $u(x)$ solves (5)–(6) with $F = 0$.

Whenever (5)–(6) has smooth solutions with $F = 0$, the threshold $F_s(d)$ vanishes. Finding a characterization of the sources g for which this is possible is an open problem. We conjecture that $F_s(d) > 0$ generically, though it may become zero in pathological cases. In [8, 11] asymptotic techniques are used to obtain exponentially small predictions of $F_s(d)$ as $d \to \infty$ for some sources. Numerical tests carried out for $g(u) = -\sin(u)$ and $g(u) = u(1-u^2)$ support those asymptotic predictions and suggest that $F_s(d) > 0$ up to d fairly large. Nevertheless, this question cannot be answered numerically. A rigorous proof of the asymptotic predictions is needed.

3. Travelling wave fronts

Maximum principles [10, 2] imply that wave fronts cannot propagate in (1) as long as stationary wave fronts exist. Travelling waves are expected when the source becomes asymmetric enough, depending on d [2]:

Theorem 2. *For any $d > 0$, there exists a threshold $F_d(d) \in [F_s(d), F_{\max})$ such that:*

1. *If $F_d(d) < F < F_{\max}$, travelling wave front solutions $u_n(t) = u(n - ct)$ of (1) with $c > 0$ exist. The wave profile $u(x)$ solves an eigenvalue problem for a differential-difference equation:*

$$-cu_x(x) = d(u(x + 1) - 2u(x) + u(x - 1)) + g(u(x)) + F, \quad x \in \mathbb{R} \tag{7}$$

$$\lim_{x \to -\infty} u(x) = U_3(F), \quad \lim_{x \to \infty} u(x) = U_1(F) \tag{8}$$

   and is unique up to translations.
2. *$F_d(d) = F_s(d)$ and the speed $c \to 0$ as $F \to F_s(d)^+$.*

Existence of travelling waves can be proved using fixed point and homotopy techniques [18] or continuation methods [2, 12]. Theorem 2 relies on a perturbation approach. First, we find families of sources g^* for which solutions (c^*, u^*) of the eigenvalue problem (7)–(8) exist. This can be done by choosing odd profiles u^* that decrease from a constant value to another and tend exponentially to such constants at infinity. Given $c^* > 0$, we set $F^* = 0$ and define:

$$g^*(x) = -c^* u_x^*(x) - d(u^*(x + 1) - 2u^*(x) + u^*(x - 1)). \tag{9}$$

Then, we need a perturbation and comparison result: If $g(u) + F < g^*(u)$ and the C^1 norm of the difference is small, a solution u of (7)–(8) still exists with speed $c \geq c^*$. Finally, we prove that $g(u) + F < g^*(u)$ holds when F is large enough.

4. Depinning transitions

Theorems 1 and 2 provide a qualitative picture of wave front propagation and pinning in spatially discrete reaction-diffusion equations: a threshold $F_s(d)$ is found such that smooth travelling waves propagate above threshold whereas discrete stationary fronts exist below it. We want to understand the mathematical nature of the transition taking place at $F_s(d) > 0$, the so-called depinning transition [6, 4].

Let us examine the evolution of the different solutions as we approach $F_s(d)$ from above and from below. When $F \to F_s(d)^+$, the profiles of the travelling waves develop a sequence of steps, see Figure 1(c)–(d). At $F = F_s(d)$, the profiles become discontinuous and fail to propagate. When $F \to F(d)^-$, the smallest eigenvalue λ_1 of the linearized problem about the stable stationary wave front s_n^F tends to zero. At $F = F_s(d)$, $\lambda_1 = 0$ and the stable and unstable branches of stationary wave fronts collide. The normal form of the bifurcation at $F = F_s(d)$ reveals a saddle-node bifurcation:

$$\phi' = A(F - F_s(d)) + B\phi^2, \quad A > 0, B > 0. \tag{10}$$

The coefficients A and B have explicit expressions involving a normalized positive eigenfunction associated to the zero eigenvalue, see [4]. To compute the normal form, we may replace the infinite system with a finite system. Its size depends on how fast the stationary wave fronts reach their constant values. When d is small, we are left with a one-dimensional system.

We conjecture that the depinning transition is a global bifurcation in the system, locally of saddle-node type. From a practical point of view, our understanding of the depinning transition yields an accurate prediction of the speed and shape of the wave fronts near the threshold. Integrating the normal form we find solutions blowing up in finite time:

$$\phi(t) \sim \sqrt{\frac{A(F - F_s(d))}{B}} \, \tan\left(\sqrt{AB\left(F - F_s(d)\right)} \, (t - t_0)\right). \tag{11}$$

This solution is very small most of the time but it blows up when the argument of the tangent function approaches $\pm\pi/2$. Let us see how this information can be used to understand the shape of the trajectories depicted in Figure 1(d). First, notice that at $F = F_s(d)$ we have a one parameter family of shifted stationary wave fronts: s_{n+k}^F. As long as (11) is small, the trajectories $u_n(t)$ of the travelling wave remain near one of the stationary solutions. When (11) blows up, the pattern advances and gets trapped about the next stationary solution. This process is iterated and explains the steps in the trajectories. Each step in a trajectory corresponds to waiting period near a stationary solution. The motion of these waves is 'saltatory': the pattern does not move for a certain time and then jumps abruptly. The speed of the wave front is the reciprocal of the time spent in one step of the trajectory, which is computed using the blow up time of the normal form:

$$c \sim \frac{\sqrt{A B (F - F_s(d))}}{\pi}. \tag{12}$$

Far from the depinning threshold, the wave front profiles become smooth. For large speeds, these waves can be approximated by continuous travelling wave solutions of a nonlinear heat equation, as explained in [10], see also [9]. When $F_s(d) = 0$, no depinning transition is observed. As in the continuous heat equation (2), we just have a family of smooth travelling waves whose speeds change sign at $F = 0$.

This description of the depinning transition is quite robust and still holds in two-dimensional models for dislocation motion [3] in crystalline solids:

$$u'_{ij} = u_{i+1,j} - 2u_{ij} + u_{i-1,j} + A[\sin(u_{i,j+1} - u_{ij}) + \sin(u_{i,j-1} - u_{i,j+1})]. \tag{13}$$

The control parameter F enters these models through the boundary conditions at infinity. Below a threshold $F_s(d)$, stationary waves corresponding to static dislocations exist. Above threshold, travelling waves $u_{ij}(t) = u(i - ct, j)$ can be numerically constructed, corresponding to moving dislocations. The profiles are shown in Figure 2(a).

5. Models with inertia

Atomic models for propagation of cracks in brittle materials or dislocations in crystals lead precisely to the study of wave fronts in discrete wave equations:

$$m\frac{d^2 u_n}{dt^2} + \alpha\frac{du_n}{dt} = d(u_{n+1} - 2u_n + u_{n-1}) + g(u_n) + F, \ m > 0, \ \alpha > 0. \tag{14}$$

Pinned cracks or dislocations are identified with stationary wave fronts. Moving cracks or dislocations are identified with travelling wave fronts. For a discontinuous piecewise linear source, $g(u) = -u - 1$, $u < 0$, $g(u) = -u + 1$, $u \geq 0$, wave front solutions can be explicitly constructed [15, 5]. This construction points out new features, compared to the overdamped case $m = 0$. First, the two thresholds $F_s(d)$ and $F_d(d)$ for existence of stationary and travelling wave fronts differ: $F_d(d) < F_s(d)$. These thresholds $F_s(d)$ and $F_d(d)$ are identified with the static and dynamic stresses for defect motion. Moreover, several families of travelling waves may coexist, with different speeds and profiles. The wave profiles become oscillatory as α decreases. However, only one particular class of oscillatory waves appears to be stable [5]: fast waves with a monotone leading edge and an oscillatory wake. These waves persist when we replace piecewise linear sources with smooth sources. Figure 2(b) shows an oscillatory travelling wave for a conservative model, constructed by a numerical continuation procedure. A rigorous existence proof for these waves is still missing. Variational techniques have been successful to prove existence of solitary waves [7] or periodic wave trains [17]. For oscillatory wave fronts with non decaying tails, it is not clear how to define a finite energy functional.

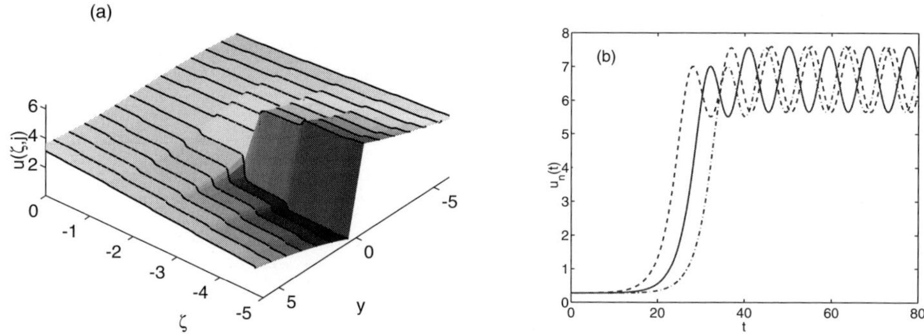

FIGURE 2. (a) Wave profiles in two-dimensional models; (b) Trajectories in models with inertia: $u_1(t)$ (dashed), $u_0(t)$ (solid), $u_{-1}(t)$ (dotted-dashed).

Again, comparison with continuum wave equations produces stricking results. A simple phase plane argument shows that:

$$m\, u_{tt}(x,t) = d\, u_{xx}(x,t) + g(u(x,t)) + F, \quad n \in \mathbb{N}, t > 0 \tag{15}$$

has no travelling wave fronts unless $F = 0$. The situation we expect for (14) is completely different: pinned waves when F is small and travelling waves when F is large. Nevertheless, higher order continuum limits recover the correct discrete qualitative picture [14, 16].

References

[1] D.G. Aronson and H.F. Weinberger, Nonlinear diffusion in population genetics, combustion and nerve pulse propagation in PDE and related topics. Lect. N. Math. **446** (1975), 5–49. Springer, Berlin.

[2] A. Carpio, S.J. Chapman, S. Hastings, J.B. McLeod, Wave solutions for a discrete reaction-diffusion equation, Eur. J. Appl. Math. **11** (2000), 399–412.

[3] A. Carpio, L.L. Bonilla, Edge dislocations in crystal structures considered as travelling waves of discrete models, Phys. Rev. Lett., **90** (2003), 135502; **91** (2003), 029901.

[4] A. Carpio, L.L. Bonilla, Depinning transitions in spatially discrete reaction-diffusion equations, SIAM J. Appl. Math., **63** (2003), 1056–1082.

[5] A. Carpio, Nonlinear stability of oscillatory wave fronts in chains of coupled oscillators, Phys. Rev. E, **69** (2004), 046601.

[6] G. Fáth, Propagation failure of travelling waves in a discrete bistable medium, Physica D **116** (1998), 176–190.

[7] G. Friesecke, J. Wattis, Existence theorem for solitary waves on lattices, Commun. Math. Phys., **161** (1994), 391–418.

[8] V. Hakim and K. Mallick, Exponentially small splitting of separatrices, matching in the complex plane and Borel summation, Nonlinearity, **6** (1993), 57–70.

[9] S. Heinze, G. Papanicolau, A. Stevens, Variational principles for propagation speeds in inhomogeneous media, SIAM J. Appl. Math. **62** (2001), 129–150.

[10] J. P. Keener, Propagation and its failure in coupled systems of discrete excitable cells, SIAM J. Appl. Math. **47** (1987), 556–572.

[11] J.R. King and S.J. Chapman, Asymptotics beyond all orders and Stokes lines in nonlinear differential-difference equations, Eur. J. Appl. Math. **12** (2001), 433–463.

[12] J. Mallet-Paret, The global structure of travelling waves in spatially discrete dynamical systems. J. Dyn. Diff. Eq. **11** (1999), 49–127.

[13] I. Mitkov, K. Kladko and J.E. Pearson, Tunable pinning of bursting waves in extended systems with discrete sources, Phys. Rev. Lett. **81** (1998), 5453–5456.

[14] P. Rosenau, Hamiltonian dynamics of dense chains and lattices or how to correct the continuum, Phys. Lett. A, **311** (2003), 39–52.

[15] L.I. Slepyan, Dynamics of a crack in a lattice, Sov. Phys. dokl. **26** (1981), 538-540 [dokl. Akad. Nauk SSSR **258** (1981), 561–564].

[16] O. Kresse, L. Truskinovsky, Mobility of lattice defects: discrete and continuum approaches, J. Mech. Phys. Sol., **51** (2003), 1305–1332.

[17] A.M. Filip and S. Venakides, Existence and modulation of travelling waves in particle chains, Comm. Pure Appl. Math. **52** (1999), 693–735.

[18] B. Zinner, Existence of travelling wave front solutions for the discrete Nagumo equation, J. Diff. Eqs. **96** (1992), 1–27.

Ana Carpio
Departamento de Matemática Aplicada
Universidad Complutense de Madrid
E-28040 Madrid, Spain

Progress in Nonlinear Differential Equations
and Their Applications, Vol. 63, 135–138

A Solution of the Heat Equation with a Continuum of Decay Rates

Thierry Cazenave, Flávio Dickstein and Fred B. Weissler

Abstract. In this paper, we prove the existence of a solution of the heat equation on $\mathbb{R}^N$ which decays at different rates along different time sequences going to infinity. In fact, all decay rates $t^{-\frac{\sigma}{2}}$ with $0 < \sigma < N$ are realized by this solution.

Mathematics Subject Classification (2000). 35K05, 35B40.

Keywords. heat equation, asymptotic behavior, decay rate.

In the study of the long-time asymptotic behavior of global solutions of evolution equations, the first step is often to establish the decay rate of the solution with respect to a suitable norm. Afterwards, one hopes to study the finer structure of the asymptotic behavior using a dilation adapted to the decay rate.

It has already been shown [1, 4] that solutions with a fixed temporal decay rate do not always have a definite spatial asymptotic form. In these papers, solutions are constructed which have different asymptotic limits (with respect to the same rescaling) along different time sequences $t_n \to \infty$.

The purpose of this note is to point out that even the search for an appropriate time decay rate may sometimes fail. It turns out that already for the linear heat equation on $\mathbb{R}^N$ there exist solutions with multiple decay rates.

Theorem 1. *There exists* $u \in C_0(\mathbb{R}^N) \cap C^\infty(\mathbb{R}^N)$, $u \geq 0$ *such that for all* $0 < \sigma < N$ *and all* $0 \leq c < \infty$, *there exists a sequence* $t_n \to \infty$ *such that* $t_n^{\frac{\sigma}{2}} \|e^{t_n \Delta} u\|_{L^\infty} \to c$ *as* $n \to \infty$.

Before giving the proof of Theorem 1, we make the following observations.

Remark 2. The range of σ is optimal.

(i) Since $u \in C_0(\mathbb{R}^N)$, $\|e^{t\Delta} u\|_{L^\infty} \to 0$ as $t \to \infty$ (i.e., $\sigma = 0$ is not achieved). This follows from the density of $L^1(\mathbb{R}^N) \cap C_0(\mathbb{R}^N)$ in $C_0(\mathbb{R}^N)$.

(ii) $t^{\frac{N}{2}} \|e^{t\Delta} u\|_{L^\infty} \to \infty$ as $t \to \infty$ (i.e., $\sigma = N$ is not achieved). Indeed, since $u \geq 0$, if $t_n^{\frac{N}{2}} \|e^{t_n \Delta} u\|_{L^\infty}$ is bounded for some sequence $t_n \to \infty$, then $u \in L^1(\mathbb{R}^N)$.

On the other hand, we know that $u \notin L^1(\mathbb{R}^N)$, because otherwise we would have $t^{\frac{\sigma}{2}} \|e^{t\Delta} u\|_{L^\infty} \to 0$ as $t \to \infty$ for all $\sigma < N$.

Remark 3. For each decay rate, there exists an asymptotic limit to the solution rescaled in the corresponding variables. More precisely, the proof will show (see (13), (14) and (16)) that for all $0 < \sigma < N$ and all $0 < c < \infty$, there exists a sequence $\lambda_n \to \infty$ such that $\lambda_n^\sigma u(\lambda_n \cdot) \to c\delta_0$ in $\mathcal{S}'(\mathbb{R}^N)$ as $n \to \infty$, where δ_0 denotes the Dirac mass at 0. Consequently, setting $t_n = \sqrt{\lambda_n}$ and $\varphi(x) = (4\pi)^{-\frac{N}{2}} e^{-\frac{|x|^2}{4}}$ (so that $\varphi = e^\Delta \delta_0$), we see that $t_n^{\frac{\sigma}{2}} [e^{t_n \Delta} u](\cdot \sqrt{t_n}) \to c\varphi$ in $C_0(\mathbb{R}^N)$ as $n \to \infty$. (See (17).)

Remark 4. Since $u \geq 0$, this asymptotic behavior cannot be the result of cancellations.

Remark 5. In articles [2, 3], we combine the ideas of the present paper with the ideas in [1]. More precisely, we construct an initial value u for which the solution of the heat equation has a continuum of decay rates. Furthermore, there exists a dense set D of decay rates such that if $\sigma \in D$ and $\varphi \in C_0(\mathbb{R}^N)$, there exists a sequence $t_n \to \infty$ such that $t_n^{\frac{\sigma}{2}} [e^{t_n \Delta} u](\cdot \sqrt{t_n}) \to \varphi$ in $C_0(\mathbb{R}^N)$ as $n \to \infty$.

Proof of Theorem 1. Consider a function $\theta \in C_c^\infty(\mathbb{R}^N)$ such that

$$\begin{cases} \theta \geq 0, \\ \|\theta\|_{L^1} = 1, \\ \operatorname{supp} \theta \subset \{1/2 < |x| < 1\}, \end{cases} \tag{1}$$

and set

$$M = \|\theta\|_{L^\infty}. \tag{2}$$

Let the sequence $(a_j)_{j \geq 1}$ be defined by

$$\begin{cases} a_1 = e^e, \\ a_{j+1} = \exp(\exp a_j) \quad j \geq 1, \end{cases} \tag{3}$$

and set

$$f_j = (\log a_j)^{-1} \xrightarrow[j \to \infty]{} 0. \tag{4}$$

We note that by (3),

$$a_{j+1} > 2a_j \quad j \geq 1. \tag{5}$$

Let $u \geq 0$ be defined by

$$u(x) = \sum_{j=1}^{\infty} f_j \theta(x/a_j). \tag{6}$$

It follows from (1) and (5) that all the terms in the sum (6) have disjoint support, so that $u \in C^\infty(\mathbb{R}^N)$. Moreover, it follows from (4) that $u(x) \to 0$ as $|x| \to \infty$, so that $u \in C_0(\mathbb{R}^N)$.

We now fix $c > 0$ and $0 < \sigma < N$ and define the dilation operator $D_\lambda^\sigma u(x) = \lambda^\sigma u(\lambda x)$ for $\lambda > 0$. Given a sequence $\lambda_n \uparrow \infty$, we write $D_{\lambda_n}^\sigma u = u^n + v^n + w^n$ for $n \geq 2$, where

$$\begin{cases} u^n(x) = \displaystyle\sum_{j=1}^{n-1} \lambda_n^\sigma f_j \theta(\lambda_n x / a_j), \\[2mm] v^n(x) = \lambda_n^\sigma f_n \theta(\lambda_n x / a_n), \\[2mm] w^n(x) = \displaystyle\sum_{j=n+1}^{\infty} \lambda_n^\sigma f_j \theta(\lambda_n x / a_j). \end{cases}$$

Since $f_j \leq f_1$, it follows from (1)–(2) that

$$u^n(x) \leq M f_1 \lambda_n^\sigma 1_{\{|x| < a_{n-1}/\lambda_n\}},$$

so that

$$\|u^n\|_{L^1} \leq M K_N f_1 \lambda_n^{-(N-\sigma)} a_{n-1}^N, \tag{7}$$

where K_N is the volume of the unit sphere of $\mathbb{R}^N$. Moreover, since f_j is decreasing,

$$\|w^n\|_{L^\infty} = M f_{n+1} \lambda_n^\sigma. \tag{8}$$

Moreover,

$$\operatorname{supp} v^n \subset \{|x| < a_n/\lambda_n\}, \tag{9}$$

and

$$\|v^n\|_{L^1} = f_n \lambda_n^{-(N-\sigma)} a_n^N. \tag{10}$$

We now set

$$\lambda_n = c^{-\frac{1}{N-\sigma}} f_n^{\frac{1}{N-\sigma}} a_n^{\frac{N}{N-\sigma}}, \tag{11}$$

so that by (10)

$$\|v^n\|_{L^1} = c. \tag{12}$$

It follows from (7), (11), (4) and (3) that

$$\|u^n\|_{L^1} \leq c M K_N f_1 (\log a_n) a_n^{-N} a_{n-1}^N \xrightarrow[n\to\infty]{} 0. \tag{13}$$

Also, it follows from (8), (11), (4) and (3) that

$$\|w^n\|_{L^\infty} = M c^{-\frac{\sigma}{N-\sigma}} (a_n^N / \log a_n)^{\frac{\sigma}{N-\sigma}} e^{-a_n} \xrightarrow[n\to\infty]{} 0. \tag{14}$$

Next, we deduce from (4), (11) and (3) that

$$\frac{a_n}{\lambda_n} = [c a_n^{-\sigma} \log a_n]^{\frac{1}{N-\sigma}} \xrightarrow[n\to\infty]{} 0. \tag{15}$$

We note in particular that by (9), (15) and (12),

$$v^n \xrightarrow[n\to\infty]{} c\delta_0, \tag{16}$$

in $\mathcal{S}'(\mathbb{R}^N)$, where δ_0 is the Dirac measure at 0. We now claim that

$$D_{\lambda_n}^\sigma e^{\lambda_n^2 \Delta} u = e^\Delta D_{\lambda_n}^\sigma u \xrightarrow[n\to\infty]{} c e^\Delta \delta_0, \tag{17}$$

in $C_0(\mathbb{R}^N)$. Indeed, the equality $D^{\sigma}_{\lambda_n} e^{\lambda_n^2 \Delta} = e^{\Delta} D^{\sigma}_{\lambda}$ follows from a straightforward computation. Next, since $e^{\Delta}(u^n + w^n) \to 0$ by (13) and (14) (and the continuity of $e^{\Delta} : L^1(\mathbb{R}^N) \to C_0(\mathbb{R}^N)$ and $C_0(\mathbb{R}^N) \to C_0(\mathbb{R}^N)$), we need only show that $e^{\Delta} v^n \to ce^{\Delta}\delta_0$. It follows from (16) that $e^{\Delta} v^n \to ce^{\Delta}\delta_0$ in $\mathcal{S}'(\mathbb{R}^N)$. Thus it remains to show that $e^{\Delta} v^n$ is relatively compact in $C_0(\mathbb{R}^N)$. Relative compactness on bounded sets follows from (12) and the smoothing properties of the heat semigroup. Hence, we need only show that $e^{\Delta} v^n(x) \to 0$ as $|x| \to \infty$, uniformly in n. This is immediate because for n sufficiently large, $\operatorname{supp} v^n \subset \{|x| < 1\}$ by (9) and (15), so that

$$e^{\Delta} v^n(x) = (4\pi)^{-\frac{N}{2}} \int_{\{|y|<1\}} e^{-\frac{|x-y|^2}{4}} v^n(y)\, dy \leq C \sup_{|y|<1} e^{-\frac{|x-y|^2}{4}}, \qquad (18)$$

where we used (12) in the last inequality. It is clear that the right-hand side of (18) converges to 0 as $|x| \to \infty$. This proves the claim (17). Setting $t_n = \lambda_n^2$, we get from (17) that

$$t_n^{\sigma/2} \|e^{t_n \Delta} u\|_{L^\infty} \xrightarrow[n\to\infty]{} c(4\pi)^{-\frac{N}{2}}.$$

This shows the result for $c > 0$. The case $c = 0$ can be obtained by applying the above result with $c' = 1$ and $\sigma' > \sigma$. $\qquad\qquad\square$

References

[1] Cazenave T., Dickstein F. and Weissler F.B. Universal solutions of the heat equation on $\mathbb{R}^N$, Discrete Contin. Dynam. Systems **9** (2003), 1105–1132.

[2] Cazenave T., Dickstein F. and Weissler F.B. Multiscale asymptotic behavior of a solution of the heat equation in $\mathbb{R}^N$, preprint, 2005.

[3] Cazenave T., Dickstein F. and Weissler F.B. A solution of the heat equation in $\mathbb{R}$ with exceptional asymptotic properties, preprint, 2005.

[4] Vázquez J.L. and Zuazua E. Complexity of large time behaviour of evolution equations with bounded data, Chinese Ann. Math. Ser. B **23** (2002), 293–310.

Thierry Cazenave
Laboratoire Jacques-Louis Lions, UMR CNRS 7598
B.C. 187, Université Pierre et Marie Curie
4, place Jussieu, F-75252 Paris Cedex 05, France
e-mail: `cazenave@ccr.jussieu.fr`

Flávio Dickstein
Instituto de Matemática, Universidade Federal do Rio de Janeiro
Caixa Postal 68530, 21944–970 Rio de Janeiro, R.J., Brazil
e-mail: `flavio@labma.ufrj.br`

Fred B. Weissler
LAGA UMR CNRS 7539, Institut Galilée–Université Paris XIII
99, Avenue J.-B. Clément, F-93430 Villetaneuse, France
e-mail: `weissler@math.univ-paris13.fr`

Progress in Nonlinear Differential Equations
and Their Applications, Vol. 63, 139–146

Finite Volume Scheme for Semiconductor Energy-transport Model

Claire Chainais-Hillairet and Yue-Jun Peng

Abstract. In this paper, we propose a finite volume scheme for the Chen energy transport model. We present numerical results obtained for the simulation of a one-dimensional n^+nn^+ ballistic diode.

1. Introduction

In the modelling of semiconductor devices, there exists a hierarchy of models ranging from the kinetic transport equations (microscopic model) to the drift-diffusion equations (macroscopic model), see for instance [11, 10].

The drift-diffusion model is well adapted for micrometer size devices because in this case it displays both computational efficiency and physical consistency. This model consists of continuity equations for the density of charges (electrons and holes) which are coupled to the Poisson equation for the electrostatic potential. A lot of numerical algorithms for solving the drift-diffusion system have already been proposed. Following the work by F. Brezzi, L. Marini and P. Pietra [1], A. Jüngel and P.Pietra [9] proposed a mixed exponential fitting finite element method for the approximation of the nonlinear drift-diffusion system. The efficiency of the scheme is established but there is no proof of convergence. In a former paper [2], we defined a finite volume scheme for this system and, using the technique developed by R. Eymard, T. Gallouët and R. Herbin [5], we proved its convergence, which simultaneously proves the existence of solutions.

The energy-transport models are used in the modelling of submicron semiconductor devices. They are more complex than the drift-diffusion models since they include an energy equation. Indeed, they consist of the conservation laws of mass and energy, together with constitutive relations for the particle and energy currents. With such systems, it is possible to model hot electron effects in submicron devices.

The discretization of the energy-transport equations has already been studied in many papers: extensions of Scharfetter-Gummel schemes in [12], ENO schemes in [8], finite element schemes in [4], [7], high-order compact difference schemes in [6]... In this paper, we construct a finite volume scheme for an energy transport model (the Chen model, see [3]) and we present some numerical results for the case of an n^+nn^+ ballistic diode.

2. Formulation of the model

The energy transport models for semiconductor devices can be derived either from the hydrodynamic models by neglecting certain convection terms or from the Boltzmann equation. In [3], D. Chen *et al.* derived such a model which is now called the Chen model. It consists of (stationary) continuity equations for charge and energy, coupled with a Poisson equation for the electrostatic potential. We consider here the transient version of this system.

2.1. The system written in the physical variables

Let us denote by q the elementary charge of electrons, ε_S the permittivity constant of the material, μ_0 the mobility constant, τ_0 the energy relaxation time, T_0 the ambient temperature, k_B the Boltzmann constant and U_T the thermal voltage at T_0 defined by $qU_T = k_BT_0$. Then, the transient Chen model writes:

$$\partial_t N - \frac{1}{q}\mathrm{div}(J_N) = 0 \text{ in } \Omega\times]0, T[, \tag{1}$$

$$\partial_t U - \mathrm{div}(J_U) = -J_N \cdot \nabla V + W(N, T) \text{ in } \Omega\times]0, T[, \tag{2}$$

$$\varepsilon_S \Delta V = q(N - C) \text{ in } \Omega\times]0, T[. \tag{3}$$

where N is the density of electrons, U is the density of energy, V is the electrostatic potential, J_N is the current density of electrons, J_U is the current density of energy, T is the temperature of the lattice, $W(N, T)$ is the energy relaxation term, $J_N \cdot \nabla V$ is the Joule heating term and C is the initial doping. The current densities may be written:

$$J_N = qU_T \left(\mu_0\nabla N - \frac{q\mu_0}{k_B}N\frac{\nabla V}{T} \right), \tag{4}$$

$$J_U = qU_T \left(\frac{\mu_0}{q}\nabla U - \frac{\mu_0}{k_B}U\frac{\nabla V}{T} \right). \tag{5}$$

The density of energy, U, the density of electrons, N and the temperature T are linked by

$$U = \frac{3}{2}k_B NT$$

and the energy relaxation term is given by

$$W(N, T) = \frac{3}{2}k_B N\frac{T - T_0}{\tau_0}.$$

Equations (1)–(5) are supplemented with initial data N_0, U_0 and boundary conditions. On the Ohmic contacts, we assume that the total space charge $N - C$ vanishes, that the temperature is the ambient temperature and that the applied voltage is given: it yields Dirichlet boundary conditions for N, U and V. On the insulating boundary segments, we assume that the fluxes of J_N, J_U and V vanish: it yields homogeneous Neumann boundary conditions.

2.2. The scaled system

We denote by C_m and L some characteristic values for the doping and the size of the domain. The scaling is the following:

$$C \to C_m C, \ N \to C_m N, \ x \to Lx, \ T \to T_0 T, \ t \to \frac{L^2}{U_T \mu_0} t, \ V \to U_T V,$$

$$U \to q U_T C_m U, \ W \to \frac{q\mu_0 U_T^2 C_m}{L^2} W, \ J_N \to \frac{q\mu_0 U_T C_m}{L} J_N, \ J_U \to \frac{q\mu_0 U_T^2 C_m}{L} J_U.$$

The system in the scaled variables becomes:

$$\partial_t N - \mathrm{div}(J_N) = 0,$$

$$\partial_t U - \mathrm{div}(J_U) = -J_N \cdot \nabla V + W(N,T)$$

$$\lambda^2 \Delta V = N - C \quad \text{where} \quad \lambda^2 = \frac{\varepsilon_S U_T}{q C_m L^2},$$

with

$$J_N = \nabla N - N\frac{\nabla V}{T}, \quad J_U = \nabla U - U\frac{\nabla V}{T}, \quad U = \frac{3}{2} NT,$$

$$W(N,T) = c_1 N - c_2 U \quad \text{with} \quad c_1 = \frac{3}{2}\frac{L^2}{\tau_0 \mu_0 U_T}, \quad c_2 = \frac{L^2}{\tau_0 \mu_0 U_T}.$$

3. The finite volume scheme

For the sake of simplicity, we describe the finite volume scheme in the 1-D case. Then a mesh of $\Omega =]0,1[$ is given by L control volumes $(K_i)_{i=1...L}$ such that $K_i =]x_{i-\frac{1}{2}}, x_{i+\frac{1}{2}}[$ with $x_i = \dfrac{x_{i-\frac{1}{2}} + x_{i+\frac{1}{2}}}{2}$ and

$$0 = x_0 = x_{\frac{1}{2}} < x_1 < \cdots < x_{i-\frac{1}{2}} < x_i < x_{i+\frac{1}{2}} < \cdots < x_L < x_{L+\frac{1}{2}} = x_{L+1} = 1.$$

We set $h_i = x_{i+\frac{1}{2}} - x_{i-\frac{1}{2}}$ and $h_{i+\frac{1}{2}} = x_{i+1} - x_i$ for $1 \leq i \leq L$. The size of the mesh is then defined by $h = \max h_i$. Let k be the time step and set $t^n = nk$. First of all, the initial conditions and the doping profile are approximated by $(N_i^0, U_i^0, C_i)_{1\leq i \leq L}$ by taking the mean values of the initial data and the doping profile on each cell K_i. We also restrict our presentation to the case of Dirichlet boundary conditions (Ohmic contacts). Then, the numerical boundary conditions $(N_0^n, U_0^n, V_0^n)_{n\geq 0}$, $(N_{L+1}^n, U_{L+1}^n, V_{L+1}^n)_{n\geq 0}$ are also given by the mean values of the Dirichlet boundary data.

The scheme on V is the classical FV-scheme for an elliptic equation:

$$\lambda^2(-dV_{i+\frac{1}{2}}^n + dV_{i-\frac{1}{2}}^n) = h_i(C_i - N_i^n), \ \forall 1 \leq i \leq L, \ \forall n \geq 0,$$

where $dV^n_{i+\frac{1}{2}}$ is the numerical approximation of $\partial_x V$ at the point $x_{i+\frac{1}{2}}$:

$$dV^n_{i+\frac{1}{2}} = \frac{V^n_{i+1} - V^n_i}{h_{i+\frac{1}{2}}} \quad \forall 0 \leq i \leq L, \ \forall n \geq 0. \tag{6}$$

For N and U, the schemes are Euler implicit in time and finite volume in space. For $1 \leq i \leq L$ and $n \geq 0$, they write:

$$h_i \frac{N^{n+1}_i - N^n_i}{k} - dN^{n+1}_{i+\frac{1}{2}} + dN^{n+1}_{i-\frac{1}{2}} + F^{n+1,C}_{i+\frac{1}{2}} - F^{n+1,C}_{i-\frac{1}{2}} = 0,$$

$$h_i \frac{U^{n+1}_i - U^n_i}{k} - dU^{n+1}_{i+\frac{1}{2}} + dU^{n+1}_{i-\frac{1}{2}} + G^{n+1,C}_{i+\frac{1}{2}}$$
$$- G^{n+1,C}_{i-\frac{1}{2}} + h_i(c_2 U^{n+1}_i - c_1 N^{n+1}_i) = -S^n_i.$$

The numerical diffusion fluxes $dN^{n+1}_{i+\frac{1}{2}}$ and $dU^{n+1}_{i+\frac{1}{2}}$ are defined in the same way of $dV^n_{i+\frac{1}{2}}$ (6). The numerical convection fluxes $F^{n+1,C}_{i+\frac{1}{2}}$ and $G^{n+1,C}_{i+\frac{1}{2}}$ are the approximations of $\dfrac{\partial_x V}{T(N,U)} N$ and $\dfrac{\partial_x V}{T(N,U)} U$ at the point $x_{i+\frac{1}{2}}$. We choose a classical upwind discretization:

$$F^{n+1,C}_{i+\frac{1}{2}} = (dV^n_{i+\frac{1}{2}})^+ \frac{N^{n+1}_i}{T(N^{n+1}_i, U^{n+1}_i)} + (dV^n_{i+\frac{1}{2}})^- \frac{N^{n+1}_{i+1}}{T(N^{n+1}_{i+1}, U^{n+1}_{i+1})}, \quad \forall 0 \leq i \leq L,$$

$$G^{n+1,C}_{i+\frac{1}{2}} = (dV^n_{i+\frac{1}{2}})^+ \frac{U^{n+1}_i}{T(N^{n+1}_i, U^{n+1}_i)} + (dV^n_{i+\frac{1}{2}})^- \frac{U^{n+1}_{i+1}}{T(N^{n+1}_{i+1}, U^{n+1}_{i+1})}, \quad \forall 0 \leq i \leq L,$$

where $s^+ = \max(s, 0)$, $s^- = \min(s, 0)$ for $s \in \mathbb{R}$.

The key point is now the approximation of the term $\displaystyle\int_{K_i} J_N \cdot \partial_x V$ by S^n_i. We propose

$$S^n_i =$$

$$\frac{h_i}{2} dV^n_{i+\frac{1}{2}} \left(-dN^{n+1}_{i+\frac{1}{2}} + (dV^n_{i+\frac{1}{2}})^+ \frac{N^{n+1}_i}{T(N^{n+1}_i, U^{n+1}_i)} + (dV^n_{i+\frac{1}{2}})^+ \frac{N^{n+1}_{i+1}}{T(N^{n+1}_{i+1}, U^{n+1}_{i+1})} \right)$$

$$+ \frac{h_i}{2} dV^n_{i-\frac{1}{2}} \left(-dN^{n+1}_{i-\frac{1}{2}} + (dV^n_{i-\frac{1}{2}})^+ \frac{N^{n+1}_{i-1}}{T(N^{n+1}_{i-1}, U^{n+1}_{i-1})} + (dV^n_{i-\frac{1}{2}})^+ \frac{N^{n+1}_i}{T(N^{n+1}_i, U^{n+1}_i)} \right).$$

4. Simulation of an $n^+ n n^+$ ballistic diode

We present here the simulation of a one-dimensional $n^+ n n^+$ ballistic diode, which is a simple model for the channel of the MOS transistor. This test case is classical: it has already been computed for instance in [4] or [6] by other kind of schemes for the steady-state Chen model.

The semiconductor domain is the interval $\Omega = (0, l^*)$ with $l^* = 0.6\mu m$. The initial doping is given by

$$C(x) = \begin{cases} c_1 = 5.10^{17} \text{ cm}^{-3} \text{ in }]0; 0.1\mu m[\, \cup \,]0.5\mu m; 0.6\mu m[\text{ (n}^+\text{-region)} \\ c_0 = 2.10^{15} \text{ cm}^{-3} \text{ elsewhere (n-region, channel).} \end{cases}$$

The boundary conditions write

$$N(0) = N(l^*) = c_1, \quad T(0) = T(l^*) = T_0, \quad V(0) = V_{\text{app}}, V(l^*) = 0,$$

where V_{app} is the applied voltage.

For the scaling, we take $C_m = c_1$ and $L = l^*$. The numerical values of the physical parameters are given in Table 1.

Parameter	Numerical value
q	$1.6 \cdot 10^{-19}$ As
ε_S	10^{-12} AsV^{-1}cm^{-1}
μ_0	$1.5 \cdot 10^3$ cm^{2}V^{-1}s^{-1}
τ_0	$0.4 \cdot 10^{-12}$ s
T_0	300 K
U_T	0.026 V
V_{app}	1.5 V

TABLE 1. Numerical values of the parameters

We compute the solution of the transient system at a large time because it becomes stationary. In Figure 1, we present the variations of the temperature and the electrostatic potential in the diode. As expected, we observe that the temperature is high in the n-channel. These numerical results obtained with the finite volume scheme can be compared to the numerical results in [4] and [6]. In Figure 2, we present the computation of the electron mean velocity $u = J_N/(qN)$ in the diode.

Figure 3 presents the current-voltage characteristics for the ballistic diode with the Chen model. As expected, the dependence of the current $I = J_N(l^*)$ with respect to the applied potential V_{app} has the form $I \sim V_{\text{app}}^\gamma$ and in the voltage range $V_{\text{app}} \in [0.5V, 1.5V]$, we compute $\gamma = 0.90$ as in [4].

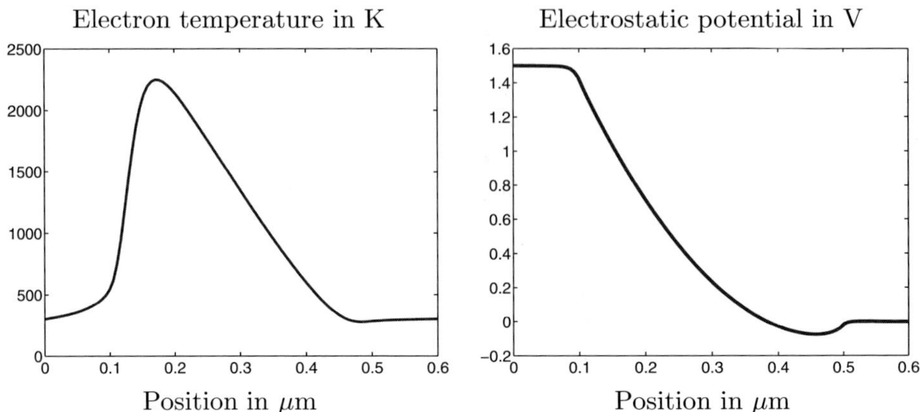

FIGURE 1. Electron temperature and electrostatic potential in the diode

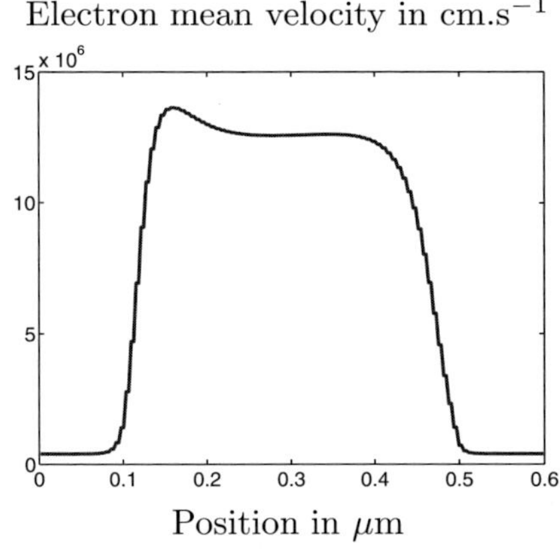

FIGURE 2. Electron mean velocity in the diode

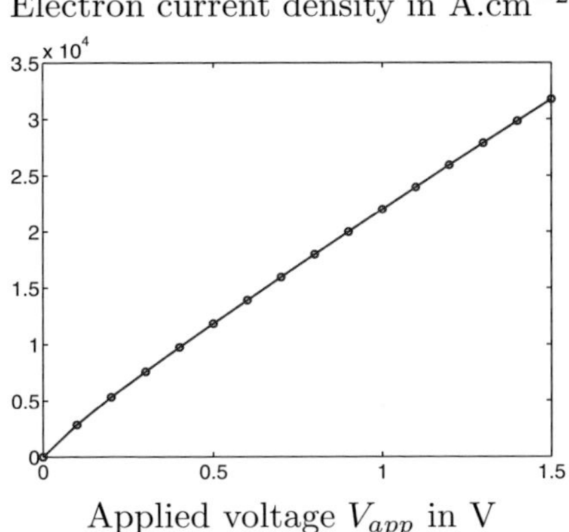

FIGURE 3. Current-voltage characteristics

References

[1] F. Brezzi, L.D. Marini, and P. Pietra. Two-dimensional exponential fitting and applications to drift-diffusion models. *SIAM J. Numer. Anal.*, 26(6):1342–1355, 1989.

[2] Claire Chainais-Hillairet and Yue-Jun Peng. Finite volume approximation for degenerate drift-diffusion system in several space dimensions. *Math. Models Methods Appl. Sci.*, 14(3):461–481, 2004.

[3] D. Chen, Kan E., Ravaioli U., Shu C.-W., and Dutton R. An improved energy transport model including nonparabolicity and non-maxwellian distribution effects. *IEE Electron Device Letters*, 13(1):26–28, 1992.

[4] Pierre Degond, Ansgar Jüngel, and Paola Pietra. Numerical discretization of energy-transport models for semiconductors with nonparabolic band structure. *SIAM J. Sci. Comput.*, 22(3):986–1007, 2000.

[5] R. Eymard, T. Gallouët, and R. Herbin. Finite volume methods. In *Handbook of numerical analysis, Vol. VII*, pages 713–1020. North-Holland, Amsterdam, 2000.

[6] M. Fournié. Numerical discretization of energy-transport model for semiconductors using high-order compact schemes. *Appl. Math. Lett.*, 15(6):721–726, 2002.

[7] Stefan Holst, Ansgar Jüngel, and Paola Pietra. A mixed finite-element discretization of the energy-transport model for semiconductors. *SIAM J. Sci. Comput.*, 24(6):2058–2075, 2003.

[8] Joseph W. Jerome and Chi-Wang Shu. Energy transport systems for semiconductors: analysis and simulation. In *World Congress of Nonlinear Analysts '92, Vol. I–IV (Tampa, FL, 1992)*, pages 3835–3846. de Gruyter, Berlin, 1996.

[9] A. Jüngel and P. Pietra. A discretization scheme for a quasi-hydrodynamic semiconductor model. *Math. Models Methods Appl. Sci.*, 7(7):935–955, 1997.

[10] Ansgar Jüngel. *Quasi-hydrodynamic semiconductor equations*. Progress in Nonlinear Differential Equations and their Applications, 41. Birkhäuser Verlag, Basel, 2001.

[11] P.A. Markowich, C.A. Ringhofer, and C. Schmeiser. *Semiconductor equations.* Springer-Verlag, Vienna, 1990.

[12] K. Souissi, F. Odeh, H.H.K. Tang, and A. Gnudi. Comparative studies of hydrodynamic and energy transport models. *COMPEL*, 13(2):439–453, 1994.

Claire Chainais-Hillairet and Yue-Jun Peng
Laboratoire de Mathématiques
UMR 6620,
Université Blaise Pascal
F-63177 Aubière Cedex, France

Progress in Nonlinear Differential Equations
and Their Applications, Vol. 63, 147–156

Asymptotic Behavior of Nonlinear Parabolic Problems with Periodic Data

Michel Chipot and Yitian Xie

Abstract. We consider parabolic problems with periodic data in space. We study the asymptotic behavior of the solution when the size of the domain on which the problem is set becomes unbounded.

1. Introduction

If $(x_1, \ldots, x_k)$ denotes the points in $\mathbb{R}^k$, Ω_n is a bounded open set in $\mathbb{R}^k$, $\tilde{T}$ is a positive constant, we consider the following problem:

$$
\begin{cases}
\partial_t u_n - \partial_{x_i}(a_{ij}(t, x, u_n)\partial_{x_j} u_n) + b(t, x, u_n) = f(t, x) & \text{in } (0, \tilde{T}) \times \Omega_n, \\[2mm]
u_n(0, \cdot) = u_0 & \text{in } \Omega_n, \\[2mm]
+ \text{ boundary conditions.}
\end{cases}
\tag{1.1}
$$

$a_{ij}(t, x, p)$, $i, j = 1, \ldots, k$, $b(t, x, p)$ are Carathéodory functions defined on $(0, \tilde{T}) \times \mathbb{R}^k \times \mathbb{R}$, i.e.,

$$(t, x) \longrightarrow a_{ij}(t, x, p), \ (resp. \ b(t, x, p)) \qquad \text{is measurable } \forall p \in \mathbb{R}, \tag{1.2}$$

$$p \longrightarrow a_{ij}(t, x, p), \ (resp. \ b(t, x, p)) \qquad \text{is continuous a.e. } (t, x) \in (0, \tilde{T}) \times \mathbb{R}^k, \tag{1.3}$$

$f(t, x)$ is a function in $(0, \tilde{T}) \times \mathbb{R}^k$. (Here we omit the summation of indices and use the Einstein convention instead.) Denote by T a positive constant. We will say that a function $h(x) \in L^2_{\text{loc}}(\mathbb{R}^k)$ is T-periodic with respect to x, if it holds that

$$h(x) = h(x + Te_j) \qquad \text{a.e. } x \in \mathbb{R}^k, \quad j = 1, \ldots, k,$$

where (e_j) is the canonical basis in $\mathbb{R}^k$. Our goal is to study, when the size of Ω_n is supposed to become infinite in all directions, whether the periodicity of the data with respect to x will force the solution to be periodic in any fixed bounded open set. For the case of elliptic operators, we refer the readers to [3] and [4].

2. The main result

We denote by Ω_n, Q the k-dimensional domains given by:

$$\Omega_n = (-nT, nT)^k, \quad Q = (0, T)^k.$$

Let V_n be a closed subspace in $H^1(\Omega_n)$ such that

$$H_0^1(\Omega_n) \subset V_n \subset H^1(\Omega_n).$$

Assume that $f(t, x)$, $a_{ij}(t, x, p)$, $i, j = 1, \ldots, k$, and $b(t, x, p)$ are T-periodic with respect to x. Assume also that $a_{ij}(t, x, p)$ satisfy (1.2) (1.3) and there exist two constants λ, Λ, such that

$$a_{ij}(t, x, p)\xi_i\xi_j \geq \lambda|\xi|^2 \quad \xi \in \mathbb{R}^k, \ \forall p \in \mathbb{R}, \ \text{a.e. } (t, x) \in (0, \tilde{T}) \times \mathbb{R}^k, \tag{2.1}$$

$$|a_{ij}(t, x, p)|, |\partial_{x_m} a_{ij}(t, x, p)| \leq \Lambda$$

$$\text{a.e. } (t, x) \in (0, \tilde{T}) \times \mathbb{R}^k, \ \forall p \in \mathbb{R}, \ i, j, m = 1, \ldots, k. \tag{2.2}$$

Furthermore, we suppose that there exists a nondecreasing positive function $\omega(x)$ defined on $\mathbb{R}^+$, such that

$$|a_{ij}(t, x, u) - a_{ij}(t, x, v)| \leq \omega(|u - v|)$$
$$\forall u, v \in \mathbb{R}, \ \text{a.e. } (t, x) \in (0, \tilde{T}) \times \mathbb{R}^k, \tag{2.3}$$

$$\int_{0+} \frac{ds}{\omega^2(s)} = +\infty. \tag{2.4}$$

In addition, we assume that there exist a constant L and a function $g(t, x) \in L^\infty((0, \tilde{T}) \times \mathbb{R}^k)$ such that:

$$|b(t, x, u) - b(t, x, v)| \leq L|u - v|g(t, x) \quad \forall u, v \in \mathbb{R}, \ \text{a.e. } (t, x) \in (0, \tilde{T}) \times \mathbb{R}^k. \tag{2.5}$$

Let us denote by u_n a weak solution to

$$\begin{cases} u_n \in H^1(0, \tilde{T}; V_n, V_n'), \\ u_n(0, \cdot) = u_0 \quad \text{in } \Omega_n, \\ \dfrac{\mathrm{d}}{\mathrm{d}t}(u_n, v)_{\Omega_n} + \displaystyle\int_{\Omega_n} a_{ij}(t, x, u_n)\partial_{x_j} u_n \partial_{x_i} v + b(t, x, u_n)v \, \mathrm{d}x = \langle f, v \rangle \\ \quad \text{in } \mathcal{D}'(0, \tilde{T}), \ \forall v \in V_n, \end{cases} \tag{2.6}$$

where V_n' denotes the dual of V_n,

$$H^1(a, b; V_n, V_n') = \{v \in L^2(a, b; V_n), v_t \in L^2(a, b; V_n')\},$$

$\langle \cdot \rangle$ denotes the duality bracket between V_n' and V_n, $(u, v)_K$ is the usual scalar product in $L^2(K)$, i.e.,

$$(u, v)_K = \int_K uv \, \mathrm{d}x.$$

Denote also by u_∞ the weak solution to

$$
\begin{cases}
u_\infty \in H^1(0,\tilde{T}; H^1_{\text{per}}(Q), H^1_{\text{per}}(Q)'), \\[2mm]
u_\infty(0,\cdot) = u_0 \quad \text{in } Q, \\[2mm]
\dfrac{\mathrm{d}}{\mathrm{d}t}(u_\infty, v)_Q + \displaystyle\int_Q a_{ij}(t,x,u_\infty)\partial_{x_j} u_\infty \partial_{x_i} v + b(t,x,u_\infty) v \, \mathrm{d}x = \langle f, v \rangle \\[2mm]
\quad \text{in } \mathcal{D}'(0,\tilde{T}), \ \forall v \in H^1_{\text{per}}(Q),
\end{cases}
\tag{2.7}
$$

where $H^1_{\text{per}}(Q) = \{ v \in H^1(Q) \mid v(x) = v(x + Te_j) \quad \forall x \in \partial Q \cap \{x_j = 0\}, j = 1, \cdots, k \}$.

Remark 1. *Let β be a constant. If we look for functions of the type $u_n e^{\beta t}$, $u_\infty e^{\beta t}$ for solutions of (2.6) and (2.7), we see that u_n is the solution to*

$$
\begin{cases}
u_n \in H^1(0,\tilde{T}; V_n, V_n'), \\[2mm]
u_n(0,\cdot) = u_0 \quad \text{in } \Omega_n, \\[2mm]
\dfrac{\mathrm{d}}{\mathrm{d}t}(u_n, v)_{\Omega_n} + \displaystyle\int_{\Omega_n} a_{ij}(t,x,u_n e^{\beta t})\partial_{x_j} u_n \partial_{x_i} v \\[2mm]
\quad + \{\beta u_n + e^{-\beta t}b(t,x,e^{\beta t}u_n) - e^{-\beta t}b(t,x,0)\}v \, \mathrm{d}x \\[2mm]
\quad = \langle e^{-\beta t}f - e^{-\beta t}b(t,x,0), v \rangle \ \text{in } \mathcal{D}'(0,\tilde{T}), \ \forall v \in V_n,
\end{cases}
\tag{2.8}
$$

and u_∞ is the solution to

$$
\begin{cases}
u_\infty \in H^1(0,\tilde{T}; H^1_{per}(Q), H^1_{per}(Q)'), \\[2mm]
u_\infty(0,\cdot) = u_0 \quad \text{in } Q, \\[2mm]
\dfrac{\mathrm{d}}{\mathrm{d}t}(u_\infty, v)_Q + \displaystyle\int_Q a_{ij}(t,x,u_\infty e^{\beta t})\partial_{x_j} u_\infty \partial_{x_i} v \\[2mm]
\quad + \{\beta u_\infty + e^{-\beta t}b(t,x,e^{\beta t}u_\infty) - e^{-\beta t}b(t,x,0)\}v \, \mathrm{d}x \\[2mm]
\quad = \langle e^{-\beta t}f - e^{-\beta t}b(t,x,0), v \rangle \ \text{in } \mathcal{D}'(0,\tilde{T}), \ \forall v \in H^1_{per}(Q),
\end{cases}
\tag{2.9}
$$

where $a_{ij}(t,x,u_n e^{\beta t})$ and $a_{ij}(t,x,u_\infty e^{\beta t})$ still satisfy (2.1)–(2.4). Since $g(t,x)$ is bounded, if we choose β large enough, it holds that

$$
\beta - Lg(t,x) \geq \lambda,
$$

where λ is the elliptic constant in (2.1). Then by the Lipschitz condition (2.5),

$$
\beta v^2 + \{e^{-\beta t}b(t,x,e^{\beta t}v) - e^{-\beta t}b(t,x,0)\}v \geq \beta v^2 - e^{-\beta t}L|e^{\beta t}v|g(t,x)|v|,
$$
$$
\geq \{\beta - Lg(t,x)\}v^2,
$$
$$
\geq \lambda v^2.
$$

If $v_1 - v_2 \geq 0$, we have also

$$\beta(v_1 - v_2) + e^{-\beta t}\{b(t, x, e^{\beta t}v_1) - b(t, x, e^{\beta t}v_2)\}$$

$$\geq \beta(v_1 - v_2) - e^{-\beta t}Lg(t, x)e^{\beta t}(v_1 - v_2) \geq \lambda(v_1 - v_2).$$

So, in the rest of this note, we will assume without loss of generality that

$$b(t, x, u)u \geq \lambda u^2, \qquad b(t, x, u) \text{ is monotone in } u, \tag{2.10}$$

and we will simply denote $b(t,x,v)$ by $b(v)$.

Under the assumptions (2.1)–(2.5) and (2.10), (2.7) admits a unique solution (see [2] for reference). (2.6) admits also a solution but it is not clear for a general V_n that the solution is unique, (see [2]).

Now we can show:

Theorem 2.1 (Convergence result). *For any $n_0 > 0$, any $\gamma > 0$, there exists a constant c, independent of n, such that*

$$|u_n - u_\infty|_{L^1(0,\tilde{T};L^1(\Omega_{n_0}))} \leq \frac{c}{n^\gamma}.$$

In the above estimate, the norm used is a particular case of the following norm,

$$|u|^p_{L^p(a,b;\mathbb{X})} = \int_a^b |u|^p_{\mathbb{X}}\, dx \qquad p < \infty,$$

$|u|_{\mathbb{X}}$ being the norm in the space $\mathbb{X}$ and

$$\Omega_{n_0} = (-n_0 T, n_0 T)^k.$$

We need several lemmas to prove this theorem. First:

Lemma 2.2 (Estimate of u_n). *Let u_n be the solution to (2.6). It holds that for some constant c, independent of n,*

$$|u_n|^2_{L^2(0,\tilde{T};H^1(\Omega_n))} \leq cn^k\{|f|^2_{L^2(0,\tilde{T};L^2(Q))} + |u_0|^2_{L^2(Q)}\}.$$

Proof. Take $v = u_n$ in (2.6), we have

$$\frac{1}{2}\frac{d}{dt}|u_n|^2_{L^2(\Omega_n)}(t) + \int_{\Omega_n} a_{ij}(t, x, u_n)\partial_{x_j}u_n\partial_{x_i}u_n + b(u_n)u_n\, dx = \int_{\Omega_n} fu_n\, dx.$$

Using the ellipticity condition (2.1) on the left-hand side and applying the Young inequality on the right-hand side, we have

$$\frac{1}{2}\frac{d}{dt}|u_n|^2_{L^2(\Omega_n)}(t) + \lambda\int_{\Omega_n} |\nabla u_n|^2 + u_n^2\, dx \leq \frac{\lambda}{2}\int_{\Omega_n} u_n^2\, dx + \frac{1}{2\lambda}\int_{\Omega_n} f^2\, dx,$$

i.e.,

$$\frac{1}{2}\frac{d}{dt}|u_n|^2_{L^2(\Omega_n)}(t) + \frac{\lambda}{2}\int_{\Omega_n} |\nabla u_n|^2 + u_n^2\, dx \leq \frac{1}{2\lambda}\int_{\Omega_n} f^2\, dx.$$

Integrating the above inequality on t leads to

$$|u_n|^2_{L^2(\Omega_n)}(\tilde{T}) + \lambda\int_0^{\tilde{T}}\int_{\Omega_n} |\nabla u_n|^2 + u_n^2\, dxdt \leq \frac{1}{\lambda}\int_0^{\tilde{T}}\int_{\Omega_n} f^2\, dxdt + |u_0|^2_{L^2(\Omega_n)}.$$

Recalling that f and u_0 are T-periodic with respect to x, we derive that for some constant c it holds that

$$\int_0^{\tilde{T}} \int_{\Omega_n} |\nabla u_n|^2 + u_n^2 \, \mathrm{d}x\mathrm{d}t \le c(2n)^k \{ |f|^2_{L^2(0,\tilde{T};L^2(Q))} + |u_0|^2_{L^2(Q)} \}.$$

The proof is complete. $\qquad\qquad\qquad\qquad\qquad\qquad\qquad\qquad\qquad\qquad\qquad\qquad\square$

Lemma 2.3. *(see* [1], [2]*) If $u \in H^1(a,b;V,V')$, then for all $v \in V$,*

$$\frac{\mathrm{d}}{\mathrm{d}t}(u,v) = \left\langle \frac{\mathrm{d}u}{\mathrm{d}t}, v \right\rangle \qquad in \ \mathcal{D}'(a,b).$$

Lemma 2.4. *Let u_∞ be the solution to (2.7). Suppose that u_∞ is extended by periodicity to $(0,\tilde{T}) \times \mathbb{R}^k$. It holds that*

$$\int_{\mathbb{R}^k} \partial_t u_\infty \phi + a_{ij}(t,x,u_\infty)\partial_{x_j} u_\infty \partial_{x_i}\phi + b(u_\infty)\phi \, \mathrm{d}x$$

$$= \int_{\mathbb{R}^k} f\phi \, \mathrm{d}x \qquad \forall \phi \in \mathcal{D}(\mathbb{R}^k) \quad in \ \mathcal{D}'(0,\tilde{T}).$$

In particular, we have also

$$\frac{\mathrm{d}}{\mathrm{d}t}(u_\infty,v) + \int_{\Omega_n} a_{ij}(t,x,u_\infty)\partial_{x_j} u_\infty \partial_{x_i}v + b(u_\infty)v \, \mathrm{d}x$$

$$= \int_{\Omega_n} fv \, \mathrm{d}x \qquad \forall v \in H_0^1(\Omega_n), \quad in \ \mathcal{D}'(0,\tilde{T}). \quad (2.11)$$

Proof. $\forall \phi \in \mathcal{D}(\mathbb{R}^k)$, set $K = \mathrm{supp}\ \phi$ (the support of ϕ). Let $(Q_\ell), \ell = 1, \cdots, m$ be a finite set of translated of Q, recovering K, i.e., such that

$$K \subset \cup_{\ell=1}^m Q_\ell.$$

Let $(\theta_\ell), \ell = 1, \cdots, m$ be a partition of the unity associated to this covering, i.e., a family of functions such that

$$\begin{cases} \theta_\ell \in \mathcal{D}(\mathbb{R}^k), 0 \le \theta_\ell \le 1, \quad \forall \ell \in \{1, \cdots, m\}, \sum_{\ell=1}^m \theta_\ell = 1 \ \text{in}\ K, \\ \mathrm{supp}\ \theta_\ell \subset Q_\ell, \quad \forall \ell \in \{1, \cdots, m\}. \end{cases}$$

One has

$$\phi = \phi \sum_{\ell=1}^m \theta_\ell = \sum_{\ell=1}^m (\phi\theta_\ell) \qquad \text{in}\ \mathbb{R}^k.$$

For any $\ell = 1, \cdots, m$, denote by $(\phi\theta_\ell)^T$ the translation of $\phi\theta_\ell$ from Q_ℓ to Q. One remarks that $(\theta_\ell)^T \in H^1_{\mathrm{per}}(Q)$, thus we have

$$\int_{\mathbb{R}^k} a_{ij}(t,x,u_\infty)\partial_{x_j} u_\infty \partial_{x_i}\phi + b(u_\infty)\phi \, \mathrm{d}x$$

$$= \sum_{\ell=1}^m \int_{\mathbb{R}^k} a_{ij}(t,x,u_\infty)\partial_{x_j} u_\infty \partial_{x_i}(\phi\theta_\ell) + b(u_\infty)(\phi\theta_\ell) \, \mathrm{d}x$$

$$= \sum_{\ell=1}^{m} \int_{Q_\ell} a_{ij}(t,x,u_\infty)\partial_{x_j}u_\infty \partial_{x_i}(\phi\theta_\ell) + b(u_\infty)(\phi\theta_\ell)\, \mathrm{d}x$$

$$= \sum_{\ell=1}^{m} \int_{Q} a_{ij}(t,x,u_\infty)\partial_{x_j}u_\infty \partial_{x_i}(\phi\theta_\ell)^T + b(u_\infty)(\phi\theta_\ell)^T \, \mathrm{d}x$$

$$= \sum_{\ell=1}^{m} \int_{Q} f(\phi\theta_\ell)^T - \partial_t u_\infty \{(\phi\theta_\ell)^T\} \, \mathrm{d}x = \sum_{\ell=1}^{m} \int_{Q_\ell} f(\phi\theta_\ell) - \partial_t u_\infty(\phi\theta_\ell)\, \mathrm{d}x$$

$$= \int_{\mathbb{R}^k} f\phi - \partial_t u_\infty \phi \, \mathrm{d}x \qquad \forall \phi \in \mathcal{D}(\mathbb{R}^k), \quad \text{in } \mathcal{D}'(0,\tilde{T}). \qquad \square$$

Proof of Theorem 2.1. We set

$$F_\epsilon(x) = \begin{cases} \dfrac{1}{I_\epsilon}\displaystyle\int_\epsilon^x \dfrac{\mathrm{d}s}{\omega^2(s)} & x > \epsilon, \\[2mm] 0 & x \le \epsilon, \end{cases}$$

where $I_\epsilon = \displaystyle\int_\epsilon^\infty \dfrac{\mathrm{d}s}{\omega^2(s)} < \infty$, (w.l.o.g. we can suppose that the integral converges).
Assume that ρ is a smooth function such that

$$\rho \equiv 1 \text{ in } \left(-\frac{1}{2}, \frac{1}{2}\right), \quad \rho \equiv 0 \text{ outside } (-1,1).$$

Let $v = F_\epsilon(u_n - u_\infty)\Pi^2 := F_\epsilon(u_n - u_\infty)\prod_{i=1}^{k} \rho^2\left(\frac{x_i}{n_1 T}\right)$, $n_1 \le n$. It is easy to check
that $v(t,\cdot) \in H_0^1(\Omega_{n_1})$, a.e. $t \in (0,\tilde{T})$ (see [2]). Plugging it into (2.6), (2.11), we
have:

$$\left\langle \frac{\mathrm{d}}{\mathrm{d}t}(u_n - u_\infty), F_\epsilon(u_n - u_\infty)\Pi^2 \right\rangle$$

$$+ \int_{\Omega_{n_1}} \{a_{ij}(t,x,u_n)\partial_{x_j}u_n - a_{ij}(t,x,u_\infty)\partial_{x_j}u_\infty\}\partial_{x_i}\{F_\epsilon \Pi^2\}\, \mathrm{d}x$$

$$+ \int_{\Omega_{n_1}} \{b(u_n) - b(u_\infty)\}F_\epsilon \Pi^2 \, \mathrm{d}x = 0,$$

i.e.,

$$\left\langle \frac{\mathrm{d}}{\mathrm{d}t}(u_n - u_\infty), F_\epsilon(u_n - u_\infty)\Pi^2 \right\rangle$$

$$+ \int_{\Omega_{n_1}^\epsilon} \{a_{ij}(t,x,u_n)\partial_{x_j}u_n - a_{ij}(t,x,u_\infty)\partial_{x_j}u_\infty\}\partial_{x_i}\Pi^2 F_\epsilon \, \mathrm{d}x$$

$$+ \int_{\Omega_{n_1}^\epsilon} \{b(u_n) - b(u_\infty)\}F_\epsilon \Pi^2 \, \mathrm{d}x$$

$$= -\int_{\Omega_{n_1}^\epsilon} \{a_{ij}(t,x,u_n)\partial_{x_j}u_n - a_{ij}(t,x,u_\infty)\partial_{x_j}u_\infty\}\partial_{x_i}F_\epsilon \Pi^2 \, \mathrm{d}x, \quad (2.12)$$

where $\Omega_{n_1}^{\epsilon} = \{x \in \Omega_{n_1},\ u_n(x) - u_\infty(x) > \epsilon\}$. Noticing that

$$\partial_{x_i} F_\epsilon = \frac{\partial_{x_i}(u_n - u_\infty)}{I_\epsilon\, \omega^2(u_n - u_\infty)},$$

the last term in (2.12) becomes:

$$-\int_{\Omega_{n_1}^{\epsilon}} \{a_{ij}(t,x,u_n)\partial_{x_j}u_n - a_{ij}(t,x,u_\infty)\partial_{x_j}u_\infty\}\partial_{x_i}F_\epsilon \Pi^2\, \mathrm{d}x$$

$$= -\int_{\Omega_{n_1}^{\epsilon}} a_{ij}(t,x,u_n)\partial_{x_j}(u_n - u_\infty)\partial_{x_i}(u_n - u_\infty)\frac{1}{I_\epsilon\, \omega^2(u_n - u_\infty)}\Pi^2\, \mathrm{d}x$$

$$\quad -\int_{\Omega_{n_1}^{\epsilon}} \{a_{ij}(t,x,u_n) - a_{ij}(t,x,u_\infty)\}\partial_{x_j}u_\infty \frac{\partial_{x_i}(u_n - u_\infty)}{I_\epsilon\, \omega^2(u_n - u_\infty)}\Pi^2\, \mathrm{d}x$$

$$\leq -\lambda\int_{\Omega_{n_1}^{\epsilon}} \frac{|\nabla(u_n - u_\infty)|^2}{I_\epsilon\, \omega^2}\Pi^2\, \mathrm{d}x + \int_{\Omega_{n_1}^{\epsilon}} \frac{|\nabla(u_n - u_\infty)|}{I_\epsilon\, \omega}|\nabla u_\infty|\Pi^2\, \mathrm{d}x$$

$$\leq -\lambda\int_{\Omega_{n_1}^{\epsilon}} \frac{|\nabla(u_n - u_\infty)|^2}{I_\epsilon\, \omega^2}\Pi^2\, \mathrm{d}x$$

$$\quad +\lambda\int_{\Omega_{n_1}^{\epsilon}} \frac{|\nabla(u_n - u_\infty)|^2}{I_\epsilon\, \omega^2}\Pi^2\, \mathrm{d}x + \frac{1}{4\lambda}\int_{\Omega_{n_1}^{\epsilon}} \frac{|\nabla u_\infty|^2}{I_\epsilon}\Pi^2\, \mathrm{d}x$$

$$\leq \frac{1}{4\lambda I_\epsilon}\int_{\Omega_{n_1}} |\nabla u_\infty|^2\, \mathrm{d}x.$$

Since $b(u)$ is monotone, we have

$$\int_{\Omega_{n_1}^{\epsilon}} \{b(u_n) - b(u_\infty)\}F_\epsilon\Pi^2\, \mathrm{d}x \geq 0.$$

Therefore we can omit the positive term on the left-hand side in (2.12) to obtain:

$$\left\langle \frac{\mathrm{d}}{\mathrm{d}t}(u_n - u_\infty),\ F_\epsilon(u_n - u_\infty)\Pi^2 \right\rangle$$

$$\quad + \int_{\Omega_{n_1}^{\epsilon}} \{a_{ij}(t,x,u_n)\partial_{x_j}u_n - a_{ij}(t,x,u_\infty)\partial_{x_j}u_\infty\}\partial_{x_i}\Pi^2 F_\epsilon\, \mathrm{d}x$$

$$\leq \frac{1}{4\lambda I_\epsilon}\int_{\Omega_{n_1}} |\nabla u_\infty|^2\, \mathrm{d}x.$$

Setting $H_\epsilon(x) = \int_0^x F_\epsilon(s)\, \mathrm{d}s$, we have:

$$\left\langle \frac{\mathrm{d}}{\mathrm{d}t}(u_n - u_\infty),\ F_\epsilon(u_n - u_\infty)\Pi^2 \right\rangle = \frac{\mathrm{d}}{\mathrm{d}t}\int_{\Omega_{n_1}^{\epsilon}} H_\epsilon(u_n - u_\infty)\Pi^2\, \mathrm{d}x.$$

So we get

$$\frac{\mathrm{d}}{\mathrm{d}t} \int_{\Omega_{n_1}^{\epsilon}} H_{\epsilon}(u_n - u_\infty)\Pi^2 \, \mathrm{d}x$$

$$+ \int_{\Omega_{n_1}^{\epsilon}} \{a_{ij}(t, x, u_n)\partial_{x_j} u_n - a_{ij}(t, x, u_\infty)\partial_{x_j} u_\infty\}\partial_{x_i}\Pi^2 F_{\epsilon} \, \mathrm{d}x$$

$$\leq \frac{1}{4\lambda I_{\epsilon}} \int_{\Omega_{n_1}} |\nabla u_\infty|^2 \, \mathrm{d}x. \quad (2.13)$$

Integrating from $(0, t)$ and letting $\epsilon \to 0$, we have

$$\int_{\Omega_{n_1}^0} (u_n - u_\infty)^+(t)\Pi^2 \, \mathrm{d}x$$

$$+ \int_0^t \int_{\Omega_{n_1}^0} \{a_{ij}(t, x, u_n)\partial_{x_j} u_n - a_{ij}(t, x, u_\infty)\partial_{x_j} u_\infty\}\partial_{x_i}\Pi^2 \, \mathrm{d}x \leq 0, \quad (2.14)$$

where v^+ denotes the positive part of v. Define

$$\bar{a}_{ij}(t, x, z) = \int_0^z a_{ij}(t, x, s) \, \mathrm{d}s,$$

we have

$$\partial_{x_m} \bar{a}_{ij}(t, x, z) = a_{ij}(t, x, z)\partial_{x_m} z + \int_0^z \partial_{x_m} a_{ij}(t, x, s) \, \mathrm{d}s,$$

$$|\bar{a}_{ij}(t, x, u_n) - \bar{a}_{ij}(t, x, u_\infty)| = |\int_{u_\infty}^{u_n} a_{ij}(t, x, s) \, \mathrm{d}s \leq \Lambda|u_n - u_\infty|.$$

Denote by w the function $(u_n - u_\infty)^+$. We derive from (2.14)

$$\int_{\Omega_{n_1}^0} (u_n - u_\infty)^+(t)\Pi^2 \, \mathrm{d}x$$

$$\leq -\int_0^t \int_{\Omega_{n_1}^0} \left\{ \partial_{x_j}\left(\bar{a}_{ij}(t, x, u_n) - \bar{a}_{ij}(t, x, u_\infty)\right) \right.$$

$$\left. - \int_{u_\infty}^{u_n} \partial_{x_j} a_{ij}(t, x, s) \, \mathrm{d}s \right\} \partial_{x_i}\Pi^2 \, \mathrm{d}x\mathrm{d}t$$

$$\leq -\int_0^t \int_{\Omega_{n_1}^0} \partial_{x_j}\{\bar{a}_{ij}(t, x, u_\infty + w) - \bar{a}_{ij}(t, x, u_\infty)\}\partial_{x_i}\Pi^2 \, \mathrm{d}x\mathrm{d}t$$

$$+ \int_0^t \int_{\Omega_{n_1}^0} \Lambda|u_n - u_\infty||\partial_{x_i}\Pi^2| \, \mathrm{d}x\mathrm{d}t$$

$$= -\int_0^t \int_{\Omega_{n_1}} \partial_{x_j}\{\bar{a}_{ij}(t, x, u_\infty + w) - \bar{a}_{ij}(t, x, u_\infty)\}\partial_{x_i}\Pi^2 \, \mathrm{d}x\mathrm{d}t$$

$$+ \int_0^t \int_{\Omega_{n_1}^0} \Lambda|u_n - u_\infty||\partial_{x_i}\Pi^2| \, \mathrm{d}x\mathrm{d}t$$

$$\leq \int_0^t \int_{\Omega_{n_1}} |\bar{a}_{ij}(t,x,u_\infty + w) - \bar{a}_{ij}(t,x,u_\infty)||\partial_{x_i x_j} \Pi^2| \, dxdt$$

$$+ \int_0^t \int_{\Omega_{n_1}^0} \Lambda |u_n - u_\infty||\partial_{x_i} \Pi^2| \, dxdt$$

$$\leq \Lambda \int_0^t \int_{\Omega_{n_1}} |u_n - u_\infty|\{|\partial_{x_i x_j}\Pi^2| + |\partial_{x_i}\Pi^2|\} \, dxdt$$

$$\leq \frac{c}{n_1} \int_0^t \int_{\Omega_{n_1}} |u_n - u_\infty| \, dxdt,$$

where c is some constant independent of n_1.

Since the same result holds for $(u_n - u_\infty)^-$, we have

$$\int_{\Omega_{n_1}} |u_n - u_\infty|(t)\Pi^2 \, dx \leq \frac{c}{n_1} \int_0^t \int_{\Omega_{n_1}} |u_n - u_\infty| \, dxdt.$$

Due to $\Pi \equiv 1$ in $\Omega_{\frac{n_1}{2}}$, we derive:

$$\int_{\Omega_{\frac{n_1}{2}}} |u_n - u_\infty|(t) \, dxdt \leq \frac{c}{n_1} \int_0^t \int_{\Omega_{n_1}} |u_n - u_\infty| \, dxdt \leq \frac{c}{n_1} \int_0^{\tilde{T}} \int_{\Omega_{n_1}} |u_n - u_\infty| \, dxdt.$$

Integrating on t from 0 to $\tilde{T}$, we have

$$\int_0^{\tilde{T}} \int_{\Omega_{\frac{n_1}{2}}} |u_n - u_\infty| \, dxdt \leq \frac{c\tilde{T}}{n_1} \int_0^{\tilde{T}} \int_{\Omega_{n_1}} |u_n - u_\infty| \, dxdt.$$

Taking $n_1 = \frac{n}{2^{\ell-1}}$ and iterating the above inequality, we have for some constant c, independent of n:

$$\begin{aligned}
\int_0^{\tilde{T}} \int_{\Omega_{\frac{n}{2^\ell}}} |u_n - u_\infty| \, dxdt \;&\leq\; \frac{c}{n} \int_0^{\tilde{T}} \int_{\Omega_{\frac{n}{2^{\ell-1}}}} |u_n - u_\infty| \, dxdt \\[2mm]
&\leq\; \frac{c}{n^\ell} \int_0^{\tilde{T}} \int_{\Omega_n} |u_n - u_\infty| \, dxdt \\[2mm]
&\leq\; \frac{c}{n^\ell} \left\{ \int_0^{\tilde{T}} \int_{\Omega_n} 1 \, dxdt \right\}^{\frac{1}{2}} |u_n - u_\infty|_{L^2(0,\tilde{T};L^2(\Omega_n))} \\[2mm]
&\leq\; \frac{c}{n^{\ell - \frac{k}{2}}} \left\{ |u_n|_{L^2(0,\tilde{T};L^2(\Omega_n))} + |u_\infty|_{L^2(0,\tilde{T};L^2(\Omega_n))} \right\}.
\end{aligned}$$

Recalling that u_∞ is periodic with respect to x, by Lemma 2.2, it holds that for some constant c, independent of n

$$\int_0^{\tilde{T}} \int_{\Omega_{\frac{n}{2^\ell}}} |u_n - u_\infty| \, dxdt \leq \frac{c}{n^{\ell-k}}.$$

Choosing $n_0 < \frac{n}{2^\ell}$, $\ell - k \geq \gamma$, the proof is complete. $\qquad\square$

Remark 2. *The convergence result above includes in particular the linear case when the operator L and the equation are:*

$$Lu := \partial_t u - \partial_{x_i}(a_{ij}(t,x)\partial_{x_j} u) + b(t,x)u = f(t,x),$$

with all the functions, a_{ij}, b, f T-periodic with respect to x and satisfying

$$\exists \lambda > 0 \qquad a_{ij}(t,x)\xi_i\xi_j \geq \lambda|\xi|^2, \quad \forall \xi \in \mathbb{R}^k, \quad a.e. \ (t,x) \in (0,\tilde{T}) \times \mathbb{R}^n,$$
$$\exists M \qquad |b(t,x)| \leq M \quad a.e. \ (t,x) \in (0,\tilde{T}) \times \mathbb{R}^n.$$

Remark that in this case, $u_n \in V_n$ solution to (2.6) is of course unique (see [2]).

Acknowledgements

The work of both authors has been supported by the Swiss Nationalfonds under the contract # 20-103300/1. We are very grateful to this institution.

References

[1] M. Chipot: ℓ *goes to plus infinity*, Birkhäuser, 2002.

[2] M. Chipot: *Element of nonlinear analysis*, Birkhäuser, 2000.

[3] M. Chipot, Y. Xie: On the asymptotic behaviour of elliptic problems with periodic data, C. R. Acad. Sci. Paris, Ser I 339 (2004), pp. 477–482.

[4] M. Chipot, Y. Xie: Elliptic problems with periodic data: an asymptotic analysis, to appear.

[5] D. Cioranescu, P. Donato: *An introduction to homogenization*, Oxford Lect. Series, vol 17, Oxford University Press, 1999.

Michel Chipot and Yitian Xie
Angewandte Mathematik
Universität Zürich
Winterthurerstr. 190
CH-8057 Zürich, Switzerland
e-mail: `m.m.chipot@math.unizh.ch`
e-mail: `yitian.xie@math.unizh.ch`

Progress in Nonlinear Differential Equations
and Their Applications, Vol. 63, 157–171

Geodesic Computations for Fast and Accurate Surface Remeshing and Parameterization

Gabriel Peyré and Laurent Cohen

Abstract. In this paper, we propose fast and accurate algorithms to remesh and flatten a genus-0 triangulated manifold. These methods naturally fits into a framework for 3D geometry modeling and processing that uses only fast geodesic computations. These techniques are gathered and extended from classical areas such as image processing or statistical perceptual learning. Using the *Fast Marching* algorithm, we are able to recast these powerful tools in the language of mesh processing. Thanks to some classical geodesic-based building blocks, we are able to derive a flattening method that exhibit a conservation of local structures of the surface.

On large meshes (more than 500 000 vertices), our techniques speed up computation by over one order of magnitude in comparison to classical remeshing and parameterization methods. Our methods are easy to implement and do not need multilevel solvers to handle complex models that may contain poorly shaped triangles.

Keywords. Remeshing, geodesic computation, fast marching algorithm, mesh segmentation, surface parameterization, texture mapping, deformable models.

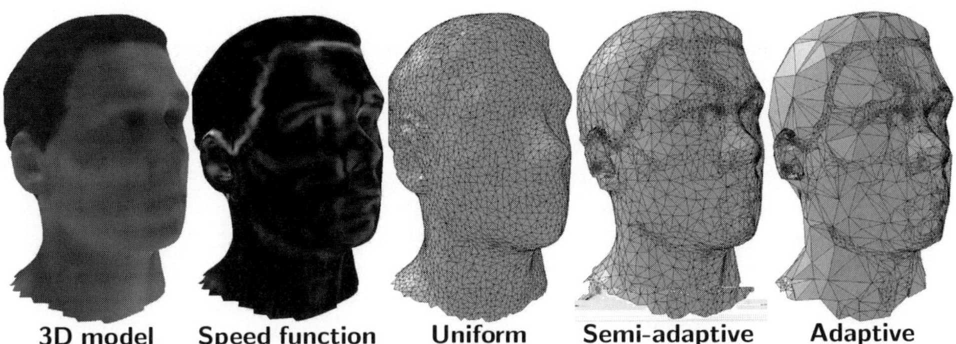

FIGURE 1. Remeshing of a 3D model using increasing weight for the speed function.

1. Introduction

The applications of 3D geometry processing abound nowadays. They range from finite element computation to computer graphics, including solving all kinds of surface reconstruction problems. The most common representation of 3D objects is the triangle mesh, and the need for fast algorithms to handle this kind of geometry is obvious. Classical 3D triangulated manifold processing methods have several well-identified shortcomings: mainly, their high complexity when dealing with large meshes, and their numerical instabilities.

To overcome these difficulties, we propose a geometry processing pipeline that relies on *intrinsic* information of the surface and not on its underlying triangulation. Borrowing from well-established ideas in different fields (including image processing and perceptual learning) we are able to process very large meshes efficiently.

1.1. Overview

In Section 2 we introduce some concepts we use in our geodesic computations. This includes basic facts about the Fast Marching algorithm, and a recently proposed greedy algorithm for manifold sampling.

To flatten each patch of a segmented surface, we will recall some recent advances in perceptual inference learning in Section 3. Combining these techniques with our geodesic computational framework will lead to an elegant solution to the flattening problem for large meshes.

In the conclusion, we will show the two algorithms in action, and see how we can texture large meshes faster than current techniques would otherwise allow. We will then give a complete study of the timings of each part of our algorithm, including a comparison with classical methods.

1.2. Related Work

Surface Remeshing and Finite Elements. Remeshing methods roughly fall into two categories:

- *Isotropic remeshing*: a surface density of points is defined, and the algorithm tries to position the new vertices to match this density. For example the algorithm of *Terzopoulos* and *Vasilescu* [Terzopoulos and Vasilescu, 1992] uses dynamic models to perform the remeshing. Remeshing is also a basic task in the computer graphics community, and [Surazhsky et al., 2003] have proposed a procedure based on local parameterization.
- *Anisotropic remeshing*: the algorithm takes into account the principal directions of the surface to align locally the newly created triangles and/or rectangles. Finite element methods make heavy use of such remeshing algorithms [Kunert, 2002]. The algorithm proposed in [Alliez et al., 2003] uses lines of curvature to build a quad-dominant mesh.

The importance of using geodesic information to perform this remeshing task is emphasized in [Sifri et al., 2003].

Greedy solutions for sampling a manifold (see Section 2.2) have been used with success in other fields such as computer vision (component grouping, [Cohen, 2001]), halftoning (void-and-cluster, [Ulichney, 1993]) and remeshing (Delaunay refinement, [Ruppert, 1995]).

Flattening and Parameterization. The flattening problem can be seen as a particular instance of parameterization. The work of [Eck et al., 1995] first introduces the harmonic formulation for the resolution of the mesh parameterization problem. Most of these classical methods come from graph-drawing theory, and [Floater et al., 2002] gives a survey of these techniques. Authors of [Desbrun et al., 2002] give an in-depth study of the various energies that can be built to flatten a mesh. The flattening algorithm of [Zigelman et al., 2002] is based on methods for finding parameters that reduce a dataset's dimensionality. Such methods have been developed for the purpose of perceptual learning [Tenenbaum et al., 2000, Roweis and Saul, 2000], and we will explain in Section 3 how to exploit these methods to handle a 3D mesh with a large amount of vertices.

2. Geodesic Remeshing

2.1. Fast Marching Algorithm

The classical Fast Marching algorithm is presented in [Sethian, 1999], and a similar algorithm was also proposed in [Tsitsiklis, 1995]. This algorithm is used intensively in computer vision, for instance it has been applied to solve global minimization problems for deformable models [Cohen and Kimmel, 1997].

This algorithm is formulated as follows. Suppose we are given a metric $P(s)\mathrm{d}s$ on some manifold $\mathcal{S}$ such that $P > 0$. If we have two points $x_0,\, x_1 \in \mathcal{S}$, the weighted geodesic distance between x_0 and x_1 is defined as

$$d(x_0,\, x_1) \stackrel{\text{def.}}{=} \min_{\gamma} \left(\int_0^1 \|\gamma'(t)\| P(\gamma(t)) \mathrm{d}t \right), \tag{1}$$

where γ is a piecewise regular curve with $\gamma(0) = x_0$ and $\gamma(1) = x_1$. When $P = 1$, the integral in (1) corresponds to the length of the curve γ and d is the classical geodesic distance. To compute the distance function $U(x) \stackrel{\text{def.}}{=} d(x_0,\, x)$ with an accurate and fast algorithm, this minimization can be reformulated as follows. The level set curve $\mathcal{C}_t \stackrel{\text{def.}}{=} \{x \setminus U(x) = t\}$ propagates following the evolution equation $\frac{\partial \mathcal{C}_t}{\partial t}(x) = \frac{1}{P(x)} \overrightarrow{n_x}$, where $\overrightarrow{n_x}$ is the exterior unit vector normal to the curve at x, and the function U satisfies the nonlinear *Eikonal* equation:

$$\|\nabla U(x)\| = P(x). \tag{2}$$

The function $F = 1/P > 0$ can be interpreted as the propagation speed of the front $\mathcal{C}_t$.

The Fast Marching algorithm on an orthogonal grid makes use of an upwind finite difference scheme to compute the value u of U at a given point $x_{i,j}$ of a grid:

$$\max(u - U(x_{i-1,j}),\, u - U(x_{i+1,j}), 0)^2$$
$$+ \max(u - U(x_{i,j-1}),\, u - U(x_{i,j+1}), 0)^2 = h^2 P(x_{i,j})^2.$$

This is a second-order equation that is solved as detailed for example in [Cohen, 2001]. An optimal ordering of the grid points is chosen so that the whole computation only takes $O(N \log(N))$, where N is the number of points.

In [Kimmel and Sethian, 1998], a generalization to an arbitrary triangulation is proposed. This allows performing front propagations on a triangulated manifold, and computing geodesic distances with a fast and accurate algorithm. The only issue arises when the triangulation contains obtuse angles. The numerical scheme presented above is not monotone anymore, which can lead to numerical instabilities. To solve this problem, we follow [Kimmel and Sethian, 1998] who propose to "unfold" the triangles in a zone where we are sure that the update step will work. Figure 2 shows the calculation of a geodesic path computed using a gradient descent of the distance function.

FIGURE 2. Front Propagation (on the left), level sets of the distance function and geodesic path (on the right).

2.2. A Greedy Algorithm for Uniformly Sampling a Manifold

A new method for sampling a 3D mesh was recently proposed in [Peyré and Cohen, 2003] that follows a farthest point strategy based on the weighted distance obtained through Fast Marching on the initial triangulation. This is related to the method introduce in [Cohen, 2001]. A similar approach was proposed independently and simultaneously in [Moenning and Dodgson, 2003]. It follows the *farthest point* strategy, introduced with success for image processing in [Eldar et al., 1997] and related to the remeshing procedure of [Chew, 1993].

This approach iteratively adds new vertices based on the geodesic distance on the surface. Figure 3 shows the first steps of our algorithm on a square. The

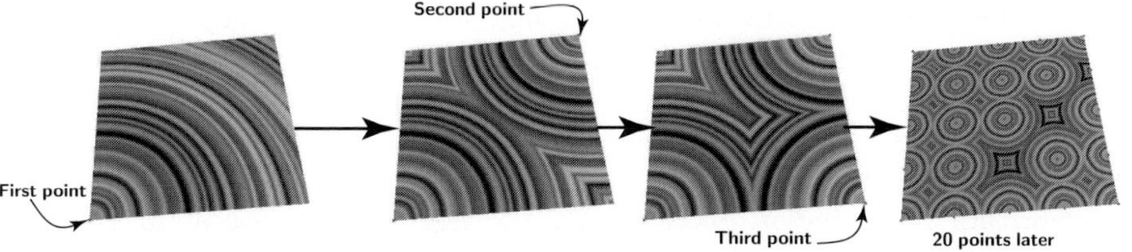

FIGURE 3. An overview of the greedy sampling algorithm.

result of the algorithm gives a set of vertices uniformly distributed on the surface according to the geodesic distance.

Once we have found enough points, we can link them together to form a *geodesic Delaunay triangulation*. This is done incrementally during the algorithm, and leads to a powerful remeshing method.

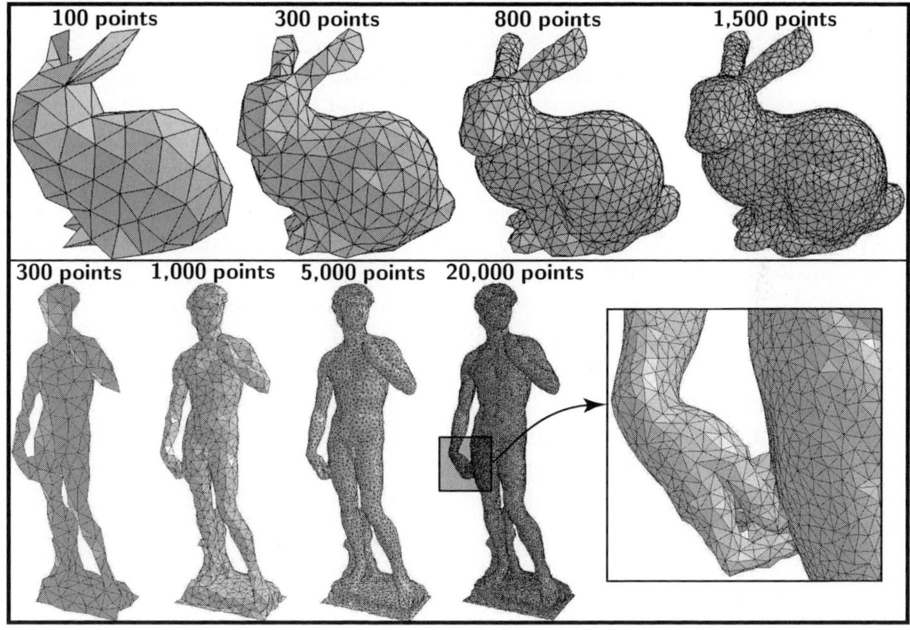

FIGURE 4. Geodesic remeshing with an increasing number of points.

Figure 4 shows progressive remeshing of the bunny and the David. In order to have a valid triangulation, the sampling of the manifold must be dense enough (for example 100 points is not enough to capture the geometry of the ears of the bunny). A theoretical proof of the validity of geodesic Delaunay triangulation can be found in [Leibon and Letscher, 2000], and more precise bound on the number of points is derived in [Onishi and Itoh, 2003]. Note that our algorithm works with manifolds with boundaries, of arbitrary genus, and with multiple connected components.

2.3. Adaptive Remeshing

In the algorithm presented in Sections 2.2, the fronts propagate at a constant speed which results in uniformly spaced mesh. To introduce some adaptivity in the sampling performed by this algorithm, we use a speed function $F = 1/P$ (which is the right-hand side of the Eikonal equation (2)) that is not constant across the surface. Figure 5 shows the progressive sampling of a square using a speed function with two different values. The colors show the level sets of the distance function U to the set of points.

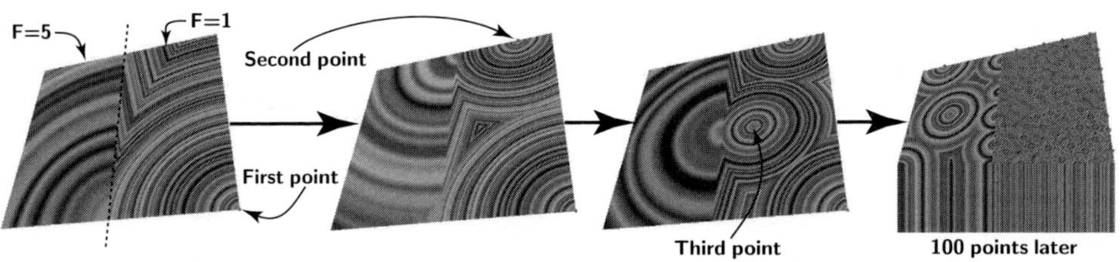

FIGURE 5. Iterative insertion of points in a square.

When a mesh is obtained from range scanning, a picture I of the model can be mapped onto the 3D mesh. Using a function F of the form $F(x) = (1 + \mu|\nabla(I(x))|)^{-1}$, where μ is a user-defined constant, one can refine regions with high variations in intensity. On Figure 1, one can see a 3D head remeshed with various μ ranging from $\mu = 0$ (uniform) to $\mu = 20/\max(|\nabla(I(x))|)$ (highly adaptive).

The local density of vertices can also reflect some geometric properties of the surface. The most natural choice is to adapt the mesh in order to be finer in regions where the local curvature is larger. The evaluation of the curvature tensor is a vast topic. We used a robust construction proposed recently in [Cohen-Steiner and Morvan, 2003]. Let us denote by $\tau(x) \overset{\text{def.}}{=} |\lambda_1| + |\lambda_2|$ the total curvature at a given point x of the surface, where λ_i are the eigenvalues of the second fundamental form. We can introduce two speed functions $F_1(x) \overset{\text{def.}}{=} 1 + \varepsilon\tau(x)$ and $F_2(x) \overset{\text{def.}}{=} \frac{1}{1+\mu\tau(x)}$, where ε and μ are two user-defined parameters. Figure 6 (a) shows that by using function F_1, we avoid putting more vertices in regions of the surface with high curvature. The speed function F_1 can be interpreted as an "edge repulsive" function. On the other hand, function F_2 could be called "edge attractive" function, since it forces the sampling to put vertices in region with high curvature such as mesh corners and edges. Figure 6 (b) shows that this speed function leads to very good results for the remeshing of a surface with sharp

features, which is obviously not the case for the "edge repulsive" speed function
(Figure 6 (a)).

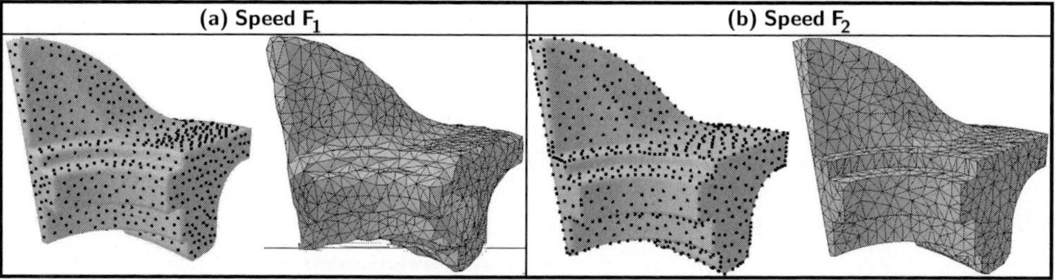

FIGURE 6. Uniform versus curvature-based sampling and remeshing.

3. Fast Geodesic Parameterization

The flattening problem can be seen as a particular instance of the more generic
problem of mesh parameterization. Given a genus-0 triangulated manifold S home-
omorphic to a disc, it consists in finding a map $f : S \to U$, where $U \subset \mathbb{R}^2$ is a
planar domain.

3.1. Geodesic Flattening and IsoMap

Recently, some nonlinear algorithms for dimensionality reduction have appeared
in the community of perceptual manifold learning. The most notable are *IsoMap*
[Tenenbaum et al., 2000] and *Locally Linear Embedding* (LLE) [Roweis and Saul,
2000].

Interestingly, the only echo of these techniques in the computer graphics
community seems to be the multi-dimensional scaling approach to flattening of
[Zigelman et al., 2002]. This method is closely related to IsoMap, and we will see
that it shares its main drawbacks.

We start with a given set of points $\{x_1, \ldots, x_n\}$ on our manifold, and we seek
$f(x_i) = \widetilde{x}_i \in \mathbb{R}^2$ such that the mapping minimizes some measure of distortion. The
most natural constraint is to try to keep the same distance between points, which
is exactly what IsoMap is doing by requiring that $d(x_i, x_j) \approx \|\widetilde{x}_i - \widetilde{x}_j\|$, where d
stands for (some approximation of) the geodesic distance on the manifold. The
method of [Zigelman et al., 2002] is very close to this approach, since it uses
the geodesic distance d computed via the Fast Marching algorithm presented in
Section 2.1.

The major bottleneck of this method is that it needs to compute all pairwise
distances $d(x_i, x_j)$. To overcome this difficulty, the authors of [Zigelman et al.,
2002] proposed to restrict the computations to a small set of points, which gives
rise to three questions:

- What should be done to speed up computation?
- How should we choose this small set of base points?
- How should we extend the map f from this small set of points to the rest of the mesh?

In the next subsection, we will show how the LLE algorithm can bring a important speed improvement that answers the first question. The answer to the two last questions will be given in Subsections 3.3 and 3.4 respectively, with an extension of LLE to triangulated manifolds.

3.2. Speeding Up Computation with LLE

The LLE algorithm is explained in detail in [Roweis and Saul, 2000]. The goal of the algorithm is to find a low-dimensional embedding in $\mathbb{R}^d$ of a set of points $\{x_1, \ldots, x_n\}$ in $\mathbb{R}^s$, $s > d$. The only parameter of this algorithm is an integer K that measures the size of the neighborhood of each point. We will denote by N_i the K-neighborhood of x_i, that is to say $N_i \overset{\text{def.}}{=} \{x_{m(1)}, \ldots, x_{m(K)}\}$, where $x_{m(j)}$ is the jth closest point to x_i for the Euclidean metric. We will briefly recall the two main steps of the procedure:

Step 1: First, for each point x_i, we are looking for some weights $w_{i,j}$ that locally best reconstruct the manifold, from the set N_i only, by minimizing

$$E_1\left(\{w_{i,j}\}_j\right) = \left\|x_i - \sum_j w_{i,j} x_j\right\|^2. \tag{3}$$

We further enforce that $w_{i,j} = 0$ if $x_j \notin N_i$, and that $\sum_j w_{i,j} = 1$. This imposes that the reconstruction is both local and invariant under affine transformations. In a Euclidean setting, the minimization (3) requires the introduction of the Gram matrix $C(x)$ defined by

$$(C(x))_{i,j} \overset{\text{def.}}{=} \left\langle x - x_{m(i)}, \, x - x_{m(j)} \right\rangle \tag{4}$$

The solution of the minimization (3) is then

$$w_{i,j} = \mathbf{1}_{N_i}(x_j) \frac{\sum\limits_q (C(x_i)^{-1})_{k,q}}{\sum\limits_{p,q} (C(x_i)^{-1})_{p,q}}, \quad \text{where} \quad x_j = x_{m(k)}.$$

The value of $\mathbf{1}_{N_i}(x)$ is equal to 1 if $x \in N_i$, and 0 otherwise.

Step 2: To reconstruct the manifold in low dimension (here in 2D), we want to solve a global minimization procedure, for $\widetilde{x}_i \in \mathbb{R}^2$:

$$\text{minimize} \quad E_2\left(\{\widetilde{x}_i\}\right) = \sum_i \left\|\widetilde{x}_i - \sum_j w_{i,j} \widetilde{x}_j\right\|^2.$$

subject to $\sum \widetilde{x}_i = 0$ and $\sum \|\widetilde{x}_i\|^2 = 1$ to avoid a degenerate solution. To solve this problem, we need to form the matrix $M \overset{\text{def.}}{=} (W - \text{Id})^{\text{T}}(W - \text{Id})$, where W is a sparse matrix containing all the weights. The eigenvector of M with lowest eigenvalue it the constant vector $\mathbf{1}$ which should be discarded. The d following eigenvectors give us the coordinates of our embedding in $\mathbb{R}^d$ for each point.

The fact that we only need to perform computations on sparse matrices allows an improvement of one order of magnitude over dense procedures such as in [Zigelman et al., 2002].

3.3. Geodesic LLE

In the previous section we saw the classical LLE algorithm in a Euclidean setting. To solve the flattening problem for a mesh, we need to extend these computations to the manifold setting. The following modifications allow such an extension.

Modification 1: The points $\{x_1, \ldots, x_n\}$ should be sampled as uniformly as possible on $\mathcal{S}$. That is why we use the greedy sampling algorithm of Section 2.2 to select these points. To get an adaptive sampling, one could use a varying speed function, as shown in Figure 2.2 (see also [Peyré and Cohen, 2003] for a curvature-based adaptation).

Modification 2: The K-neighborhood N_i of each point should be computed using the geodesic distance and not the Euclidean one. This can be done very quickly using a local front propagation.

Modification 3: The matrix $C(x)$ of equation (4) can not be computed anymore using dot products. Instead, following [Roweis and Saul, 2000] (pairwise LLE), we propose the following formula

$$-2C(x)_{i,j} \stackrel{\text{def.}}{=} d(x_{m(i)}, x_{m(j)})^2 - \frac{1}{K} \sum_{k=1}^{K} d(x_{m(i)}, x_{m(k)})^2$$

$$- \frac{1}{K} \sum_{k=1}^{K} d(x_{m(k)}, x_{m(j)})^2 + \frac{1}{K^2} \sum_{k,l=1}^{K} d(x_{m(k)}, x_{m(l)})^2,$$

that only uses geodesic distance information. This formula is equivalent to (4) in the Euclidean setting.

3.4. Extending the Map

The three modifications proposed in the previous section allow us to find the location of $\widetilde{x}_i = f(x_i) = (f_1(x_i), f_2(x_i))^{\mathrm{T}} \in \mathbb{R}^2$ for each base point x_i. To compute the whole map f, we need to interpolate the location of $f(x) = (f_1(x), f_2(x))^{\mathrm{T}}$ for each point $x \in \mathcal{S}$, using the known locations $f(x_i)$.

This problem has been addressed very recently in [Bengio et al., 2003], by recasting it into a unified framework of eigenvector learning, common to many dimensionality reduction methods.

To extend f, we use the fact that vectors $\{f_1(x_i)\}_{i=1}^n$ and $\{f_2(x_i)\}_{i=1}^n$ are eigenvectors of the symmetric matrix $M = (W - \mathrm{Id})^{\mathrm{T}}(W - \mathrm{Id})$ (with eigenvalues λ_1 and λ_2). In the continuous setting, this matrix becomes a symmetric kernel

$\widetilde{M}(x, y)$ for each point x, y in $\mathcal{S}$. Matrix multiplication by M is then replaced by

$$\varphi \mapsto \widetilde{M}\varphi(x) \stackrel{\text{def.}}{=} \int_{\mathcal{S}} \varphi(y)\widetilde{M}(x, y)\mathrm{d}y, \tag{5}$$

where φ is any mapping from $\mathcal{S}$ to $\mathbb{R}$. Using this remark, it is natural to suppose that the continuous maps f_1 and f_2 are eigenfunctions of the operator defined by equation (5) for the same eigenvalues λ_1 and λ_2. This implies that we can compute them using a Nyström-like formula

$$f_1(x) = \frac{1}{\lambda_1} \int_{\mathcal{S}} f_1(y)\widetilde{M}(x, y)\mathrm{d}y \approx \frac{1}{n\lambda_1} \sum_{i=1}^{n} f_1(x_i)\widetilde{M}(x_i, y), \tag{6}$$

and similarly for f_2.

Since the $f_1(x_i)$ are known, we just need to setup our kernel $\widetilde{M}$. The only constraint is that for all $y \in \mathcal{S}$, $\widetilde{M}(x_i, y)$ should be easy to compute, e.g., it should only involve already computed distances such as $d(x_j, y)$ for $x_j \in N_i$. This can be done in a straightforward manner by first setting the weights for y:

$$w(x_i, y) \stackrel{\text{def.}}{=} \mathbf{1}_{N_i}(y) \frac{\sum\limits_{q} (C(x_i)^{-1})_{k,q}}{\sum\limits_{p,q} (C(x_i)^{-1})_{p,q}} \quad \text{with} \quad y = x_{m(k)},$$

and then defining the kernel:

$$\widetilde{M}(x, y) \stackrel{\text{def.}}{=} w(x, y) + w(y, x) - \sum_{k} w(x_k, x)w(x_k, y),$$

We can check that for base points, we retrieve the original matrix up to a substraction of the identity, i.e., $\widetilde{M}(x_i, x_j) = \delta_{i,j} - M_{i,j}$. This shift is only here to avoid a singularity along the diagonal and does not modify the computation.

This shows that we can extend the map f to a new point x using only some local distance information between x and its neighborhood in $\{x_1, \ldots, x_n\}$. Furthermore, most of the time, this information is already available from previous front propagations performed to flatten $\{x_1, \ldots, x_n\}$.

Figure 7 shows the flattening of one half of a human head. Even with a large patch that contains holes, our method gives very good results (no face flip) with only 100 base vertices.

Figure 8 shows the influence of the number of base points on the flattening. Even with only 20 points, the resulting embedding is nearly smooth except at the border of the mesh, and with 100 points, we get a perfectly smooth flattening.

4. Results and Discussion

Texturing of a Complex Model. To perform texture mapping on a complex 3D mesh, a segmentation step is required to first cut the model into disk-shaped charts. Although the study of this step is outside the scope of this paper, we note that the notion of Voronoi cells is often to perform mesh partition, as introduced

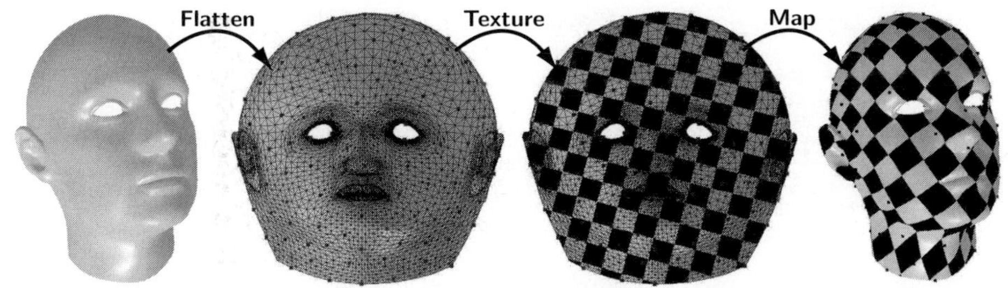

FIGURE 7. The original model, texture on the flattened domain, and on the 3D mesh.

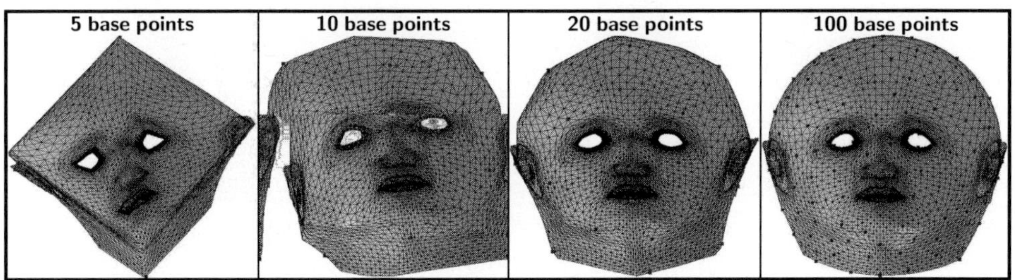

FIGURE 8. Influence of the number of base points. The original model is shown on the left of Figure 7

in [Eck et al., 1995]. We choose to use a scheme based on a weighted geodesic distance [Peyré and Cohen, 2004], since its continuous nature is clearly related to our flattening approach.

On Figure 9 one can see the whole pipeline in action. This includes first a centroidal tessellation of the mesh, then the extraction and flattening of each cell, and lastly the texturing of the model.

Computation Times. For our tests of the flattening procedure, we have chosen to use a fixed number of points (200 points), since the geometric complexity of the meshes was almost constant. The parameterization of [Desbrun et al., 2002] is implemented using the boundary-free formulation (Neumann condition).

Table 1 shows the complexity of the algorithms mentioned in the paper, for a mesh of 10k vertices. The constant A is the number of steps in the gradient descent for the localization of the intrinsic center of mass, which is about $A = 8$ for 10k vertices. The constant B represent the number of base points, which is $n/100$ in our tests. This clearly shows the speed up that Geodesic LLE can bring over global methods such as [Zigelman et al., 2002]. This is confirmed by the running times reported in table 2. For large meshes, the stability of our method is an advantage over the approaches based on large linear system such as [Desbrun et al., 2002], for which it is difficult to ensure the convergence of the conjugate gradient.

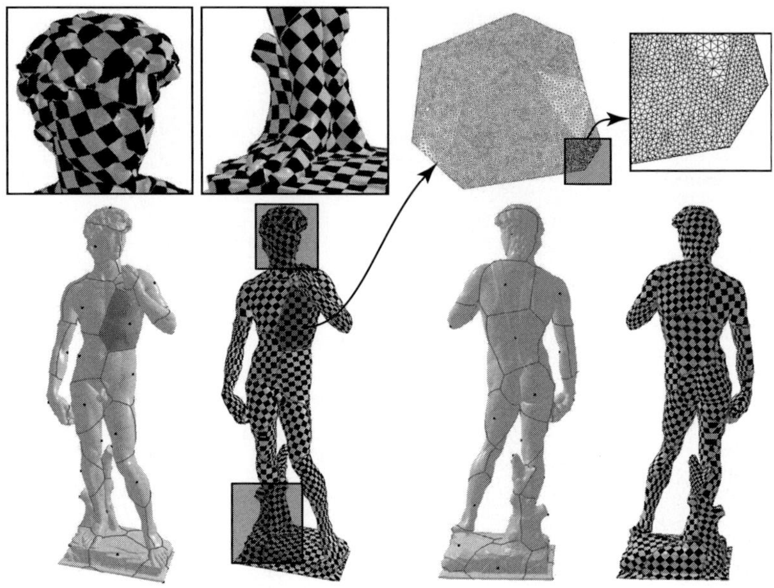

FIGURE 9. Texturing of the David.

	F. Marching	Greedy sampl.	1 Lloyd iter.	Zigelman02	Geodesic LLE
Complexity	$n \log(n)$	$n \log(n)^2$	$An \log(n)$	$Bn \log(n) + B^3$	$n \log(n) + B^2$
Times	$2s$	$10s$	$6s$	$55s$	$28s$

TABLE 1. Complexity of the algorithms

Nbr.vertices	[Zigelman et al., 2002]	[Desbrun et al., 2002]	Geodesic LLE
1,000	$7s$	$3s$	$5s$
10,000	$55s$	$25s$	$28s$
100,000	$440s$	$210s$	$150s$
700,000	$2160s$	$1320s$	$740s$

TABLE 2. Comparison of flattening algorithms

Discussion. The complete texturing of the David mesh (700,000 vertices) shown on Figure 9 clearly enlightens the strengths of our approach:

- The resulting flattening map is smooth, with no face flip (at least on this model). This can be seen on the close-up of the flattened domain.
- The whole texturing procedure takes $740s$, which shows an important speed up with respect to previous methods.
- Our scheme is more local than the flattening procedure of [Zigelman et al., 2002], but it does not reach the per-vertex resolution of classical methods such as [Desbrun et al., 2002]. This enables both fast computations and respect of

	M_1	M_2	M_3
[Desbrun et al., 2002] (conformal)	$E_c = 0.9$ $E_a = 1.2$	$E_c = 1.5$ $E_a = 3$	$E_c = 2.5$ $E_a = 10.4$
[Desbrun et al., 2002] (authalic)	$E_c = 1.4$ $E_a = 0.6$	$E_c = 3.0$ $E_a = 1.1$	$E_c = 8.3$ $E_a = 3.5$
[Zigelman et al., 2002] (MDS)	$E_c = 0.8$ $E_a = 0.9$	The flattening is not valid	
Our scheme (GeodesicLLE)	$E_c = 1.1$ $E_a = 0.9$	$E_c = 1.7$ $E_a = 1.6$	$E_c = 6.5$ $E_a = 5.5$

TABLE 3. Area and angular distortion for various schemes

the small scale variations (bumps or noise), which is not the case of [Zigelman et al., 2002].

Notice that there is no theoretical guarantee on the validity of the flattening. The only cases where face flips occurs is on patches with huge isoperimetric distortion. We believe that this is not a real issue since such degenerate cases can be easily detected and fixed (for example by subdividing the region). It is important to note that classical methods also face similar problems. In [Levy et al., 2002] a cut is performed to accelerate the convergence of the system resolution and ease the parameterization in regions with a sock-like shape. In practice however, our segmentation algorithm ensures that patches that need to be flattened do not contain high curvature variations and the whole process performs very well with no face flip.

Distorsion Measures. To support our claim that our flattening scheme performs a trade-off between conservation of area and conservation of angle, we have performed some test (see table 3). We used 3 finger-like meshes M_1, M_2 and M_3 with increasing isoperimetric distortion. On each face x we compute the eigenvalues (s_1, s_2) of of the Jacobian of the parameterization map (linearly evaluated). Locally, conformality is characterized by $s_1 = s_2$ and conservation of area by $s_1 s_2 = 1$. As a conformal metric, we use

$$E_c(M)^2 = \frac{1}{A} \sum_x \left| \frac{s_1(x)}{s_2(x)} + \frac{s_2(x)}{s_1(x)} - 2 \right|^2 A(x)$$

and as an equi-areal metric we use

$$E_a(M)^2 = \frac{1}{A} \sum_x \left| s_1(x)s_2(x) + \frac{1}{s_1(x)s_2(x)} - 2 \right|^2 A(x)$$

where $A(x)$ is the area of a face x, A is the total area.

5. Conclusion

We have described new algorithms to perform the remeshing and the flattening of a genus-0 triangulated manifold. The main tool that allows having a fast algorithm is the fast marching on a triangulated mesh, together with some improvements we added. We have presented a fast algorithm for remeshing of a surface with a uniform or adaptive distribution. This is based on iteratively choosing the farthest point according to a weighted distance on the surface. We introduced a geodesic version of Locally Linear Embedding that is able to perform fast computations on a given set of points, and to extend the embedding to the rest of the mesh in a transparent manner. The resulting flattening is smooth and achieves a desirable trade-off between conservation of angle and area.

References

[Alliez et al., 2003] Alliez, P., D. Cohen-Steiner, O. Devillers, B. Levy, and M. Desbrun: 2003, 'Anisotropic Polygonal Remeshing'. *ACM Transactions on Graphics. Special issue for SIGGRAPH conference* pp. 485–493.

[Bengio et al., 2003] Bengio, Y., J.-F. Paiement, and P. Vincent: 2003, 'Out-of-Sample Extensions for LLE, Isomap, MDS, Eigenmaps, and Spectral Clustering'. *Proc. NIPS 2003*.

[Chew, 1993] Chew, L. P.: 1993, 'Guaranteed-Quality Mesh Generation for Curved Surfaces'. *Proc. of the Ninth Symposium on Computational Geometry* pp. 274–280.

[Cohen, 2001] Cohen, L.: 2001, 'Multiple Contour Finding and Perceptual Grouping Using Minimal Paths'. *Journal of Mathematical Imaging and Vision* **14**(3), 225–236.

[Cohen and Kimmel, 1997] Cohen, L.D. and R. Kimmel: 1997, 'Global Minimum for Active Contour Models: A Minimal Path Approach'. *International Journal of Computer Vision* **24**(1), 57–78.

[Cohen-Steiner and Morvan, 2003] Cohen-Steiner, D. and J.-M. Morvan: 2003, 'Restricted Delaunay Triangulations and Normal Cycles'. *Proc. 19th ACM Sympos. Comput. Geom.* pp. 237–246.

[Desbrun et al., 2002] Desbrun, M., M. Meyer, and P. Alliez: 2002, 'Intrinsic Parameterizations of Surface Meshes'. *Eurographics conference proceedings* **21**(2), 209–218.

[Eck et al., 1995] Eck, M., T. DeRose, T. Duchamp, H. Hoppe, M. Lounsbery, and W. Stuetzle: 1995, 'Multiresolution Analysis of Arbitrary Meshes'. *Computer Graphics* **29**(Annual Conference Series), 173–182.

[Eldar et al., 1997] Eldar, Y., M. Lindenbaum, M. Porat, and Y. Zeevi: 1997, 'The Farthest Point Strategy for Progressive Image Sampling'. *IEEE Trans. on Image Processing* **6**(9), 1305–1315.

[Floater et al., 2002] Floater, M. S., K. Hormann, and M. Reimers: 2002, 'Parameterization of Manifold Triangulations'. *Approximation Theory X: Abstract and Classical Analysis* pp. 197–209.

[Kimmel and Sethian, 1998] Kimmel, R. and J. Sethian: 1998, 'Computing Geodesic Paths on Manifolds'. *Proc. Natl. Acad. Sci.* **95**(15), 8431–8435.

[Kunert, 2002] Kunert, G.: 2002, 'Towards Anisotropic Mesh Construction and Error Estimation in the Finite Element Method'. *Numerical Methods in PDE* **18**, 625–648.

[Leibon and Letscher, 2000] Leibon, G. and D. Letscher: 2000, 'Delaunay triangulations and Voronoi diagrams for Riemannian manifolds'. *ACM Symposium on Computational Geometry* pp. 341–349.

[Levy et al., 2002] Levy, B., S. Petitjean, N. Ray, and J. Maillot: 2002, 'Least Squares Conformal Maps for Automatic Texture Atlas Generation'. In: ACM (ed.): *Special Interest Group on Computer Graphics – SIGGRAPH'02, San-Antonio, Texas, USA*.

[Moenning and Dodgson, 2003] Moenning, C. and N.A. Dodgson: 2003, 'Fast Marching Farthest Point Sampling'. *Proc. EUROGRAPHICS 2003*.

[Onishi and Itoh, 2003] Onishi, K. and J. Itoh: 2003, 'Estimation of the necessary number of points in Riemannian Voronoi diagram'. *Proc. CCCG*.

[Peyré and Cohen, 2003] Peyré, G. and L.D. Cohen: 2003, 'Geodesic Remeshing Using Front Propagation'. *Proc. IEEE Variational, Geometric and Level Set Methods 2003*.

[Peyré and Cohen, 2004] Peyré, G. and L.D. Cohen: 2004, 'Surface Segmentation Using Geodesic Centroidal Tesselation'. *Proc. 3D Data Processing Visualization Transmission 2004*.

[Roweis and Saul, 2000] Roweis, S. and L. Saul: 2000, 'Nonlinear Dimensionality Reduction by Locally Linear Embedding'. *Science* **290**(5500), 2323–2326.

[Ruppert, 1995] Ruppert, J.: 1995, 'A Delaunay Refinement Algorithm for Quality 2-Dimensional Mesh Generation'. *Journal of Algorithms* **18**(3), 548–585.

[Sethian, 1999] Sethian, J.: 1999, *Level Sets Methods and Fast Marching Methods*. Cambridge University Press, 2nd edition.

[Sifri et al., 2003] Sifri, O., A. Sheffer, and C. Gotsman: 2003, 'Geodesic-based Surface Remeshing'. *Proc. 12th International Meshing Roundtable* pp. 189–199.

[Surazhsky et al., 2003] Surazhsky, V., P. Alliez, and C. Gotsman: 2003, 'Isotropic Remeshing of Surfaces: a Local Parameterization Approach'. *Proc. 12th International Meshing Roundtable*.

[Tenenbaum et al., 2000] Tenenbaum, J.B., V. de Silva, and J.C. Langford: 2000, 'A Global Geometric Framework for Nonlinear Dimensionality Reduction'. *Science* **290**(5500), 2319–2323.

[Terzopoulos and Vasilescu, 1992] Terzopoulos, D. and M. Vasilescu: 1992, 'Adaptive Meshes and Shells: Irregular Triangulation, Discontinuities, and Hierarchical Subdivision'. In: *Proc. IEEE CVPR '92*. Champaign, Illinois, pp. 829–832.

[Tsitsiklis, 1995] Tsitsiklis, J.: 1995, 'Efficient Algorithms for Globally Optimal Trajectories'. *IEEE Trans. on Automatic Control*.

[Ulichney, 1993] Ulichney, R.: 1993, 'The Void-and-Cluster Method for Generating Dither Arrays'. *Proc. IS&T Symposium on Electronic Imaging Science & Technology, San Jose, CA* **1913**(9), 332–343.

[Zigelman et al., 2002] Zigelman, G., R. Kimmel, and N. Kiryati: 2002, 'Texture Mapping Using Surface Flattening via Multi-dimensional Scaling'. *IEEE Trans. on Visualization and Computer Graphics* **8**(1), 198–207.

Gabriel Peyré
CMAP, École Polytechnique, UMR CNRS 7641
e-mail: `peyre@cmapx.polytechnique.fr`

Laurent Cohen
CEREMADE, Université Paris Dauphine, UMR CNRS 7534
e-mail: `cohen@ceremade.dauphine.fr`

Progress in Nonlinear Differential Equations
and Their Applications, Vol. 63, 173–177

On the Newton Body Type Problems

M. Comte

Abstract. We study solutions of Newton's type problems. More precisely we
minimize functionals

$$\int_\Omega F(|\nabla u(x)|)dx \tag{1}$$

where F is first concave and then convex, look at critical points of the functional, prove existence of particular ones and describe their behavior.

In his Principia Mathematica I. Newton considered, in 1685 ([13]), one of the
pioneering papers of the Calculus of Variations: to find the shape of a symmetrical
revolution body moving in a fluid with minimal resistance to motion. As a matter
of fact, the problem was already suggested by Galilee in his famous *Discursi* in
1638 (for a detailed history see Goldstine [9]).

In fact Newton works in the framework of "a rare medium consisting of equal
particles freely disposed at equal distances". Thus particles do not interact with
each other. Under this assumption, the shape of the front part of the required
body can be described by the graph of a function $u : \Omega \to \mathbb{R}$, where $\Omega \subset \mathbb{R}^2$ is a
given bounded domain. This problem is explained with more details in [4]. Since
the length of the body is finite the quantity $\max u - \min u$ does not exceed a given
constant $M > 0$; since the resistance does not change by adding a constant to
u (translation of the body), it is assumed without loss of generality that u takes
values in $[0, M]$. Even with these restrictions, the computation of the effective
resistance of the body could be very complicated. Some particles hit the body,
transmitting momentum; this quantity depends only on the impact angle and
therefore on $|\nabla u(x)|$ at the contact point. But after reflection, some particles
may hit the body a second time; therefore it is necessary to restrict the class
of admissible functions in order to exclude this event. If each particle hit the
body at most once, then the effective resistance of the body can be expressed (in
appropriate units) by:

$$F(u) = \int_\Omega \frac{dx}{1 + |\nabla u(x)|^2}. \tag{2}$$

One of the simplest way to ensure the single-impact assumption is to require that u
is concave. This is indeed the case considered first by Newton. However, this is very

restrictive and appears to be mathematical artifact. Therefore, many authors tried to remove this limitation. It was already noticed by Legendre in 1786 [11] that the solution of Newton does not minimize the functional among all radial functions; for, if you consider wildly oscillating functions, the value of the functional becomes arbitrarily small. However, many authors objected that it does not make sense from the physical statement of the problem: indeed for these functions, the resistance is not expressed by $F(u)$, since the single impact assumption is not taken into account.

Some different approaches are explained in [3]. But the least restrictive one, for this expression of the resistance, is to require only that any particle hitting the graph of u at x with vertical velocity does not hit it again, as explained in Section 5 of [3]. Under this assumption we considered in a joined paper with T. Lachand-Robert [7], a set of minimization CCM containing the set of concave functions and we looked at the problem:

$$\inf_{u \in CCM} F(u) \quad \text{where } F(u) := \int_\Omega \frac{dx}{1 + |\nabla u(x)|^2}. \tag{3}$$

We then obtained

Theorem 0.1. *Let $u \in CCM$ be regular on the boundary . Then u is not a minimizer for* (3).

For more details see [7].

On the other hand in the case of Newton, that is when Ω is the unit ball and when the infimum is searched in the radial functions of CCM, we have the existence of minimizers. In [6] we proved

Theorem 0.2. *There exists a number M^* such that, if $M \geq M^*$, then there is exactly one local minimizer of F; if $M < M^*$, there is an infinite number of local minimizers for F, and the corresponding set is not compact in $W^{1,p}$.*

As mentioned in Armanini ([1]), in the same historical book Newton considered also other resistance assumptions leading to different power expressions of the type

$$\int_\Omega \frac{1}{1 + |\nabla u(x)|^n} dx \tag{4}$$

with $n \geq 1$.

It is not strange that the Newton resistance functional led need some suitable corrections when the present fluid mechanics theory is applied in the context of an ideal or viscous fluid and so, for instance, in [14] it was proposed a resistance functional of the type

$$\int_\Omega \frac{1}{1 + |\nabla u(x)|^n} dx + \int_\Omega p(x, u(x)) dx.$$

Nevertheless, it is worth mentioning that even though the Newton's resistance model is only a crude approximation, it appears to provide good results in many contexts dealing, for instance, with a rarified gas in hypersonic aerodynamics, having been considered by many distinguished specialists of this area as, for instance, von Karman, Ferrari, Lightill and Sears (see the exposition made in the NASA report [8] and the book [12]).

We can remark that the expression under the integral in functionals (3) (or (4)) is not globally convex on ∇u (although it is a convex function when $|\nabla u(x)| \geq \alpha$ for some suitable $\alpha > 0$) and that it is not coercive (in fact it converges to zero when $|\nabla u(x)| \to +\infty$). Those two facts arise quite often in many other special (but relevant) problems of the Calculus of Variations (see, for instance, some other classical and recent examples mentioned in [2]). This is our motivation with J.I. Diaz to consider a general class of functional (being invariant by symmetrical changes of coordinates) of the form

$$\int_\Omega F(|\nabla u(x)|)dx \tag{5}$$

In fact, we did not deal with the associated minimization problem but with the study of the associated stationary points. So, with J.I. Diaz, we considered a class of quasi-linear obstacle problems which can be formulated as follows

$$(OP) \begin{cases} -div(A(|\nabla u|)\nabla u) + \beta(u) \ni 0 & \text{in } \Omega, \\ u = 0 & \text{on } \partial\Omega, \end{cases} \tag{6}$$

where β is the maximal monotone graph given by

$$\begin{cases} \beta(u) = \{0\} & \text{if } u < M \\ \beta(M) = [0, +\infty) \\ \beta(u) = \phi & \text{if } u > M \end{cases}$$

with $A \in C^1(0, +\infty)$ satisfying the following set of assumptions: there exists $\alpha_A \geq 0$ such that

$$\begin{cases} \text{the function } t \to tA(t) \text{ is decreasing on} & (0, \alpha_A) \\ \text{and is increasing on} & (\alpha_A, +\infty), \end{cases} \tag{7}$$

$$A < 0 \quad \text{on} \quad [0, +\infty) \quad \text{and} \quad \lim_{t \to +\infty} tA(t) = 0, \tag{8}$$

$$\lim_{a \to +\infty} A(a) \int_{\alpha_A}^a \frac{d\tau}{\tau A(\tau)} < 1. \tag{9}$$

Notice that the formulation corresponding to the stationary points of functional (5) leads to $A(|\nabla u|) = \frac{F'(|\nabla u|)}{|\nabla u|}$ and that for the classical Newton minimal resistance problem we have $F(t) = (1 + t^2)^{-1}$ and thus $A(t) = \frac{-2}{(1+t^2)^2}$ which satisfies the first set of assumptions with $\alpha_A = \frac{1}{\sqrt{3}}$ and

$$\lim_{a \to +\infty} A(a) \int_\alpha^a \frac{d\tau}{\tau A(\tau)} = \frac{1}{4}.$$

We shall deal with solutions of the obstacle problem (OP) in the class of functions such that

$$u \in H_0^1(\Omega) \text{ and } |\nabla u(x)| \geq \alpha_A \text{ if } u(x) < M. \tag{10}$$

In the radial case, that is $\Omega = B(0, R)$ and u radially symmetric satisfying (OP) and (10), it is easy to see that the problem reduces to the study of the one-dimensional free boundary value problem

$$\begin{cases} -\frac{1}{r}\frac{d}{dr}(rA(|u'|)u') = 0 & \text{in } (\rho, R) \\ u(R) = 0, \ u(\rho) = M \\ -u'(R^-) > \alpha_A \end{cases} \tag{11}$$

where $\rho \in [0, R)$ must be determined. In [5] we obtained

Theorem 0.3. *Let $R > 0$ be given. For every $M > 0$ there exists $\rho_M \in (0, R)$ such that for any $m \in (\alpha_A, +\infty)$ there exists $u(r) = u(r : m)$ solution of the obstacle problem satisfying*

   i) *$u \equiv M$ in $[0, \rho_m]$ for some $\rho_m \in [\rho_M, R)$,*
   ii) *$-u'(\rho_m) = m$*
   iii) *u is strictly concave in (ρ_m, R) and $u \in W^{1,\infty}(0, R)$.*

Finally, the map $M \to \rho_M$ is decreasing and concave.

In [5] we also obtained results about the coincidence set (the flat region of the body) and results with different kind of assumptions on the function A.

References

[1] E. Armanini, *Sulla superficie di minima resistenza*, Ann. mat. pura appl. (3) **4** (1900), pp. 131–148.

[2] B. Botteron and P. Marcellini, *A general approach to the existence of minimizers of one-dimensional non-coercive integrals of the calculus of variations*, Ann. Inst. Henri Poincaré, **8** (1991), pp. 197–223.

[3] G. Buttazzo, B. Kawohl, *On Newton's problem of Minimal Resistance*, Mathematical Intelligencer, **15** (1993), pp. 7–12.

[4] G. Buttazzo, V. Ferone, B. Kawohl, *Minimum Problems over Sets of Concave Functions and Related Questions*, Math. Nachrichten, **173** (1993), pp. 71–89.

[5] M. Comte and J.I. Díaz, *On the Newton partially flat minimal resistance body type problems*, in preparation.

[6] M. Comte, T. Lachand-Robert, *Newton's problem of the body of minimal resistance under a single-impact assumption*, Calc. Var. 12, (2001), pp. 173–211.

[7] M. Comte, T. Lachand-Robert, *Existence of minimizers for the Newton's problem of the body of minimal resistance under a single-impact assumption*, J. Anal. Math., 83, 2001, p. 313–335.

[8] A.J. Eggers Jr., M.M. Resnikoff and D.H. Dennis, *Bodies of revolution having minimum drag at high supersonic airspeeds*, NASA Report 1306, 1958.

[9] H.H. Goldstine, *A History of the Calculus of Variations from the 17th through the 19th Century*, Springer-Verlag, Heidelberg, 1980.

[10] D. Kinderlehrer and G. Stampacchia, An *Introduction to Variational Inequalities and Their Applications*, Academic Press, New York, 1990.

[11] A.-M. Legendre, *Sur la manière de distinguer les maxima des minima dans le calcul des variations*, Mém. Acad. Sci., Paris (1786) éd 1788, pp. 7–37.

[12] A. Miele, Theory of Optimum Aerodynamic Shapes. Academic Press, London 1965.

[13] I. Newton, *Philosophiae Naturalis Principia Mathematica*. 1686.

[14] A. Wagner, *A Remark on Newton's Resistance Formula*, Z. Angew. math. Mech. ZAMM, **79** (1999), pp. 423–427.

M. Comte
Université Pierre et Marie Curie
Laboratoire Jacques-Louis Lions
F-75252 Paris Cedex 05, France
e-mail: comte@ann.jussieu.fr
URL: www.ann.jussieu.fr

Progress in Nonlinear Differential Equations
and Their Applications, Vol. 63, 179–188

Some Open Problems on Water Tank Control Systems

Jean-Michel Coron

It is a great pleasure and honor for me to write a paper in this volume to celebrate Haïm Brezis's 60th birthday. Haïm, as every one knows, is not only an outstanding mathematician with an immense far-reaching influence: he is also an outstanding Ph.D. adviser. It has been a great chance for me to have the opportunity to do my Ph.D. under his supervision and to work after with him. One of the numerous Haïm's qualities which make him a terrific adviser is his ability to propose seminal and fascinating open problems. In order to celebrate Haïm's birthday, I try to follow my master and friend and propose some open problems (on a nonlinear partial differential equation modelling a water tank control system).

1. Modelling equations of a water tank control system

We consider a 1-D tank containing an inviscid incompressible irrotational fluid. The tank is subject to one-dimensional horizontal moves. We assume that the horizontal acceleration of the tank is small compared to the gravity constant and that the height of the fluid is small compared to the length of the tank. This motivates the use of the Saint-Venant equations [12] (also called shallow water equations) to describe the motion of the fluid; see, e.g., [8, Section 4.2]. After suitable scaling arguments, the length of the tank and the gravity constant can be taken to be equal to 1; see [4]. Then the dynamics equations considered are, see [9] and [4],

$$H_t\left(t, x\right) + \left(Hv\right)_x\left(t, x\right) = 0, \tag{1}$$

$$v_t\left(t, x\right) + \left(H + \frac{v^2}{2}\right)_x\left(t, x\right) = -u\left(t\right), \tag{2}$$

$$v(t, 0) = v(t, 1) = 0, \tag{3}$$

$$\frac{\mathrm{d}s}{\mathrm{d}t}\left(t\right) = u\left(t\right), \qquad \frac{\mathrm{d}D}{\mathrm{d}t}\left(t\right) = s\left(t\right), \tag{4}$$

where (see Figure 1)

- $H(t, x)$ is the height of the fluid at time t and for $x \in [0, 1]$,
- $v(t, x)$ is the horizontal water velocity of the fluid *in a referential attached to the tank* at time t and for $x \in [0, 1]$ (in the shallow water model, all the points on the same vertical have the same horizontal velocity),
- u is the horizontal acceleration of the tank in the absolute referential,
- s is the horizontal velocity of the tank,
- D is the horizontal displacement of the tank.

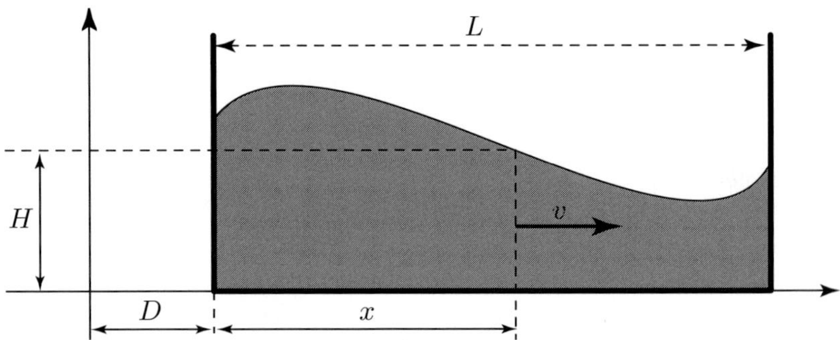

FIGURE 1. Fluid in the 1-D tank

This is a control system, denoted Σ, where

- the state is $Y = (H, v, s, D)$,
- the control is $u \in \mathbb{R}$.

By scaling arguments we may assume that $g = L = 1$ and that, for every steady state, H, which is then a constant function, is equal to 1; see [4]. One is interested in the local controllability of the control system Σ around the equilibrium point

$$(Y_e, u_e) := ((1, 0, 0, 0), 0).$$

Of course, the total mass of the fluid is conserved so that, for every solution of (1) to (3),

$$\frac{\mathrm{d}}{\mathrm{d}t} \int_0^1 H(t, x)\, \mathrm{d}x = 0. \tag{5}$$

(One gets (5) by integrating (1) on $[0, 1]$ and by using (3) together with an integration by parts.) Moreover, if H and v are of class $\mathcal{C}^1$, it follows from (2) and (3) that

$$H_x(t, 0) = H_x(t, 1) \quad (= -u(t)). \tag{6}$$

Therefore we introduce the vector space E of functions

$$Y = (H, v, s, D) \in \mathcal{C}^1([0, 1]) \times \mathcal{C}^1([0, 1]) \times \mathbb{R} \times \mathbb{R}$$

such that

$$H_x(0) = H_x(1), \; v(0) = v(1) = 0, \tag{7}$$

and consider the affine subspace $\mathcal{Y} \subset E$ of $Y = (H, v, s, D) \in E$ satisfying

$$\int_0^1 H(x)\mathrm{d}x = 1. \tag{8}$$

With these notations, we can define a trajectory of the control system Σ.

Definition of a trajectory. *Let T_1 and T_2 be two real numbers satisfying $T_1 \leqslant T_2$. A function $(Y, u) = ((H, v, s, D), u) : [T_1, T_2] \to \mathcal{Y} \times \mathbb{R}$ is a trajectory of the control system Σ if*

   (i) *the functions H and v are of class C^1 on $[T_1, T_2] \times [0, 1]$,*
   (ii) *the functions s and D are of class C^1 on $[T_1, T_2]$ and the function u is continuous on $[0, T]$,*
   (iii) *the equations (1) to (4) hold for every $(t, x) \in [T_1, T_2] \times [0, 1]$.*

2. Controllability: results and open problems

We are interested in the local controllability of Σ around (Y_e, u_e). Since one is looking for a *local* result, one first studies the linearized control system around the trajectory (Y_e, u_e). Indeed, if this linearized control system is controllable, one can expect to get the local controllability of Σ by using the inverse mapping theorem. (In fact, for nonlinear partial differential control systems, this is not so direct when some "loss of derivatives" appear; see [1, 4].) The linearized control system around (Y_e, u_e) is

$$(\Sigma_{\text{lin}}) \left\{ \begin{array}{l} h_t + v_x = 0, \; v_t + h_x = -u\,(t)\,, \; v(t, 0) = v(t, 1) = 0, \\ \frac{\mathrm{d}s}{\mathrm{d}t}\,(t) = u\,(t)\,, \; \frac{\mathrm{d}D}{\mathrm{d}t}\,(t) = s\,(t)\,, \end{array} \right. \tag{9}$$

where the state is $(h, v, s, D) \in \mathcal{Y}_{\text{lin}}$, with

$$\mathcal{Y}_{\text{lin}} := \left\{ (h, v, s, D) \in E; \; \int_0^L h\mathrm{d}x = 0 \right\},$$

and the control is $u \in \mathbb{R}$. It has been proved by F. Dubois, N. Petit and P. Rouchon in [9] that the linear control system (9) is not controllable (see also [11]). This noncontrollability property can be seen by noticing that (9) implies that, if

$$h(0, 1 - x) = -h(0, x) \text{ and } v(0, 1 - x) = v(0, x) \; \forall x \in [0, 1],$$

then

$$h(t, 1 - x) = -h(t, x) \text{ and } v(t, 1 - x) = v(t, x) \; \forall x \in [0, 1], \; \forall t.$$

Even if the control system (9) is not controllable, this control system, as it is proved in [9], is steady-state controllable, which means that one can move, from every steady state $(h_0, v_0, s_0, D_0) := (0, 0, 0, D_0)$ to every steady state $(h_1, v_1, s_1, D_1) := (0, 0, 0, D_1)$ for this control system (see also [11] when the tank has a non-straight bottom). This does not imply that the related property (move from $(1, 0, 0, D_0)$ to

$(1, 0, 0, D_1))$ also holds for the nonlinear control system Σ (even if D_0 and D_1 are small), as shown by the following example.

Example 1. *Consider the two following control systems*

$$\dot{x}_1 = x_2^3, \ \dot{x}_2 = x_1 + u, \dot{s} = u, \dot{D} = s, \tag{10}$$

$$\dot{x}_1 = x_2^2, \ \dot{x}_2 = x_1 + u, \dot{s} = u, \dot{D} = s, \tag{11}$$

where, for both systems, the state is $x = (x_1, x_2, s, D) \in \mathbb{R}^4$ and the control is $u \in \mathbb{R}$. The set of steady states of these two control systems is the set of the $((0, 0, 0, a), 0)$ with $a \in \mathbb{R}$. These two control systems have the same linearized control system around $(x_e, u_e) := ((0, 0, 0, 0), 0)$, namely

$$\dot{x}_1 = 0, \ \dot{x}_2 = x_1 + u, \ \dot{s} = u, \ \dot{D} = s. \tag{12}$$

The linear control system (12) is not controllable, but this linear control is steady-state controllable: for (12), one can move from $(0, 0, 0, a_1)$ to $(0, 0, 0, a_2)$ whatever are $a_1 \in \mathbb{R}$ and $a_2 \in \mathbb{R}$. But

- *For control system (10), one can move from $(0, 0, 0, a_1)$ to $(0, 0, 0, a_2)$ whatever are $a_1 \in \mathbb{R}$ and $a_2 \in \mathbb{R}$*
- *For control system (11), one can never move from $(0, 0, 0, a_1)$ to $((0, 0, 0, a_2)$ whatever are $a_1 \in \mathbb{R}$ and $a_2 \in \mathbb{R}$ with $a_1 \neq a_2$.*

Let us introduce our definition of local controllability $C(T)$ and our definition of steady-state local controllability $S(T)$. For $w \in C^1([0, L])$, let

$$|w|_1 := \text{Max}\{|w(x)| + |w_x(x)|; \ x \in [0, L]\}.$$

The definition of $C(T)$ is the following one.

Definition of $C(T)$. *Let $T > 0$. The control system Σ satisfies the property $C(T)$ if, for every ϵ, there exists $\eta > 0$ such that, for every $Y_0 = (H_0, v_0, s_0, D_0) \in \mathcal{Y}$, and for every $Y_1 = (H_1, v_1, s_1, D_1) \in \mathcal{Y}$ such that*

$$|H_0 - 1|_1 + |v_0|_1 + |H_1 - 1|_1 + |v_1|_1 + |s_0| + |s_1| + |D_0| + |D_1| < \eta,$$

there exists a trajectory

$$(Y, u) : [0, T] \ \to \mathcal{Y} \times \mathbb{R}, \ t \mapsto ((H(t), v(t), s(t), D(t)), u(t))$$

of the control system Σ such that

$$Y(0) = Y_0 \text{ and } Y(T) = Y_1, \tag{13}$$

and, for every $t \in [0, T]$,

$$|H(t) - 1|_1 + |v(t)|_1 + |s(t)| + |D(t)| + |u(t)| < \epsilon. \tag{14}$$

The definition of our steady-state local controllability property $S(T)$ is the following one.

Definition of $S(T)$**.** *Let $T > 0$. The control system Σ satisfies the property $S(T)$ if, for every ϵ, there exists $\eta > 0$ such that, for every $D_0 \in \mathbb{R}$, and for every $D_1 \in \mathbb{R}$ such that*

$$|D_0| + |D_1| < \eta,$$

there exists a trajectory

$$(Y, u) : [0, T] \; \to \; \mathcal{Y} \times \mathbb{R}, \; t \mapsto ((H(t), v(t), s(t), D(t)), u(t))$$

of the control system Σ such that

$$Y(0) = (0, 0, 0, D_0) \; and \; Y(T) = (0, 0, 0, D_1), \tag{15}$$

and, for every $t \in [0, T]$,

$$|H(t) - 1|_1 + |v(t)|_1 + |s(t)| + |D(t)| + |u(t)| < \epsilon.$$

Clearly $C(T)$ implies $S(T)$, and, for $T \leq T'$, $C(T)$ implies $C(T')$ and $S(T)$ implies $S(T')$. Using the characteristics of the hyperbolic system (1)–(2) one easily sees that $S(T)$ does not hold if $T < 1$. The following theorem is proved in [4].

Theorem 2. *Property $C(T)$ holds for T large enough.*

The proof of Theorem 2 given in [4] relies on the return method, a method that we have introduced in [2] for a stabilisation problem in finite dimension and first used in infinite dimension for the controllability of the Euler equations in [3, 5]. This method allows in some cases to get the local controllability at an equilibrium of a nonlinear control system even if the linearized control system at the equilibrium is not controllable. The idea of the return consists in the following one. If one can find a trajectory of the nonlinear control system such that

(i) it starts and ends at the equilibrium,

(ii) the linearized control system around this trajectory is controllable,

then, in general, the inverse mapping theorem allows to conclude that one can go from any state close to the equilibrium to any other state close to the equilibrium. This method, as used in [4], requires $T > 2$ at least.

Our first open problem is

Open Problem 3. *What is the value of*

$$T_c := Inf\,\{T > 0; \; C(T) \; holds\}. \tag{16}$$

and our second open problem is

Open Problem 4. *What is the value of*

$$T_s := Inf\,\{T > 0; \; S(T) \; holds\}. \tag{17}$$

Our guess is that

$$T_c = T_s = 2. \tag{18}$$

Note that, for the linearized control system (9), it is proved that the steady-state controllability holds for every $T > 1$. So, if our guess (18) holds, the nonlinearity which has helped us in order to get the local controllability, would impose an extra time for the steady-state local controllability.

3. A toy model

In this section, we give the main ideas of the proof of Theorem 2 on a very simple finite-dimensional control system which shares some properties with our control system Σ. This will be our toy model, denoted by $\mathcal{T}$. We also propose on this toy model a method which could be tried on Σ in order to get an upper bound on T_c.

For a function $w : [0,1] \to \mathbb{R}$, we denote by w^{ev} "the even part" of w and by w^{od} the odd part of w:

$$w^{\mathrm{ev}}(x) := \frac{1}{2}(w(x) + w(1-x)), \; w^{\mathrm{od}}(x) := \frac{1}{2}(w(x) - w(1-x)).$$

Then, if $h := 1 - H$, one has from (1) to (4)

$$\begin{cases} h_t^{\mathrm{od}} + v_x^{\mathrm{ev}} = -(h^{\mathrm{ev}}v^{\mathrm{ev}} + h^{\mathrm{od}}v^{\mathrm{od}})_x, \; v_t^{\mathrm{ev}} + h_x^{\mathrm{od}} = -u(t) - (v^{\mathrm{ev}}v^{\mathrm{od}})_x, \\ v^{\mathrm{ev}}(t,0) = v^{\mathrm{ev}}(t,1) = 0, \; \frac{\mathrm{d}s}{\mathrm{d}t}(t) = u(t), \; \frac{\mathrm{d}D}{\mathrm{d}t}(t) = s(t), \end{cases} \tag{19}$$

$$\begin{cases} h_t^{\mathrm{ev}} + v_x^{\mathrm{od}} = -(h^{\mathrm{ev}}v^{\mathrm{od}} + h^{\mathrm{od}}v^{\mathrm{ev}})_x, \; v_t^{\mathrm{od}}(t,x) + h_x^{\mathrm{ev}} = -\frac{1}{2}((v^{\mathrm{ev}})^2 + (v^{\mathrm{od}})^2)_x, \\ v^{\mathrm{od}}(t,0) = v^{\mathrm{od}}(t,1) = 0, \end{cases} \tag{20}$$

together with the initial conditions

$$h^{\mathrm{od}}(0,x) = h_0^{\mathrm{od}}(x), \; v^{\mathrm{ev}}(0,x) = v_0^{\mathrm{ev}}(x), \; s_1(0) = s_1, \; D_1(0) = D_1, \tag{21}$$

$$h^{\mathrm{ev}}(0,x) = h_0^{\mathrm{ev}}(x), \; v^{\mathrm{od}}(0,x) = v_0^{\mathrm{od}}(x). \tag{22}$$

The linearized control system of the part (19) is

$$\begin{cases} h_t^{\mathrm{od}} + v_x^{\mathrm{ev}} = 0, \; v_t^{\mathrm{ev}} + h_x^{\mathrm{od}} = -u(t), \; v^{\mathrm{ev}}(t,0) = v^{\mathrm{ev}}(t,1) = 0, \\ \frac{\mathrm{d}s}{\mathrm{d}t}(t) = u(t), \; \frac{\mathrm{d}D}{\mathrm{d}t}(t) = s(t). \end{cases} \tag{23}$$

Let us consider (23) as a control system where the control is u and where the state $(h^{\mathrm{od}}, v^{\mathrm{ev}}, s, D) \in C^1([0,1]) \times C^1([0,1]) \times \mathbb{R} \times \mathbb{R}$ satisfies

$$h^{\mathrm{od}}(1-x) = -h^{\mathrm{od}}(x), \; v^{\mathrm{ev}}(L-x) = v^{\mathrm{ev}}(x), \; v^{\mathrm{ev}}(0) = v^{\mathrm{ev}}(L) = 0.$$

Note that, by [9], this control system is controllable (in every time $T > 1$). When one linearizes the first two equations of (19) and the two equations (20), one gets the usual wave equations. A natural analogous of the wave equation in finite dimension is the oscillator equation. Hence a natural analogous to our control system (19)–(20) is

$$(\mathcal{T}) \quad \dot{x}_1 = x_2, \; \dot{x}_2 = -x_1 + x_2 x_3 + u, \; \dot{x}_3 = x_4, \; \dot{x}_4 = -x_3 + 2x_1 x_2, \; \dot{s} = u, \; \dot{D} = s \tag{24}$$

where the state is $x = (x_1, x_2, x_3, x_4, s, D) \in \mathbb{R}^6$ and the control is $u \in \mathbb{R}$. (The quadratic terms in (24) need some specific properties but could be much more general than the one chosen here.) This control system is our toy model.

The linearized control system of $\mathcal{T}$ around $(0,0) \in \mathbb{R}^6 \times \mathbb{R}$ is

$$\dot{x}_1 = x_2, \; \dot{x}_2 = -x_1 + u, \; \dot{x}_3 = x_4, \; \dot{x}_4 = -x_3, \; \dot{s} = u, \; \dot{D} = s. \tag{25}$$

This linear control system is not controllable (look at (x_3, x_4)). But, as one easily checks, (25) is steady-state controllable for arbitrary time T, that is, for every $(D_0, D_1) \in \mathbb{R}^2$ and for every $T > 0$, there exists a trajectory

$$((x_1, x_2, x_3, x_4, s, D), u) : [0, T] \to \mathbb{R}^6 \times \mathbb{R}$$

of the linear control system (25) such that

$$x_1(0) = x_2(0) = x_3(0) = x_4(0) = 0, \ s(0) = 0, \ D(0) = D_0,$$
$$x_1(T) = x_2(T) = x_3(T) = x_4(T) = 0, \ s(T) = 0, \ D(T) = D_1.$$

But the same does not hold for the nonlinear control system $\mathcal{T}$: one can check that one needs $T > 2\pi$ in order to have the steady-state local controllability for the nonlinear control system $\mathcal{T}$.

Let us now prove the following large time local controllability of $\mathcal{T}$.

Proposition 5. *There exists $T > 0$ and $\delta > 0$ such that, for every $a \in \mathbb{R}^6$ and every $b \in \mathbb{R}^6$ with $|a| < \delta$ and $|b| < \delta$, there exists $u \in L^\infty(0, T)$ such that, if $x = (x_1, x_2, x_3, x_4, s, D) : [0, T] \to \mathbb{R}^6$ is the solution of the Cauchy problem*

$$\dot{x}_1 = x_2, \quad \dot{x}_2 = -x_1 + x_2 x_3 + u, \quad \dot{x}_3 = x_4,$$
$$\dot{x}_4 = -x_3 + 2x_1 x_2, \quad \dot{s} = u, \quad \dot{D} = s, \ x(0) = a,$$

then $x(T) = b$.

Let us prove this proposition by using the return method. In order to use this method one needs, at least, to know trajectories of the control system $\mathcal{T}$ such that the linearized control systems around these trajectories are controllable. The simplest trajectories one can consider are the trajectories

$$((x_1^\gamma, x_2^\gamma, x_3^\gamma, x_4^\gamma, s^\gamma, D^\gamma), u^\gamma) = ((\gamma, 0, 0, 0, \gamma t, \gamma t^2/2), \gamma), \tag{26}$$

where $\tau_1 > 0$ is fixed, γ is any real number different from 0 and $t \in [0, \tau_1]$. The linearized control system around the trajectory

$$(x^\gamma, u^\gamma) := ((x_1^\gamma, x_2^\gamma, x_3^\gamma, x_4^\gamma, s^\gamma, D^\gamma), u^\gamma)$$

is the linear control system

$$\dot{x}_1 = x_2, \ \dot{x}_2 = -x_1 + u, \ \dot{x}_3 = x_4, \ \dot{x}_4 = -x_3 + 2\gamma x_2, \ \dot{s} = u, \ \dot{D} = s. \tag{27}$$

Using the usual Kalman condition for controllability, one easily checks that this linear control system is controllable if and only if $\gamma \neq 0$. Let us now choose $\gamma \neq 0$. Then, since the linearized control system around (x^γ, u^γ) is controllable, there exists $\delta_1 > 0$ such that for every $a \in B(x^\gamma(0), \delta_1) := \{x \in \mathbb{R}^6; |x - x^\gamma(0)| < \delta_1\}$ and for every b in $B(x^\gamma(\tau_1), \delta_1) := \{x \in \mathbb{R}^6; |x - x^\gamma(\tau_1)| < \delta_1\}$ there exists $u \in L^\infty([0, \tau_1]; \mathbb{R})$ such that

$$\left(\dot{x}_1 = x_2, \ \dot{x}_2 = -x_1 + x_2 x_3 + u, \ \dot{x}_3 = x_4, \right.$$

$$\left. \dot{x}_4 = -x_3 + 2x_1 x_2, \ \dot{s} = D, \ \dot{D} = u, \ x(0) = a \right) \Rightarrow (x(T_0) = b).$$

Hence, by the continuity of the solutions of the Cauchy problem with respect to the initial condition, in order to prove Proposition 5 it suffices to check that

(i) there exists $\tau_2 > 0$ and a trajectory $(\tilde{x}, \tilde{u}) : [0, \tau_2] \to \mathbb{R}^6 \times \mathbb{R}$ of the control system $\mathcal{T}$ such that $\tilde{x}(0) = 0$ and $|\tilde{x}(\tau_2) - x^\gamma(0)| < \delta_1$.

(ii) there exists $\tau_3 > 0$ and a trajectory $(\hat{x}, \hat{u}) : [0, \tau_3] \to \mathbb{R}^6 \times \mathbb{R}$ of the control system $\mathcal{T}$ such that $\hat{x}(\tau_3) = 0$ and $|\hat{x}(0) - x^\gamma(\tau_1)| < \delta_1$.

In order to prove (i) we consider quasi-static deformations. Let $g \in \mathcal{C}^2([0,1];\mathbb{R})$ be such that

$$g(0) = 0, \ g(1) = 1. \tag{28}$$

Let $\tilde{u} : [0, 1/\epsilon] \to \mathbb{R}$ be defined by $\tilde{u}(t) = \gamma g(\epsilon t)$. Let $\tilde{x} := (\tilde{x}_1, \tilde{x}_2, \tilde{x}_3, \tilde{x}_4, \tilde{s}, \tilde{D}) : [0, 1/\epsilon] \to \mathbb{R}^6$ be defined by

$$\dot{\tilde{x}}_1 = \tilde{x}_2, \ \dot{\tilde{x}}_2 = -\tilde{x}_1 + \tilde{x}_2\tilde{x}_3 + \tilde{u}, \ \dot{\tilde{x}}_3 = \tilde{x}_4,$$

$$\dot{\tilde{x}}_4 = -\tilde{x}_3 + 2\tilde{x}_1\tilde{x}_2, \ \dot{\tilde{s}} = \tilde{u}, \ \dot{\tilde{D}} = \tilde{s}, \ \tilde{x}(0) = 0.$$

One easily checks that

$$\tilde{x}(1/\epsilon) \to (\gamma, 0, 0, 0, 0, 0) \text{ as } \epsilon \to 0,$$

which ends the proof of (i).

In order to get (ii) one needs just to modify a little bit the above construction.

Remark 6. *The linearized system of*

$$\dot{x}_1 = x_2, \ \dot{x}_2 = -x_1 + x_2 x_3 + \gamma, \ \dot{x}_3 = x_4, \ \dot{x}_4 = -x_3 + 2x_1 x_2,$$

at the equilibrium $(\gamma, 0, 0, 0)$ has i and $-i$ for eigenvalues. This is why the quasi-static deformation is so easy to perform. If this linearized would have eigenvalues with strictly positive real part, it is still possible to perform in some cases the quasi-static deformation by stabilizing the equilibriums by suitable feedbacks. For an application to a partial differential equation, see [7].

The method which we have used in order to prove Proposition 5 has an important drawback: due to the quasi-static deformation parts it leads to quite bad estimates on the time T for controllability. Let us now propose another method which a good estimate on the time for local controllability. This method is classical in finite dimension – see for example [10] and the references therein – and has been used in infinite dimension for a KdV control system in [6]. It consists in looking for "higher order variations" which allows to move in the directions which are missed by the controllability of the linearized control system. These directions are $\pm(0, 0, 1, 0, 0, 0)$ and $\pm(0, 0, 0, 1, 0, 0)$ for our toy control system $\mathcal{T}$. For the control system Σ, these directions are given by $(h, v, 0, 0) \in \mathcal{Y}_{\text{lin}}$, with $h(1 - x) = h(x)$ and $v(1 - x) = -v(x)$. Let us describe this method on $\mathcal{T}$ in order to get that Proposition 5 holds for every $T > 2\pi$.

One first look to the linearized control system around 0, i.e., the linear control system (25). Let $T > 0$ and let $(e_i)_{i \in \{1,\dots,6\}}$ be the usual basis of $\mathbb{R}^6$. One easily sees that, for every $i \in \mathcal{I}_c := \{1, 2, 5, 6\}$, there exists $u_i \in L^\infty(0, T)$ such that

$$\Big(\dot{x}_1 = x_2, \ \dot{x}_2 = -x_1 + u_i, \ \dot{x}_3 = x_4,$$
$$\dot{x}_4 = -x_3, \ \dot{s} = u_i, \ \dot{D} = s, \ x(0) = 0 \Big) \Rightarrow (x(T) = e_i).$$

Let us assume for the moment being that, for every $i \in \mathcal{I}_u := \{3, 4\}$, there exist $u_i^\pm \in L^\infty(0, T)$ such that

$$\Big(\dot{x}_1 = x_2, \ \dot{x}_2 = -x_1 + u_i^\pm, \ \dot{x}_3 = x_4, \ \dot{x}_4 = -x_3 + 2x_1 x_2,$$
$$\dot{s} = u_i^\pm, \ \dot{D} = s, \ x(0) = 0 \Big) \Rightarrow (x(T) = \pm e_i). \tag{29}$$

Note that in the left-hand side of (29), we have put $\dot{x}_2 = -x_1 + u_i^\pm$ and not $\dot{x}_2 = -x_1 + x_2 x_3 + u_i^\pm$. The reason is that the x_i with $i \in \mathcal{I}_c$ and u are considered of order 1, and the x_i with $i \in \mathcal{I}_u$ are considered to be of order 2. With this choice of scaling, the left-hand side of (29) is the approximation of order 2 of the control system $\mathcal{T}$. Then, let $b := \sum_{i=1}^6 b_i e_i$. Let, for $i \in \mathcal{I}_u$,

$$\text{if } b_i \geq 0, u_i := u_i^+ \text{ and if } b_i < 0, u_i := u_i^-.$$

Let $u \in L^\infty(0, T)$ be defined by

$$u := \sum_{i \in \mathcal{I}_c} b_i u_i + \sum_{i \in \mathcal{I}_u} |b_i|^{1/2} u_i.$$

Let $x : [0, T] \to \mathbb{R}^6$ be the solution of the Cauchy problem

$$\dot{x}_1 = x_2, \ \dot{x}_2 = -x_1 + x_2 x_3 + u, \ \dot{x}_3 = x_4, \ \dot{x}_4 = -x_3 + 2x_1 x_2, \ \dot{s} = u, \ \dot{D} = s, \ x(0) = 0.$$

Then straightforward estimates lead to

$$x(T) = b + o(b) \text{ as } b \to 0.$$

Hence using the Brouwer fixed point theorem (and standard estimates on ordinary differential equations) one gets the local controllability of $\mathcal{T}$ (around 0) in time T (and therefore Proposition 5). It remains to prove the existence of $u_i^\pm \in L^\infty(0, T)$ for every $i \in \mathcal{I}_u := \{3, 4\}$. Easy computations show that, if

$$\dot{x}_1 = x_2, \ \dot{x}_3 = x_4, \ \dot{x}_4 = -x_3 + 2x_1 x_2,$$

then

$$x_3(T) = \int_0^T x_1^2(t) \cos(T - t) dt, \ x_4(T) = x_1^2(T) - \int_0^T x_1^2(t) \sin(T - t) dt. \tag{30}$$

From (30), it is not hard to get that the existence of $u_i^\pm \in L^\infty(0, T)$ for every $i \in \mathcal{I}_u$ holds if (and only if) $T > 2\pi$.

References

[1] K. Beauchard, Local controllability of a 1d Schrödinger equation. *Preprint of the Université de Paris-Sud*, 2004.

[2] J.-M. Coron, Global asymptotic stabilization for controllable systems without drift, *Math. Control Signals Systems*, 5 (1992) pp. 295–312.

[3] J.-M. Coron, Contrôlabilité exacte frontière de l'équation d'Euler des fluides parfaits incompressibles bidimensionnels, *C.R. Acad. Sci. Paris*, 317 (1993) pp. 271–276.

[4] J.-M. Coron, Local controllability of a 1-D tank containing a fluid modeled by the shallow water equations, *ESAIM: COCV*, 8 (2002) pp. 513–554.

[5] J.-M. Coron, On the controllability of 2-D incompressible perfect fluids, *J. Math. Pures & Appliquées*, 75 (1996) pp. 155–188.

[6] J.-M. Coron and E. Crépeau, Exact boundary controllability of the nonlinear KdV equation for critical lengths, *J. European Mathematical Society*, 6 (2003) pp. 367–398.

[7] J.-M. Coron and E. Trélat, Global steady-state controllability of 1-D semilinear heat equations, *SIAM J. Control Optim.*, 43 (2004) pp. 549–569.

[8] L. Debnath, *Nonlinear water waves*, Academic Press, San Diego (1994).

[9] F. Dubois, N. Petit and P. Rouchon, Motion planning and nonlinear simulations for a tank containing a fluid, ECC 99.

[10] M. Kawski, High-order small time local controllability, in: *Nonlinear Controllability and Optimal Control* (H.J. Sussmann, ed.), Monogr. Textbooks Pure Appl. Math. **113**, Dekker, New York, 1990, pp. 431–467.

[11] N. Petit and P. Rouchon, Dynamics and solutions to some control problems for water-tank systems, *IEEE Transactions on Automatic Control*, 47 (2002) pp. 594–609.

[12] A.J.C.B. de Saint-Venant, Théorie du mouvement non permanent des eaux, avec applications aux crues des rivières et à l'introduction des marées dans leur lit, *C.R. Acad. Sci. Paris*, 53 (1871) pp. 147–154.

Jean-Michel Coron
Université Paris-Sud et Institut universitaire de France
Département de Mathématique
Bâtiment 425
F-91405 Orsay, France
e-mail: `Jean-Michel.Coron@math.u-psud.fr`

Progress in Nonlinear Differential Equations
and Their Applications, Vol. 63, 189–205

Hölder Estimates for Solutions to a Singular Nonlinear Neumann Problem

Juan Dávila and Marcelo Montenegro

Abstract. We consider the elliptic equation $-\Delta u + u = 0$ in a bounded, smooth domain Ω in $\mathbb{R}^n$ subject to the nonlinear singular Neumann condition $\frac{\partial u}{\partial \nu} = -u^{-\beta} + f(x, u)$. Here $0 < \beta < 1$ and $f \geq 0$ is C^1. We prove estimates for solutions to the same equation with $\frac{\partial u_\varepsilon}{\partial \nu} = -\frac{u_\varepsilon}{(u_\varepsilon + \varepsilon)^{1+\beta}} + f(x, u_\varepsilon)$ on the boundary, uniformly in ε.

1. Introduction

This note is intended as a complement of previous work by the authors [2]. We study the regularity of solutions of the following nonlinear boundary value problem

$$\begin{cases} -\Delta u + u = 0 & \text{in } \Omega \\ \quad u \geq 0 & \text{in } \Omega \\ \dfrac{\partial u}{\partial \nu} = -u^{-\beta} + f(x, u) & \text{on } \partial\Omega \cap \{u > 0\}, \end{cases} \tag{1}$$

where $\Omega \subset \mathbb{R}^n$, $n \geq 2$, is a bounded domain with smooth boundary, $0 < \beta < 1$ and ν is the exterior unit normal vector to $\partial\Omega$. We assume that

$$f : \partial\Omega \times \mathbb{R} \to \mathbb{R} \text{ is } C^1 \text{ and } f \geq 0. \tag{2}$$

By a solution of (1) we mean a function $u \in H^1(\Omega) \cap C(\overline{\Omega})$ satisfying

$$\int_\Omega \nabla u \cdot \nabla \varphi + u\varphi = \int_{\partial\Omega \cap \{u>0\}} (-u^{-\beta} + f(x, u))\varphi, \quad \forall \varphi \in C_0^1(\Omega \cup (\partial\Omega \cap \{u > 0\})). \tag{3}$$

One natural approach to prove existence of solutions of (1) is the following: take $\varepsilon > 0$ and consider

$$\begin{cases} -\Delta u + u = 0 & \text{in } \Omega \\ \dfrac{\partial u}{\partial \nu} = -\dfrac{u}{(u + \varepsilon)^{1+\beta}} + f(x, u) & \text{on } \partial\Omega. \end{cases} \tag{4}$$

It is not difficult to show that under the additional assumption

$$\lim_{u \to \infty} \frac{f(x,u)}{u} = 0 \quad \text{uniformly for } x \in \Omega \tag{5}$$

(4) has a maximal solution $\overline{u}^{\varepsilon}$. In [2] we proved that this maximal solution satisfies an estimate of the form

$$|\nabla \overline{u}^{\varepsilon}| \leq C(\overline{u}^{\varepsilon})^{-\beta} \quad \text{in } \Omega,$$

with C independent of ε. This was an essential step in proving that the limit $\lim_{\varepsilon \to 0} u^{\varepsilon}$ exists and is a solution of (1). Nevertheless there could exist other solutions of (4). For instance assuming (2) and (5) problem (4) admits also a minimal nonnegative solution $\underline{u}^{\varepsilon}$ (it could be zero but assuming $f(\cdot, 0) \not\equiv 0$ guarantees $\underline{u}^{\varepsilon} \not\equiv 0$). Assuming some growth conditions on f, any critical point of Φ_{ε} is also a solution with

$$\Phi_{\varepsilon}(u) = \frac{1}{2}\int_{\Omega}(|\nabla u|^2 + u^2) - \int_{\partial \Omega} G^{\varepsilon}(x,u), \tag{6}$$

where

$$G^{\varepsilon}(x,u) = \int_0^u g^{\varepsilon}(x,t)\,dt, \quad \text{and } g^{\varepsilon}(x,u) = -\frac{u}{(u+\varepsilon)^{1+\beta}} + f(x,u).$$

In this note we prove the following result concerning any kind of solution to (4).

Theorem 1.1. *Suppose f satisfies (2). Then for any bounded solution u of (4) we have*

$$|\nabla u| \leq Cu^{-\beta} \quad \text{in } \Omega,$$

where C is independent of ε, and depends on Ω, n, β, f and $\|u\|_{L^\infty(\Omega)}$.

A consequence of the previous gradient estimate is the following convergence result (the proof is exactly as in [2]).

Corollary 1.2. *Assume (2) and let $\varepsilon_k \to 0$ and u^{ε_k} be a sequence of solutions of (4) with*

$$\|u^{\varepsilon_k}\|_{L^\infty(\Omega)} \leq C,$$

where C is independent of k. Then up to a subsequence $u^{\varepsilon_k} \to u$ in $C^\mu(\overline{\Omega})$ for any $0 < \mu < \frac{1}{1+\beta}$ and u is a solution of (1).

This result enables us to consider other type of nonlinearities than in [2]. For example

Theorem 1.3. *Assume that $n \geq 3$ and $1 < p < \frac{n}{n-2}$. Then there exists a nontrivial solution to*

$$\begin{cases} -\Delta u + u = 0 & \text{in } \Omega \\ \quad\ u \geq 0 & \text{in } \Omega \\ \dfrac{\partial u}{\partial \nu} = -u^{-\beta} + u^p & \text{on } \partial\Omega \cap \{u > 0\}. \end{cases} \tag{7}$$

By Theorem 1.1 this solution is $C^{\frac{1}{1+\beta}}(\overline{\Omega})$.

Previous work with a singular Neumann condition include [3] where the authors study the evolution equation $u_t = u_{xx}$ in $(0,1)$ with Neumann conditions $u_x(0,t) = 0$, $u_x(1,t) = -u(1,t)^{-\beta}$. The initial condition is $u(x,0) = u_0(x) > 0$ and sufficiently smooth. They prove that the solution exists up to a quenching time $0 < T < \infty$ with $\lim_{t \nearrow T} u(1,t) = 0$ and they provide estimates of the type $C_1 \leq (1-x)^{\frac{1}{\beta+1}} u(x,T) \leq C_2$.

In higher dimensions a similar evolution problem was addressed in [6] with a positive unbounded nonlinearity such as $1/(1-u)$, but the authors only work with a time interval $[0,T)$ where $0 \leq u(t) < 1$.

As mentioned earlier this work is a continuation of previous work of the authors. For this reason not all proofs are supplied here and we refer to [2].

2. Preliminaries

There are two important key points in the proof of Theorem 1.1. First there is a construction of a local subsolution. The second ingredient is a Hardy type inequality, which roughly speaking asserts that a solution that stays above the local subsolution is locally a minimum of the related energy. To make this more precise we rescale the problem to a small ball. It is convenient at this point to introduce some notation. Let $\tau_0 > 0$ be small enough to be fixed in Proposition 2.1 below. For $0 < \tau < \tau_0$ and $x_0 \in \partial\Omega$ let us write $\partial(B_\tau(x_0) \cap \Omega) = \Gamma^e \cup \Gamma^i$ where

$$\Gamma^i = \partial B_\tau(x_0) \cap \Omega, \quad \Gamma^e = B_\tau(x_0) \cap \partial\Omega$$

are the internal and external boundaries. We also decompose $\Gamma^e = \Gamma^1 \cup \Gamma^2$ with

$$\Gamma^1 = \varphi^{-1}(B_{\tau/2}(0)) \cap \partial\Omega, \quad \Gamma^2 = \Gamma^e \setminus \Gamma^1, \tag{8}$$

where φ is a smooth diffeomorphism which flattens the boundary of Ω near x_0. This means that $\varphi : W \subset \mathbb{R}^n \to B_{\tau_0}(0)$ is smooth with W an open set containing the ball $B_{\tau_0}(x_0)$ and $\varphi(W \cap \Omega) = B_{\tau_0}(0) \cap H$, $\varphi(W \cap \partial\Omega) = B_{\tau_0}(0) \cap \partial H$, $\varphi(W \setminus \overline{\Omega}) = B_{\tau_0}(0) \setminus \overline{H}$, where

$$H = \{(x', x_n) : x' \in \mathbb{R}^{n-1}, x_n > 0\}.$$

Let us introduce the rescaled domains which allow us to work in balls of unit size:

$$B_\tau^+ = \frac{1}{\tau}(B_\tau(x_0) \cap \Omega - x_0) = B_1(0) \cap \frac{1}{\tau}(\Omega - x_0), \quad \Omega_\tau = \frac{1}{\tau}(\Omega - x_0)$$

$$\Gamma_\tau^i = \frac{1}{\tau}(\Gamma^i - x_0), \quad \Gamma_\tau^e = \frac{1}{\tau}(\Gamma^e - x_0), \quad \Gamma_\tau^k = \frac{1}{\tau}(\Gamma^k - x_0), \quad k = 1, 2. \tag{9}$$

Given $x_0 \in \partial\Omega$ and $0 < \tau < \tau_0$ we let v_τ be the solution of the linear equation

$$\begin{cases} -\Delta v_\tau + \tau^2 v_\tau = 0 & \text{in } B_\tau^+, \\[2mm] \dfrac{\partial v_\tau}{\partial \nu}(y) = -\operatorname{dist}(y, \Gamma_\tau^2)^{-\frac{\beta}{1+\beta}} & y \in \Gamma_\tau^1, \\[2mm] v_\tau(y) = 0 & y \in \Gamma_\tau^2, \\[2mm] v_\tau(y) = s\operatorname{dist}(y, \partial\Omega_\tau) & y \in \Gamma_\tau^i. \end{cases} \tag{10}$$

For large s its solution will be called a local subsolution because of the next lemma.

Proposition 2.1. *There exist $\tau_0 > 0$ and $s_0 > 0$ such that if $0 < \tau < \tau_0$ and $s \geq s_0$ the solution of* (10) *is positive in B_τ^+ and satisfies*

$$v_\tau(y) \geq cs\,\mathrm{dist}(y, \Gamma_\tau^2)^{\frac{1}{1+\beta}}, \quad \forall y \in \Gamma_\tau^1, \tag{11}$$

where $c > 0$ is independent of x_0, τ and s (c depends only on Ω, n, β). In particular, choosing s_0 larger if necessary

$$\frac{\partial v_\tau}{\partial \nu} \leq -v_\tau^{-\beta} \quad on\ \Gamma_\tau^1. \tag{12}$$

We will not include the proof of the statements in this section. They can be found in [2].

Next we state a Hardy type inequality.

Proposition 2.2. *There exists a constant C_h such that*

$$\int_{\Gamma_\tau^1} \frac{\varphi^2}{\mathrm{dist}(y, \Gamma_\tau^2)} \leq C_h \int_{B_\tau^+} |\nabla \varphi|^2, \quad \forall \varphi \in C_0^\infty(B_\tau^+ \cup \Gamma_\tau^1). \tag{13}$$

The constant C_h can be taken independent of τ and $x_0 \in \partial\Omega$ if $0 < \tau < \tau_0$.

Finally we mention some lemmas on linear equations with a Neumann boundary condition. Again, the proofs can be found in [2].

This is a sort of Harnack inequality.

Lemma 2.3. *Let $a \in L^\infty(\Omega_\tau \cap B_3)$, $a \geq 0$ and suppose that $u \in H^1(\Omega_\tau \cap B_3)$, $u \geq 0$ satisfies*

$$\begin{cases} -\Delta u + a(y)u = 0 & in\ \Omega_\tau \cap B_3 \\[2mm] \dfrac{\partial u}{\partial \nu} \leq N & on\ \Gamma_\tau^e, \end{cases}$$

where N is a constant. Then there is a constant $c_k > 0$ such that

$$u(y) \geq c_k\,\mathrm{dist}(y, \Gamma_\tau^e)(c_k u(y_1) - N), \quad \forall y \in B_\tau^+\ and\ \forall y_1 \in B_\tau^+ \cap B_{1/2}.$$

The constant c_k can be chosen independent of $x_0 \in \partial\Omega$ and of $0 < \tau < \tau_0$.

These last two estimates are standard in the theory of L^p regularity theory, see for instance [9].

Lemma 2.4. *Let $a \in L^\infty(B_\tau^+)$. Suppose $u \in H^1(B_\tau^+)$ satisfies*

$$\begin{cases} -\Delta u + a(x)u = 0 & in\ B_\tau^+ \\[2mm] \dfrac{\partial u}{\partial \nu} = g & on\ \Gamma_\tau^e, \end{cases}$$

where $g \in L^p(\Gamma_\tau^e)$ and $p \geq 1$. Let $1 \leq r < \frac{np}{n-1}$. Then there exists C independent of g and u such that

$$\|u\|_{W^{1,r}(\Omega_\tau \cap B_{3/4})} \leq C\left(\|g\|_{L^p(\Gamma_\tau^e)} + \|u\|_{L^1(B_\tau^+)}\right).$$

Lemma 2.5. *Let $a \in L^\infty(B_\tau^+)$ and suppose that $u \in H^1(B_\tau^+)$, $u \geq 0$ satisfies*

$$\begin{cases} -\Delta u + a(x)u \geq 0 & \text{in } B_\tau^+ \\[2mm] \dfrac{\partial u}{\partial \nu} \geq -N & \text{on } \Gamma_\tau^e, \end{cases}$$

where N is a constant. Then there is a constant $C > 0$ independent of u, N such that

$$\int_{B_{3/4} \cap B_\tau^+} u \leq C(u(x) + N) \quad \forall x \in B_{1/2} \cap B_\tau^+.$$

3. Proof of Theorem 1.1

Let u be a bounded nontrivial solution of equation (4) and write

$$M = \max\left(\sup_{x \in \partial \Omega} f(x, u(x)), \max_{\overline{\Omega}} u \right).$$

Let τ_0 and s_0 be the constants in Proposition 2.1 and fix $\widetilde{C} > 0$ such that

$$s_0 < \frac{1}{2} c_k^2 \widetilde{C}, \tag{14}$$

$$M^{1+\beta} < \tau_0 \widetilde{C}^{1+\beta}, \tag{15}$$

$$M^{1+\beta} < \frac{1}{2} c_k \widetilde{C}^{1+\beta}. \tag{16}$$

Next we fix C_0 large enough such that

$$\left(\frac{C_0}{\widetilde{C}} \right)^{1+\beta} \geq 6. \tag{17}$$

Let x_1 be a point in Ω. We distinguish two cases.

Case 1. Assume $u(x_1) \leq C_0 \operatorname{dist}(x_1, \partial \Omega)^{\frac{1}{1+\beta}}$. Consider the scaling about the point x_1 given by $\tilde{u}(y) = \tau^{-\frac{1}{1+\beta}} u(\tau y + x_1)$, with $\tau = \frac{1}{2} \operatorname{dist}(x_1, \partial \Omega)$. Then $-\Delta \tilde{u} + \tau^2 \tilde{u} = 0$ in $B_1(0)$, $\tilde{u} \geq 0$ in $B_1(0)$ and $\tilde{u}(0) \leq 2^{\frac{1}{1+\beta}} C_0$. Since $\tilde{u} \geq 0$, by elliptic estimates we have $|\nabla \tilde{u}(0)| \leq C(n, \beta) C_0$, where $C(n, \beta)$ depends only on n, β. This implies $|\nabla u(x_1)| \leq C(n, \beta) C_0 \tau^{-\frac{\beta}{1+\beta}} \leq C(n, \beta) C_0^{1-\beta} u(x_1)^{-\beta}$. Thus

$$|\nabla u(x_1)| \leq C(n, \beta) C_0^{1-\beta} u(x_1)^{-\beta}. \tag{18}$$

We keep the explicit dependence on C_0 for future reference.

Case 2. Assume

$$u(x_1) > C_0 \operatorname{dist}(x_1, \partial \Omega)^{\frac{1}{1+\beta}}. \tag{19}$$

Let

$$x_0 \in \partial \Omega, \quad \operatorname{dist}(x_1, \partial \Omega) = |x_0 - x_1|. \tag{20}$$

Our first task is to show that u satisfies an inequality such as (19) on all points on the line segment

$$[x_0, x_1] = \left\{ x_0 + t\frac{x_1 - x_0}{|x_1 - x_0|} : 0 \le t \le \bar{t} \right\},$$

where $\bar{t} = |x_1 - x_0|$.

Lemma 3.1. *Choosing C_0 larger if necessary (only depending on n, β and $\widetilde{C}$ as in (17)) we have*

$$u(x) \ge C_0 \operatorname{dist}(x, \partial\Omega)^{\frac{1}{1+\beta}} \quad \forall x \in [x_0, x_1]. \tag{21}$$

Proof. For the sake of notation we write

$$x_t = x_0 + t\frac{x_1 - x_0}{|x_1 - x_0|} \quad 0 \le t \le \bar{t},$$

and observe that $\operatorname{dist}(x_t, \partial\Omega) = |x_t - x_0| = t$. Suppose that (21) fails. Then

$$t_0 = \sup\{t \in [0, \bar{t}] : u(x_t) \le C_0 t^{\frac{1}{1+\beta}}\}$$

is well defined, $t_0 > 0$ and by (19) we have $t_0 < \bar{t}$. Define $g(t) = u(x_t)$. Using the same argument as in case 1, see (18), we have that

$$g'(t) \le C(n, \beta)C_0^{1-\beta}g(t)^{-\beta} \quad \text{whenever } g(t) \le C_0 t^{\frac{1}{1+\beta}}. \tag{22}$$

Let $h(t) = C_0 t^{\frac{1}{1+\beta}}$, so that $h'(t) = \frac{C_0^{1+\beta}}{1+\beta}h(t)^{-\beta}$. Then we have $g(t_0) = h(t_0)$ and by (22)

$$g'(t_0) \le C(n, \beta)C_0^{1-\beta}g(t_0)^{-\beta} = C(n, \beta)\frac{1+\beta}{C_0^{2\beta}}h'(t_0).$$

Choose C_0 larger so that $C(n, \beta)\frac{1+\beta}{C_0^{2\beta}} < \frac{1}{2}$. Then $g(t) > h(t)$ for $t \in (t_0 - \sigma, t_0)$ for some $\sigma > 0$. This is impossible. $\qquad\square$

Define τ_1 by

$$\tau_1 = \left(\frac{u(x_1)}{\widetilde{C}}\right)^{1+\beta} \tag{23}$$

and observe that by (15) we have

$$\tau_1 < \tau_0.$$

We look now at the rescaled function u around the point $x_0 \in \partial\Omega$ given by (20): for $0 < \tau < \tau_0$ and $x_0 \in \partial\Omega$ define

$$u_\tau(y) = \tau^{-\frac{1}{1+\beta}}u(\tau y + x_0), \qquad y \in \Omega_\tau = \frac{1}{\tau}(\Omega - x_0). \tag{24}$$

At this point it is convenient to replace f with a C^1 function $\bar{f} : \partial\Omega \times \mathbb{R} \to \mathbb{R}$ with $\bar{f} \ge 0$ and $f, \frac{\partial f}{\partial u}$ bounded, and such that $f(x, u) = \bar{f}(x, u)$ for all $x \in \partial\Omega$ and

$0 \leq u \leq M$. Then u solves (4) with f replaced by $\bar{f}$ and therefore u_τ is a solution of

$$\begin{cases} -\Delta u_\tau + \tau^2 u_\tau = 0 & \text{in } \Omega_\tau, \\ \dfrac{\partial u_\tau}{\partial \nu} = g_\tau^\varepsilon(y, u_\tau) & \text{on } \partial\Omega_\tau. \end{cases} \tag{25}$$

where g_τ^ε is given by

$$g_\tau^\varepsilon(y, w) = \tau^{\frac{\beta}{1+\beta}} g^\varepsilon\left(\tau y + x_0, \tau^{\frac{1}{1+\beta}} w\right), \tag{26}$$

and

$$g^\varepsilon(x, u) = -\frac{u}{(u+\varepsilon)^{1+\beta}} + \bar{f}(x, u). \tag{27}$$

Observe that we have changed the definition of g^ε and g_τ^ε from the one given in the introduction replacing f by $\bar{f}$.

We will see that as a consequence of (21) u_τ has to be suitably large on the internal boundary Γ_τ^i.

Lemma 3.2. *For $0 < \tau \leq \tau_1$ we have*

$$u_\tau(y) \geq s_0 \operatorname{dist}(y, \partial\Omega_\tau) \quad \forall y \in \Gamma_\tau^i.$$

Proof. Let $z_\tau = \frac{1}{2}\frac{x_1 - x_0}{|x_1 - x_0|} \in B_\tau^+ \cap B_{1/2}$. By (21) and the definition of u_τ we have

$$u_\tau(z_\tau) = \tau^{-\frac{1}{1+\beta}} u(\tau z_\tau + x_0) \geq \frac{C_0}{2} \geq \widetilde{C}, \tag{28}$$

where the last inequality is a consequence of (17). Using Harnack's Lemma 2.3 and (28) we obtain

$$u_\tau(y) \geq c_k \operatorname{dist}(y, \partial\Omega_\tau)\left(c_k \widetilde{C} - \sup_{\Gamma_\tau^e} \frac{\partial u_\tau}{\partial \nu}\right), \quad \forall y \in B_\tau^+. \tag{29}$$

From the boundary condition in (25) and the definition of M

$$\sup_{\Gamma_\tau^e} \frac{\partial u_\tau}{\partial \nu} \leq \tau^{\frac{\beta}{1+\beta}} M.$$

Notice that from (16) we deduce $u(x_1)^\beta \leq \frac{c_k \widetilde{C}^{1+\beta}}{2M}$ which is the same as

$$M\left(\frac{u(x_1)}{\widetilde{C}}\right)^\beta \leq \frac{1}{2} c_k \widetilde{C}.$$

Thus

$$\tau^{\frac{\beta}{1+\beta}} M \leq \tau_1^{\frac{\beta}{1+\beta}} M = \left(\frac{u(x_1)}{\widetilde{C}}\right)^\beta M \leq \frac{1}{2} c_k \widetilde{C}.$$

Inserting this in (29) and recalling (14) we find

$$u_\tau(y) \geq \frac{1}{2} c_k^2 \widetilde{C} \operatorname{dist}(y, \partial\Omega_\tau) \geq s_0 \operatorname{dist}(y, \partial\Omega_\tau) \quad \forall y \in \Gamma_\tau^i. \qquad \square$$

The main step that we shall prove in the sequel is the following:

Proposition 3.3. *For all $0 < \tau \leq \tau_1$ we have*

$$u_\tau \geq v_\tau \quad in \ B_\tau^+. \tag{30}$$

For the proof of Proposition 3.3 we consider the nonlinear problem

$$\begin{cases} -\Delta w + \tau^2 w = 0 & \text{in } B_\tau^+ \\ \quad\; w = u_\tau & \text{on } \Gamma_\tau^i \cup \Gamma_\tau^2 \\ \quad\; \dfrac{\partial w}{\partial \nu} = g_\tau^\varepsilon(x, w) & \text{on } \Gamma_\tau^1 \end{cases} \tag{31}$$

where we regard u_τ as data and w as the unknown. Observe that u_τ is a solution of (31).

The solutions of (31) are the critical points of the functional

$$\psi_\tau(w) = \frac{1}{2} \int_{B_\tau^+} (|\nabla w|^2 + \tau^2 w^2) - \int_{\Gamma_\tau^1} G_\tau^\varepsilon(x, w)$$

on the set

$$E_\tau = \{ w \in H^1(B_\tau^+) \mid w = u_\tau \text{ on } \Gamma_\tau^i \cup \Gamma_\tau^2 \},$$

where

$$G_\tau^\varepsilon(y, w) = \int_0^w g_\tau^\varepsilon(y, r)\, dr,$$

and g_τ^ε defined in (26).

We remark that any nontrivial solution u of the regularized problem (4) is positive by the strong maximum principle, the fact that $f \geq 0$ and Hopf's lemma. This implies that $u_\tau \to \infty$ in B_τ^+ as $\tau \to 0$, more precisely $u_\tau \sim \tau^{-\frac{1}{1+\beta}} u(x_0)$ in B_τ^+. As a consequence, for fixed $\varepsilon > 0$ as $\tau \to 0$ problem (31) is less singular and we have

Lemma 3.4. *For $\tau > 0$ small enough problem (31) has a unique solution.*

How small τ has to be may depend on ε.

Proof. Suppose that there exists a sequence $\tau_j \to 0$ and solutions $w_j^1, w_j^2 \in H^1(\Omega_\tau)$ to equation (31) with $w_j^1 \neq w_j^2$.

Since $w_j^1 = w_j^2 = u_{\tau_j}$ on $\Gamma_\tau^i \cup \Gamma_\tau^2$ we have $w_j^i \leq \tau_j^{-\frac{1}{1+\beta}} M$ on $\Gamma_\tau^i \cup \Gamma_\tau^2$, $i = 1, 2$. Also, $\frac{\partial w_j^i}{\partial \nu} \leq \bar{f}_{\tau_j}(y, w_j^i)$ on Γ_τ^1 where

$$\bar{f}_{\tau_j}(y, w) = \tau_j^{\frac{\beta}{1+\beta}} \bar{f}(\tau_j y + x_0, \tau_j^{\frac{1}{1+\beta}} w) \leq C\tau_j^{\frac{\beta}{1+\beta}},$$

since $\bar{f}$ is bounded. By the maximum principle we have

$$w_j^i \leq C\tau_j^{-\frac{1}{1+\beta}} \quad \text{on } B_{\tau_j}^+. \tag{32}$$

with C independent of j.

Let $w_j = w_j^1 - w_j^2$. Then w_j satisfies

$$\begin{cases} -\Delta w_j + \tau_j^2 w_j = 0 & \text{in } B_{\tau_j}^+ \\ w_j = 0 & \text{on } \Gamma_{\tau_j}^i \cup \Gamma_{\tau_j}^2 \\ \dfrac{\partial w_j}{\partial \nu} = b_j(x) w_j & \text{on } \Gamma_{\tau_j}^1, \end{cases} \tag{33}$$

where

$$b_j(x) = \frac{\partial g_{\tau_j}^\varepsilon}{\partial w}(x, \xi(x))$$

for some $\xi(x) \in [w_j^1(x), w_j^2(x)]$ (we use the notation $[a, b] = [\min(a, b), \max(a, b)]$).
Now we estimate

$$b_j(x) = \frac{\partial g_{\tau_j}^\varepsilon}{\partial w}(x, \xi(x)) = \tau_j^{\frac{2}{1+\beta}} \frac{\partial g^\varepsilon}{\partial w}(\tau_j x + x_0, \tau_j^{\frac{1}{1+\beta}} \xi(x)),$$

where g^ε is defined in (27). By (32) we see that $\tau_j^{\frac{1}{1+\beta}} \xi(x) \leq C$ and since g^ε is C^1
we thus conclude that

$$b_j \to 0 \quad \text{uniformly on } \Gamma_{\tau_j}^1.$$

Thus, for j large enough the operator in (33) becomes coercive and hence $w_j = 0$
if j is large. Indeed, multiplying (33) by w_j and integrating we find

$$\int_{B_{\tau_j}^+} |\nabla w_j|^2 + \tau_j^2 \int_{B_{\tau_j}^+} w_j^2 = \int_{\Gamma_{\tau_j}^1} b_j w_j^2$$

Since $w_j = 0$ in $\Gamma_{\tau_j}^2 \cup \Gamma_{\tau_j}^i$ we have by the Sobolev trace inequality

$$\int_{B_{\tau_j}^+} |\nabla w_j|^2 + \tau_j^2 \int_{B_{\tau_j}^+} w_j^2 \leq C \|b_j\|_{L^\infty(\Gamma_{\tau_j}^1)} \int_{B_{\tau_j}^+} |\nabla w_j|^2,$$

which shows that $w_j \equiv 0$ for j large enough. $\qquad\square$

Lemma 3.5. *Fix $s = s_0$ in Proposition (2.1) and let v_τ be the solution of (10).
Assume $w, v \in E_\tau$ are subsolutions of (31) such that*

$$v \geq v_\tau \quad \text{on} \quad \Gamma_\tau^1, \quad \text{and} \quad v \leq w \quad \text{on} \quad \Gamma_\tau^i \cup \Gamma_\tau^2.$$

Then

$$\psi_\tau(\max(w, v)) \leq \psi_\tau(w) + \left(\frac{C}{s_0^{1+\beta}} + C\tau - \frac{1}{2} \right) \int_{B_\tau^+ \cap \{v > w\}} |\nabla(v - w)|^2,$$

where C is independent of ε, s_0, τ, v and w.

Proof. We derive first some estimates for the nonlinear terms. The functions
$G^\varepsilon(x, u)$, $G_\tau^\varepsilon(x, w)$ are given by

$$G^\varepsilon(x, u) = \int_0^u g^\varepsilon(x, s)\, ds = \frac{(u + \varepsilon)^{-\beta}(\varepsilon + \beta u) - \varepsilon^{1-\beta}}{\beta(-1 + \beta)} + \overline{F}(x, u),$$

where $\overline{F}(x, u) = \int_0^u \bar{f}(x, s)\, ds$, and

$$G_\tau^\varepsilon(x, w) = \tau^{\frac{-1+\beta}{1+\beta}} G^\varepsilon(\tau x + x_0, \tau^{\frac{1}{1+\beta}} w).$$

Note that

$$-u^{-\beta} + \bar{f}(x, u) \le g^\varepsilon(x, u) \le \bar{f}(x, u)$$

and hence we have the estimates

$$-\frac{u^{1-\beta}}{1-\beta} + \overline{F}(x, u) \le G^\varepsilon(x, u) \le \overline{F}(x, u)$$

and

$$-\frac{w^{1-\beta}}{1-\beta} + \tau^{\frac{-1+\beta}{1+\beta}} \overline{F}(\tau x + x_0, \tau^{\frac{1}{1+\beta}} w) \le G_\tau^\varepsilon(x, w) \le \tau^{\frac{-1+\beta}{1+\beta}} \overline{F}(\tau x + x_0, \tau^{\frac{1}{1+\beta}} w).$$

Let $W = \max(w, v)$. Then W satisfies

$$\begin{cases} -\Delta W + \tau^2 W \le 0 & \text{in } B_\tau^+, \\ W \le u_\tau & \text{on } \Gamma_\tau^i \cup \Gamma_\tau^2 \\ \dfrac{\partial W}{\partial \nu} \le g_\tau^\varepsilon(x, W) & \text{on } \Gamma_\tau^1. \end{cases} \tag{34}$$

We have the equality

$$\begin{aligned} \psi_\tau(W) - \psi_\tau(w) = &-\frac{1}{2} \int_{B_\tau^+} \left(|\nabla(W - w)|^2 + \tau^2 (W - w)^2 \right) \\ &+ \int_{B_\tau^+} \left(\nabla W \cdot \nabla(W - w) + \tau^2 W(W - w) \right) \\ &- \int_{\Gamma_\tau^1} \left(G_\tau^\varepsilon(x, W) - G_\tau^\varepsilon(x, w) \right). \end{aligned} \tag{35}$$

Next we multiply (34) by $W - w \ge 0$ and integrate by parts. Note that $W - w = 0$ on $\Gamma_\tau^i \cup \Gamma_\tau^2$ so that

$$\begin{aligned} \int_{B_\tau^+} \nabla W \cdot \nabla(W - w) + \tau^2 W(W - w) &\le \int_{\Gamma_\tau^1} \frac{\partial W}{\partial \nu}(W - w) \\ &\le \int_{\Gamma_\tau^1} g_\tau^\varepsilon(x, W)(W - w). \end{aligned} \tag{36}$$

Combining (35) and (36) we obtain

$$\begin{aligned} \psi_\tau(W) - \psi_\tau(w) \le &-\frac{1}{2} \int_{B_\tau^+} |\nabla(W - w)|^2 \\ &- \int_{\Gamma_\tau^1} \left(G_\tau^\varepsilon(x, W) - G_\tau^\varepsilon(x, w) - g_\tau^\varepsilon(x, W)(W - w) \right). \end{aligned} \tag{37}$$

We claim that

$$-[G_\tau^\varepsilon(x, W) - G_\tau^\varepsilon(x, w) - g_\tau^\varepsilon(x, W)(W - w)] \le C(\tau + W^{-1-\beta})(W - w)^2, \tag{38}$$

where C is a constant independent of ε.

To verify (38) we consider first the case $W \leq 2w$. By Taylor's theorem

$$-[G_\tau^\varepsilon(x, W) - G_\tau^\varepsilon(x, w) - g_\tau^\varepsilon(x, W)(W - w)] = \frac{1}{2}\frac{\partial g_\tau^\varepsilon}{\partial w}(x, \xi)(W - w)^2,$$

for some $w < \xi < W$. A computation shows that

$$\frac{\partial g_\tau^\varepsilon}{\partial w}(x, w) = \tau\frac{\beta\tau^{\frac{1}{1+\beta}}w - \varepsilon}{\left(\tau^{\frac{1}{1+\beta}}w + \varepsilon\right)^{2+\beta}} + \tau\bar{f}_u(\tau x + x_0, \tau^{\frac{1}{1+\beta}}w)$$

and therefore

$$\frac{\partial g_\tau^\varepsilon}{\partial w}(x, w) \leq \tau\beta\left(\tau^{\frac{1}{1+\beta}}w + \varepsilon\right)^{-1-\beta} + K\tau \leq \beta w^{-1-\beta} + K\tau, \tag{39}$$

where $K = \sup_{x,u}|\bar{f}_u(x, u(x))| < \infty$. Hence

$$-[G_\tau^\varepsilon(x, W) - G_\tau^\varepsilon(x, w) - g_\tau^\varepsilon(x, W)(W - w)] \leq (\beta\xi^{-1-\beta} + K\tau)(W - w)^2.$$

But $\xi^{-\beta} \leq w^{-\beta} \leq (W/2)^{-\beta}$ and we obtain

$$-[G_\tau^\varepsilon(x, W) - G_\tau^\varepsilon(x, w) - g_\tau^\varepsilon(x, W)(W - w)] \leq C(\tau + W^{-1-\beta})(W - w)^2.$$

For the case $W > 2w$ observe that

$$-\left[G_\tau^\varepsilon(x, W) - G_\tau^\varepsilon(x, w) - g_\tau^\varepsilon(x, W)(W - w)\right]$$
$$= -G_\tau^\varepsilon(x, W) + G_\tau^\varepsilon(x, w) + g_\tau^\varepsilon(x, W)(W - w)$$
$$\leq \frac{W^{1-\beta}}{1 - \beta} + \tau^{\frac{-1+\beta}{1+\beta}}\left[\overline{F}(\tau x + x_0, \tau^{\frac{1}{1+\beta}}W)\right.$$
$$- \overline{F}(\tau x + x_0, \tau^{\frac{1}{1+\beta}}w)$$
$$\left. + \tau^{\frac{1}{1+\beta}}\bar{f}(\tau x + -x_0, \tau^{\frac{1}{1+\beta}}W)(W - w)\right].$$

But for $W > 2w$ we have

$$\frac{W^{1-\beta}}{1 - \beta} = \frac{1}{1 - \beta}W^{-1-\beta}W^2 \leq \frac{4}{1 - \beta}W^{-1-\beta}(W - w)^2$$

and

$$\left|\overline{F}(\tau x + x_0, \tau^{\frac{1}{1+\beta}}W) - \overline{F}(\tau x + x_0, \tau^{\frac{1}{1+\beta}}w) + \tau^{\frac{1}{1+\beta}}\bar{f}(\tau x + x_0, \tau^{\frac{1}{1+\beta}}W)(W - w)\right|$$
$$= \frac{1}{2}\tau^{\frac{2}{1+\beta}}|\bar{f}_u(\tau x + x_0, \tau^{\frac{1}{1+\beta}}\xi)|(W - w)^2,$$

for some ξ. Thus

$$-[G_\tau^\varepsilon(x, W) - G_\tau^\varepsilon(x, w) - g_\tau^\varepsilon(x, W)(W - w)] \leq (CW^{-1-\beta} + K\tau)(W - w)^2.$$

Using estimate (38) in (37) we find

$$\psi_\tau(W) - \psi_\tau(w) \leq -\frac{1}{2}\int_{B_\tau^+}|\nabla(W - w)|^2 + C\int_{\Gamma_\tau^1}(W^{-1-\beta} + \tau)(W - w)^2.$$

But $W \geq v_\tau \geq cs_0 \operatorname{dist}(y, \Gamma_\tau^2)$ by (11) and therefore

$$\psi_\tau(W) - \psi_\tau(w) \leq -\frac{1}{2} \int_{B_\tau^+} |\nabla(W - w)|^2$$
$$+ C \int_{\Gamma_\tau^1} \left(s_0^{-1-\beta} \operatorname{dist}(\Gamma_\tau^2)^{-1-\beta} + \tau \right) (W - w)^2.$$

By Hardy's (Proposition 2.2) and Sobolev's inequality

$$\psi_\tau(W) - \psi_\tau(w) \leq \left(\frac{C}{s_0^{1+\beta}} + C\tau - \frac{1}{2} \right) \int_{B_\tau^+} |\nabla(W - w)|^2. \tag{40}$$

$\square$

Proof of Proposition 3.3. For $\tau > 0$ sufficiently small (31) has a unique solution. Therefore for τ small u_τ is the solution of (31) and the minimizer of ψ_τ.

We claim that if w is any minimizer of ψ_τ then $w \geq v_\tau$ in B_τ^+. Indeed take $v = v_\tau$ in Lemma 3.5 and observe that since $w = u_\tau$ on Γ_τ^i, we have by Lemma 3.2 $w \geq v_\tau$ on Γ_τ^i. Thus we can apply Lemma 3.5. Let us look at (40). We can choose s_0 larger and τ_0 smaller if necessary in order to make $\frac{C}{s_0^{1+\beta}} + C\tau - \frac{1}{2} < 0$. Thus $\psi_\tau(\max(w, v_\tau)) < \psi_\tau(w)$ unless $\max(w, v_\tau) \equiv w$, which is equivalent to assert $v_\tau \leq w$ in B_τ^+.

Let us see now that for $0 < \tau \leq \tau_1$ ψ_τ has a unique minimizer. Indeed, consider w_1, w_2 minimizers of ψ_τ. By the previous claim they satisfy $w_j \geq v_\tau$, $j = 1, 2$. Then from Lemma 3.5 it follows that $w_1 = w_2$. From now on w_τ denotes the unique minimizer of ψ_τ. We claim that the operator $D^2\psi_\tau(w_\tau)$ is coercive on the space $E_\tau = \{w \in H^1(B_\tau^+) \mid w = 0 \text{ on } \Gamma^i \cup \Gamma^2\}$ in the sense that

$$\int_{B_\tau^+} (|\nabla\varphi|^2 + \tau^2\varphi^2) - \int_{\Gamma_\tau^1} \frac{\partial g_\tau^\varepsilon}{\partial u}(x, w_\tau)\varphi^2 \geq \sigma \int_{B_\tau^+} |\nabla\varphi|^2 \tag{41}$$

for some $\sigma > 0$ independent of $0 < \tau \leq \tau_1$ and all $\varphi \in H^1(B_\tau^+)$ with $\varphi = 0$ on $\Gamma_\tau^i \cup \Gamma_\tau^2$. This follows from the behavior of $\frac{\partial g_\tau^\varepsilon}{\partial u}$ as given in (39), the estimate $w_\tau \geq v_\tau \geq cs_0 \operatorname{dist}(y, \Gamma_\tau^2)^{\frac{1}{1+\beta}}$ and Hardy's inequality, Proposition 2.2. We will use this to show that u_τ is the minimizer of ψ_τ. We know that this is true for small $\tau > 0$. Assume this fails for some $0 < \tau < \tau_1$ and set

$$\mu = \inf\{\tau \in (0, \tau_1) \mid u_\tau \text{ is not the minimizer of } \psi_\tau\}.$$

Then by continuity u_μ is the minimizer of ψ_μ. Thus $D^2\psi_\mu(u_\mu)$ is coercive in the sense above. On the other hand, for a sequence (τ_j) such that $\mu < \tau_j < \tau_1$, $\tau_j \to \mu$ there are at least two solutions of (31), one being u_τ and the other one the minimizer w_τ of ψ_τ. Both of them are uniformly bounded as $\tau_j \to \mu$. Set

$$z_j = \frac{u_{\tau_j} - w_{\tau_j}}{\|u_{\tau_j} - w_{\tau_j}\|_{L^2(B_{\tau_j}^+)}}.$$

Then

$$\begin{cases} -\Delta z_j + \tau^2 z_j = 0 & \text{in } B_{\tau_j}^+ \\ \qquad\quad z_j = 0 & \text{on } \Gamma_{\tau_j}^i \cup \Gamma_{\tau_j}^2 \\ \dfrac{\partial z_j}{\partial \nu} = \dfrac{\partial g_{\tau_j}^\varepsilon}{\partial u}(y, \xi_j(y))z_j & \text{on } \Gamma_{\tau_j}^1, \end{cases}$$

where ξ_j is between u_{τ_j} and w_{τ_j}. Multiplying by z_j and integrating we find

$$\int_{B_{\tau_j}^+} (|\nabla z_j|^2 + \tau_j^2 z_j^2) = \int_{\Gamma_{\tau_j}^1} \frac{\partial g_{\tau_j}^\varepsilon}{\partial u}(y, \xi_j(y))z_j^2.$$

Since z_j is bounded in $L^2(B_{\tau_j}^+)$ and for fixed $\varepsilon > 0$ $\frac{\partial g_{\tau_j}^\varepsilon}{\partial u}(y, \xi_j(y))$ is continuous and bounded, we see that z_j is bounded in $H^1(B_{\tau_j}^+)$. Thus we can extract a subsequence for which $z_j \rightharpoonup z$ weakly in $H^1(B_{\tau_j}^+)$ and strongly in $L^2(B_{\tau_j}^+)$. In particular $\|z\|_{L^2(B_\mu^+)} = 1$ which shows that $z \not\equiv 0$. Taking $j \to \infty$ we find

$$\int_{B_\mu^+} (|\nabla z|^2 + \mu^2 z^2) \leq \int_{\Gamma_\mu^1} \frac{\partial g_\mu^\varepsilon}{\partial u}(y, u_\mu(y))z^2,$$

and since $z \not\equiv 0$ we have a contradiction with (41). $\qquad\qquad\square$

Finally let us show that estimate (30) is enough to obtain the desired result.

Proposition 3.6. *Let $x_1 \in \Omega$ and assume we are in Case 2, i.e., (20) holds. Then*

$$|\nabla u(x_1)| \leq Cu(x_1)^{-\beta},$$

with a constant that depends on Ω, n, β, f and $\|u\|_{L^\infty(\Omega)}$.

Proof. Recall x_0 given by (20), the definition of τ_1 in (23) and u_{τ_1}, c.f. (24). Let $y_1 = \frac{1}{\tau_1}(x_1 - x_0)$ which satisfies

$$|y_1| \leq \frac{1}{6} \tag{42}$$

by (17), (19), (20). A direct calculation shows that it is sufficient to establish

$$|\nabla u_{\tau_1}(y_1)| \leq C. \tag{43}$$

By (30) and (11) we have the estimate

$$u_{\tau_1}(y) \geq cs_0 \operatorname{dist}(y, \Gamma_{\tau_1}^2)^{\frac{1}{1+\beta}} \quad \forall y \in \Gamma_{\tau_1}^1. \tag{44}$$

Using this in the boundary condition in (31) we deduce that

$$\left|\frac{\partial u_{\tau_1}}{\partial \nu}\right| \leq C \operatorname{dist}(y, \Gamma_{\tau_1}^2)^{-\frac{\beta}{1+\beta}} + \tau^{\frac{\beta}{1+\beta}} M \quad \text{on } \Gamma_{\tau_1}^1, \tag{45}$$

and therefore, on a smaller set we obtain an estimate

$$\left|\frac{\partial u_{\tau_1}}{\partial \nu}\right| \leq C \quad \text{on } B_{1/3} \cap \partial\Omega_{\tau_1}, \tag{46}$$

with a constant C independent of ε.

Let us prove (43). For this purpose choose $p > n$ and take $n < r < \frac{np}{n-1}$. By Lemma 2.4

$$\|u_{\tau_1}\|_{W^{1,r}(B_{1/4}\cap\Omega_{\tau_1})} \leq C \left(\left\|\frac{\partial u_{\tau_1}}{\partial \nu}\right\|_{L^p(B_{1/3}\cap\partial\Omega_{\tau_1})} + \|u_{\tau_1}\|_{L^1(B_{1/3}\cap\Omega_{\tau_1})} \right),$$

and by the embedding $W^{1,r} \subset C^\mu$ we have for some $0 < \mu < 1$

$$\|u_{\tau_1}\|_{C^\mu(B_{1/4}\cap\Omega_{\tau_1})} \leq C \left(\left\|\frac{\partial u_{\tau_1}}{\partial \nu}\right\|_{L^p(B_{1/3}\cap\partial\Omega_{\tau_1})} + \|u_{\tau_1}\|_{L^1(B_{1/3}\cap\Omega_{\tau_1})} \right).$$

By the assumption (2) and the lower bound (44) we see that the right-hand side of the boundary condition in (31) satisfies

$$\|g_\tau^\varepsilon(y, u_{\tau_1})\|_{C^\mu(B_{1/4}\cap\partial\Omega_{\tau_1})} \leq C \left(\left\|\frac{\partial u_{\tau_1}}{\partial \nu}\right\|_{L^p(B_{1/3}\cap\partial\Omega_{\tau_1})} + \|u_{\tau_1}\|_{L^1(B_{1/3}\cap\Omega_{\tau_1})} \right).$$

Using Schauder estimates (see, e.g., [8]) we deduce

$$\|u_{\tau_1}\|_{C^{1,\mu}(B_{1/5}\cap\Omega_{\tau_1})} \leq C \left(\left\|\frac{\partial u_{\tau_1}}{\partial \nu_{\tau_1}}\right\|_{L^p(B_{1/3}\cap\partial\Omega_{\tau_1})} + \|u_{\tau_1}\|_{L^1(B_{1/3}\cap\Omega_{\tau_1})} \right).$$

Recalling that $|y_1| \leq \frac{1}{6}$ by (42) we obtain

$$|\nabla u_{\tau_1}(y_1)| \leq C \left(\left\|\frac{\partial u_{\tau_1}}{\partial \nu}\right\|_{L^p(B_{1/3}\cap\partial\Omega_{\tau_1})} + \|u_{\tau_1}\|_{L^1(B_{1/3}\cap\Omega_{\tau_1})} \right).$$

By (46) we can assert that

$$\left\|\frac{\partial u_{\tau_1}}{\partial \nu}\right\|_{L^p(B_{1/3}\cap\partial\Omega_{\tau_1})} \leq C$$

with C independent of ε. It suffices then to find an estimate for $\|u_{\tau_1}\|_{L^1(B_{1/3}\cap\Omega_{\tau_1})}$. Using (45) we see that

$$\left|\frac{\partial u_{\tau_1}}{\partial \nu}\right| \leq C \quad \text{on } B_{5/12} \cap \partial\Omega_{\tau_1}$$

and therefore, using Lemma 2.5 we find

$$\int_{B_{1/3}\cap\Omega_{\tau_1}} u_{\tau_1} \leq C(u_{\tau_1}(y) + 1), \quad \forall y \in B_{1/2} \cap \Omega_{\tau_1}. \tag{47}$$

Remark that by the choice of τ_1 (cf. 23) we have

$$u_{\tau_1}(y_1) = \widetilde{C}.$$

Thus, selecting $y = y_1$ in (47) (recall (42)) we obtain the desired conclusion. $\qquad\square$

4. Proof of Theorem 1.3

We consider the approximating scheme (4) with $f(x, u) = u^p$ and $1 < p < \frac{n}{n-2}$:

$$\begin{cases} -\Delta u + u = 0 & \text{in } \Omega \\ \dfrac{\partial u}{\partial \nu} = -\dfrac{u}{(u + \varepsilon)^{1+\beta}} + u^p & \text{on } \partial\Omega. \end{cases} \tag{48}$$

Let Φ_ε be defined as in (6) with

$$g^\varepsilon(u) = \begin{cases} -\dfrac{u}{(u+\varepsilon)^{1+\beta}} + u^p & \text{if } u \geq 0 \\ |u|^p & \text{if } u < 0. \end{cases}$$

We will show that for fixed $\varepsilon > 0$ (48) has a nontrivial solution, using the mountain pass theorem of Ambrosetti and Rabinowitz [1, 10] in the space $H^1(\Omega)$ with the usual norm $\|u\|_{H^1}^2 = \int_\Omega |\nabla u|^2 + u^2$. We have

$$g^\varepsilon(u)u \geq \theta G^\varepsilon(u) \qquad \forall u \geq u_0$$

for some $\theta > 2$ and some $u_0 > 0$ and this together with the subcritical exponent $1 < p < \frac{n}{n-2}$ implies that the Palais-Smale condition holds for Φ_ε. Also, if $\|u\|_{H^1} = \rho$ we have by the trace embedding theorem

$$\int_{\partial\Omega} G^\varepsilon(u) \leq C \int_{\partial\Omega} |u|^{p+1} \leq a \int_{\partial\Omega} u^2 + C_a \int_{\partial\Omega} |u|^{\frac{2(n-1)}{n-2}}$$
$$\leq Ca\|u\|_{H^1}^2 + C_a\|u\|_{H^1}^{p+1}$$

with $a > 0$ as small as we like. Thus if $\|u\|_{H^1} = \rho$ then

$$\Phi_\varepsilon(u) \geq \frac{1}{2}\rho^2 - Ca\rho^2 - C_a\rho^{p+1} \geq \alpha > 0$$

choosing $\rho >$ small. Notice that ρ and $\alpha > 0$ are independent of ε. Let u_ε denote the mountain pass solution to (48). We will show that $\|u_\varepsilon\|_{L^\infty(\Omega)} \leq C$ for some C independent of ε employing the blow-up method of [4]. Suppose that for a sequence $\varepsilon \to 0$ we have $m_\varepsilon \equiv \|u_\varepsilon\|_{L^\infty(\Omega)} \to \infty$ and let x_ε be a point where the maximum of u_ε in $\overline{\Omega}$ is attained. Then necessarily $x_\varepsilon \in \partial\Omega$ and we can assume that $x_\varepsilon \to x_0 \in \partial\Omega$. Define

$$v_\varepsilon(y) = \frac{1}{m_\varepsilon}u(m_\varepsilon^{1-p}y + x_\varepsilon).$$

Then $\Delta v_\varepsilon + m_\varepsilon^{2(1-p)}v_\varepsilon = 0$ in the domain $\Omega_\varepsilon \equiv (\Omega - x_\varepsilon)/m_\varepsilon^{1-p}$ and

$$\frac{\partial v_\varepsilon}{\partial \nu} = -m_\varepsilon^{-p-\beta}v_\varepsilon^{-\beta} + v_\varepsilon^p \qquad \text{on } \partial\Omega_\varepsilon.$$

The proof of Theorem 1.1 can be adapted to yield a uniform Hölder estimate locally for v_ε:

$$\|v_\varepsilon\|_{C^\gamma(\overline{\Omega}_\varepsilon \cap B_R)} \leq C \qquad \forall \varepsilon > 0$$

for some constant C depending on R but independent of ε. For a subsequence we find that $v_\varepsilon \to v$ uniformly on compact sets with v a nontrivial, nonnegative solution to the problem

$$\begin{cases} \Delta v = 0 & \text{in } \mathbb{R}^n_+ \\ \dfrac{\partial v}{\partial \nu} = v^p & \text{on } \partial\mathbb{R}^n_+, \end{cases}$$

where $\mathbb{R}^n_+$ is a half-space. But this is impossible, see, e.g., [5] and also [7]. This shows that u_ε is uniformly bounded in $L^\infty(\Omega)$. Corollary 1.2 implies that $u = \lim_{\varepsilon \to 0} u_\varepsilon$ is a solution to (7). This solution is nontrivial because $\Phi_\varepsilon(u_\varepsilon) \geq \alpha > 0$ for all $\varepsilon > 0$.

Acknowledgement

J. Dávila was partially supported by Fondecyt 1020815. He would like also to thank H. Brezis and the organizers of the *Fifth European Conference on Elliptic and Parabolic Problems: A special tribute to the work of Haim Brezis* for the kind invitation to participate in this event.

References

[1] A. Ambrosettiy, P.H. Rabinowitz, *Dual variational methods in critical point theory and applications*. J. Functional Analysis **14** (1973), 349–381.

[2] J. Dávila, M. Montenegro, *Nonlinear problems with solutions exhibiting a free boundary on the boundary*. To appear in Ann. Inst. H. Poincaré Anal. Non Linéaire.

[3] M. Fila, H.A. Levine, *Quenching on the boundary*. Nonlinear Anal. **21** (1993), 795–802.

[4] B. Gidas, J. Spruck, *A priori bounds for positive solutions of nonlinear elliptic equations*. Comm. Partial Differential Equations **6** (1981), 883–901.

[5] B. Hu, *Nonexistence of a positive solution of the Laplace equation with a nonlinear boundary condition*. Differential Integral Equations **7** (1994), 301–313.

[6] H.A. Levine, G.M. Lieberman, *Quenching of solutions of parabolic equations with nonlinear boundary conditions in several dimensions*. J. Reine Angew. Math. **345** (1983), 23–38.

[7] Y. Li, M. Zhu, *Uniqueness theorems through the method of moving spheres*. Duke Math. J. **80** (1995), 383–417.

[8] G.M. Lieberman, *Boundary regularity for solutions of degenerate elliptic equations*. Nonlinear Anal. **12** (1988), 1203–1219.

[9] J.L. Lions and E. Magenes, *Problemi ai limiti non omogenei. V*. Ann. Scuola Norm Sup. Pisa **16** (1962), 1–44.

[10] P.H. Rabinowitz, Minimax methods in critical point theory with applications to differential equations. American Mathematical Society, Providence, RI, 1986.

Juan Dávila
Departamento de Ingeniería Matemática
CMM (UMR CNRS)
Universidad de Chile
Casilla 170/3, Correo 3
Santiago, Chile
e-mail: jdavila@dim.uchile.cl

Marcelo Montenegro
Universidade Estadual de Campinas IMECC
Departamento de Matemática,
Caixa Postal 6065, CEP 13083-970
Campinas, SP, Brasil
e-mail: msm@ime.unicamp.br

Progress in Nonlinear Differential Equations
and Their Applications, Vol. 63, 207–215

Asymptotic Analysis of the Neumann Problem for the Ukawa Equation in a Thick Multi-structure of Type 3:2:2

U. De Maio and T.A. Mel'nyk

Abstract. We propose two different approaches for asymptotic analysis of the Neumann boundary-value problem for the Ukawa equation in a thick multi-structure Ω_ε, which is the union of a domain Ω_0 and a large number N of ε−periodically situated thin annular disks with variable thickness of order $\varepsilon = \mathcal{O}(N^{-1})$, as $\varepsilon \to 0$. In the first approach, using some special extension operator, the convergence theorem is proved as $\varepsilon \to 0$. In the second one, the leading terms of the asymptotic expansion for the solution are constructed and the corresponding estimates in the Sobolev space $H^1(\Omega_\varepsilon)$ are proved.

Mathematics Subject Classification (2000). 3540, 35B27, 35J25, 35C20, 35B25.

Keywords. homogenization, asymptotic expansion, thick multi-structure.

1. Background and objective of the study

A thick multi-structure (or thick junction) Ω_ε of type $k : p : d$ is a domain in $\mathbb{R}^n$, which consists of some domain Ω_0 and a large number of ε-periodically situated thin domains along some manifold on the boundary of Ω_0. This manifold is called the joint zone and the domain Ω_0 is called the junction's body. Here ε is a small parameter, which characterizes the distance between the neighboring thin domains and their thickness. In general, the junction's body and the joint zone can be depend on ε as well. The type $k : p : d$ of the thick junction refers respectively to the limiting dimensions of the body, the joint zone, and each of the attached thin domains. This classification was given by T.A. Mel'nyk and S.A. Nazarov in the papers [7]–[11]. It was shown in these papers that scheme of investigation and qualitative properties of solutions to boundary value problems in thick multi-structures essentially depend on the junction type.

Thick junctions are prototypes of widely used engineering constructions such as long bridges on supports, frameworks of houses, industrial installations, spaceship grids as well as many other physical and biological systems with very distinct characteristic scales.

Despite the enormous growth in computational power, it is often impossible to represent a complete system at the finest scale for which the various constitutive elements may suitably be represented. Increase in the size of computational domains for thick multi-structures naturally leads to longer computing times and makes it very difficult to maintain an acceptable level of accuracy.

Thus, asymptotic analysis of problems in such domains is an important task for applied mathematics. The aim of the analysis is to develop rigorous asymptotic methods for boundary value problems in thick junctions of different types as $\varepsilon \to 0$, i.e., when the number of attached thin domains infinitely increases and their thickness tends to zero.

1.1. Statement of the problem and the main result

Let l, a_0 and a_1 be positive real numbers and $a_0 < a_1$. Consider a model thick multi-structure Ω_ε of type $3 : 2 : 2$ that consists of the cylinder

$$\Omega_0 = \{x \in \mathbb{R}^3 : 0 < x_2 < l, \quad r := \sqrt{x_1^2 + x_3^2} < a_0\}$$

and a large number N of thin annular disks

$$G_j(\varepsilon) = \{x \in \mathbb{R}^3 : |x_2 - \varepsilon(j + 1/2)| < \varepsilon h_0(r)/2, \quad a_0 \le r < a_1\},$$

$$j = 0, 1, \ldots, N - 1,$$

i.e., $\Omega_\varepsilon = \Omega_0 \cup G(\varepsilon)$, $G(\varepsilon) = \cup_{j=0}^{N-1} G_j(\varepsilon)$. Here h_0 is a piecewise smooth function on the segment $[a_0, a_1]$ and $0 < h_0(r) < 1$ for $r \in [a_0, a_1]$; the number of the thin disks is equal to a large even integer N, therefore, $\varepsilon = l/N$ is a small parameter, which characterizes the distance between the neighboring thin disks and their thickness. These values respectively equal $\varepsilon(1 - h_0)$ and εh_0. The cross-section of Ω_ε in the plane $\mathbb{R}^2_{x_2 x_3}$ is shown in Figure 1.

FIGURE 1. The cross-section of the thick junction Ω_ε of type 3:2:2.

In Ω_ε we consider the following Neumann boundary-value problem

$$\begin{aligned}
-\Delta_x u_\varepsilon(x) + u_\varepsilon(x) &= f_\varepsilon(x), & x \in \Omega_\varepsilon, \\
\partial_\nu u_\varepsilon(x) &= 0, & x \in \partial\Omega_\varepsilon,
\end{aligned} \tag{1}$$

where $\partial_\nu = \partial/\partial\nu$ is the outward normal derivative.

We can regard without loss of generality that the right-hand side f_ε belongs to $L_2(\Omega_1)$, where $\Omega_1 = \{x : \ 0 < x_2 < l, \ r < a_1\}$.

It follows from the theory of boundary-value problems that for any fixed ε there exists a unique weak solution $u_\varepsilon \in H^1(\Omega_\varepsilon)$ to problem (1), which satisfies the integral identity

$$\int_{\Omega_\varepsilon} (\nabla u_\varepsilon \cdot \nabla\varphi + u_\varepsilon\varphi)\, dx = \int_{\Omega_\varepsilon} f_\varepsilon\, \varphi\, dx \qquad \forall\, \varphi \in H^1(\Omega_\varepsilon). \tag{2}$$

The aim of our research is to study the asymptotic behavior of the weak solution to problem (1) as $\varepsilon \to 0$, i.e., when the number of the attached thin disks infinitely increases and their thickness tends to zero.

1.2. Principal components of analysis

The extensive reviews on asymptotic analysis ob boundary-value problems in thick multi-structures are presented in [7]–[11]. Here we will mention principal components of analysis and new moments for our problem.

Extension operators play an important role in proofs of convergence theorems for solutions to boundary-value problems in domains depending on a small parameter ϵ. Usually such operators act from the Sobolev space $H^1(D_\epsilon)$ into the Sobolev space $H^1(D_0)$ and they allow to pass into a domain D_0 which does not depend on the parameter ϵ. It is very essential that such operators be uniformly bounded with respect to ϵ in H^1. Therefore, the uniform boundedness of extension operators is the necessary condition in statements of some problems (see [12]). If D_ϵ is an ϵ-periodically perforated domain by small holes, then the existence of uniformly bounded extension operators was proved in [4]. However, in case when D_ϵ is a thick junction, there are no extension operators that would be bounded uniformly in ϵ. This is one of the main difficulties in investigations of boundary-value problems in thick multi-structures. But, it turned out that for solutions of such problems it is possible to construct extensions whose norms in H^1 are uniformly bounded in ϵ. The general approach to construct such operators was proposed in [9, 10].

It should be emphasized here that there is essential difference between the asymptotic investigation of boundary-value problems in thick multi-structures and in domains with rapidly oscillating boundaries [3]. The extension in [3] was constructed without conservation of the class of the space (only in $H^1_{\mathrm{loc}}(\Omega_1^+)$, where $\Omega_1^+ \subset \mathbb{R}^2$ is a domain that in the limit is filled up by the oscillating boundary) and under the assumption that the right-hand side $f \in H^1$. In addition, a function h, which defines the oscillating boundary in [3], must be a continuously differentiable periodic function, that is the boundary is smooth, and the reciprocal functions of h on some intervals have to be provided to construct an extension operator. These conditions do not hold for thick junctions, which have the Lipschitz boundaries and the periodical structure of the joining of the thin domains (the thin domains can have various lengths). Therefore, the scheme of construction of extension operators in [3] is not applicable for thick multi-structures. Moreover, we consider more weak assumption for the right-hand side of problem (1) (see (3)).

Regarding the approximations of solutions it should be mentioned the paper [1, 2], where the corrector for the solution to the Laplace equation in a plane thick junction of type $2 : 1 : 1$ was constructed outside a layer of width 2ε in the joint zone. But for applied problems, it is very important to construct the asymptotic expansion for the solution and to prove the asymptotic estimates all over the thick junction since the solution has singularities exactly in the joint zone. Such asymptotic constructions were made for different boundary-value problems in [7]–[11] with the help of the method of matched asymptotic expansions and asymptotic methods for thin domains.

It should be stressed that the corresponding limit problem is derived from limit problems for each domain forming the thick junction with the help of the solutions to the junction-layer problems in the joint zone. However, the junction-layer solutions behave as powers (or logarithm) at infinity and do not decrease exponentially. Therefore, they influence directly the leading terms of the asymptotics. The cause of this effect is in the modification of the geometrical structure of the domain, where junction-layer problems are considered, and this domain depends on the type of a thick junction.

Only boundary-value problems in thick junctions with attached thin domains whose thickness are unvarying or problems in thick junctions with the smooth boundaries (see [3]) were considered till now. Our thick junction Ω_ε has the type $3 : 2 : 2$ and the Lipschitz boundary. In addition, the thickness of each thin disk is equal to the value $\varepsilon h_0(r)$, $r \in [a_0, a_1]$. Because of this the coefficients of the corresponding limit problem depend on h_0.

The results of this paper and their detail proofs are printed in the preprints [5, 6].

2. The convergence theorem

In this section we will assume that

$$f_\varepsilon \to f_0 \quad \text{in} \quad L_2(\Omega_1) \quad \text{as} \quad \varepsilon \to 0 \tag{3}$$

and there exist positive constants C_1, ε_0 such that for all $\varepsilon \in (0, \varepsilon_0)$

$$\int_{\widehat{\Omega}_\varepsilon} \left(\mathbf{F}_\varepsilon(x)\right)^2 \, dx \leq C_1, \tag{4}$$

where $\mathbf{F}_\varepsilon(x) = \varepsilon^{-1}\left(\widehat{f}_\varepsilon(x + \varepsilon \bar{e}_2) - \widehat{f}_\varepsilon(x)\right)$, $\ (\bar{e}_2 = (0, 1, 0)\,)$. Here and further we interpret symbols with hats $(\widehat{Y}, \widehat{f}_\varepsilon, \widehat{\Omega}_\varepsilon \ldots)$ as follows: if Y is a set, then $\widehat{Y}$ is the union of Y and of its image, symmetric with respect to the plane $\{x : x_2 = 0\}$; if Y is a function, then $\widehat{Y}$ is its even extension into the relevant domain with respect to the plane $\{x : x_2 = 0\}$. Condition (4) means that f_ε has not big scattering of values on the thin disks.

Theorem 1. *If the conditions* (3) *and* (4) *hold, then there exists an extension operator*

$$\mathbf{P}_\varepsilon : H^1(\Omega_\varepsilon) \mapsto H^1(\Omega_1)$$

such that for the solution u_ε to problem (1) *we have*

$$\| \mathbf{P}_\varepsilon u_\varepsilon \|_{H^1(\Omega_1)} \leq C_0 \left(\|\mathbf{F}_\varepsilon\|_{L_2(\widehat{\Omega}_\varepsilon)} + \|f_\varepsilon\|_{L_2(\Omega_\varepsilon)} \right) \leq C_1. \tag{5}$$

Let us describe the main steps of the proof. At first, using condition (4) and taking into account that the even extension of problem (1) is invariant with respect to the shift by ε along the axis x_2, we prove that the scattering of the values of u_ε on the neighboring thin disks is small in some sense. Then, with the help of the "linear matching", we construct an extension of u_ε in domains between the neighboring thin cut disks and estimate the norm of the extension. Finally, we use Lemma 2.1 ([9]) to extend our solution into thin tori.

Theorem 2. *Under the assumptions for the right-hand side f_ε the extension $\mathbf{P}_\varepsilon u_\varepsilon$ of the solution u_ε weakly in $H^1(\Omega_1)$ converges as $\varepsilon \to 0$ to a unique weak solution of the following limit problem*

$$-\Delta_x v_0^+(x) + v_0^+(x) = f_0(x), \qquad\qquad x \in \Omega_0, \tag{6}$$

$$-\mathrm{div}_{\widetilde{x}}\big(h_0(r)\nabla_{\widetilde{x}} v_0^-(x)\big) + h_0(r)v_0^-(x) = h_0(r)\, f_0(x), \qquad x \in D,$$

$$\partial_r v_0^-(x) = 0, \qquad\qquad r = a_1, \quad x_2 \in (0,l),$$

$$\partial_{x_2} v_0^+(x) = 0, \qquad\qquad x \in S^{(0)} \cup S^{(l)},$$

$$v_0^+(x) = v_0^-(x), \qquad\qquad r = a_0, \quad x_2 \in (0,l),$$

$$\partial_r v_0^+(x) = h_0(a_0)\,\partial_r v_0^-(x), \quad r = a_0, \quad x_2 \in (0,l).$$

Here $\widetilde{x} = (x_1, x_3)$, $D = \{x : \ 0 < x_2 < l, \ a_0 < r < a_1\}$, $\partial_r = \partial/\partial r$ is the derivative with respect to polar radius r; $S^{(0)} = \{x \in \partial\Omega_0 : \ x_2 = 0\}$, $S^{(l)} = \{x \in \partial\Omega_0 : \ x_2 = l\}$.

It should be stressed here that we have the differential equation only with respect to x_1 and x_3 in the domain D, which is filled up by the thin disks in the limit passage and there are no any boundary conditions on the vertical parts of ∂D.

Sketch of the proof. Because of conditions (3), (4) and Theorem 1, the sequences $\{\chi_\varepsilon \partial_{x_i}(\mathbf{P}_\varepsilon u_\varepsilon)\}_{\varepsilon > 0}$, $i = 1, 2, 3$, are bounded in $L_2(D)$ and the sequence $\{\mathbf{P}_\varepsilon u_\varepsilon\}_{\varepsilon > 0}$ is bounded in $H^1(\Omega_1)$. Therefore, we can choose a subsequence of $\{\varepsilon\}$ (still denoted by $\{\varepsilon\}$) such that $\chi_\varepsilon \partial_{x_i}(\mathbf{P}_\varepsilon u_\varepsilon) \to \gamma_i$ weakly in $L_2(D)$, $i = 1, 2, 3$; and

$$\mathbf{P}_\varepsilon u_\varepsilon \to \begin{cases} v_0^+(x), & x \in \Omega_0, \\ v_0^-(x), & x \in D, \end{cases} \quad \text{weakly in} \quad H^1(\Omega_1).$$

Here χ_ε is the characteristic function of the set $G(\varepsilon)$. Next we find that $\gamma_2 \equiv 0$ and $\gamma_i(x) = h_0(r)\,\partial_{x_i}v_0^-(x)$, $x \in D$, $i = 1$, $i = 3$. Finally, passing to the limit in the integral identity (2), we get the following identity

$$\int_{\Omega_0} \left(\nabla v_0^+ \cdot \nabla \varphi + v_0^+ \varphi \right) dx + \int_D h_0(r) \left(\partial_{x_1} v_0^- \partial_{x_1}\varphi + \partial_{x_3} v_0^- \partial_{x_3}\varphi + v_0^- \varphi \right) dx$$

$$= \int_{\Omega_0} f_0(x)\varphi(x)\,dx + \int_D h_0(r)\,f_0(x)\varphi(x)\,dx, \quad \forall\,\varphi \in H^1(\Omega_1). \quad (7)$$

The identity (7) is the corresponding integral identity for the limit problem (6). Using standard Hilbert space methods, we prove that there exists a unique function from the anisotropic Sobolev space $\mathcal{V} = \{v \in L_2(\Omega_1) : \partial_{x_1}v \in L_2(\Omega_1), \partial_{x_3}v \in L_2(\Omega_1), \partial_{x_2}v \in L_2(\Omega_0)\}$ with the scalar product

$$(u,v)_\mathcal{V} = \int_{\Omega_0} \left(\nabla u \cdot \nabla v + uv \right) dx + \int_D h_0(r) \left(\partial_{x_1}u\,\partial_{x_1}v + \partial_{x_3}u\,\partial_{x_3}v + uv \right) dx,$$

which satisfies the identity (7). This function is called the weak solution to problem (6).

3. Asymptotic approximation

Usually to construct the asymptotic extensions for solutions of perturbed problems we need more stronger assumptions for the right-hands. For our problem we assume that f_ε has the following form $f_\varepsilon(x) = f_0(x) + \varepsilon f_1(x,\varepsilon)$, $x \in \Omega_1$, where f_0 is a smooth function in $\overline{\Omega_1}$ that vanishes on $S^{(0)}$ and $S^{(l)}$, $f_1 \in L_2(\Omega_1)$ and $\|f_1(\cdot,\varepsilon)\|_{L_2(\Omega_1)} = \mathcal{O}(1)$ as $\varepsilon \to 0$. Obviously, from this assumption it follows (3) and (4). Also we assume that h_0 is locally constant in some sufficiently small neighborhood of the point a_0.

We seek the leading terms of the asymptotics for the solution u_ε, restricted to Ω_0, in the form

$$u_\varepsilon(x) \approx v_0^+(x) + \sum_{k=1}^\infty \varepsilon^k v_k^+(x,\varepsilon), \quad (8)$$

and, restricted to $G_j(\varepsilon)$, in the form

$$u_\varepsilon(x) \approx v_0^-(x) + \sum_{k=1}^\infty \varepsilon^k v_k^-(x,\xi_2 - j), \quad \xi_2 = \varepsilon^{-1}x_2. \quad (9)$$

The expansions (8) and (9) are usually called *outer expansions*.

In a neighborhood of $\Gamma_0 = \{x \in \partial\Omega : r = a_0\}$ we consider the Laplace operator in the cylindrical coordinates r, φ, x_2, where $r = \sqrt{x_1^2 + x_3^2}$, $\tan(\varphi) = x_3/x_1$, and then pass to the "rapid" coordinates $\xi = (\xi_1,\xi_2)$, where $\xi_1 = -\varepsilon^{-1}(r - a_0)$ and $\xi_2 = \varepsilon^{-1}x_2$. After that the Laplace operator takes the following form

$$\varepsilon^{-2}\left(\partial_{\xi_1\xi_1}^2 + \partial_{\xi_2\xi_2}^2\right) - \varepsilon^{-1}\left(a_0 - \varepsilon\xi_1\right)^{-1}\partial_{\xi_1} + \left(a_0 - \varepsilon\xi_1\right)^{-2}\partial_{\varphi\varphi}^2. \quad (10)$$

We seek the leading terms of the inner expansion in a neighborhood of Γ_0 in view

$$u_\varepsilon(x) \approx v_0^+(x)|_{r=a_0} + \varepsilon\Big(Z_1(\xi)\,(\partial_{x_2} v_0^+(x))|_{r=a_0} + Z_2(\xi)\,(\partial_r v_0^+(x))|_{r=a_0}\Big) + \cdots . \quad (11)$$

Substituting (11) in (10) and in the corresponding boundary conditions of problem (1) and collecting the coefficients of the same power of ε, we arrive junction-layer problems for the functions Z_1 and Z_2:

$$
\begin{array}{rcll}
-\Delta_{\xi_1\xi_2} Z_i(\xi) &=& 0, & \xi \in \Pi, \\
\partial_{\xi_1} Z_i(0,\xi_2) &=& 0, & (0,\xi_2) \in \partial\Pi^+ \setminus I_h, \\
\partial_{\xi_2} Z_i(\xi) &=& -\delta_{1i}, & \xi \in \partial\Pi^- \setminus I_h, \\
\partial_{\xi_2}^k Z_i(\xi_1,0) &=& \partial_{\xi_2}^k Z_i(\xi_1,1), & \xi_1 > 0, \quad k = 0,1.
\end{array}
\quad (12)
$$

Here Π is the union of semi-infinite strips $\Pi^+ = (0,+\infty) \times (0,1)$ and $\Pi^- = (-\infty,0] \times I_h$, where $I_h = \big((1-h)/2,\,(1+h)/2\big)$, the constant h is equal to $h_0(a_0)$. The last periodic conditions in (12) due to the periodicity of the thin rings $\{G_j(\varepsilon) : j = 0, \ldots, N-1\}$.

The main asymptotic relations for $\{Z_i\}$ can be obtained from general results about the asymptotic behavior of solutions to elliptic problems in domains with different exits to infinity. However, using the symmetry of the domain Π, we can define more exactly the asymptotic relations and detect other properties of the junction-layer solutions Z_1, Z_2 similarly as in the papers [8, 9].

Statement 1.([8]) *There exist solutions $Z_i \in H^1_{loc,\sharp}(\Pi)$, $i = 1,2$, of problems (12), which have the following differentiable asymptotics*

$$
Z_1(\xi) = \begin{cases} \mathcal{O}(\exp(-2\pi\xi_1)), & \xi_1 \to +\infty, \\ -\xi_2 + \tfrac{1}{2} + \mathcal{O}(\exp(\pi h^{-1}\xi_1)), & \xi_1 \to -\infty; \end{cases}
$$

$$
Z_2(\xi) = \begin{cases} -\xi_1 + c_h + \mathcal{O}(\exp(-2\pi\xi_1)), & \xi_1 \to +\infty, \\ -h^{-1}\xi_1 + \mathcal{O}(\exp(\pi h^{-1}\xi_1)), & \xi_1 \to -\infty. \end{cases}
$$

In addition, the function Z_1 is odd in ξ_2 and Z_2 is even in ξ_2 with respect to $1/2$. Here $H^1_{loc,\sharp}(\Pi) = \{u : \Pi \to \mathbb{R} \mid u(\xi_1,0) = u(\xi_1,1) \text{ for any } \xi_1 > 0, \ u \in H^1(\Pi_R) \text{ for any } R > 0\}$, where $\Pi_R = \Pi \cap \{\xi : -R < \xi_1 < R\}$.

Substituting the outer expansions (8) and (9) in problem (1), collecting the coefficients of the same powers of ε and matching with the inner expansion (10), we deduce that the function $v_0^\pm$ must satisfies the relations of the limit problem (6).

Next we construct an approximation function $R_\varepsilon \in H(\Omega_\varepsilon)$

$$R_\varepsilon(x) := R_\varepsilon^+(x) = v_0^+(x) + \varepsilon\chi_0(r)\mathcal{N}^+\Big(-\frac{r-a_0}{\varepsilon},\frac{x_2}{\varepsilon},\varphi,x_2\Big),$$

$$x \in \Omega_0, \quad (13)$$

$$R_\varepsilon(x) := R_\varepsilon^-(x) = v_0^-(x) + \varepsilon\Big(Y\Big(\frac{x_2}{\varepsilon}\Big)\partial_{x_2} v_0^-(x) + \chi_0(r)\mathcal{N}^-\Big(-\frac{r-a_0}{\varepsilon},\frac{x_2}{\varepsilon},\varphi,x_2\Big)\Big),$$

$$x \in D, \quad (14)$$

where χ_0 is a cut-off function that is equal to 1 in a neighborhood of the contact zone Γ_0, $Y(\xi_2) = -\xi_2 + \frac{1}{2} + [\xi_2]$ ($[\xi_2]$ is the integral part of ξ_2),

$$\mathcal{N}^+(\xi, \varphi, x_2) = Z_1(\xi)\left(\partial_{x_2} v_0^+(x)\right)_{|r=a_0} + \left(Z_2(\xi) + \xi_1\right)\left(\partial_r v_0^+(x)\right)_{|r=a_0},$$

$$\mathcal{N}^-(\xi, \varphi, x_2) = \left(Z_1(\xi) - Y(\xi_2)\right)\left(\partial_{x_2} v_0^+(x)\right)_{|r=a_0} + \left(Z_2(\xi) + h^{-1}\xi_1\right)\left(\partial_r v_0^+(x)\right)_{|r=a_0}.$$

Observing that $[\Delta_{\widetilde{x}}, \chi_0(r)]\Phi(x) = \nabla_{\widetilde{x}} \cdot \left(\Phi(x)\nabla_{\widetilde{x}} \chi_0(r)\right) + \nabla_{\widetilde{x}}\Phi(x) \cdot \nabla_{\widetilde{x}} \chi_0(r)$, where $[\mathbf{A}, \mathbf{B}] = \mathbf{AB} - \mathbf{BA}$ is the commutator of two operators $\mathbf{A}$ and $\mathbf{B}$, and taking into account the form (10) of Laplace's operator, we get

$$\begin{aligned}
&-\Delta_x R_\varepsilon^+(x) + R_\varepsilon^+(x) - f_\varepsilon(x) \\
&= -\chi_0(r)\left(\partial_{x_2\xi_2}^2 \mathcal{N}^+(\xi, \varphi, x_2) - r^{-1}\partial_{\xi_1}\mathcal{N}^+(\xi, \varphi, x_2)\right) \\
&\quad - \nabla_\xi \mathcal{N}^+(\xi, \varphi, x_2) \cdot \nabla_{\widetilde{x}} \chi_0(r) \\
&\quad - \varepsilon\left(\nabla_{\widetilde{x}} \cdot \left(\mathcal{N}^+\nabla_{\widetilde{x}} \chi_0(r)\right)\right) \\
&\quad + \chi_0(r)\left(-\mathcal{N}^+ + \partial_{x_2 x_2}^2 \mathcal{N}^+(\xi, \varphi, x_2) + r^{-2}\partial_{\varphi\varphi}^2 \mathcal{N}^+(\xi, \varphi, x_2)\right) + f_1(x, \varepsilon)\right),
\end{aligned}$$

$$\xi_1 = -\frac{r - a_0}{\varepsilon}, \quad \xi_2 = \frac{x_2}{\varepsilon}, \quad r = \sqrt{x_1^2 + x_3^2}, \quad \tan(\varphi) = \frac{x_3}{x_1}, \qquad x \in \Omega_0. \tag{15}$$

Similarly, we obtain

$$-\Delta_x R_\varepsilon^-(x) + R_\varepsilon^-(x) - f_\varepsilon(x)$$

$$= \nabla_{\widetilde{x}}(\ln h_0) \cdot \nabla_{\widetilde{x}} v_0^- + \chi_0(r)\left(r^{-1}\partial_{\xi_1}\mathcal{N}^- - \partial_{\xi_2}\mathcal{N}^-\right) - \nabla_\xi \mathcal{N}^- \cdot \nabla_{\widetilde{x}} \chi_0(r)$$

$$-\varepsilon\partial_{x_2}\left(Y\left(\frac{x_2}{\varepsilon}\right)\partial_{x_2 x_2}^2 v_0^- + \chi_0(r)\left(\partial_{x_2}\mathcal{N}^-\right)_{|\xi_2=x_2/\varepsilon}\right) - \varepsilon\left(Y\left(\frac{x_2}{\varepsilon}\right)\left(\Delta_{\widetilde{x}}\left(\partial_{x_2} v_0^-\right) - v_0^-\right)\right.$$

$$\left. + \nabla_{\widetilde{x}} \cdot \left(\mathcal{N}^-\nabla_{\widetilde{x}} \chi_0\right) + \chi_0\left(-\mathcal{N}^- + r^{-2}\partial_{\varphi\varphi}^2 \mathcal{N}^-\right) + f_1\right), \qquad x \in G(\varepsilon). \tag{16}$$

Estimating the right-hand sides in (15), (16) and finding residuals from R_ε in the boundary condition on $\partial\Omega_\varepsilon$, we deduce the following theorem.

Theorem 3. *For any $\delta > 0$ the difference between the solution u_ε to problem (1) and the approximation function R_ε, which is defined by (13) and (14), satisfies the estimate*

$$\|u_\varepsilon - R_\varepsilon\|_{H^1(\Omega_\varepsilon)} \leq c_1(\delta)\,\varepsilon^{1-\delta}. \tag{17}$$

References

[1] Y. Amirat and O. Bodart, Boundary layer corrector for the solution of Laplace equation in a domain with oscillating boundary, Zeitschrift für Analysis und ihre Anwendungen, 20 (2001), No. 4, 929–940.

[2] Y. Amirat, O. Bodart, U. De Maio and A. Gaudiello, Asymptotic approximation of the solution of the Laplace equation in a domain with highly oscillating boundary, (to appear in SIAM, J. Math. Anal.)

[3] R. Brizzi and J.P. Chalot, Homogenization and Neumann Boundary Value Problem, *Ric. Mat.* **46** (1997), 347–387.

[4] D. Cioranescu and J. Saint Jean Paulin, Homogenization in open sets with holes, *J. Math. Anal. and Appl.* **71** (1979), 590–607.

[5] U. De Maio and T.A. Mel'nyk, Homogenization of the Neumann problem in a thick multi-structures of type 3 : 2 : 2, *Preprint* No. **11**, Department of Information Engineering and Applied Mathematics, University of Salerno, October, 2003.

[6] U. De Maio and T.A. Mel'nyk, Asymptotic solution to a mixed boundary value problem in a thick multi-structure of type 3:2:2, *Preprint* No. **13**, Department of Information Engineering and Applied Mathematics, University of Salerno, October, 2003.

[7] T.A. Mel'nyk and S.A. Nazarov, Asymptotic structure of the spectrum of the Neumann problem in a thin comb-like domain, *C.R. Acad. Sci. Paris*, **319** (1994), Serie 1, 1343–1348.

[8] T.A. Mel'nyk and S.A. Nazarov, Asymptotics of the Neumann spectral problem solution in a domain of "thick comb" type, *Trudy Seminara imeni I.G. Petrovskogo, Moscow University*, **19** (1996), 138–173 (in Russian); English transl. in: *Journal of Mathematical Sciences*, **85** (1997), No. 6, 2326–2346.

[9] T.A. Mel'nyk, Homogenization of the Poisson equation in a thick periodic junction, *Zeitschrift für Analysis und ihre Anwendungen*, **18** (1999), No. 4, 953–975.

[10] T.A. Mel'nyk, Asymptotic analysis of a spectral problem in a periodic thick junction of type 3:2:1, *Mathematical Methods in the Applied sciences*, **23** (2000), No. 4, 321–346.

[11] T.A. Mel'nyk and S.A. Nazarov, Asymptotic analysis of the Neumann problem of the junction of a body and thin heavy rods, *St. Petersburg Math.J.* **12** (2001), No. 2, 317–351.

[12] V.V. Zhikov, S.M. Kozlov, and O.A. Oleinik, *Homogenization of Differential Operators and Integral Functionals*, Springer, Berlin, 1994.

U. De Maio
Dipartimento di Matematica e Applicazioni
Università degli Studi di Napoli Federico II
Complesso Monte S. Angelo – Edificio "T"
Via Cintia
I-80126 Napoli, Italia
e-mail: `udemaio@unina.it`

T.A. Mel'nyk
National Taras Shevchenko University of Kyiv
Faculty of Mathematics and Mechanics
Volodymyrska str. 64
01033 Kyiv, Ukraine
e-mail: `melnyk@imath.kiev.ua`

Progress in Nonlinear Differential Equations
and Their Applications, Vol. 63, 217–234

On the Haïm Brezis Pioneering Contributions on the Location of Free Boundaries

J.I. Díaz

1. Introduction

Starting in the seventies, and simultaneously to his beautiful results on the existence and regularity of solutions of many nonlinear PDEs, Haïm Brezis produced a series of papers in which, in a pioneering way, he rigorously found new qualitative phenomena as, for instance, the compactness of the support of the solution of suitable problems posed on unbounded domains and, more generally, on the location of this type of free boundaries (sometimes unexpected from the original formulation).

In this paper, we shall recall some of his results indicating their great impact in the literature which remains being relevant and useful thirty years later.

Our presentation starts by making mention to his results on the support of the solution of Variational Inequalities, specially on some ones arising in Fluid Mechanics (Section 2). Some of his results on the support of the solution of semilinear equations are collected in Section 3. Finally, in Section 4, we shall recall his works connecting compact support properties and the abstract theory of monotone operators.

As Haïm Brezis commented at the official dinner of the Gaeta meeting, this set of results looks like a set of geological, or archeological, layers (almost the first ones among the generated by him) in his very vast production. Nevertheless, as in Geology, the time and the life use to fracture such set of initially well-ordered layers producing unexpected changes and mixtures. Something similar is produced also in Mathematics and so, for instance, the study of some special obstacle problem became of great interest to understand some limit behavior in the Ginzburg-Landau model in superconductivity (see Sandir and Serfaty [62]).

Research partially supported by project MTM2004-07590-C03-01 of the DGISGPI (Spain) and RTN HPRN-CT-2002-00274 of the EC..

2. The support of the solution of a variational inequality in fluid mechanics

Starting in 1973, Haïm Brezis and Guido Stampacchia studied in a series of papers (see [31], [32], [20], [68] and the presentation made in [56]) a very classical problem of the Fluid Mechanics introducing a new approach. They considered the problem of a flow past a given profile with prescribed velocity at the infinity.[1]

At the beginning of the seventies, the literature on the problem was very vast, with important contributions by many authors: P. Molenbroeck (1890), S.A. Chaplying (1902), J. Leray (1935), H. Bateman (1938), T. von Karman (1941), R. Courant and K.O. Friedrich (1948), L. Crocco (1951), L. Bers (1954), P. Germain (1954), M.J. Lighthill (1955), R. Finn and D. Gilbarg (1957), R. Finn and J. Serrin (1958) (see a larger and detailed list of references in the book by L. Bers [16]). From the mathematical point of view, the study of the incompressible case was essentially complete after the works by R. Finn. The situation was entirely different for the study of the compressible fluids.

Besides of studying the compressible case, another goal of the works by Brezis and Stampacchia was to get some sharp estimates on the maximum velocity by means of some method leading to some easy application of numerical algorithms (Stampacchia mentioned in [68] the suggestion received from the Instituto per le Applicazioni del Calcolo dall'Instituto di Meccanica Razionale del Politecnico di Torino). In fact, with their works, they initiated the development of the study of solutions with compact support on unbounded domains which would be extended later to a general class of semilinear and quasilinear partial differential equations.

The new approach by Brezis and Stampacchia was to show, rigorously, how the study of the associate hodograph plane (in the study of steady subsonic flow for a non viscous fluid, past a given symmetric convex profile in the plane) leads to a suitable obstacle problem on an unbounded domain.

They considered a closed convex profile $\mathcal{P}$ in $\mathbb{R}^2$, symmetric with respect to the x-axis. They assumed the fluid to be irrotational and so, the velocity $\mathbf{q} = (u, v)$ verifies the equations

$$\operatorname{div}(\rho\mathbf{q}) = 0, \ \operatorname{rot}(\mathbf{q}) = \mathbf{0}$$

where ρ denotes the density of the fluid (a constant in the incompressible case). It is also assumed that $\mathbf{q} \to \mathbf{q}_\infty = (q_\infty, 0)$ as $|(x, y)| \to +\infty$ and $\mathbf{q} \cdot \mathbf{n} = 0$ on $\partial\mathcal{P}$. Then, it is possible to define the stream function ψ given by

$$\psi_x = -\rho v \ \ \psi_y = \rho u.$$

Using Bernoulli's equation, there exists a decreasing function $\rho = h(q)$ relating ρ with $q = |\mathbf{q}|$ which depends on the physical properties of the fluid (for instance, $h(q) = (1 - Cq^2)^{1/(\gamma-1)}$ for barotopic gases). Then q can be considered as a function

---

[1]This subject already attracted the attention of scientists and artists (as for instance, Leonardo da Vinci (1452–1519)) since the beginnings of our culture.

of ψ_x and ψ_y and we get the equation

$$(1 - \frac{u^2}{a^2(q)})\psi_{xx} + (1 - \frac{v^2}{a^2(q)})\psi_{yy} - \frac{2uv}{a^2(q)}\psi_{xy} = 0, \tag{1}$$

where

$$a^2(q) = -q\frac{h(q)}{h'(q)},$$

$a(q)$ is the local speed of sound. In particular, (1) reduces to $\Delta\psi = 0$ when the fluid is incompressible. The boundary condition along $\partial\mathcal{P}$ is $\psi = 0$. Equation (1) is a mixed type quasilinear equation which is elliptic in the subsonic range ($q < q_c$) and hyperbolic in the supersonic range ($q > q_c$). Here q_c is the speed of sound, solution of $a(q_c) = q_c$.

It is well known that if we consider ψ as a function of $\mathbf{q}$ instead of (x, y) (*the hodograph plane*) then equation (1) becomes linear in the new variables. More precisely, the *hodograph transform*, in polar coordinates, $\mathcal{T} : (x, y) \to (u, v) \to (\theta, q)$

$$tg\ \theta = \frac{v}{u},$$

leads (1) to the Chaplying equation, which, by introducing

$$\sigma = \int_q^{q_c} \frac{h(\tau)}{\tau}d\tau$$

and

$$k(q) = \frac{1}{h^2(q)}(1 - \frac{q^2}{a^2(q)}) = k(\sigma)$$

can be written as

$$\psi_{\sigma\sigma} + k\psi_{\theta\theta} = 0. \tag{2}$$

This becomes the Tricomi equation when $k(q)$ is replaced by a linear function near $\sigma = 0$. Notice that $k(\sigma) > 0$ in the subsonic range ($\sigma > 0$) and $k(\sigma) < 0$ in the supersonic one ($\sigma < 0$).

Although the main interest of the hodograph transform lies in the fact that we deal with a linear equation, this equation has to be solved on a domain which is a priori unknown (the image of the profile $\mathcal{P}$ under $\mathcal{T}$ is not known since we do not know the distribution of velocities along $\mathcal{P}$). Because of the symmetry, we have $\psi = 0$ along the x-axes and it is sufficient to study the problem in the upper half plane where $\psi > 0$. Assuming that the flow is *totally subsonic*, the hodograph transform leads the profile $\mathcal{P}$ into a curve Γ (*a free boundary*) contained in the region $[\sigma > 0]$. If we denote by $\sigma = l(\theta)$ to this free boundary, it was shown in Ferrari and Tricomi [51] that the boundary conditions satisfied by ψ along Γ are the following

$$\frac{\partial\psi}{\partial\sigma} = -\frac{R(\theta)q(\sigma)}{1 + k(\sigma)(\frac{dl}{d\theta})^2} \quad \text{and} \quad \frac{\partial\psi}{\partial\theta} = -\frac{R(\theta)q(\sigma)\frac{dl}{d\theta}}{1 + k(\sigma)(\frac{dl}{d\theta})^2},$$

with $R(\theta)$ the radius of curvature of $\mathcal{P}$ at the point $P \in \mathcal{P}$ where the tangent makes an angle θ with the x-axis (we take $R(\theta) < 0$ since $\mathcal{P}$ is convex).

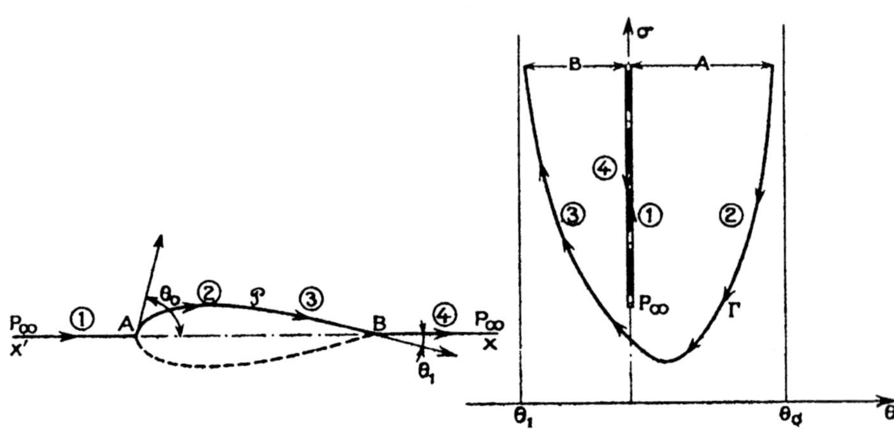

FIGURE 1

Inspired by the work of C. Baiocchi [6] on a different hydrodynamics problem, Brezis and Stampacchia introduced the change of unknown

$$u(\theta, \sigma) = \int_{l(\theta)}^{\sigma} \frac{k(\tau)}{q(\tau)} \psi(\theta, \tau) d\tau, \tag{3}$$

for $\sigma > l(\theta)$ and $\theta_1 < \theta < \theta_0$. In order to identify the properties satisfied by $u(\theta, \sigma)$ it is useful to introduce the set

$$\mathcal{D} = \{(\theta, \sigma) : \theta_1 < \theta < \theta_0, \ \sigma > l(\theta)\} \backslash \{(0, \sigma) : \sigma \geq \sigma_\infty\}$$

where

$$\sigma_\infty = \int_{q_\infty}^{q_c} \frac{h(\tau)}{\tau} d\tau$$

(q_∞ being the x-component of the prescribed velocity at the infinity). Then, they show (see the exposition made in [20]) that u verifies $u > 0$ in $\mathcal{D}$ and

$$\begin{cases} -\frac{1}{q^2}\left(\frac{q^2}{k} u_\sigma\right)_\sigma - u_{\theta\theta} - u = R & \text{in } \mathcal{D}, \\ u = 0 & \text{on } \Gamma, \\ \nabla u = \mathbf{0} & \text{on } \Gamma, \\ u(0, \sigma) = \text{Constant} = H_\mathcal{P} & \sigma \geq \sigma_\infty, \end{cases}$$

where $2H_\mathcal{P}$ coincides with the height of the profile. To get a *complementary formulation* (i.e., without any explicit mention to the free boundary Γ) they introduce the set $\Omega = \{(\theta, \sigma) : \theta_1 < \theta < \theta_0, \ \sigma > 0\}$ and extend u to Ω by choosing $u(0, \sigma) = 0$ for $0 < \sigma \leq l(\theta)$. Then, they show that u satisfies an obstacle problem by introducing the functional space

$$V = \{v : qv \in L^2(\Omega), \ qv_\theta \in L^2(\Omega), \ \frac{q}{\sqrt{k}} v_\sigma \in L^2(\Omega), \ v = 0 \text{ on } \partial\Omega\}$$

with the canonical norm and the closed convex subset

$$K_H = \{v \in V : v \geq 0 \text{ on } \Omega \text{ and } u(0, \sigma) = H_{\mathcal{P}} \text{ for } \sigma \geq \sigma_\infty\}.$$

Then, they define the bilinear form

$$a(u, v) = \int_\Omega (\frac{1}{k} u_\sigma v_\sigma + u_\theta v_\theta - uv) q^2(\sigma) d\theta d\sigma.$$

After proving that $a(u, v)$ is coercive on K_H, i.e.,

$$\lim_{\substack{\|u\| \to \infty \\ u \in K_H}} \frac{a(u, u)}{\|u\|_V} = \infty,$$

they conclude that *function u defined by (3) is the unique solution of the variational inequality*

$$\begin{cases} u \in K_H \\ a(u, v - u) \geq \int_\Omega R(\theta) v q^2(\sigma) d\theta d\sigma & \text{for all } v \in K_H. \end{cases} \tag{4}$$

Having solved (4), if we denote by

$$D^+ = \{(\theta, \sigma) \in \Omega : u(\theta, \sigma) > 0\},$$

when $\overline{D^+}$ does not intersect the axis $\{\sigma = 0\}$, the curve Γ, boundary of D^+, represents the distribution of velocities along $\mathcal{P}$. If $\overline{D^+}$ intersects the axis $\{\sigma = 0\}$ we conclude that q_∞ is too large and there exists no totally subsonic flow past $\mathcal{P}$.

In this way, their treatment[2] allows to apply, in an automatic way, well-known algorithms for the numerical approximation of u (see, for instance [53]).

But this nice results were not entirely complete since in order to estimate the maximum of the speed $q_{max} := \max q$ it was needed to get some lower estimate on the location of the free boundary Γ. They proved that if $q_A \geq q_\infty$ is the solution of the equation

$$\frac{H}{R_m} - 1 = \frac{q_A}{q_\infty} [-1 + \frac{1}{h(q_\infty)} \int_{q_\infty}^{q_c} \frac{h(\tau)}{\tau} d\tau] \tag{5}$$

with $R_m := \min_\theta |R(\theta)| > 0$ and if $q_A \leq q_c$ then, the maximum velocity satisfies that $q_{max} \leq q_A$. To do that, they construct the auxiliary function

$$\phi(\sigma) = \begin{cases} R_m q_A \int_A^\sigma \frac{k(\tau)}{q(\tau)} (\tau - A) d\tau & \text{if } A \leq \sigma \leq \sigma_\infty, \\ 0 & \text{if } 0 \leq \sigma \leq A, \end{cases}$$

where

$$A = \int_{q_A}^{q_c} \frac{h(\tau)}{\tau} d\tau,$$

---

[2] In my modest opinion, this new approach to such a classical problem has many common intellectual points with some other cultural creations of the value as, for instance, the *Rhapsody on a Theme of Paganini, Op.* 43 by Sergei Vasilyevich Rachmaninov or *Les Demoiselles d'Avignon.*(1907) by Pablo Picasso (oeuvre in which many people find some motivations on *The Visitation* (1610–14) by Domenicos Theotocopoulos "El Greco").

and they prove that it is a supersolution of problem (4). They also proved that the comparison principle holds for this problem and so the inequality $u \leq \phi$ leads to a lower estimate of the free boundary, $D^+ \subset [\sigma > A]$, and, finally, to the conclusion $q_{\max} \leq q_A$.

In the incompressible case, equation (5) reduces to

$$\frac{H}{R_m} - 1 = \frac{q_A}{q_\infty}[-1 + \log \frac{q_A}{q_\infty}],$$

and, in the particular case of an sphere ($H = R_m$ and $\log \frac{q_A}{q_\infty} = 1$) it is obtained that $q_{\max} \leq eq_\infty$ (some explicit computation shows that $q_{\max} = 2q_\infty$).

Before passing to recall other results by Brezis on the location of free boundaries, we must mention some other papers on the study of subsonic flows inspired by the articles by Brezis and Stampacchia. The previous study was extended to the case in which the flow presents a free boundary S (the *sillage*, boundary of a wake) where $q = q_S$ in Brezis and Duvaut [26]). They proved that if $q_S < q_\infty$ then the wake disappears at a finite distance of the profile but that when $q_S = q_\infty$ the free boundary converges to $(0, +\infty)$ as $|(x, y)| \to +\infty$. The problem was later developed, from the numerical point of view, in Bourgat and Duvaut [17]. Some sharper estimates on the location of the free boundary in the hodograph plane were obtained in [37] and [41]. The problem concerning an obstacle in a channel was considered in Tomarelli [69] (see also Bruch and Dormiani [33]). The case of non-symmetric convex profiles was studied in Hummel [55] and later extended by Shimborsky [66] to plane channels, Venturi tubes and flow around a Joukowski airfoil. A careful study of the convergence of solutions and free boundaries was given in Santos [63], [64] (see also the presentation made in Rodrigues [61]). Many references on the collision of two jets of compressible fluids can be found in the books Friedman [52] and Antontsev, Díaz and Shmarev [4].

3. The support of the solution of semilinear (multivalued or sublinear) second order equations

Simultaneously to his works with Stampacchia on the above fluid mechanics problem (the paper [32] was received on June 28, 1975), Brezis found that the support of the solution of other variational inequalities (of obstacle type) for a general second order elliptic operator verifies also similar compactness properties. So, in Brezis [18] (see also [19]) he studied the compactness of the support of the solution of the multivalued semilinear equation

$$\begin{cases} Lu + \beta(u) \ni f & \text{in } \Omega, \\ u = \varphi & \text{on } \partial\Omega, \end{cases} \tag{6}$$

where Ω is a smooth *unbounded* domain of $\mathbb{R}^N$, L is a second-order elliptic operator

$$L = -\sum_{i,j} a_{ij}\frac{\partial^2}{\partial x_i \partial x_j} + \sum_i a_i \frac{\partial}{\partial x_i} + a$$

and β is a maximal monotone graph in $\mathbb{R}^2$ such that $0 \in \beta(0)$. He assumed that

$$a_{ij} \in C^1(\overline{\Omega}) \cap L^\infty(\Omega); \quad a_i, a \in L^\infty(\Omega),$$

$$\begin{cases} \text{for every } r > 0 \text{ there is } \alpha(r) > 0 \text{ such that} \\ \sum_{i,j} a_{ij}\xi_i\xi_j \geq \alpha(r)|\xi|^2 \text{ for } x \in \Omega, \ |x| \leq r, \ \xi \in \mathbb{R}^N, \end{cases}$$

$$a(x) \geq \delta > 0 \text{ for } x \in \Omega.$$

It is clear that if problem (6) has a solution with compact support then, necessarily φ has also compact support and

$$\beta^-(0) \leq f(x) \leq \beta^+(0) \text{ for } |x| \text{ large,}$$

where $[\beta^-(0), \beta^+(0)]$ denotes the interval $\beta(0)$. These conditions are not sufficient but he proved in [18] that they "almost" sufficient. More precisely, he proved that if

$$\varphi \in C^2(\partial\Omega), \ \varphi \text{ has compact support and } \beta^0(\varphi) \in L^\infty(\partial\Omega),$$

$$f \in L^\infty_{loc}(\overline{\Omega}) \text{ and } \beta^-(0) < \lim_{|x|\to\infty} \inf ess f(x) \leq \lim_{|x|\to\infty} \sup ess f(x) < \beta^+(0), \quad (7)$$

then (6) has a unique solution with compact support, $u \in W^{2,p}(\Omega)$ for all $p < \infty$. The proof was based in the explicit construction of suitable radially symmetric super and subsolutions defined in the whole space $\mathbb{R}^N$. Besides to study the optimality of assumption (7), by particularizing β as different multivalued maximal monotone graphs in $\mathbb{R}^2$, Brezis stated, as corollaries, the existence and uniqueness of a solution with compact support to some minimization problems of the type

$$\underset{\substack{u \in H^1_0(\Omega), \ u \geq 0 \\ u=\varphi \text{ on } \partial\Omega, \text{ supp } u \text{ compact}}}{Min} \int (\frac{1}{2}|\nabla u|^2 + \frac{1}{2}|u|^2 - fu)dx,$$

and

$$\underset{\substack{u \in H^1(\Omega)\cap L^1(\Omega) \\ u=\varphi \text{ on } \partial\Omega}}{Min} \int (\frac{1}{2}|\nabla u|^2 + \frac{1}{2}|u|^2 + |u|)dx.$$

In that paper, he wrote the following remark:

> It has been shown by several authors that some nonlinear variational problems have a solution with compact support (see [5], [15], [59]). It would be of interest to unify these various results.

He added a footnote to this remark:

> A new result in that direction has been obtained very recently by M. Crandall.

At this time he also knew the results on the support of the solutions of the porous media equation by Oleinik, Kalahnikov and Yui Lin, Barenblatt, Aronson, Peletier and many others[3].

---

[3]As a matter of fact, the study of this subject was one of the several points suggested by Haïm Brezis to this author as thesis subjects, during their first meeting, on April 1974. Roughly speaking I could summarize a large part of my own scientific production as an attempt of to give an answer to the above mentioned remark by Brezis. To be more specific, the reader is sent to the monographs Díaz [38] and Antontsev, Díaz and Shmarev [4].

The interest of Brezis on the support of solutions of variational inequalities was extended to the parabolic case in his paper with A. Friedman [27]. They study the obstacle problem

$$\begin{cases} u_t - \Delta u + \beta(u) \ni 0 & \text{in } (0,\infty) \times \mathbb{R}^N, \\ u(x,0) = u_0(x) & \text{on } \mathbb{R}^N. \end{cases} \tag{8}$$

with β the maximal monotone graph in $\mathbb{R}^2$ given by

$$\beta(r) = \begin{cases} \phi & \text{if } r < 0, \\ (-\infty, 0] & \text{if } r = 0, \\ 0 & \text{if } r > 0. \end{cases}$$

Besides proving the compactness of the support of the solution $u(t,.)$, for any fixed $t > 0$ (once u_0 has a compact support), they proved, by first time in the literature, the property of *support shrinking* (concerning positive initial data u_0 such that $u_0(x) \to 0$ when $|x| \to \infty$). They also give fine estimates on the support of $u(t,.)$ and prove the *extinction in finite time* (i.e., the existence of $t^* < \infty$ such that $u(x,t) \equiv 0$, on $\mathbb{R}^N$, for any $t \geq t^*$). This paper was the inspiration of many subsequent researches by different authors (Tartar, Evans, Knerr, Veron, J.I. Díaz, Herrero, Vázquez, G. Díaz, Gilding, Kersner and many others: see, e.g., references in the monographs [38] and [4]). We must mention also the study of first order hyperbolic Variational Inequalities made in Bensoussan and J.L. Lions [11] for linear operators and Díaz and Veron [50] for nonlinear balance laws.

In a paper with A. Bensoussan and A. Friedman [12], Brezis reconsidered the question of the location of the free boundary for variational and quasi variational inequalities but now by means of the construction of local supersolutions which, in particular, allows to get estimates on the support of the solution also for bounded domains. This technique was extended to a very general class of nonlinear equations in [39] and [38].

We cannot end this part of the section dealing with multivalued equations without making mention to the results on qualitative properties of solutions (independently of his deep results on the regularity of the solution) obtained by Haïm Brezis on other different (but typical) free boundary problems. This was the case of the *dam problem* (considered firstly under general geometry conditions in Brezis, Kinderlehrer, Stampacchia [28] and later improved by Brezis' students J. Carrillo and M. Chipot [35]). Brezis returned on this problem in [24].

The interest of Brezis on mathematical problems suggested by the Environment was recently illustrated with the organization (jointly to this author) of the meeting between the Académie des Sciences and the Real Academia de Ciencias on Mathematics and Environment held at Paris, 23–24 May, 2002 ([25]). The meeting was additionally an occasion to render homage to the memory of Jacques-Louis Lions.

Another different problem studied by him was the magnetic confinement of a plasma in a Tokamaks. In collaboration with H. Berestycki [13], he introduced some variations to a previous formulation by Mercier and Temam giving many qualitative properties for the solutions. The problem was latter considered by many authors: Ambrosetti, Mancini, Damlamian, Caffarelli, Friedman, Kinderlehrer, Nirenberg, Stakgold, Bandle, Marcus, Sermonge, Mossino, Rakotoson, Blum, Gallouet and Simon, among them. Let us mention that the modelling of other types of magnetic confinement plasma fusion machines, the so called Stellarators (as, for instance, the TJ-II of the CIEMAT, Madrid) presents important differences with respect to the usual model for Tokamaks (see Díaz and Rakotoson [49]).

As a natural continuation of the Brezis result on the multivalued semilinear problem (6) and in connection with the above mentioned footnote of his paper, he studied with Ph. Benilan and M.G. Crandall, the support of the solution of the equation

$$-\Delta u + \beta(u) \ni f \quad \text{in } \mathbb{R}^N,$$

when $f \in L^1(\mathbb{R}^N)$ improving his results of [18] and considering also the case in which f has a compact support. They proved that the necessary and sufficient condition on β in order to get a solution with compact support is that

$$\int_0 \frac{ds}{\sqrt{j(s)}} < +\infty \tag{9}$$

where j is the convex primitive of β (i.e., such that $\partial j = \beta$). This criterion was extended to the case of quasilinear problems of the type

$$-\Delta_p u + \beta(u) \ni f \quad \text{in } \mathbb{R}^N, \tag{10}$$

in Díaz and Herrero ([43], [44]) where $\Delta_p u := \text{div}(|\nabla u|^{p-2} \nabla u)$, $p > 1$, to the criterion

$$\int_0 \frac{ds}{\sqrt[p]{j(s)}} < +\infty \tag{11}$$

which, for instance, now applies to Lipschitz functions $\beta(u)$ if $p > 2$. The above results were extended in many directions in the literature. For instance, the study of the semilinear elliptic equation (6), but now on a bounded domain Ω and with $f = 0$ on Ω and $\varphi = 1$ on the boundary, was studied by Bandle, Sperb and Stakgold [8] (see also [42]) showing that condition (9) is, again, the necessary and sufficient condition on β for the formation of a internal free boundary (the boundary of the *dead core*). The most general result in connection with the necessity of condition (11) was due to Vázquez who extended, in [70], the *Höpf strong maximum principle*. Many other contributions on this subject were produced by many authors (Veron, Serrin, Lanconelli, Díaz, Saa, Thiel, Kamin, Pucci, Zou,...: we send the reader to the monographs [38] and [4], and the recent survey Pucci and Serrin [57] for detailed references).

In seems interesting to point out that in Brezis and Nirenberg [30] the authors use the transformation $u = e^{-v}$ to study the singularity of v, solution of $-\Delta v + |\nabla v|^2 = h^2(v)$ for a suitable function $h^2(v)$, by analyzing the vanishing at a single point of u, solution of a semilinear equation of the type (6).

In collaboration with E. Lieb [29], Brezis also studied the support of a (vector) solution $\mathbf{u}$ of some nonlinear elliptic systems arising in the study of the Minimum Action to some Vector Field Equations. They proved that, under suitable conditions, $|\mathbf{u}|$ is a nonnegative subsolution of a semilinear equations similar to (6). The study of the support of solutions of nonlinear systems and higher order equations was carried out by many authors: (Bidaut-Veron, Bernis, Antontsev, Bertsch, Dal Passo, Shiskhov, Andreucci, Tedev, Cirmi, ...: see [38] and [4] for detailed references).

We briefly mention here that besides the use of the super and subsolutions method we also know other useful tools to this purpose such as appropriate energy methods [4], the application of rearrangement techniques leading to measure estimates on the dead core and coincidence sets ([38], [48], [9]), etc.

4. Compact support properties and the abstract theory of monotone operators

The fundamental contributions of Haïm Brezis to the abstract theory of maximal monotone operators on Hilbert spaces (and accretive operators in Banach spaces) are well known (see, for instance [22]). Even in that period of full dedication to that line of research he also was interested in many different applications to non linear partial differential equations (see, for instance, his lecture at the Vancouver *International Congress of Mathematicians* [23]). This abstract theory allows to get, also, general results for the numerical analysis of difficult problems generating a free boundary (see, for instance [14]) and can be applied to show the connections on the behavior of the free boundaries associated to some parabolic problems and the ones associated to the family of elliptic problems generated by time-implicit discretization [1].

But which I would like to illustrate here is the way in which such special problem, as the flow past a given profile mentioned in Section 2, seems to have been the starting point of an abstract result in the framework of the maximal operators in Hilbert spaces.

Although it was not explicitly said anywhere, it seems to me that his results on the support of the solution of second order elliptic variational inequalities could be the motivation for the study of the abstract Cauchy problem

$$\begin{cases} \frac{du}{dt}(t) + Au(t) \ni f(t) & \text{in } X, \\ u(0) = u_0, \end{cases}$$

in the case in which $X = H$ is a Hilbert space and $A : D(A) \to \mathcal{P}(H)$ a maximal monotone operator multivalued at 0 (with $0 \in int D(A)$). So, in a pioneering way, he obtain in [23] the first abstract result on the finite extinction time property. He proved that if we assume $f(t)$ such that

$$B(f(t), \epsilon) \subset A0, \text{ for a.e. } t \geq t_f, \text{ for some } \epsilon > 0 \text{ and } t_f \geq 0, \tag{12}$$

then the property of *finite extinction time* holds (there exists $t^* \in [t_f, +\infty)$ such that $u(t) \equiv 0$, in H, for any $t \geq t^*$) in a similar way to his results with A. Friedman on the semilinear equation (8). In contrast to the use of the comparison principle made in his previous results for elliptic and parabolic partial differential equations, now he merely used the fact that A is a maximal monotone operator and assumption (12).

Brezis considered in [23] a classical pursuit problem (already proposed by Leibnitz but modelled, now, in terms of a multivalued system associated to some suitable ordinary differential equations, i.e., with $H = \mathbb{R}^N$) as a simple application of the above abstract result. It turns out that assumption (12) is difficult to be checked in order to get some possible applications to partial differential equations (where, for instance $H = L^2(\Omega)$). This was the motivation of the work [39] in which the property of finite extinction time was proved for Banach spaces X and $A : D(A) \to \mathcal{P}(X)$ a multivalued m-accretive operator. Several applications for the special case of $X = L^\infty(\Omega)$, to some parabolic problems of the type (8) with β a multivalued maximal monotone graph of $\mathbb{R}^2$ (including second-order parabolic obstacle problems) were given in that paper. By working, again, on the space $X = L^\infty(\Omega)$ and using a certain duality with some fully nonlinear parabolic equation, the above abstract result yields to the extinction in a finite time of solutions to multivalued nonlinear diffusion equations of the form

$$u_t - \Delta\beta(u) \ni f,$$

arising in several contexts ([36]).

The finite extinction property can be proved also (via this abstract result) for other nonlinear multivalued parabolic problems of the type

$$\begin{cases} u_t - \nu\Delta u - g\mathrm{div}\left(\frac{\nabla u}{|\nabla u|}\right) = f(t,x) & \text{in } Q_\infty, \\ u = 0 & \text{on } \Sigma_\infty, \\ u(0,x) = u_0(x) & \text{on } \Omega, \end{cases}$$

for $\nu \geq 0$ and $g > 0$ and $f(t,x) \neq 0$. Such formulation arises in very different applied problems (non-Newtonian fluids of Bingham type, image processing, microgranular structures: see references, for instance, in [3]). Moreover, coming back to the similarity with the unexpected mixtures of geological layers mentioned at the Introduction, it seems interesting to point out that the above multivalued operator is also related to some very old works in Differential Geometry ([60]).

A different problem which looks quite similar to the previous ones (since it deals with a multivalued operator) but for which the above abstract results does not apply directly is the multivalued hyperbolic dry friction type problem as, for instance,

$$\begin{cases} u_{tt} - u_{xx} + \beta(u_t) \ni 0 & \text{in } (0,1) \times (0,+\infty), \\ u(t,0) = u(t,1) = 0 & t \geq 0, \\ u(0,.) = u_0(.) & t > 0, \\ u_t(0,.) = v_0(.) & \text{in } (0,1), \end{cases}$$

where now β denotes the maximal monotone graph of $\mathbb{R}^2$ given by

$$\beta(u) = \{1\} \text{ if } u > 0, \ \beta(0) = [-1,1] \text{ and } \beta(u) = \{-1\} \text{ if } u < 0. \tag{13}$$

This problem was already considered by Haïm Brezis in his paper [21]. Later, he proposed to his student A. Haraux (as one of the main thesis goals) the study of the dynamics of solutions of this problem. Haraux [54] proved that $u(t,x) \to \zeta(x)$ in $H_0^1(0,1)$ as $t \to +\infty$, with ζ verifying $-1 \leq \zeta_{xx} \leq 1$ and then (at the beginnings of the seventies) Brezis proposed the conjecture that the equilibrium position ζ is reached after a finite time (*stabilization in finite time*). Although some partial results in this direction were obtained by H. Cabannes [34] (for some special initial data u_0 and v_0) the case of arbitrary initial data seems to be still an open problem.

Motivated by this, and also suggested by the numerical approach of solutions, some easier formulations were considered in the literature, as, for instance, the spatially discretized vibrating string via a finite differences. The resulting system also arises in the study of the vibration of N-particles of equal mass m. In fact, it was by passing to the limit in the number of particles (in absence of any friction) how the wave equation was obtained in 1746 by Jean Le Rond D'Alembert.

If we denote the located positions, along the interval $(0,1)$ of the x axis, by $x_i(t)$ and we assume that each particle is connected to its neighbors by two harmonic springs of strength k, then the equations of motion can be written as the vectorial problem

$$(\mathbf{P}_N) \begin{cases} m\ddot{\mathbf{x}}(t) + k\mathbf{A}\mathbf{x}(t) + \mu_\beta\mathbf{B}(\dot{\mathbf{x}}(t)) + \mu_\beta\mathbf{G}(\dot{\mathbf{x}}(t)) \ni \mathbf{0}, \\ \mathbf{x}(0) = \mathbf{x}_0, \ \dot{\mathbf{x}}(0) = \mathbf{v}_0, \end{cases}$$

where $\mathbf{x}(t) := (x_1(t), x_2(t), \ldots, x_N(t))^T$ (here $\mathbf{h}^T$ means the transposed vector of $\mathbf{h}$), $\mathbf{A}$ is the symmetric positive definite matrix of $\mathbb{R}^{N \times N}$ given by

$$\mathbf{A} = \begin{pmatrix} 2 & -1 & \ldots & 0 \\ -1 & 2 & -1 & \ldots \\ \ldots & -1 & 2 & -1 \\ 0 & \ldots & -1 & 2 \end{pmatrix},$$

and $\mathbf{B} : \mathbb{R}^N \to \mathcal{P}(\mathbb{R}^N)$ (respectively $\mathbf{G} : \mathbb{R}^N \to \mathbb{R}^N$) denotes the (multivalued) maximal monotone operator (respectively the Lipschitz continuous function) given by $\mathbf{B}(y_1, \ldots, y_N) = (\beta(y_1), \ldots, \beta(y_N))^T$ (resp. $\mathbf{G}(y_1, \ldots, y_N) = (g(y_1), \ldots, g(y_N))^T$). The term $\mu_\beta\beta(\dot{x}_i(t))$ represents the Coulomb friction and $\mu_\beta\mathbf{G}$ represents other type of frictions such as, for instance, the one due to the viscosity of an surrounding fluid. We point out that this type of friction arises very often in the applications and that its consideration was already proposed by Lord Rayleigh (see, e.g., [58]).

The study of the special case of a single oscillator, $N = 1$, without viscous friction

$$m\ddot{x} + 2kx + \mu_\beta\beta(\dot{x}) \ni 0 \tag{14}$$

can be found in many textbooks. The motion stops definitively after a finite time $T_e < +\infty$ ($x(t) \equiv x_\infty$ for any $t \geq T_e$ for some $x_\infty \in [-\frac{\mu_\beta}{2k}, \frac{\mu_\beta}{2k}]$). As in the case of the damped wave equation, it is not difficult to prove ([47]) that for any

$(\mathbf{x_0}, \mathbf{v_0}) \in \mathbb{R}^{2N}$, problem (P_N) admits a unique weak solution $x \in C^1([0, +\infty) : \mathbb{R}^N)$ and that there exists a unique equilibrium state $x_\infty \in \mathbb{R}^N$ (i.e., satisfying that $Ax_\infty \in ([-\frac{\mu_\beta}{2k}, \frac{\mu_\beta}{2k}]^N)^T)$ such that $\| \dot{\mathbf{x}}(t) \| + \| x(t) - x_\infty \| \to 0$ as $t \to +\infty$. The stabilization in a finite time, in absence of viscous friction ($\mu_g = 0$) was proved in Bamberger and Cabannes [7]. It was proved in [47] that the presence of a viscous friction (with a suitable behavior of g near 0) may originate a qualitative distinction among the orbits in the sense that the state of the system may reach an equilibrium state in a finite time or merely in an asymptotic way (as $t \to +\infty$), according the initial data $\mathbf{x}(0) = \mathbf{x_0}$ and $\dot{\mathbf{x}}(0) = \mathbf{v_0}$. This dichotomy seems to be new in the literature and contrasts with the phenomena of *finite extinction time* for first order ODEs and parabolic PDEs. More precisely, the following was proved in [47]: i) if $g(r)r \leq 0$ in some neighborhood of 0 then all solutions of (P_N) stabilize in a finite time, ii) if $g(r) = \lambda r$ with $\lambda \geq 2\sqrt{\lambda_1 mk}/(\mu_\beta \mu_g)$, where λ_1 denotes the first eigenvalue of A then there exist solutions of (P_N) which do not stabilize in any finite time, and iii) if $N = 1$, $A = 1 \in R$ and $g'(0) < 2\sqrt{mk}/(\mu_\beta \mu_g)$ any solution stabilize in finite time but if $g'(0) \geq 2\sqrt{mk}/(\mu_\beta \mu_g)$ there exist solutions which do not stabilize in any finite time.

Another dynamical question raised by Haïm Brezis concerns the study of the damped oscillator

$$m\ddot{x} + \mu |\dot{x}|^{\alpha-1} \dot{x} + kx = 0, \tag{15}$$

when now $\alpha \in (0, 1)$. Here μ and $k > 0$ are fixed parameters. In fact we can simplify the above formulation to

$$\ddot{x} + |\dot{x}|^{\alpha-1} \dot{x} + x = 0, \tag{16}$$

by dividing by k and by introducing the rescaling $\tilde{x}(\tilde{t}) = \beta^{1/(\alpha-1)} x(\lambda \tilde{t})$ where $\lambda = \sqrt{m}/\sqrt{k}$ and $\beta = \mu/(k^{(2-\alpha)/2}m^{\alpha/2})$. Notice that the x-rescaling fails for the linear case $\alpha = 1$ since there is not any defined scale for x and the equation is merely reduced to $\ddot{x} + \beta\dot{x} + x = 0$ with $\beta = \mu/(\sqrt{km})$, a parameter which characterizes the dynamics. Notice also that the limit case $\alpha \to 0$ corresponds to the Coulomb friction equation (14).

We recall that, even if the nonlinear term $|\dot{x}|^{\alpha-1} \dot{x}$ is not a Lipschitz continuous function of $\dot{x}$, the existence and uniqueness of solutions of the associate Cauchy problem

$$(P_\alpha) \begin{cases} \ddot{x} + |\dot{x}|^{\alpha-1} \dot{x} + x = 0 & t > 0, \\ x(0) = x_0, \ \dot{x}(0) = v_0 \end{cases}$$

is well known in the literature: see, e.g., Brezis [21]. The asymptotic behavior, for $t \to \infty$, of solutions of the Coulomb and linear problems (P_0) and (P_1) (limit cases when $\alpha \to 0$ or $\alpha \to 1$) was well known. In the second case the decay is exponential. In the first one, as already mentioned, given x_0 and v_0 there exist a finite time $T = T(x_0, v_0)$ and $\zeta \in [-1, 1]$ such that $x(t) \equiv \zeta$ for any $t \geq T(x_0, v_0)$. When $\alpha \in (0, 1)$ it was also well known that the solutions of (P_α) verify $(x(t), \dot{x}(t)) \to (0, 0)$ as $t \to \infty$ (see, e.g., Haraux [54]). The question of to knowing if this convergence is in fact an identity after a finite time was proposed by Brezis.

This time the answer to his question (almost thirty years later) was not as the one expected by him. In a series of papers ([45], [46] and [2]) it was shown that the generic asymptotic behavior above described for the limit case (P_0) is only exceptional for the sublinear case $\alpha \in (0,1)$ since the generic orbits $(x(t), \dot{x}(t))$ decay to $(0,0)$ in a infinite time and only two one-parameter families of them decay to $(0,0)$ in a finite time: in other words, when $\alpha \to 0$ the exceptional behavior becomes generic. For a different approach see [71].

We end by remarking that in some other nonlinear partial differential systems it arises a feature very different from the case of scalar dissipative equations: the vector solution has some components which stabilize in finite time, and others for which this phenomenon does not occur. This property occurs, for instance, for the linear heat equation with a multivalued nonlinear dynamical boundary condition (for more details and other examples see [40]).

5. Special acknowledgements

If most of the papers ends with some acknowledgements, this presentation could not finish without expressing here, in this special occasion, the deep recognition and gratitude of many Spaniards mathematicians towards Haïm Brezis by the support and encouragements received from him since 1974. It was thanks to his generous help as the panorama of the mathematics in Spain, specially in the field of the nonlinear analysis, started to enjoy an activity and recognition nonexistent before. Fortunately, this was later extended to many other fields of the mathematics.

This singular contribution was officially recognized to him, in April 2000, when he received with two days of difference the nomination as foreign member of the Real Academia de Ciencias de España and the distinction as Doctor Honoris Causa by the Universidad Autónoma de Madrid.

References

[1] L. Alvarez and J.I. Díaz: The waiting time property for parabolic problems trough the nondiffusion of the support for the stationary problems, *Rev. R. Acad. Cien. Serie A Matem. (RACSAM)* **97** (2003), 83–88.

[2] H. Amann and J.I. Díaz, A note on the dynamics of an oscillator in the presence of strong friction, *Nonlinear Anal.* **55** (2003), 209–216.

[3] F.Andreu, V. Caselles, J.I. Díaz and J.M. Mazón, Some Qualitative Properties for the Total Variation, *Journal of Functional Analysis*, **188**, 516–547, 2002.

[4] S.N. Antontsev, J.I. Díaz and S.I. Shmarev, *Energy Methods for Free Boundary Problems: Applications to Nonlinear PDEs and Fluid Mechanics*, Progress in Nonlinear Differential Equations and Their Applications, 48, Birkhäuser, Boston, 2002.

[5] J. Auchmuty and R. Beals, Variational solutions of some nonlinear free boundary problems, *Arch. Rat. Mech. Anal.* **43**, (1971), 255–271

[6] C. Baiocchi, Su un problema di frontiera libera connesso a questioni di idraulica, *Annali di Mat. Pura ed Appl.* **92** (1972), 107–127.

[7] A. Bamberger and H. Cabannes, Mouvements d'une corde vibrante soumise à un frottement solide, *C. R. Acad. Sc. Paris*, **292** (1981), 699–705.

[8] C. Bandle, R.P. Sperb and I. Stakgold, Diffusion and reaction with monotone kinetics, *Nonlinear Analysis, TMA*, **8**, (1984), 321–333.

[9] C. Bandle and J.I. Díaz, Inequalities for the Capillary Problem with Volume Constraint. In *Nonlinear Problems in Applied Mathematics: In Honor of Ivar Stakgold on his 70th Birthday* (T.S. Angell et al. ed.), SIAM, Philadelphia, 1995.

[10] Ph. Benilan, H. Brezis and M.G. Crandall, A semilinear equation in $L^1(R^N)$, *Ann. Scuola Norm. Sup. Pisa* **4**, 2 (1975), 523–555.

[11] A. Bensoussan and J.L. Lions, On the support of the solution of some variational inequalities of evolution, *J. Math. Soc. Japan*, **28** (1976), 1–17.

[12] A. Bensoussan, H. Brezis and A. Friedman, Estimates on the free boundary for quasi variational inequalities. *Comm. PDEs* **2** (1977), no. 3, 297–321.

[13] H. Berestycki and H. Brezis, On a free boundary problem arising in plasma physics, *Nonlinear Analysis*, **4** (1980), 415–436.

[14] A. Berger, H. Brezis and J.C.W. Rogers, A numerical method for solving the problem *RAIRO Anal. Numér.*, **13** (1979), no. 4, 297–312.

[15] L. Berkowitz and H. Pollard, A non classical variational problem arising from an optimal filter problem, *Arch. Rat. Mech. Anal.*, **26** (1967), 281–304.

[16] L. Bers, *Mathematical aspects of subsonic and transient gas dynamics*, Chapman and Hall, London, 1958.

[17] J.F. Bourgat and G. Duvaut, Numerical analysis of flow with or without wake past a symmetric two-dimensional profile without incidence, *Int. Journal for Num. Math. In Eng.* **11** (1977), 975–993.

[18] H. Brezis, Solutions of variational inequalities with compact support. *Uspekhi Mat. Nauk.*, **129**, (1974) 103–108.

[19] H. Brezis, Solutions à support compact d'inequations variationelles, *Séminaire Leray*, Collège de France, 1973–74, pp. III.1–III.6

[20] H. Brezis, A new method in the study of subsonic flows, In, *Partial Differential equations and related topics,* J. Goldstein, ed., Lecture Notes in Math. Vol. 446, Springer, 1977, 50–64.

[21] H. Brézis, Problèmes unilatéraux, *J. Math. Pures Appl.* **51**, (1972), 1–168 .

[22] H. Brezis, *Opérateurs maximaux monotones et semigroupes de contractions dans les espaces de Hilbert*, North-Holland, Amsterdam 1972.

[23] H. Brezis, Monotone operators, nonlinear semigroups and applications. *Proceedings of the International Congress of Mathematicians* (Vancouver, B. C., 1974), Vol. **2**, Canad. Math. Congress, Montreal, Que., 1975, 249–255.

[24] H. Brezis, The dam problem revisted, in *Free Boundary Problems*, Proc. Symp. Montecatini, A. Fasano and M. Primicerio eds., Pitman, 1983.

[25] H. Brezis and J.I. Diaz (eds.), *Mathematics and Environment*, Proceedings of the meeting between the Académie des Sciences and the Real Academia de Ciencias, Paris, 23–24 May, 2002. Special volume of *Rev. R. Acad. Cien.Serie A Matem. (RACSAM)* **96**, n° 2, (2003).

[26] H. Brezis and G. Duvaut, Écoulements avec sillages autour d'un profil symmétrique sans incidence, *C.R. Acad. Sci.*, **276** (1973), 875–878.

[27] H. Brezis and A. Friedman, Estimates on the support of solutions of parabolic variational inequalities, *Illinois J. Math.*, **20** (1976), 82–97.

[28] H. Brezis, D. Kinderlehrer and G. Stampacchia, Sur une nouvelle formulation du problème de l'écoulement à travers une digue, *C.R. Acad. Sci. Paris,* **287** (1978), 711–714.

[29] H. Brezis and E. Lieb, Minimum action solutions of some vector field equations, *Comm. Math. Phys.*, **96** (1984), 97–113.

[30] H. Brezis and L. Nirenberg, Removable singularities for nonlinear elliptic equations, *Topol. Methods Nonlinear Anal.*, **9** (1997), 201–219.

[31] H. Brezis and G. Stampacchia, Une nouvelle méthode pour l'étude d'écoulements stationnaires, *C. R. Acad. Sci.*, **276** (1973), 129–132.

[32] H. Brezis and G. Stampacchia, The Hodograph Method in Fluid-Dynamics in the Light of Variational Inequalities, *Arch. Rat. Mech.Anal.*, **61** (1976), 1–18.

[33] J. Bruch and M. Dormiani, Flow past a symmetric two-dimensional profile with a wake in a channel, in *Nonlinear Problems,* vol. 2, C. Taylor, O.R. Oden, E. Hinton eds., Pineridge Press, Swanzea, UK, 1987.

[34] H. Cabannes, Mouvement d'une corde vibrante soumise a un frottement solide, *C. R. Acad. Sci. Paris Ser.* A-B **287** (1978), 671–673.

[35] J. Carrillo and M. Chipot, On the Dam Problem, *J. Diff. Eq.* **45** (1982), 234–271.

[36] G. Díaz, J.I. Díaz, Finite extinction time for a class of non linear parabolic equations, *Comm. in Partial Differential Equations,* **4** (1979) No 11, 1213–1231.

[37] J.I. Díaz, *Técnica de supersoluciones locales para problemas estacionarios no lineales. Aplicación al estudio de flujos subsónicos.* Memorias de la Real Academia de Ciencias Exactas, Físicas y Naturales. Serie de Ciencias Exactas, Tomo XVI, 1982.

[38] J.I. Díaz, *Nonlinear Partial Differential Equations and Free Boundaries.* Research Notes in Mathematics, 106, Pitman, London 1985.

[39] J.I. Díaz, Anulación de soluciones para operadores acretivos en espacios de Banach. Aplicaciones a ciertos problemas parabólicos no lineales. *Rev. Real. Acad. Ciencias Exactas, Físicas y Naturales de Madrid,* Tomo **LXXIV** (1980), 865–880.

[40] J.I. Díaz, Special finite time extinction in nonlinear evolution systems: dynamic boundary conditions and Coulomb friction type problems. To appear in *Nonlinear Elliptic and Parabolic Problems: A Special Tribute to the Work of Herbert Amann,* Zurich, June, 28–30, 2004 (M. Chipot, J. Escher eds.).

[41] J.I. Díaz and A. Dou, Sobre flujos subsónicos alrededor de un obstáculo simétrico. *Collectanea Mathematica,* (1983), 142–160.

[42] J.I. Díaz and J. Hernández, On the existence of a free boundary for a class of reaction diffusion systems, *SIAM J. Math. Anal.* **15**,N° 4, (1984), 670–685.

[43] J.I. Díaz and M.A. Herrero, Proprietés de support compact pour certaines équations elliptiques et paraboliques non linéaires. *C.R. Acad. Sc. Paris,* **286**, Série I, (1978), 815–817.

[44] J.I. Díaz and M.A. Herrero, Estimates on the support of the solutions of some non linear elliptic and parabolic problems. *Proceedings of the Royal Society of Edinburgh*, **98A** (1981), 249–258.

[45] J.I. Díaz and A. Liñán, On the asymptotic behavior of solutions of a damped oscillator under a sublinear friction term: from the exceptional to the generic behaviors. In *Proceedings of the Congress on non linear Problems* (Fez, May 2000), Lecture Notes in Pure and Applied Mathematics (A. Benkirane and A. Touzani. eds.), Marcel Dekker, New York, 2001, 163–170.

[46] J.I. Díaz and A. Liñán, On the asymptotic behaviour of solutions of a damped oscillator under a sublinear friction term, *Rev. R. Acad. Cien. Serie A Matem.* (RACSAM), **95** (2001), 155–160.

[47] J.I. Díaz and V. Millot, Coulomb friction and oscillation: stabilization in finite time for a system of damped oscillators. *CD-Rom Actas XVIII CEDYA / VIII CMA*, Servicio de Publicaciones de la Univ. de Tarragona 2003.

[48] J.I. Díaz and J. Mossino, Isoperimetric inequalities in the parabolic obstacle problems. *Journal de Mathématiques Pures et Appliqueés*, **71** (1992), 233–266.

[49] J.I. Díaz and J.M. Rakotoson, On a nonlocal stationary free boundary problem arising in the confinement of a plasma in a Stellarator geometry, *Archive for Rational Mechanics and Analysis*, **134** (1996), 53–95.

[50] J.I. Díaz and L. Veron, Existence, uniqueness and qualitative properties of the solutions of some first order quasilinear equations, *Indiana University Mathematics Journal*, **32**, No3, (1983), 319–361.

[51] C. Ferrari and F. Tricomi, *Aerodinamica transonica*, Cremonese, Rome,1962.

[52] A. Friedman, *Variational Principles and Free Boundary Problems*, Wiley, New York, 1982.

[53] R. Glowinski, J.L. Lions and R. Tremolieres, *Analyse Numérique des Inéquations Variationnelles*, 2 volumes, Dunod, París, 1976.

[54] A. Haraux, Comportement à l'infini pour certains systèmes dissipatifs non linéaires, *Proc. Roy. Soc. Edinburgh, Sect. A* **84A** (1979), 213–234.

[55] R.A. Hummel, The Hodograph Method for Convex Profiles, *Ann. Scuola Norm. Sup. Pisa* **9** IV, (1982), 341–363.

[56] D. Kinderlehrer and G. Stampacchia: *An introduction to variational inequalities and their applications.* Academic Press, New York 1980 (SIAM, Philadelphia, PA, 2000).

[57] P. Pucci and J. Serrin, The strong maximum principle revisted, *J. Diff. Equations*, **196** (2004), 1–66.

[58] J.W. Rayleigh, B. Strutt, *The theory of sound*, Dover Publications, New York, 2d ed., 1945.

[59] R. Redheffer, On a nonlinear functional of Berkovitz and Pollard, *Arch. Rat. Mech. Anal.* **50** (1973), 1–9.

[60] B. Riemann: Über die Fläche vom kleinsten Inhalt bei gegebener Begrenzung, *Abh. Königl. Ges. d. Wiss. Göttingen, Mathem. Cl.* **13** (1867), 3–52.

[61] J.F. Rodrigues, *Obstacle Problems in Mathematical Physics*, North-Holland, Amsterdam, 1987.

[62] E. Sandier and S. Serfaty, A rigorous derivation of a free-boundary problem arising in superconductivity, *Ann. Sci. Ecole Norm. Sup.* **4**, 33, (2000), 561–592.

[63] L. Santos, Variational convergences of a flow with a wake in a channel past a profile, *Bolletino U.M.I.*, **7**, 2-B, (1988), 788–792.

[64] L. Santos, Variational limit of compressible to incompressible fluid. In *Energy Methods in Continuum Mechanics*, S.N. Antontsev, J.I. Díaz and S.I. Shmarev eds., Kluwer, Dordrecht, 1996, 126–144.

[65] J. Serrin: Mathematical Principles of Classical Fluid Mechanics, in *Handbuch der Physik*, **8**, Springer-Verlag, Berlin 1959, 125–263.

[66] E. Shimborsky, Variational Methods Applied to the Study of Symmetric Flows in Laval Nozzles, *Comm. PDEs*, **4** (1979), 41–77.

[67] E. Shimborsky, Variational Inequalities arising in the theory of two-dimensional potential flows, *Nonlinear Anal.*, **5** (1981), 434–444.

[68] G. Stampacchia, Le disequazioni variazionali nella dinamica dei fluidi, In *Metodi Valuativi nella fisica-matemática*, Accad. Naz. Lincei, Anno CCCLXXII, Quaderno 217, (1975), 169–180.

[69] F. Tomarelli, Hodograph method and variational inequalities in fluid-dynamics, *Inst. Nat. Alta Mat.* Vol **I-II**, Roma, 1980, 565–574.

[70] J.L. Vázquez: A strong maximum principle for some quasilinear elliptic equations, *Appl. Math. Optim.* **12** (1984), 191–202.

[71] J.L. Vázquez, The nonlinearly damped oscillator, *ESAIM Control Optim. Calc. Var.* **9** (2003), 231–246.

J.I. Díaz
Departamento de Matemática Aplicada
Facultad de Matemáticas
Univiversidad Complutense de Madrid
E-28040 Madrid, Spain
e-mail: ji_diaz@mat.ucm.es

Progress in Nonlinear Differential Equations
and Their Applications, Vol. 63, 235–242

Fractal Conservation Laws: Global Smooth Solutions and Vanishing Regularization

Jérôme Droniou

Abstract. We consider the parabolic regularization of a scalar conservation law in which the Laplacian operator has been replaced by a fractional power of itself. Using a splitting method, we prove the existence of a solution to the problem and, thanks to the Banach fixed point theorem, its uniqueness and regularity. We also show that, as the regularization vanishes, the solution converge to the entropy solution of the scalar conservation law. We only present here the outlines of the proofs; we refer the reader to [4] and [5] for the details.

Mathematics Subject Classification (2000). 35L65, 35S30, 35A35, 35B65.

Keywords. regularization of scalar conservation laws, pseudo-differential operator, vanishing regularization.

1. Introduction

1.1. The equation and its motivations

The scalar conservation law

$$\begin{cases} \partial_t u(t,x) + \operatorname{div}(f(u))(t,x) = 0 & t > 0,\, x \in \mathbb{R}^N, \\ u(0,x) = u_0(x) & x \in \mathbb{R}^N, \end{cases} \tag{1}$$

where $f \in C^\infty(\mathbb{R}; \mathbb{R}^N)$ and $u_0 \in L^\infty(\mathbb{R}^N)$, is a well-known equation. S.N. Krushkov introduced in [6] a notion of solution for which existence and uniqueness holds (the entropy solution). A way to prove the existence of such entropy solutions is to consider the parabolic regularization of (1):

$$\begin{cases} \partial_t u^\varepsilon(t,x) + \operatorname{div}(f(u^\varepsilon))(t,x) - \varepsilon \Delta u^\varepsilon(t,x) = 0 & t > 0,\, x \in \mathbb{R}^N, \\ u^\varepsilon(0,x) = u_0(x) & x \in \mathbb{R}^N \end{cases} \tag{2}$$

(for which existence, uniqueness and regularity of solutions is classical), to establish so-called entropy inequalities (see Subsection 1.2), and to pass to the limit $\varepsilon \to 0$.

We are interested here in the case where we replace $-\Delta$ in the parabolic regularization (2) by a fractional power $(-\Delta)^{\lambda/2}$ of the Laplacian; precisely, we consider

$$\begin{cases} \partial_t u(t,x) + \operatorname{div}(f(u))(t,x) + g[u(t,\cdot)](x) = 0 & t > 0\,,\ x \in \mathbb{R}^N\,, \\ u(0,x) = u_0(x) & x \in \mathbb{R}^N\,, \end{cases} \tag{3}$$

where the operator g is defined through Fourier transform by

$$\mathcal{F}(g[v])(\xi) = |\xi|^{\lambda} \mathcal{F}(v)(\xi) \quad \text{with } \lambda \in \,]1,2]. \tag{4}$$

The motivation for the study of this problem comes from a question of P. Clavin; he shows in [2] that, in some cases of gas detonation, the wave front satisfies an equation which is close to (3) but with $\lambda = 1$; numerical tests indicate that shocks can occur in this case. The question was: if $\lambda > 1$, do we have for (3) the same regularization effect as for (2)? Curiously enough, the regularity of the solutions to (3) is quite easy to obtain; their global existence, on the other hand, is much harder (see Subsection 1.2). Some other motivations for (3) appear in [9].

1.2. Main difficulty

Some partial existence results for (3) can be found in [1], but they are either limited to the case $N = 1$ and $f(u) = u^2$ (and with quite regular initial data), or to results of local existence in time.

The main problem when considering (3) is the lack of *a priori* estimates (which would allow to pass from local existence to global existence). If we consider this equation as a regularization of (1), a natural space for the solutions is L^{∞}. Let us briefly recall how L^{∞} estimates are obtained on the solutions to (2): if η is a convex function and $\phi' = \eta' f'$, multiplying (2) by $\eta'(u^{\varepsilon})$ and taking into account (thanks to the convexity of η)

$$\Delta(\eta(u^{\varepsilon})) = \eta''(u^{\varepsilon})|\nabla u^{\varepsilon}|^2 + \eta'(u^{\varepsilon})\Delta u^{\varepsilon} \geq \eta'(u^{\varepsilon})\Delta u^{\varepsilon}$$

leads to

$$\partial_t \eta(u^{\varepsilon})(t,x) + \operatorname{div}(\phi(u^{\varepsilon}))(t,x) - \varepsilon\Delta(\eta(u^{\varepsilon}))(t,x) \leq 0. \tag{5}$$

Then, taking $\eta \equiv 0$ on $[-||u_0||_{\infty}, ||u_0||_{\infty}]$ and $\eta > 0$ outside $[-||u_0||_{\infty}, ||u_0||_{\infty}]$, the integration of (5) gives $||u^{\varepsilon}(t)||_{\infty} \leq ||u_0||_{\infty}$ for all $t > 0$.

Such a manipulation cannot be made if Δ is replaced by g. Thus, to obtain L^{∞} bound on the solution to (3), we use a totally different method.

2. Existence of a global solution

The semi-group generated by g is quite easy to understand: passing to Fourier transform, we see that the solution to $\partial_t v + g[v] = 0$ with initial datum $v(0) = v_0$ is given by $v(t,x) = K(t,\cdot) * v_0(x)$, where the kernel K is defined by

$$K(t,x) = \mathcal{F}^{-1}(\xi \to e^{-t|\xi|^{\lambda}}).$$

A result of [8] states that K is nonnegative, so that $||K(t)||_{L^1(\mathbb{R}^N)} = \mathcal{F}(K)(0) = 1$. As a consequence, we see that

$$||v(t)||_{L^\infty(\mathbb{R}^N)} \le ||v_0||_{L^\infty(\mathbb{R}^N)}\,,\ ||v(t)||_{L^1(\mathbb{R}^N)} \le ||v_0||_{L^1(\mathbb{R}^N)}\,,$$
$$|v(t)|_{BV(\mathbb{R}^N)} \le |v_0|_{BV(\mathbb{R}^N)}.$$

Hence, g "behaves well" (any interesting norm is preserved by g).

It is well known that the same holds for $\partial_t v + \operatorname{div}(f(v)) = 0$: if v evolves according to this scalar conservation law, its L^∞, L^1 and BV norms do not increase.

Hence, since each operator $\partial_t + g$ and $\partial_t + \operatorname{div}(f(\cdot))$ behaves well, we can let them evolve on separate time intervals and, afterwards, try to mix them together in order to get $\partial_t + \operatorname{div}(f(\cdot)) + g$. This idea is well known in numerical analysis, where it is called "splitting", but to our knowledge it has never been used before in order to prove the existence of a solution to a continuous problem.

We take $u_0 \in L^1(\mathbb{R}^N) \cap L^\infty(\mathbb{R}^N) \cap BV(\mathbb{R}^N)$ and, for $\delta > 0$, we define a function $U^\delta : [0, \infty[\times\mathbb{R}^N \to \mathbb{R}$ by (we omit the space variable):

- On $[0, \delta[$, U^δ is the solution to $\partial_t U^\delta + 2g[U^\delta] = 0$ with initial datum $U^\delta(0) = u_0$.
- On $[\delta, 2\delta[$, U^δ is the solution to $\partial_t U^\delta + 2\operatorname{div}(f(U^\delta)) = 0$ with initial datum $U^\delta(\delta)$ obtained in the first step.
- On $[2\delta, 3\delta[$, U^δ is the solution to $\partial_t U^\delta + 2g[U^\delta] = 0$ with initial datum $U^\delta(2\delta)$ given by the preceding step.
- etc...

That is to say, on half of the time – but in a set spread throughout $[0, \infty[$ – U^δ evolves according to $\partial_t + 2g = 0$ and, on the other half, it evolves according to $\partial_t + 2\operatorname{div}(f(\cdot)) = 0$; the factors "2" come from the fact that each of this operator only appears on half of the time: if we want to recover $\partial_t + \operatorname{div}(f(\cdot)) + g = 0$ on the whole of $[0, \infty[$ at the end, we must give a double weight to the operators on each half of $[0, \infty[$.

Thanks to the preceding considerations on both operators, we see that the L^∞, L^1 and BV norms of $U^\delta(t)$ are bounded by the corresponding norms of u_0. In particular, by Helly's Theorem, $\{U^\delta(t)\,;\ \delta > 0\}$ is relatively compact in $L^1_{\mathrm{loc}}(\mathbb{R}^N)$ for each $t \ge 0$. It is possible to prove that $\{U^\delta\,;\ \delta > 0\}$ is equicontinuous $[0, \infty[\to L^1(\mathbb{R}^N)$ and thus, up to a subsequence and as $\delta \to 0$, that U^δ converges in $C([0, T]; L^1_{\mathrm{loc}}(\mathbb{R}^N))$ to some u. Multiplying by $\varphi \in C^\infty_c([0, \infty[\times\mathbb{R}^N)$ the equations satisfied by U^δ and integrating, we can show that u satisfies (3) in a weak sense:

$$\int_0^\infty \int_{\mathbb{R}^N} u\partial_t\varphi + f(u) \cdot \nabla\varphi - ug[\varphi]\, dt dx + \int_{\mathbb{R}^N} u_0\varphi(0)\, dx = 0.$$

We have thus proved that, if u_0 is regular enough, (3) has a solution in a weak sense; moreover, this solution is bounded by $||u_0||_{L^\infty(\mathbb{R}^N)}$.

3. Regularity and uniqueness of the solution

3.1. Definition of solution

Another way to handle (3) is to consider that $\mathrm{div}(f(u))$ is a lower order term, and therefore to write $\partial_t u + g[u] = -\mathrm{div}(f(u))$. Since the semi-group generated by g is known, Duhamel's formula then gives

$$u(t,x) = K(t,\cdot) * u_0(x) - \int_0^t K(t-s,\cdot) * \mathrm{div}(f(u(s,\cdot)))(x)\,ds$$

and the properties of the convolution lead to

$$u(t,x) = K(t,\cdot) * u_0(x) - \int_0^t \nabla K(t-s,\cdot) * f(u(s,\cdot))(x)\,ds. \qquad (6)$$

This suggests the following definition.

Definition 3.1. *Let $u_0 \in L^\infty(\mathbb{R}^N)$. A solution to (3) is $u \in L^\infty(]0,\infty[\times\mathbb{R}^N)$ which satisfies (6) for a.e. $(t,x) \in]0,\infty[\times\mathbb{R}^N$.*

By the definition of K, it is obvious that $K(t,x) = t^{-N/\lambda}K(1,t^{-1/\lambda}x)$; hence, $\|\nabla K(t)\|_{L^1(\mathbb{R}^N)} = C_0 t^{-1/\lambda}$ and the integral term in (6) is defined as soon as u is bounded.

It is then easy, by a Banach fixed point theorem, to prove the existence of a solution on a small time interval $[0,T]$ (and its uniqueness on any time interval); but, due to the lack of estimates on this solution, nothing ensures that it can be extended to $[0,\infty[$. However, using its integrability properties, it is possible to prove that the weak solution constructed by a splitting method in Section 2 is also a solution in the sense of Definition 3.1. Hence, we have the existence of a global solution when the initial datum is regular enough, and its uniqueness for any bounded initial condition.

3.2. Regularization effect

The regularity of the solution is not very difficult to obtain. Assume that $u_0 \in L^\infty(\mathbb{R}^N)$ and take u a solution to (6) on $[0,T_0]$ (not necessarily the one constructed before, since we have not assumed that u_0 is integrable and has bounded variation). Since $\|\nabla K(t)\|_{L^1(\mathbb{R}^N)} = C_0 t^{-1/\lambda}$, the idea is to apply a Banach fixed point theorem on (6) in the space

$$E_T = \{v \in C_b(]0,T[\times\mathbb{R}^N) \mid t^{1/\lambda}\nabla v \in C_b(]0,T[\times\mathbb{R}^N;\mathbb{R}^N)\}.$$

For T small enough and $u_0 \in L^\infty(\mathbb{R}^N)$, we are able to prove the existence of a solution to (6) in E_T; since the solution is unique in $L^\infty(]0,T[\times\mathbb{R}^N)$, this proves that the given solution u is C^1 in space on $]0,T[$; this reasoning can be done from any initial time t_0 (not only $t_0 = 0$), which proves that u is C^1 in space on $]0,T_0[\times\mathbb{R}^N$.

A bootstrap technique, based on integral equations satisfied by the derivatives of u, allows to extend this method and to prove that u is C^∞ in space, and that all its spatial derivatives are bounded on $]t_0,T_0[\times\mathbb{R}^N$, for all $t_0 > 0$, by some constant depending on t_0 and $||u||_{L^\infty(]0,T_0[\times\mathbb{R}^N)}$. It is then possible to give a meaning to $g[u]$ (we prove that, if $2m > N + \lambda$, there exists integrable functions g_1 and g_2 such that $g[u] = g_1 * u + g_2 * \Delta^m u$) and to show that (3) is satisfied in the classical sense; this proves that u is also regular in time.

Thus, even if the initial datum is only bounded, the solution is regular and we have a bound on its derivatives which only depends on a bound on the solution itself. Let $u_0 \in L^\infty(\mathbb{R}^N)$; we can approximate it (a.e. and in L^∞ weak-$*$) by regular data u_0^n, for which we have proven the existence of solutions u^n (Section 2); these solutions are bounded by $\sup_n ||u_0^n||_{L^\infty(\mathbb{R}^N)} < +\infty$, which gives a bound on their derivatives; this proves that, up to a subsequence, u^n converge a.e. to some bounded u; it is then easy to pass to the limit in (6), with (u_0^n, u^n) instead of (u_0, u), to see that u is a solution to (3).

3.3. Main result

To sum up, we have obtained the following theorem.

Theorem 3.1. *If $f \in C^\infty(\mathbb{R}; \mathbb{R}^N)$ and $u_0 \in L^\infty(\mathbb{R}^N)$, then (3) has a unique solution in the sense of Definition 3.1. Moreover, this solution u satisfies*

  i) *$u \in C^\infty(]0,\infty[\times\mathbb{R}^N)$ and, for all $t_0 > 0$, all the derivatives of u are bounded on $[t_0,\infty[\times\mathbb{R}^N$,*

  ii) *for all $t > 0$, $||u(t)||_{L^\infty(\mathbb{R}^N)} \le ||u_0||_{L^\infty(\mathbb{R}^N)}$,*

  iii) *as $t \to 0$, we have $u(t) \to u_0$ in $L^p_{\text{loc}}(\mathbb{R}^N)$ for all $p < \infty$ and in $L^\infty(\mathbb{R}^N)$ weak-$*$.*

Remark 3.1. *The construction via the splitting method proves that the solution to (3) has more properties than the one stated above: any property which is satisfied by both equations $\partial_t + g = 0$ and $\partial_t + \text{div}(f(\cdot)) = 0$ is also satisfied by (3); for example: the solution takes its values between the essential lower and upper bounds of u_0, and there is a L^1-contraction principle for (3).*

Remark 3.2. *Since Theorem 3.1 only relies on the nonnegativity of K and the integrability properties of K and ∇K, it is also valid for more general g's, such as sums of operators (4) or anisotropic operators of the kind*

$$g = \sum_{j=1}^{N}(-\partial_j^2)^{\frac{\lambda_j}{2}}, \quad i.e., \quad \mathcal{F}(g[v])(\xi) = \left(\sum_{j=1}^{N}|\xi_j|^{\lambda_j}\right)\mathcal{F}(v)(\xi), \quad \text{with } \lambda_j \in]1,2].$$

The same holds for Theorem 4.1 and, in some cases, Theorem 4.2.

4. Vanishing regularization

Since (3) has been considered as a possible regularization of (1), it seems natural to wonder if, aside from the regularizing effect which has just been proved, the solutions to this equation stay close to the solution of the scalar conservation law when the weight on g is small. Precisely, if we consider

$$\begin{cases} \partial_t u^\varepsilon(t,x) + \mathrm{div}(f(u^\varepsilon))(t,x) + \varepsilon g[u^\varepsilon(t,\cdot)](x) = 0 & t > 0\,,\ x \in \mathbb{R}^N\,, \\ u^\varepsilon(0,x) = u_0(x) & x \in \mathbb{R}^N\,, \end{cases} \tag{7}$$

is it true that, as in the case of the parabolic regularization, u^ε converges as $\varepsilon \to 0$ to the entropy solution of (1)?

The answer is not obvious if we recall that some higher-order regularizations of conservation laws can generate too many oscillations, as the regularization vanishes, to allow the convergence towards the entropy solution; an example of this phenomenon, the KdV equation $\partial_t u^\varepsilon + \partial_x((u^\varepsilon)^2) = \varepsilon\partial_x^3 u^\varepsilon$, is mentioned in [3].

The convergence of the parabolic regularization (2) to the conservation law (1) is strongly based on the entropy inequality (5). If we want to prove the convergence of (7) to (1), we need to prove an entropy inequality for the non-local regularization g, and we are back to the problem mentioned in Subsection 1.2.

4.1. Entropy inequality

Therefore, we use again the splitting method. Let η be a convex function, $\phi' = \eta' f'$ and U^δ be the function constructed in Section 2 (with εg instead of g and for u_0 regular enough). On $I_\delta = \cup_{p\ \mathrm{odd}}[p\delta, (p+1)\delta]$, U^δ is the (entropy) solution of a scalar conservation law $(^1)$, and thus, for a nonnegative $\varphi \in C_c^\infty([0,\infty[\times\mathbb{R}^N)$,

$$\int_{I_\delta}\int_{\mathbb{R}^N} \eta(U^\delta)\partial_t\varphi + 2\phi(U^\delta)\cdot\nabla\varphi\,dtdx = \sum_{p\ \mathrm{odd}} a_{p+1} - a_p = -a_0 + \sum_{p\ \mathrm{even}} a_p - a_{p+1}, \tag{8}$$

where $a_p = \int_{\mathbb{R}^N} \eta(U^\delta(p\delta))\varphi(p\delta)\,dx$.

On $[p\delta, (p+1)\delta]$ for p even, U^δ satisfies $\partial_t U^\delta + 2\varepsilon g[U^\delta] = 0$ and thus $U^\delta(t) = K(2\varepsilon(t-p\delta)) * U^\delta(p\delta)$. Since η is convex and $K(2\varepsilon(t-\delta))$ is nonnegative with total mass 1, Jensen's inequality gives $\eta(U^\delta(t)) \le K(2\varepsilon(t-p\delta)) * \eta(U^\delta(p\delta))$; hence, φ being nonnegative,

$$a_{p+1} - a_p \le \int_{\mathbb{R}^N} K(2\varepsilon\delta) * \eta(U^\delta(p\delta))\varphi((p+1)\delta)\,dx - \int_{\mathbb{R}^N} \eta(U^\delta(p\delta))\varphi(p\delta)\,dx. \tag{9}$$

But $t \to K(2\varepsilon t) * \eta(U^\delta(p\delta))$ is solution to $\partial_t v + 2\varepsilon g[v] = 0$ with initial datum $\eta(U^\delta(p\delta))$, thus

$$\int_{\mathbb{R}^N} K(2\varepsilon\delta) * \eta(U^\delta(p\delta))\varphi((p+1)\delta)\,dx - \int_{\mathbb{R}^N} \eta(U^\delta(p\delta))\varphi(p\delta)\,dx \tag{10}$$

$$= \int_{p\delta}^{(p+1)\delta}\int_{\mathbb{R}^N} K(2\varepsilon(t-p\delta)) * \eta(U^\delta(p\delta))(\partial_t\varphi - 2\varepsilon g[\varphi])\,dtdx.$$

Since the L^∞, L^1 and BV norms of $U^\delta(s)$ and $\eta(U^\delta(s))$ are bounded independently of δ and s, and since $K(t)_{t>0}$ is an approximate unit as $t \to 0$ $(^2)$, we have, for

---

^{1}In fact, for δ small enough, U^δ is regular on I_δ.

^{2}This comes from the fact that $K(t)$ is nonnegative with mass 1, and that $K(t,x) = t^{-N/\lambda}K(1, t^{-1/\lambda}x)$.

$t \in]p\delta, (p+1)\delta]$,

$$\|K(2\varepsilon(t-p\delta)) * \eta(U^\delta(p\delta)) - \eta(U^\delta(p\delta))\|_{L^1(\mathbb{R}^N)} \le \omega_1(\delta)$$

$$\|\eta(U^\delta(t)) - \eta(U^\delta(p\delta))\|_{L^1(\mathbb{R}^N)} \le C\|U^\delta(t) - U^\delta(p\delta)\|_{L^1(\mathbb{R}^N)} \le \omega_2(\delta),$$

where $\omega_j(\delta) \to 0$ as $\delta \to 0$ (recall that $U^\delta(t) = K(2\varepsilon(t-p\delta)) * U^\delta(p\delta)$); therefore,

$$\|K(2\varepsilon(t-p\delta)) * \eta(U^\delta(p\delta)) - \eta(U^\delta(t))\|_{L^1(\mathbb{R}^N)} \le \omega_1(\delta) + \omega_2(\delta) = \omega_3(\delta)$$

and (9) and (10) give

$$a_{p+1} - a_p \ \le \ \int_{p\delta}^{(p+1)\delta} \int_{\mathbb{R}^N} \eta(U^\delta(t))(\partial_t\varphi - 2\varepsilon g[\varphi])\, dt dx$$

$$+ \omega_3(\delta) \int_{p\delta}^{(p+1)\delta} \|\partial_t\varphi(t)\|_{L^\infty(\mathbb{R}^N)} + 2\varepsilon\|g[\varphi(t)]\|_{L^\infty(\mathbb{R}^N)}\, dt.$$

Summing on even p's and coming back to (8), we find

$$\int_{I_\delta} \int_{\mathbb{R}^N} \eta(U^\delta)\partial_t\varphi + 2\phi(U^\delta)\cdot\nabla\varphi\, dt dx + \int_{\mathbb{R}^+\backslash I_\delta} \int_{\mathbb{R}^N} \eta(U^\delta)\partial_t\varphi - 2\varepsilon\eta(U^\delta)g[\varphi]\, dt dx$$

$$+ \int_{\mathbb{R}^N} \eta(u_0)\varphi(0)\, dx \ge -C(\varphi)\omega_3(\delta).$$

We can then pass to the limit $\delta \to 0$ (recall that $U^\delta \to u^\varepsilon$); since the characteristic functions of I_δ and $\mathbb{R}^+\backslash I_\delta$ weakly converge to $1/2$, we obtain

$$\int_0^\infty \int_{\mathbb{R}^N} \eta(u^\varepsilon)\partial_t\varphi + \phi(u^\varepsilon)\cdot\nabla\varphi - \varepsilon\eta(u^\varepsilon)g[\varphi]\, dt dx + \int_{\mathbb{R}^N} \eta(u_0)\varphi(0)\, dx \ge 0, \quad (11)$$

which is the entropy inequality for (7). This relation has been obtained in the case of regular initial data, but it can easily be extended to the case of general bounded initial data by the same idea as in the end of Subsection 3.2.

4.2. Convergence results

Once the entropy inequality for (7) has been obtained, a comparison between u^ε and u can be obtained by means of the doubling variable technique of S.N. Krushkov: we write the entropy inequality (11) with $\eta(u^\varepsilon) = |u^\varepsilon - u(s,y)|$ (s and y fixed) and φ depending on (s,y), we integrate on (s,y), we do the same with the entropy inequality satisfied by u (exchanging the roles of u^ε and u) and we sum the results. Taking φ which forces s to be near t and y to be near x, the term $|u^\varepsilon(t,x) - u(t,x)|$ appears up to an error which can be controlled, and we obtain the following result.

Theorem 4.1. *If $u_0 \in L^\infty(\mathbb{R}^N)$, then the solution to (7) converges, as $\varepsilon \to 0$ and in $C([0,T]; L^1_{\mathrm{loc}}(\mathbb{R}^N))$ for all $T > 0$, to the entropy solution of (1).*

If we assume more regularity on the initial data, then the error terms which appear in the doubling variable technique can be estimated more precisely and, as in [7] for the parabolic approximation, an optimal rate of convergence can be proved.

Theorem 4.2. *Assume that $u_0 \in L^1(\mathbb{R}^N) \cap L^\infty(\mathbb{R}^N) \cap BV(\mathbb{R}^N)$; let u^ε be the solution to (7) and u be the entropy solution to (1). Then, for all $T > 0$, $\|u^\varepsilon - u\|_{C([0,T];L^1(\mathbb{R}^N))} = \mathcal{O}(\varepsilon^{1/\lambda})$.*

Remark 4.1. *We notice that, for $\lambda < 2$, the convergence is better than in the case of parabolic approximation. This is due to the fact that, for small times* [3], *g is less diffusive than Δ; this comes from the homogeneity property $K(t,x) = t^{-N/\lambda}K(1, t^{-1/\lambda}x)$ of the kernel of g, which is to be compared with the homogeneity property $G(t,x) = t^{-N/2}G(1, t^{-1/2}x)$ of the heat kernel.*

On the contrary, and because of the same homogeneity properties, g is more diffusive than Δ for large times.

References

[1] P. Biler, T. Funaki and W.A. Woyczynski, *Fractal Burgers Equations*, J. Diff. Eq. **148** (1998), 9–46.

[2] P. Clavin, *Instabilities and nonlinear patterns of overdriven detonations in gases*, H. Berestycki and Y. Pomeau (eds.), Nonlinear PDE's in Condensed Matter and Reactive Flows, Kluwer, 2002, 49–97.

[3] R.J. DiPerna, *Measure-valued solutions to conservation laws.* Arch. Rational Mech. Anal. **88** (1985), no. 3, 223–270.

[4] J. Droniou, T. Gallouët and J. Vovelle, *Global solution and smoothing effect for a non-local regularization of an hyperbolic equation*, Journal of Evolution Equations, Vol **3**, No 3 (2003), pp. 499–521.

[5] J. Droniou, *Vanishing non-local regularization of a scalar conservation law*, Electron. J. Differential Equations 2003 (2003), no. 117, 1–20.

[6] S.N. Krushkov, *First Order quasilinear equations with several space variables.* Math. USSR. Sb., **10** (1970), 217–243.

[7] N.N. Kuznecov, *The accuracy of certain approximate methods for the computation of weak solutions of a first-order quasilinear equation*, Ž. Vyčisl. Mat. i Mat. Fiz., **16** (1976), pp. 1489–1502, 1627.

[8] P. Lévy, Calcul des Probabilités, 1925.

[9] W.A. Woyczyński, *Lévy processes in the physical sciences.* Lévy processes, 241–266, Birkhäuser Boston, Boston, MA, 2001.

Jérôme Droniou
IM^3, UMR CNRS 5149, CC 051
Université Montpellier II
Place Eugène Bataillon
F-34095 Montpellier cedex 5, France
e-mail: `droniou@math.univ-montp2.fr`

---

[3]The presence of ε entails that εg acts in finite time as g on small time.

Progress in Nonlinear Differential Equations
and Their Applications, Vol. 63, 243–258

Stationary and Self-similar Solutions for Coagulation and Fragmentation Equations

M. Escobedo

To Pr. H. Brezis in his 60th anniversary, with respect and gratitude.

1. Introduction

We are interested in self similar and stationary solutions of homogeneous coagulation and fragmentation equations. Although some explicit solutions of that kind where known for particular examples of such equations, no general existence result was known. Such results have recently been obtained (cf. [FL1], [EMR], [EM]), independently by two slightly different methods. The results exposed below are mainly taken from the references [EMR] and [EM], although similar results may be found in[FL1] and [FL2].

The coagulation and fragmentation equations describe the evolution of the size density function of a system of particles which undergo coagulation and/or fragmentation events. In order to describe these equations in more detail, consider a system of particles, denoted $\{y\}$, that are identified by their mass $y \in \mathbb{R}^+$. We denote $f(t, y)$ the density of particles of mass $y \in \mathbb{R}^+$ at time $t > 0$. These particles may agglomerate and fragment following certain rules, that we are going to describe, and one is interested in the evolution of the density function f.

The general coagulation fragmentation that we shall consider reads:

$$\frac{\partial f}{\partial t}(t, y) = \varepsilon_1 C(f)(t, y) + \varepsilon_2 L(f)(t, y), \quad t > 0, \, y > 0 \tag{1}$$

with $\varepsilon_i \in \{0, 1\}$ for $i = 1, 2$, $\varepsilon_1 + \varepsilon_2 \geq 1$. The coagulation operator C is given by:

$$C(f)(y) = \frac{1}{2}\int_0^y a(y', y - y')f(x, y')f(x, y - y')dy' - \int_0^\infty a(y, y') \, f(x, y) \, f(x, y') \, dy'$$

and the fragmentation operator L is:

$$L(f)(y) = \int_y^\infty b(y'', y)f(y'')dy'' - f(y)\int_0^\infty \frac{y'}{y}b(y, y')dy'.$$

When $\varepsilon_2 = 0$, only the term $C(f)$ is present in the right-hand side of (1):

$$\frac{\partial f}{\partial t}(t, y) = C(f)(t, y), \quad t > 0, \ y > 0 \tag{2}$$

and this is called the coagulation equation.

If $\varepsilon_1 = 0$, we are left only with the term $L(f)$. The equation is called the fragmentation equation, and it reads:

$$\frac{\partial f}{\partial t}(t, y) = L(f)(t, y), \quad t > 0, \ y > 0 \tag{3}$$

The fragmentation process may schematically be described as follows: for each particle $\{y\}$ of mass $y \in \mathbb{R}^+$:

$$\{y\} \overset{b(y,y')}{\longrightarrow} \{Y\}$$

where,

$$Y = (y_i)_{i \in \mathbb{N}^*}; \quad y_1 \geq \cdots y_i \geq \cdots \geq 0; \quad y = \sum_1^\infty y_i.$$

The rate of fragmentation of particles of mass y into particles of mass y' is given by $b(y, y')$. We shall consider rates of the form,

$$b(y, y') = y^\gamma B(\frac{y'}{y}), \quad \gamma \geq -1$$

where B is a measure such that,

$$B \geq 0, \qquad \mathrm{supp} B \subset [0, 1], \qquad \int_0^1 y \, dB(y) < +\infty.$$

The particular case $B(z) = B(1 - z)$ corresponds to binary fragmentation in which every particle of mass y gives rise to two particles of mass $y/2$.

On the other hand, the coagulation process may be described as

$$\{y\} + \{y'\} \overset{a(y,y')}{\longrightarrow} \{y + y'\}$$

where a is the rate of occurrence of the aggregation of two particles of mass y and y'. When two particles, of masses y and y', are sufficiently close, they merge into a new particle of mass $y + y'$. It is generally assumed that:

$$\begin{cases} a(y, y') = a(y', y), \\ a(r\, y, r\, y') = r^\lambda \, a(y, y'), \ \lambda \geq 0. \end{cases}$$

For example:

$$a(y, y') = (y^\alpha + y'^\alpha)^p |y^\beta \pm y'^\beta|^q,$$

for α, β, p, q real numbers. For the sake of clearness we shall only consider in all the following the case

$$a(y, y') = y^\alpha y'^\beta + y^\beta y'^\alpha, \quad \alpha + \beta = \lambda \in (0, 1). \tag{4}$$

These models may be obtained in different ways and arise in very different situations: astrophysics (formation of planetary systems), chemical physics (colloidal suspensions, aerosols, polymers, combustion), etc. ... (see for example the recent review monograph by F. Leyvraz [L], and also [D], [LM]).

The most important mathematical feature of the coagulation fragmentation equation (1) is the formal conservation of the total mass, i.e.,

$$\frac{d}{dt} \int_0^\infty x\, f(x,t) dx = 0, \quad \forall t > 0.$$

This property formally follows multiplying the equation (1) by y and using Fubini's theorem. It is known to be true under some general assumptions as it has been proved in [La], [LM1] and [ELMP]. For instance, when $\varepsilon_1\varepsilon_2 = 1$, $0 \le \lambda \le 1$ and $\gamma > \lambda - 2$ for an initial data f_{in} such that $(1 + y) f(y) \in L^1$, as it is proved in [ELMP]. But this property may be false in some cases, where the formal calculation can not be actually performed. That may be the case for instance when $\varepsilon_1\varepsilon_2 = 1$, $\lambda > 1$ and $\gamma < \lambda - 2$, or when $\varepsilon_2 = 0$ and $\lambda > 1$, even if $(1 + y) f(y) \in L^1$, (cf. [J], [EMP] and [ELMP]). In both cases, it may be shown that, for some positive $T > 0$ we have:

$$M_1(t) := \int_0^\infty x\, f(t,x) dx = \int_0^\infty x\, f_{in}(x) dx = M_1(0), \quad \forall t \in (0, T),$$

$$\text{and,} \qquad M_1(t) < M_1(0), \quad \forall t > T,$$

i.e., there is a loss of mass in finite time. This is the well-known gelation phenomenon (see [J], [EMP] and [L] for a detailed description and references). The conservation of mass is also known to break down in finite time when $\varepsilon_1 = 0$ and $\gamma \le -1$. In that case the loss of mass is due to what is called "shattering phenomenon" (see for more details [GZ], [B2], [B4]).

We shall only be concerned here with the case where the conservation of mass holds for all time. This property is essential in all the results exposed below.

2. Stationary and self similar solutions

Stationary and self similar solutions are interesting particular solutions which moreover may describe the behavior of general solutions of the Cauchy problem. We briefly introduce them in each of the three different cases of equation (1), (2) and (3).

(i) Stationary solutions for the coagulation-fragmentation equation.

In the case of binary fragmentation, a well-known condition to have stationary solutions for the coagulation fragmentation is the detailed balance property: there exists a function M such that:

$$a(y, y')M(y)M(y') = b(y + y', y)M(y + y'). \tag{5}$$

Obviously, such a function M is an equilibrium and then a stationary solution of the equation (1). Conditions on a and b for which such detailed balance condition is satisfied may be found in [LM]. We are interested in the existence of stationary solutions without the detailed balance condition, or in other words, functions satisfying $C(f) + L(f) = 0$ but not necessarily (5).

(ii) Self similar solutions.

The coagulation and the fragmentation equations, have only trivial stationary solutions. In the case of the coagulation equation this is $f \equiv 0$. For the fragmentation equation these are: $yf = A\delta$ for all real A. We look then for self similar solutions.

2.1. The coagulation equation

Explicit examples of self-similar solutions where already known for some particular cases of coagulation equations. We only mention two of them, the interested reader may consult the reference [A]:

- For $a(y, y') = 1$, (M. von Smoluchowski 1916)

$$f(y, t) = 4t^{-2} e^{-\frac{2x}{t}}, \quad t > 0.$$

- For $a(y, y') = y + y'$, (Z.A. Melzak 1953)

$$f(t, y) = (2\pi)^{-1/2} e^{-t} x^{-3/2} e^{-e^{-2t}x/2}.$$

We are interested in self similar solutions in more general situations, for kernels of the form given in (4). These self similar solutions are defined as follows. For all $\lambda \in [0, 1)$, we first define:

$$f_\mu(t, y) = \mu^{\frac{2}{1-\lambda}} f(\mu t, \mu^{\frac{1}{1-\lambda}} y) \quad \forall \mu > 0.$$

It is straightforward to check that, if f is a solution of (2), so is f_μ for any $\mu > 0$ and moreover:

$$\int_0^\infty y \, f_\mu(t, y) dy = \int_0^\infty y \, f(t, y) dy, \quad \forall \mu > 0.$$

For $\lambda = 1$ the scaling is different and that case is not considered here.

The self similar solutions (2) are those for which such $f \equiv f_\mu$ for all $\mu > 0$. It is trivial to see that f is self similar in that sense if and only if,

$$f(t, y) = t^{-\frac{2}{1-\lambda}} g\left(t^{-\frac{1}{1-\lambda}} y\right),$$

where the function $g(z) \equiv f(1, z)$ satisfies the stationary equation

$$2g + z\frac{\partial g}{\partial z} + (1 - \lambda)Q(g) = 0.$$

For $\lambda > 1$, self similar solutions of that kind do not exist due to the gelation phenomenon. Assuming, without loss of generality that the gelation time $T = 1$, the relevant scaling in that case is given by:

$$f_\mu(x, t) = \mu^{-\alpha} f(\mu^{-\beta} x, 1 + \mu(t - 1)), \quad \mu > 0,$$

with α and β satisfying:

$$\alpha + 1 = \beta(1 + \lambda) \tag{6}$$

in order for f_μ to satisfy the coagulation equation. Self similar solutions are then of the form,

$$f(x,t) = (1-t)^\alpha g(x(1-t)^\beta).$$

Equation (6) is not enough to determine uniquely the values of α and β. On the other hand, we can not, as before, impose the conservation of mass to the functions f_μ to the gelation phenomenon. To find the self similar solutions in this case is then a kind of non linear eigenvalue problem where the quotient α/β is the behavior of the profile $g(z)$ as $z \to +\infty$ (see [L] for this very interesting open question and [EMV] for a related problem).

2.2. The fragmentation equation

If $\gamma > -1$ and f solves the fragmentation equation, so does the function

$$f_\mu(t,y) = \mu^{\frac{2}{1+\gamma}} f(\mu t, \mu^{\frac{1}{1+\gamma}} y) \quad \forall \mu > 0$$

and moreover,

$$\int_0^\infty y\, f_\mu(t,y)dy = \int_0^\infty y\, f(t,y)dy, \quad \forall \mu > 0.$$

The self similar solutions of the fragmentation equation when $\gamma > -1$ have then the form:

$$f(t,y) = t^{\frac{2}{1+\gamma}} g(t^{\frac{1}{1+\gamma}} y),$$

where $g(y) = f(1,y)$ satisfies

$$2g + z\frac{\partial g}{\partial z} - (1+\gamma)L(g) = 0.$$

For $\gamma = -1$ the scaling is different and will not be considered here. When $\gamma < -1$ the solutions may loose mass instantaneously due to the so called "shattering" phenomenon (see for instance [GZ]). This makes impossible the existence of such self similar solutions.

In the mathematical literature, the study of self similar solutions for the fragmentation equation has been considered using probabilistic methods in recent works by Bertoin [B3], [B4], and [B1], [B2] for the asymptotic behavior. The convergence to the equilibrium state for the coagulation-fragmentation equation with detailed balance condition is considered in [LM2] (see also the references therein), and without this condition in [DS], [FM]. Finally, the self similar solutions and asymptotic behavior for coagulation equation have been considered by probabilistic methods in [B5], [DT], and in [MP1], [MP2], for $a(y,y') = 1$ and $a(y,y') = y + y'$, using deterministic analytic methods.

Self similar solutions have of course been widely considered in the physic's literature for instance by van Dongen and Ernst in [DE1], [DE2] but the interested reader may consult the references therein and the recent monograph [L] by Leyvraz. Among the great deal of information contained in these references, let us only mention the following concerning the coagulation equation and obtained in [DE1] and [DE2].

First, these authors remarked that the function

$$g(x) = x^{-\lambda-1}$$

is a singular weak self similar solution to the coagulation equation, but which is not of finite mass. It is then considered without physical meaning, although it seems to give the behavior of the self similar solutions of the coagulation equation as $x \to 0$ (see below). On the other hand, assuming that self similar solutions of finite mass do exist, and using formal asymptotics and numerics, the behavior of these solutions has been described as follows.

1. If $\alpha > 0$,

$$g(x) \sim Ax^{-(1+\lambda)}, \quad \text{as } x \to 0$$

for some positive and explicit constant depending only on the kernel a.

2. If $\alpha = 0$,

$$g(x) \sim x^{\tau}, \quad \text{as } x \to 0;$$

$$\tau = 2 - \Gamma \int_0^\infty x^\lambda g(x)dx < 1 + \lambda$$

where Γ is a positive explicit constant which only depends on a.

3. If $\alpha < 0$:

$$g(x) \sim Ax^{-2} \exp\left(-Bx^{-|\alpha|}\right) \quad \text{as } x \to 0, \qquad B = \frac{\Gamma}{|\alpha|} \int_0^\infty y^{\lambda-\alpha} G(y)\, dy.$$

3. The new results

In order to state our results we need first to introduce some notations. We denote by L^1_{loc} the space of integrable functions $f : (0, \infty) \to \mathcal{R}$ on any compact $[\varepsilon, 1/\varepsilon]$, $\varepsilon \in (0, 1)$ and by M^1_{loc} the associated measures spaces. For any given continuous function $\varphi : (0, \infty) \to (0, \infty)$, we define:

$$M^1_\varphi := \{\mu \in M^1_{loc}, \text{ such that } M_\varphi(|\mu|) < \infty\}, \quad L^1_\varphi := M^1_\varphi \cap L^1_{loc},$$

where, for any measure $0 \le \nu \in M^1_{loc}$, the generalized moment $M_\varphi(\nu)$ is defined as

$$M_\varphi(\nu) := \int_0^\infty \varphi(y)\, d\nu(y).$$

In order to shorten notations we also (abusively) denote, for any $k \in \mathcal{R}$,

$$\dot{M}^1_k = M^1_{y^k}, \quad \dot{L}^1_k = L^1_{y^k}, \quad M^1_k = M^1_{1+y^k}, \quad L^1_k = L^1_{1+y^k}.$$

Finally, we define

$$\dot{BV}_1 := \{f \in L^1_{loc}, \text{ such that } f' \in \dot{M}^1_1\}.$$

The same construction is made on $(0, 1)$. In that case, $L^1_{loc}(0, 1)$ is the set of measurable functions, integrable on $[\varepsilon, 1]$ for any $\varepsilon > 0$. Theses spaces are always denoted indicating the interval $(0, 1)$, like for instance $\dot{M}^1_k(0, 1)$, $\dot{L}^1_k(0, 1)$, $M^1_k(0, 1)$, $L^1_k(0, 1)$, $\dot{BV}_1(0, 1)$. Let us emphasize that all these are Banach spaces.

On the other hand, given a non negative measure μ_{in} in a suitable space of measures X, a weak solution of the coagulation fragmentation equation (1) with initial data μ_{in} is a measure $\mu \in L^\infty(0, T, X)$ such that $M_1(t) = M(0)$ for all $t \geq 0$ and

$$\int_0^\infty \int_0^\infty \mu \, \partial_t \psi \, dy dt + \int_0^\infty \mu^{in} \, \psi(0, .) \, dy + \int_0^\infty \, < \tilde{Q}(\mu), \psi > \, dt = 0$$

for each $\psi \in C_0^\infty([0, +\infty) \times (0, \infty))$, where

$$< \tilde{Q}(\mu), \psi > := < L(\mu), \psi > + < C(\mu), \psi >$$

and the terms at the right hand side are suitably defined

3.1. Stationary solutions for the coagulation fragmentation equation

Theorem 1. *Assume $B \in L^\infty(0, 1)$ and $\alpha \leq 0$ or $\gamma = -1$. Then for each $\rho > 0$ there exists at least one solution*

$$f \in \bigcap_{k \geq 1} \left(L_k^1 \cap L^{k+1} \right), \quad f \in L_{-r}^1, \ \forall r \in [0, 1),$$

of the stationary coagulation fragmentation equation $Q(f) + L(f) = 0$ such that

$$\int_0^\infty y f(t, y) dy = \rho, \quad \text{for all } t > 0.$$

Moreover, if $\alpha < 0$, this solution satisfies:

$$f \in L_{-k}^1, \quad \text{for all } k > 0,$$

3.2. Self similar solutions for the coagulation equation

Theorem 2. *Assume $\beta \in [0, 1)$ and $\alpha \in [-\beta, \beta] \cap [-\beta, 1 - \beta)$.*

1. *Assume $\alpha < 0$. Then, for any $\rho > 0$ there exists at least one self-similar profile of mass ρ such that $g \in C^\infty((0, \infty))$ such that, for all $y > 0$,*

$$e^{-a y^\alpha} \mathbf{1}_{y \leq 1} + e^{-b y} \mathbf{1}_{y \geq 1} \leq g(y) \leq e^{-A y^\alpha} \mathbf{1}_{y \leq 1} + e^{-B y} \mathbf{1}_{y \geq 1}$$

   for some positive constants a, b, A, B.

2. *Assume $\alpha \geq 0$. Then, for any $\rho > 0$ there exists at least one self-similar profile $g \in C((0, \infty))$, of mass ρ such that, for every $\varepsilon > 0$ there exist two positive constants $b_\varepsilon, B_\varepsilon$ for which the following holds.*

$$\forall y \in (\varepsilon, \infty) \quad e^{-b_\varepsilon y} \leq g(y) \leq e^{-B_\varepsilon y}$$

   Moreover,

$$g \, y^k \in L^\infty(0, 1) \ \forall k > 1 + \lambda,$$

   and, if $\alpha > 0$, $g \, y^k \notin L^\infty(0, 1) \ \forall k < 1 + \lambda$.

Remark 1. The results on the regularity and behavior of the self similar solutions obtained in Theorem 2 are not so precise as those obtained by formal asymptotic arguments and quoted in the introduction. The method of the proof nevertheless show that the conclusion of Theorem 2 holds for more general coagulation kernels a as soon as their behavior is similar, in a suitable way, to that of the kernel (4).

250 M. Escobedo

Let us mention nevertheless that for the precise kernel (4) with $\alpha = 0$ the precise behavior is proved in [FL].

3.3. Self similar solutions for the fragmentation equation

Theorem 3. *Assume*

$$B \in BV_1(0,1) \cap L^1_m, \quad m \le \gamma + 1.$$

Then for each $\rho > 0$ there exists a unique self similar solution

$$f(t,y) = t^{\frac{2}{1+\gamma}} g(t^{\frac{1}{1+\gamma}} y),$$

$$g \in \bigcap_{k \ge m} L^1_k, \quad g \in BV_1(\mathbb{R}^+),$$

of the fragmentation equation $\partial_t f = L(f)$ such that

$$\int_0^\infty y g(y) dy = \rho, \quad \text{for all } t > 0.$$

Moreover, if $\alpha < 0$, this solution satisfies:

$$g \in L^1_{-k}, \quad \text{for all } k > 0$$

3.4. Asymptotic behavior of the solutions to the fragmentation equation

Theorem 4. *Assume*

$$B \in BV_1(0,1) \cap L^1_m, \quad m \le \gamma + 1$$

and

$$f_{in} \in BV_1 \cap L^1_1 \cap L^1_{-1}.$$

Then, there exists a unique solution

$$f \in C\left([0,+\infty); L^1_1 \cap BV_1\right) \cap L^1(0,T; L^1_{\gamma+2}),$$

to the fragmentation equation. Moreover:

$$\lim_{t \to +\infty} \int_0^\infty y |f(t,y) - t^{\frac{2}{1+\gamma}} g(t^{\frac{1}{1+\gamma}} y)| dy = 0.$$

Remark. By the very definition of the fragmentation equation, one easy believes that, for any solution f with total mass M the following holds:

$$y f(t,y) \to M\delta, \quad \text{as } t \to +\infty.$$

The asymptotic behavior contained in Theorem 4 precises this information. It says, in precise terms, that this asymptotic delta-formation is described by the approximation of the identity:

$$\left(t^{\frac{2}{1+\gamma}} y\, g(t^{\frac{1}{1+\gamma}} y) \right)_{t>1}.$$

4. Main ingredients of the proofs

All the existence part of our results are proved with the same method. To obtain a priori estimates on the solutions of the evolution equation and use the following fixed point result.

4.1. A fixed point theorem

The following result follows easily from Tykhonov's fixed point Theorem.

Theorem. *Let Y be a Banach space and $(S_t)_{t\geq 0}$ be a continuous semigroup on Y. Assume that S_t is weakly (sequentially) continuous for any $t > 0$ and that there exists $\mathcal{Z}$ a nonempty convex and weakly (sequentially) compact subset of Y which is invariant under the action of S_t (that is $S_t z \in \mathcal{Z}$ for any $z \in \mathcal{Z}$ and $t \geq 0$). Then, there exists $z_0 \in \mathcal{Z}$ which is stationary under the action of S_t (that is $S_t z_0 = z_0$ for any $t \geq 0$).*

Proof. For any $t > 0$, thanks to the Tykhonov's point fixed theorem (see [E]), there exists $z_t \in \mathcal{Z}$ such that $S_t z_t = z_t$. On the one hand,

$$S_{i\,2^{-m}} z_{2^{-n}} = z_{2^{-n}} \quad \text{for any} \quad i, n, m \in \mathbf{N}, \quad m \leq n.$$

On the other hand, by weak compactness of $\mathcal{Z}$, we may extract a subsequence $(z_{2^{-n_k}})_k$ which converges weakly to a limit $z_0 \in \mathcal{Z}$. By weak continuity of S_t we may pass to the limit $n_k \to \infty$ and we obtain $S_t z_0 = z_0$ for any dyadic time $t \geq 0$. We conclude that z_0 is stationary by continuity of $t \mapsto S_t$ and density of the dyadic real numbers in the real line. $\qquad\square$

The proof of the existence of steady solutions using this abstract result is a slight modification of the method used in [GPV] for granular flows equations.

4.2. A priori estimates

The a priori estimates are obtained in a very similar way in each of the three cases of equations (1), (2) and (3). For the sake of brevity we only present here the case of the coagulation fragmentation (1). The corresponding arguments to obtain the self similar solutions for the coagulation and for the fragmentation equations are very similar (see Remark 3).

Moreover, we only present the formal calculations. One has first to construct a suitable sequence of regularised approximating equations on which such calculations are allowed and then pass to the limit in the estimates. For the sake of brevity we do not give the details.

Consider then the equation:

$$\frac{\partial f}{\partial t} = Q(f) + L(f).$$

After multiplication by $\Phi \in L^{\infty}(\mathbb{R}^+)$ we obtain the basic identity:

$$\frac{d}{dt} \int_0^{\infty} f(t,y)\Phi(y)dy = \frac{1}{2} \int_0^{\infty} \int_0^{\infty} a(y,y')f(t,y)f(t,y')\tilde{\Phi}(y,y')dydy'$$
$$+ \int_0^{\infty} f(t,y)y^{\gamma} \int_0^y B(\frac{y}{y'}) \left(\Phi(y') - \frac{y'}{y}\Phi(y) \right) dydy'$$

where

$$\tilde{\Phi}(y,y') = \Phi(y+y') - \Phi(y) - \Phi(y').$$

We first estimate the moments of the solution and chose then $\Phi(y) = y^k$:

$$\frac{d}{dt} \int_0^{\infty} f(t,y)y^k dy = \frac{1}{2} \int_0^{\infty} \int_0^{\infty} \Lambda_k(y,y')f(t,y)f(t,y')dydy'$$
$$+ C_{k,\gamma}(1-k) \int_0^{\infty} y^{\gamma+k+1} f(t,y)dy'$$

with

$$C_{k,\gamma} = \frac{1+\gamma}{1-k} \int_0^1 B(\sigma)(\sigma^k - \sigma)d\sigma > 0,$$

$$\Lambda_k(y,y') = (k-1)(y^{\alpha}(y')^{\beta} + y^{\beta}(y')^{\alpha})((y+y')^k - y^k - (y')^k) \geq 0.$$

We deduce that for $k > 1$ there exists some positive constants $C_{k,1}, C_{k,2}, C_{\lambda,1}$ and $C_{\lambda,2}$ independent of f such that

$$\frac{d}{dt}M_k \leq C_{k,1} M_{\beta-1+k} M_{1+\alpha} - C_{k,2}M_{1+\gamma+k},$$

and

$$\frac{d}{dt}M_{\lambda} \leq C_{\lambda,1} M_{1+\gamma+\lambda} - C_{\lambda,2} M_{\lambda}^2,$$

Using that $\alpha \leq 0$ and $\gamma \geq -1$ we finally obtain, for some positive constants C_1 and C_2 independent of f,

$$\frac{d}{dt}(M_{\lambda} + M_{2-\beta}) + (M_{\lambda} + M_{2-\beta})^q \leq C_1 + C_2 (M_{\lambda} + M_{2-\beta}).$$

Using Gronwall's Lemma we deduce:

$$\sup_{t \geq 0}(M_{\lambda}(t) + M_{2-\beta}(t)) \leq \max\left(C_0, M_{\lambda}(0) + M_{2-\beta}(0)\right)$$

We may then estimate M_k for $k > 2 - \beta$ as

$$\frac{d}{dt}M_k \leq C_1 M_k^{\theta_1} - C_2 M_k^{\theta_2},$$

with $0 \leq \theta_1 \leq 1$ and $\theta_2 > \theta_1, \theta_2 \geq 1$. Using again the Gronwall's Lemma we conclude

$$\sup_{t \geq 0} M_k(t) \leq \max\left(C, M_k(0)\right), \text{ for all } k > 2 - \beta.$$

A similar argument shows that

$$\sup_{t \geq 0} M_{-r}(t) \leq \max(C, M_{-r}(0)), \forall r \in [0,1).$$

In order to perform the final approximation argument, we need some weak compactness of the trajectories of the solutions f. This is obtained by means of suitable L^p estimates. We chose then, $\Phi(y) = f^{p-1}(y)$. The coagulation integral gives

$$\int_0^\infty Q(f)f^{p-1}dy = \int_0^\infty \int_0^\infty y^\alpha y'^\beta f f' \left((f'')^{p-1} - f^{p-1} - (f')^{p-1}\right) dy dy'$$

$$\leq -M_\beta \int_0^\infty y^\alpha f^p dy.$$

On the other hand, the fragmentation integral gives

$$\int_0^\infty L(f)f^{p-1}dy = \int_0^\infty y^\gamma f \int_0^y B(\frac{y'}{y}) \left((f')^{p-1} - \frac{y'}{y}f^p\right) dy' dy$$

$$\leq ||B||_\infty M_\gamma \int_0^\infty f^{p-1}dy - \int_0^1 yB(y)dy \int_0^\infty y^{\gamma+1}f^p dy.$$

Combining the two terms we obtain:

$$\frac{d}{dt}\int_0^\infty f^p dy + C_1 \int_0^\infty (y^\alpha + y^{\gamma+1})f^p dy \leq C_2 \left(\int_0^\infty f^p dy\right)^{\frac{p-2}{p-1}},$$

and using that $\alpha \leq 0$ or $\gamma + 1 = 0$ we deduce:

$$\frac{d}{dt}\int_0^\infty f^p dy + C_1 \int_0^\infty f^p dy \leq C_2 \left(\int_0^\infty f^p dy\right)^{\frac{p-2}{p-1}}.$$

We finally deduce that, for all $p \geq 2$ there exists a positive constant $C_p > 0$, such that:

$$\sup_{t \geq 0} ||f(t)||_p \leq \max(C_p, ||f_{in}||_p).$$

Using the a priori estimates, we may then apply the fixed point theorem proved above as follows. The Banach space is

$$Y = L^1_{2\alpha} \cap L^1_m, \quad m = \max(2\beta, 2 - \beta, \gamma + 1),$$

$(S_t)_{t>0}$ is the semigroup of the coagulation fragmentation equation $S_t : Y \to Y$ and

$$\mathcal{A}_k = \left\{f, M_1(f) = \rho, ||f||_2 \leq \mu_0, M_k^1(f(t)) \leq \mu_k\right\}.$$

The a priori estimates show that S_t sends A_k into itself for μ_k sufficiently large, and is weakly sequentially continuous for any $t > 0$. Finally the non empty convex set is defined as define:

$$\mathcal{Z}_k = \bigcap_{\ell=1}^k A_k.$$

We obtain a sequence $(G_k)_{k>1}$ of steady states such that $G_k \in \mathcal{Z}_k$. A compactness argument gives a fixed point $g \in \bigcap_{k=1}^\infty \mathcal{Z}_k$.

Remark 2. It is possible to obtain the existence of self similar profiles under weaker assumptions on a and B. The regularity obtained for the profile may be then less

                           M. Escobedo

regular. For instance, if we only assume that $B \in M_m$ for some $m \leq 2\alpha$, we only obtain a profile $g \in \bigcap_{k \geq 1} M_k$.

Remark 3. In order to prove the existence of self similar solutions to the coagulation or the fragmentation equation we apply the same method to the following evolution equations:

$$\frac{\partial g}{\partial t} = 2g + z\frac{\partial g}{\partial z} + (1 - \lambda)C(g), \tag{7}$$

$$\frac{\partial g}{\partial t} + 2g + z\frac{\partial g}{\partial z} = (1 + \gamma)L(g). \tag{8}$$

Notice that the self similar profiles g of the coagulation equation (2) are the stationary solutions of (7), and the self similar profiles g of the fragmentation equation (3) are the stationary solutions of (8).

To prove the regularity of the self similar profiles for the coagulation equation stated in Theorem 2, we use that the equation of these profiles may be written as

$$\partial_y(y^2 g) = (1 - \lambda)C(g) \quad \text{in} \quad \mathcal{D}'(0, \infty), \tag{9}$$

where

$$C(g)(z) := \int_0^z g\, y \left\{ \int_{z-y}^{\infty} a\, g'\, dy' \right\} dy$$

$$= (y^{\alpha+1}\, g) \star_z L_\beta(g) + (y^{\beta+1}\, g) \star_z L_\alpha(g),$$

and:

$$z \mapsto h \star_z g := \int_0^z h(y)\, g(z - y)\, dy, \quad L_\nu(g)(u) := \int_u^{\infty} v^\nu\, g(v)\, dv.$$

It is then possible to use a bootstrap type argument in equation (9) to obtain regularity properties of the solutions.

Finally, the main argument in the proof of the uniform lower bound on the profiles is the following. Since the function g is a steady solution of equation (7), we have

$$\partial_t g - 2g - yg_y + g\int_0^{\infty} a\, g'\, dy' = \frac{1}{2}\int_0^y a(y - y', y')\, g(y - y')\, g(y')\, dy'.$$

and then

$$\partial_t g - 2g - yg_y + g\, \ell_0 \geq 0,$$

where

$$\ell_0(y) := M\,(y^\alpha + y^\beta) \geq \int_0^{\infty} a\, g'\, dy', \quad M := \max(M_\alpha(g), M_\beta(g)).$$

By a simple comparison argument, the profile g is then greater than the solutions of the linear equation

$$\partial_t h = D\, h - \ell_0\, h, \qquad h(0) = f_{in}.$$

The desired estimates follow using a final iteration argument.

5. Asymptotic behavior of the solutions to the fragmentation equation

The uniqueness of the self similar or stationary solutions, for a given mass, is in general an open question. Nevertheless, a uniqueness result is proved in Theorem 3 above for the self similar solutions of the fragmentation equation. Actually, a general uniqueness result for solutions of the coagulation fragmentation is proved in [EMR] but under some strong regularity assumptions of the solutions, that are not necessarily satisfied by the solutions obtained in the existence results. It turns out that the self similar solutions for the fragmentation equation obtained in Theorem 3 fulfills the regularity assumptions and are then unique.

This uniqueness allows us to prove Theorem 4 on the asymptotic behavior, as $t \to +\infty$, of the general solutions to the Cauchy problem associated to the fragmentation equation. Let us briefly sketch the proof. Take then $f_{in} \in C_0(\mathbb{R}^+)$ such that $M_1(f_{in}) = \rho > 0$ and let g_ρ be the unique self similar profile of the equation of mass ρ. Suppose, to avoid a trivial situation, that $f_{in} \neq g_\rho$. Let f be the unique solution of the Cauchy problem:

$$\partial_t f = L(f), \quad f(0, x) = f_{in}.$$

As we already indicated in the introduction, it is the easy to check that the function

$$g(\tau, y) = e^{-\frac{2\tau}{1+\gamma}} f(e^\tau - 1, e^{-\frac{\tau}{1+\gamma}} y)$$

is the unique solution to

$$\begin{cases} \dfrac{\partial g}{\partial \tau} = -2g - y\dfrac{\partial g}{\partial y} + (1+\gamma)L(g), & y > 0, t > 0, \\[2mm] g(0, y) = f_{in}(y), & y > 0. \end{cases} \tag{10}$$

Consider now the following Lyapunov functional:

$$H(g(\tau)) = \int_0^\infty y|g(\tau, y) - g_\rho(y)|dy.$$

A straightforward calculation gives:

$$\frac{d}{d\tau}H(g(\tau)) = D(h(\tau))$$

with $h(\tau, y) = g(\tau, y) - g_\rho(y)$ and

$$D(h) = \int_0^\infty (-2h + yh_y + (1+\gamma)L(h))\, sign(h(y))dy$$

$$= \int_0^\infty \int_0^y b(y, y')(h(y)sign(h(y') - |h(y)|)y'dy'dy \le 0.$$

The function $\tau \to H(g(\tau))$ is then non increasing along the trajectory of g. Let us show that it is strictly decreasing. Assume on the contrary that, for $0 < \tau_1 < \tau_2$,

$H(g(\tau_1)) = H(g(\tau_2))$. Then:

$$\int_{\tau_1}^{\tau_2} D(h(s))ds = 0,$$

from where, using that the integrand is non positive,

$$D(h(s)) = 0, \quad \forall s \in (\tau_1, \tau_2).$$

We deduce

$$h(s,y)sign(h(s,y')) - |h(s,y)| = 0, \ \forall s \in (\tau_1, \tau_2)$$

and then,

$$\operatorname{sign} h(s,y) = \operatorname{sign} h(s,y'), \ \text{for a.e. } y, y'.$$

This implies that $signh(s,y)$ is constant for $y \in \mathbb{R}^+$ and $s \in (\tau_1, \tau_2)$. Since

$$\int_0^\infty yh(s,y)dy = \int_0^\infty y(g(\tau,y) - g_\rho(y))dy = 0$$

this is impossible. The function $H(g(\tau))$ is then a strict Lyapunov functional on L_1^1.

On the other hand, using the a priori estimates described above, one easily sees that if $f_{in} \in L_1^1 \cap BV_1$, the solution g of the (10) satisfies $g(\tau) \in L_1^1 \cap BV_1$ for all $\tau > 0$. The trajectory $(g(\tau))_{\tau>0}$ is then in a compact subset of L_1^1. This implies

$$\lim_{\tau \to +\infty} H(g(\tau)) = 0$$

or, equivalently

$$\lim_{\tau \to +\infty} \int_0^\infty y|e^{-\frac{2\tau}{1+\gamma}} f(e^\tau - 1, e^{-\frac{\tau}{1+\gamma}} y) - g_\rho(y)|dy = 0$$

$$\Leftrightarrow$$

$$\lim_{t \to +\infty} \int_0^\infty y|f(t-1,y) - t^{\frac{2}{1+\gamma}} g_\rho(t^{\frac{1}{1+\gamma}} y)|dy = 0. \qquad \square$$

Remark 4. We point point out that the proof of the asymptotic behavior result for the fragmentation equation sketched above is exactly the same as that for the asymptotic behavior for scalar conservation laws in $L^1(\mathbb{R}^N)$.

References

[A]   Aldous, D.J., *Deterministic and stochastic models for coalescence (aggregation, coagulation): a review of the mean-field theory for probabilists*, Bernoulli **5** (1999), 3–48.

[B1]   Bertoin, J., *On small masses in self-similar fragmentations*, Stochastic Process. Appl. 109 (2004), no. 1, 13–22.

[B2]   Bertoin, J., *The asymptotic behavior of fragmentation processes*, J. Eur. Math. Soc. (JEMS) 5 (2003), no. 4, 395–416.

[B3]   Bertoin, J., *Self-similar fragmentations*, Ann. Inst. H. Poincaré Probab. Statist. 38 (2002), no. 3, 319–340.

[B4]   Bertoin, J., *Homogeneous fragmentation processes*, Probab. Theory Related Fields 121 (2001), no. 3, 301–318.

[B5]   Bertoin, J., *Eternal solutions to Smoluchowski's coagulation equation with additive kernel and their probabilistic interpretations*, Ann. Appl. Probab. 12 (2002), no. 2, 547–564.

[DS]   Dubovskiĭ, P.B., Stewart, I.W., *Trend to equilibrium for the coagulation-fragmentation equation*, Math. Methods Appl. Sci. **19** (1996), 761–772.

[DE1]   van Dongen, P.G.J., Ernst, M.H.: Cluster size distribution in irreversible aggregation at large times. J. Phys. A, **18**, 2779–2793 (1985)

[DE2]   van Dongen, P.G.J., Ernst, M.H.: Scaling solutions of Smoluchowski's coagulation equation. J. Statist. Phys., **50**, 295–329 (1988)

[D]   Drake, R.L. *A general mathematical survey of the coagulation*, in: G. Hidy, J.R. Brocks (Eds.), Topics in Current Aerosol Research 3 (Part 2), Pergamon Press, Oxford (1972).

[E]   Edwards,R.E., "Functional Analysis, Theory and Applications", Holt, Rinehart and Winston, 1965

[ELMP]   Escobedo, M., Laurençot, Ph., Mischler, S., Perthame, B.: Gelation and mass conservation in coagulation-fragmentation models, J. Differential Equations 195 (2003), no. 1, 143–174.

[EMP]   Escobedo, M., Mischler, S., Perthame, B.: Gelation in coagulation and fragmentation models. Comm. Math. Phys., **231**, 157–188 (2002)

[EMR]   Escobedo, M, Mischler, S., Ricard, M. R., *On self-similarity and stationary problem for fragmentation and coagulation models*, to appear in Annales de l'Institut Henri Poincaré.

[EMV]   Escobedo, M, Mischler, S., Velazquez, J. J. L., *On the fundamental solution of a linearised Uehling Uhlenbeck equation*, Preprint 2004.

[FG]   Fournier, N., Giet, J.-S. , *On small particles in coagulation-fragmentation equations*, J. Statist. Phys. 111 (2003), no. 5-6, 1299–1329.

[FL1]   Fournier, N., Laurençot, P., *Existence of self-similar solutions to Smoluchoski's coagulation equation*, preprint 2004

[FL2]   Fournier, N., Laurençot, P., *Local properties of self-similar solutions to Smoluchoski's coagulation equation with sum kernel*, preprint 2004

[FM]   Fournier, N., Mischler, S., *Trend to the equilibrium for the coagulation equation with strong fragmentation but with balance condition*, preprint 2003, accepted for publication in Proceedings: Mathematical, Physical and Engineering Sciences.

[GPV]   Gamba, I.M. , Panferov, V. ,Villani, C., *On the Boltzmann equation for diffusively excited granular media*, to appear in Comm. Math. Phys.

[GZ]   McGrady, E.D., Ziff, R.M., *"Shattering" Transition in Fragmentation*, Phys. Rev. Lett. **58**, 892–895, (1987)

[J]   Jeon, I.: Existence of gelling solutions for coagulation-fragmentation equations. Comm. Math. Phys., **194**, 541–567 (1998)

[La]   Laurençot, Ph.: On a class of continuous coagulation-fragmentation models. J. Differential Equations, **167**, 145–174 (2000)

[LM1]   Laurençot, Ph., Mischler, S.: The continuous coagulation-fragmentation equations with diffusion. Arch. Rational Mech. Anal., **162**, 45–99 (2002)

[LM2]   Laurençot, Ph., Mischler, S.: Convergence to equilibrium for the continuous coagulation fragmentation equation. Bull. Sci. Math., **127**, 179–190 (2003)

[LM3]   Laurençot, Ph., Mischler, S., *On coalescence equations and related models*, to appear in "Modelling and computational methods for kinetic equations", Editors P. Degond, L. Pareschi, G. Russo, in the Series Modeling and Simulation in Science, Engineering and Technology (MSSET), Birkhäuser.

[L]   Leyvraz, F.: Scaling Theory and Exactly Solved Models In the Kinetics of Irreversible Aggregation. Phys. Reports, **383**, Issues 2-3, 95–212 (2003)

[MP1]   Menon, G, Pego, R.L.: Approach to self-similarity in Smoluchowski's coagulation equation, preprint 2003.

[MP2]   Menon, G, Pego, R.L.: Dynamical scaling in Smoluchowski's coagulation equation: uniform convergence, preprint 2003.

M. Escobedo
Departamento de Matemáticas
Universidad del País Vasco / EHU
Bilbao, Spain

Progress in Nonlinear Differential Equations
and Their Applications, Vol. 63, 259–266
© 2005 Birkhäuser Verlag Basel/Switzerland

Orlicz Capacities and Applications to PDEs and Sobolev Mappings

Alberto Fiorenza

Proceedings of the V European Conference on elliptic and parabolic problems
A special tribute to the work of Haïm Brézis. Gaeta, May, 30 – June, 3, 2004

Abstract. We discuss two applications of the notion of Orlicz capacity. The first one is related to a nonexistence result of solutions for some nonlinear elliptic equations having measure data, the second one to a capacitary estimate useful for proving an extension, due to Malý, Swanson and Ziemer ([20]), of the area and co-area formulas.

Mathematics Subject Classification (2000). Primary 35J60; Secondary 46E30, 46E35, 31C45.

Keywords. Nonlinear elliptic equations, Orlicz spaces, measure data, capacity, capacitary estimate.

1. A nonexistence result

1.1. Introduction

Let us consider the problem

$$\begin{cases} -\Delta u + |u|^{q-1}u = \mu & \text{in } \Omega, \\ u = 0 & \text{on } \partial\Omega \end{cases} \tag{1.1}$$

where $\Omega \subset \mathbf{R}^N$ ($N \geq 3$) is a bounded smooth domain, $1 < q < \infty$, and μ is a bounded Radon measure on Ω.

A function $u \in L^q(\Omega)$ is called *weak* solution of (1.1) if

$$-\int_\Omega u\Delta\varphi + \int_\Omega |u|^{q-1}u\varphi = \int_\Omega \varphi d\mu \qquad \forall\varphi \in C^2(\bar\Omega), \ \varphi = 0 \text{ on } \partial\Omega.$$

It is known (see Stampacchia [24] and Brézis, Marcus and Ponce [11]) that a weak solution u belongs to $W_0^{1,q}(\Omega)$ for every $q < N/(N-1)$.

The celebrated result by Bênilan and Brézis [9] states that if μ is the Dirac mass at a point of Ω, then in the case $q < N/(N-2)$ there exists a unique weak solution. Moreover, if $q \geq N/(N-2)$, distributional solutions in $L^q_{\mathrm{loc}}(\Omega)$ do not exist. It is to be noted here that when $\mu \in L^1(\Omega)$ the problem (1.1) admits a unique solution in some appropriate class without any restriction on q.

The phenomenon of the nonexistence can be better understood using the notion of capacity (see Brézis, Marcus and Ponce [10, 11] for recent developments). Roughly speaking, given an exponent q, if the measure μ on the right-hand side is concentrated on a very "small" set, then distributional solutions do not exist. Baras and Pierre [5] (see also Gallouët and Morel [17]) were able to characterize how much such set must be small, in terms of q, in order to obtain nonexistence of distributional solutions. Namely, they proved that a distributional solution $u \in L^q(\Omega) \bigcap W_0^{1,1}(\Omega)$ exists if and only if

$$|\mu|(E) = 0 \text{ for every Borel set } E \subset \Omega \text{ with } \mathrm{cap}_{2,q'}(E) = 0$$

where $\mathrm{cap}_{2,q'}$ denotes the capacity associated to $W_0^{2,q'}$ (see next subsection). This result is consistent with that one by P. Bênilan and H. Brézis, because a point has $(2,q')$-capacity zero if and only if $q \geq N/(N-2)$ (see, e.g., Meyers [22]).

The failure of existence discussed above can be seen also from another point of view.

Suppose that $q \geq N/(N-2)$. Let us consider first the case $\mu = f \in L^1(\Omega)$ (case of existence). Let $f_n \in L^1(\Omega)$ be a sequence of functions converging to f in the sense of measures, and consider the problem (1.1) with μ replaced by f_n. Such problem admits a unique solution u_n (see Brézis and Strauss [12]), and the sequence u_n converges to u, where u is the solution when the datum is f. Consider now the case $\mu = \delta$, where δ is the Dirac mass at a point of Ω, say 0 (case of nonexistence). Setting for instance $f_n = \chi_{B(0,1/n)}/|B(0,1/n)|$, we have $f_n \to \delta$, and proceeding analogously, one gets $u_n \to 0$. Notice that the function identically zero is not a solution of (1.1) (see Brézis [9], Brézis and Veron [13] for details). The fact that in this case solutions do not exist can be roughly expressed saying that *sequences of solutions of approximating equations do not converge to a reasonable solution.*

In Section 1.3 we will present a result of nonexistence in the sense of approximations, for a general class of *nonlinear* equations, in which the key condition (only sufficient for nonexistence) is given in terms of a generalization of the concept of capacity. In the case of problem (1.1), when μ is a Dirac mass, the result is weaker than that one in Brézis [9] (we get nonexistence for $q > N/(N-2)$, loosing the case $q = N/(N-2)$); this is due to the fact that we deal with first order capacities, which are "rougher" than the second order ones (which are in fact needed for optimal results, in view of Baras and Pierre [5]).

1.2. Orlicz capacities

Let $0 < \alpha < N$ and let r be a real number, with $r > 1$. Let K be a compact subset of Ω. The (α, r)-*capacity* of K with respect to Ω is defined (see, e.g., Adams and

Hedberg [1]) as

$$\mathrm{cap}_{\alpha,r}(K) = \mathrm{cap}_{\alpha,r}(K,\Omega) = \inf\left\{ \|u\|_{W_0^{\alpha,r}(\Omega)}^r : u \in C_c^\infty(\Omega),\ u \geq \chi_K \right\},$$

where χ_K is the characteristic function of K; we will use the convention that $\inf\emptyset = +\infty$. The (α,r)-capacity of any open subset U of Ω is then defined by

$$\mathrm{cap}_{\alpha,r}(U) = \mathrm{cap}_{\alpha,r}(U,\Omega) = \sup\left\{\mathrm{cap}_{\alpha,r}(K),\ K \text{ compact},\ K \subset U\right\},$$

and the (α,r)-capacity of any set $E \subset \Omega$ by

$$\mathrm{cap}_{\alpha,r}(E) = \mathrm{cap}_{\alpha,r}(E,\Omega) = \inf\left\{\mathrm{cap}_{\alpha,r}(U),\ U \text{ open},\ E \subset U\right\}.$$

The previous definition can be generalized in the context of Orlicz spaces. For our goals it will be sufficient to limit ourselves to the case of first order capacities. Moreover, we will give the definition only for compact sets, because the definition of Orlicz capacity can be extended to any set as before. It is possible to see that the following formulation is equivalent, for our goals, to that one appearing in Aissaoui [2, 3], Aissaoui and Benkirane [4] (see Section 2.1 for the definition, Fiorenza and Prignet [15] for details).

Let us first recall some basic notions of the theory of Orlicz spaces. An N-*function* is a function Φ continuous on $[0,\infty[$, increasing, convex, and such that $\lim_{x\to 0}\Phi(x)/x = 0$, $\lim_{x\to\infty}\Phi(x)/x = +\infty$. For our purposes we will assume throughout this paper that all N-functions satisfy $\Phi \in C^1([0,\infty[)$, Φ' strictly increasing, and

$$c_1 \min(s^{q_1-1}, s^{q_2-1})\Phi'(t) \leq \Phi'(st) \leq c_2 \max(s^{q_1-1}, s^{q_2-1})\Phi'(t) \qquad (1.2)$$

for some $q_1, q_2 > 1$, $c_1, c_2 > 0$, for all $s, t > 0$. Such assumption is a way to express that the growth of Φ "lies between" two powers with exponent greater than 1 and implies that Φ is doubling.

The *complementary function* of Φ, denoted by $\widetilde{\Phi}$, is defined by

$$\widetilde{\Phi}(s) = \sup_{t \geq 0}[st - \Phi(t)] \qquad \forall s \geq 0.$$

It can be proved that if Φ is an N-function, also $\widetilde{\Phi}$ is an N-function. If Φ' is strictly increasing, $(\widetilde{\Phi})'(t) = (\Phi')^{-1}(t)\,\forall t \geq 0$.

For an N-function Φ, the *Orlicz class* $L^\Phi(\Omega)$ is defined by

$$L^\Phi(\Omega) = \left\{ f \in L^1_{\mathrm{loc}}(\Omega) : \int_\Omega \Phi(|f|)dx < +\infty \right\}.$$

The Orlicz class $L^\Phi(\Omega)$, equipped with the norm

$$\|f\|_\Phi = \inf\left\{ k > 0 : \int_\Omega \Phi\left(\frac{|f|}{k}\right) dx \leq 1 \right\}$$

becomes the so-called *Orlicz space*, which is a reflexive Banach space.

Definition 1.1. Let K be a compact subset of Ω and A be an N-function. The *A-capacity* of K with respect to Ω is defined as:

$$\text{cap}_{1,A}(K) = \inf\left\{ A(\|\,|\nabla u|\,\|_A) : u \in C_c^\infty(\Omega),\ u \geq \chi_K \right\}.$$

1.3. The main result

We are going to consider the following class of nonlinear equations, more general than that one appearing in (1.1):

$$\begin{cases} -\text{div}(a(x, \nabla u)) + \Phi''(|u|)u = \mu & \text{in } \Omega, \\ u = 0 & \text{on } \partial\Omega. \end{cases} \tag{1.3}$$

Here $a : \Omega \times \mathbf{R}^N \to \mathbf{R}^N$ be a Carathéodory function (i.e., $a(\cdot, \xi)$ is measurable on Ω for every ξ in $\mathbf{R}^N$, and $a(x, \cdot)$ is continuous on $\mathbf{R}^N$ for almost every x in Ω), such that the following holds for some p, $1 < p < N$:

$$a(x, \xi) \cdot \xi \geq \alpha\,|\xi|^p\,, \tag{1.4}$$

$$|a(x, \xi)| \leq \beta\,[b(x) + |\xi|^{p-1}]\,, \tag{1.5}$$

$$[a(x, \xi) - a(x, \eta)] \cdot (\xi - \eta) > 0\,, \tag{1.6}$$

for almost every x in Ω, for every ξ, η in $\mathbf{R}^N$, with $\xi \neq \eta$, where α and β are two positive constants, and b is a nonnegative function in $L^{p'}(\Omega)$ ($p' = p/(p-1)$). Notice that the problem (1.1) is recovered setting

$$a(x, \xi) = \xi \qquad p = 2. \tag{1.7}$$

Under assumptions (1.4), (1.5) and (1.6), $u \to -\text{div}(a(x, \nabla u))$ is a uniformly elliptic, coercive and pseudomonotone operator acting from $W_0^{1,p}$ to its dual $W^{-1,p'}$, and so it is surjective (see Leray and Lions [19]).

The function Φ appearing in (1.3) is an N-function in $C^2([0, \infty[)$ satisfying the assumptions given in Section 1.2, and such that $\Phi'(t) \leq t\Phi''(t)\ \forall t \geq 0$. Notice that the problem (1.1) is recovered setting

$$\Phi(t) = \frac{t^{q+1}}{q(q+1)}. \tag{1.8}$$

Finally, μ is a bounded Radon measure concentrated on a set E of null A-capacity, where A is an N-function whose growth is between t^p and t^N in the sense of (1.2). We say that μ is *concentrated* on E if $\mu(B) = \mu(B \cap E)$ for every Borelian subset B of Ω. Notice that the problem (1.1) with μ replaced by δ is recovered setting

$$A(t) = t^N \tag{1.9}$$

because δ is concentrated on a point, whose $(1, N)$-capacity is zero.

The main result concerning the problem (1.3), obtained in collaboration with A. Prignet in [15], can be briefly stated as follows: *if*

$$\int^{+\infty} \frac{(\tilde{A})'(t)\,\Phi^{-1}(t^{p'})}{t^{p'}}\,dt < +\infty, \tag{1.10}$$

then sequences of solutions of approximating equations of (1.3) *do not converge to a reasonable solution of* (1.3).

Notice that in the particular case of (1.7), (1.8), (1.9) the assumption (1.10) reduces to $q > N/(N-2)$.

We would need at this point to state precisely *how* we approximate the problem (1.3) and *how* we can assert that the limit of the solutions is not "reasonable". Moreover, in order to speak about "limit of solutions", since we deal with nonlinear equations, it may be convenient to have uniqueness of solutions. This is in general lost when considering distributional solutions (for positive results in this sense see, e.g., Fiorenza and Sbordone [16] and Greco, Iwaniec and Sbordone [18]), therefore a convenient approach is to use the notion of *entropy* solution (see Bénilan, Boccardo, Gallouët, Gariepy, Pierre and Vazquez [6], Boccardo, Gallouët and Orsina [8], Dal Maso, Murat, Orsina and Prignet [14]). For details and for various relevant particular cases (related to problems considered, e.g., in Orsina and Prignet [23], Boccardo and Gallouët [7]) we refer to Fiorenza and Prignet [15] and references therein.

2. A capacitary estimate

2.1. Comparison with the Hausdorff capacity

In order to fix some notation, let us recall the well-known definition of Hausdorff measure. Let $h(r)$ be an increasing function, defined ($\leq +\infty$) for $r \geq 0$, and satisfying $h(0) = 0$. Let $E \subset \mathbf{R}^N$, and consider coverings of E by countable unions of (open or closed) balls $\{B(x_i, r_i)\}_{i=1}^{\infty}$ with radii $\{r_i\}_{i=1}^{\infty}$. Then for any ρ, $0 < \rho \leq \infty$, a set function $\Lambda_h^{(\rho)}$ is defined by

$$\Lambda_h^{(\rho)}(E) = \inf \sum_{i=1}^{\infty} h(r_i)$$

where the infimum is taken over all such coverings with $\sup_{i=1} r_i \leq \rho$. Clearly $\Lambda_h^{(\rho)}(E)$ is a decreasing function of ρ, so $\lim_{\rho \to 0} \Lambda_h^{(\rho)}(E)$ exists ($\leq +\infty$), and we can define

$$\Lambda_h(E) = \lim_{\rho \to 0} \Lambda_h^{(\rho)}(E).$$

This is the *Hausdorff measure of E with respect to the function h*. If $h(r) = r^\alpha$, we write $\mathcal{H}^\alpha$ for Λ_{r^α}. The set function $\Lambda_h^{(\infty)}$ is called the *Hausdorff capacity*.

Following Aissaoui and Benkirane [4], let us now give the following definition.

Definition 2.1. *Let E be any subset of $\mathbf{R}^N$. We set*

$$C'_{\alpha, A}(E) = \inf \left\{ \|f\|_A : f \in L^A(\mathbf{R}^N), \, G_\alpha * f \geq \chi_E \right\}$$

The following theorem has been proven in Fiorenza and Prignet [15]:

Theorem 2.2. *Let* $0 < \alpha < N$, h *be an increasing function on* $[0, \infty[$ *such that* $h(0) = 0$,

$$\int_0^4 t^{\alpha-1}(\widetilde{\Phi})'\left(\frac{h(t)}{t^{N-\alpha}}\right) dt = H < \infty$$

and let $E \subset \mathbf{R}^N$ *be a set satisfying* $\Lambda_h^{(\infty)}(E) > 0$. *Then there exists a constant* $c_\Phi > 0$, *independent of* h *and* E, *such that*

$$\Lambda_h^{(\infty)}(E) \leq \Theta(c_\Phi\, C'_{\alpha,\Phi}(E))$$

where $\Theta(t)$ *is an increasing function such that* $\Theta(0+) = 0$. *In particular,*

$$C'_{\alpha,\Phi}(E) = 0 \Rightarrow \Lambda_h(E) = 0\,.$$

2.2. Application to Sobolev mappings

A consequence of Theorem 2.2 has been proven independently in Malý, Swanson and Ziemer [20, 21] (see Theorem 5.5 in [20]), in order to obtain the following extension of the area and co-area formulas for Sobolev mappings.

Let $E \subset \mathbf{R}^n$ be a measurable set and $f : \mathbf{R}^n \to \mathbf{R}^m$ be a Lipschitz continuous function. For $1 \leq m \leq n$ let us denote by $|J_m f(x)|$ the square root of the sum of the squares of the determinants of the m by m minors of the differential of f.

The well-known *co-area formula* states that

$$\int_E |J_m f(x)|dx = \int_{\mathbf{R}^m} \mathcal{H}^{n-m}(E \cap f^{-1}(y))dy \tag{2.1}$$

and the *area formula* states that

$$\mathcal{H}^n(\bar{f}(E)) = \int_E |J_n \bar{f}(x)|dx \tag{2.2}$$

where $\bar{f} : \mathbf{R}^n \to \mathbf{R}^{n+m}$ is defined by $\bar{f}(x) = (x, f(x))$.

Since we are going to deal with functions f belonging to the Sobolev class $W_{\text{loc}}^{1,1}(\Omega, \mathbf{R}^m)$, we will need the notion of functions precisely represented.

A function $f \in L_{\text{loc}}^1(\Omega)$ is said to be *precisely represented* if

$$f(x) = \lim_{r \to 0} \frac{1}{|B(x,r)|} \int_{B(x,r)} f(y)dy$$

at all points x where this limit exists. It is clear from the Lebesgue differentiation theorem that any function in $L_{\text{loc}}^1(\Omega)$ may be modified on a set of Lebesgue measure zero so as to be precisely represented. A mapping $f \in L_{\text{loc}}^1(\Omega; \mathbf{R}^m)$ is said to be precisely represented if each of its component functions is precisely represented.

We say that a measurable function f on Ω belongs to the *Lorentz space* $L^{m,1}(\Omega)$ if

$$\|f\|_{L^{m,1}(\Omega)} = \int_0^\infty |\{x \in \Omega : |f(x)| > s\}|^{1/m}ds < +\infty$$

The following theorem holds:

Theorem 2.3 ([20]). *Suppose that $1 \leq m \leq n$, that $f \in W^{1,1}_{\mathrm{loc}}(\Omega; \mathbf{R}^m)$ is precisely represented, and that $|\nabla f| \in L^{m,1}(\Omega)$. Then $f^{-1}(y)$ is countably $\mathcal{H}^{n-m}$ rectifiable for almost all $y \in \mathbf{R}^m$, the graph of f is countably $\mathcal{H}^n$ rectifiable and the co-area formula* (2.1) *and area formula* (2.2) *hold for all measurable sets $E \subset \Omega$.*

References

[1] D.R. Adams, L.I. Hedberg, *Function spaces and potential theory*, Grundlehren der mathematischen Wissenschaften, **314**, Springer-Verlag, Berlin, 1996.

[2] N. Aissaoui, *Bessel potentials in Orlicz spaces*, Rev. Mat. Univ. Complut. Madrid **10** (1997), 55–79.

[3] N. Aissaoui, *Some developments of strongly nonlinear potential theory*, Libertas Math. **19** (1999), 155–170.

[4] N. Aissaoui, A. Benkirane, *Capacités dans les espaces d'Orlicz*, Ann. Sci. Math. Québec **18** (1994), 1–23.

[5] P. Baras, M. Pierre, *Singularités éliminables pour des équations semi-linéaires*, Ann. Inst. Fourier, Grenoble **34** (1) (1984), 185–206.

[6] P. Bénilan, L. Boccardo, T. Gallouët, R. Gariepy, M. Pierre, J.L. Vazquez, *An L^1 theory of existence and uniqueness of nonlinear elliptic equations*, Ann. Scuola Norm. Sup. Pisa Cl. Sci., **22** (1995), 240–273.

[7] L. Boccardo, T. Gallouët, *Nonlinear elliptic and parabolic equations involving measure data*, J. Funct. Anal. **87** (1989), 149–169.

[8] L. Boccardo, T. Gallouët, L. Orsina *Existence and uniqueness of entropy solutions for nonlinear elliptic equations with measure data*, Ann. Inst. H. Poincaré Anal. Non Linéaire **13** (1996), 539–551.

[9] H. Brézis, *Nonlinear elliptic equations involving measures*, in Contributions to nonlinear partial differential equations (Madrid, 1981), 82–89, Res. Notes in Math., 89, Pitman, Boston, Mass.-London, 1983.

[10] H. Brézis, M. Marcus, A.C. Ponce, *A new concept of reduced measure for nonlinear elliptic equations*, C. R. Acad. Sci. Paris, Ser. I **339** (2004) 169–174

[11] H. Brézis, M. Marcus, A.C. Ponce, *Nonlinear elliptic equations with measures revisited*, to appear.

[12] H. Brézis, W. Strauss, *Semi-linear second-order elliptic equations in L^1*, J. Math. Soc. Japan **25** (1973) 565–590.

[13] H. Brézis, L. Veron, *Removable singularities for some nonlinear elliptic equations*, Arch. Rat. Mech. Anal. **75** (1980) 1–6.

[14] G. Dal Maso, F. Murat, L. Orsina, A. Prignet, *Renormalized solutions for elliptic equations with general measure data*, Ann. Scuola Norm. Sup. Pisa CL. Sci., **28** (1999), 741–808.

[15] A. Fiorenza, A. Prignet, *Orlicz capacities and applications to some existence questions for elliptic PDEs having measure data*, ESAIM: Control, Optimisation and Calculus of Variations **9** (2003), 317–341.

[16] A. Fiorenza, C. Sbordone, *Existence and uniqueness results for solutions of nonlinear equations with right-hand side in L^1*, Studia Math. **127** (3), (1998) 223–231.

[17] T. Gallouët, J.M. Morel, *Resolution of a semilinear equation in L^1*, Proc. Roy. Soc. Edinburgh, **96** (1984), 275–288.

[18] L. Greco, T. Iwaniec, C. Sbordone, *Inverting the p-harmonic operator*, Manuscripta Math. **92** (2), (1997) 249–258.

[19] J. Leray, J.-L. Lions, *Quelques résultats de Višik sur les problèmes elliptiques non linéaires par les méthodes de Minty-Browder*, Bull. Soc. Math. France, **93** (1965), 97–107.

[20] J. Malý, D. Swanson, W.P. Ziemer, *The co-area formula for Sobolev mappings*, Transactions of the A.M.S. **355** (2), (2002), 477–492.

[21] J. Malý, D. Swanson, W.P. Ziemer, *Fine behavior of functions with gradients in a Lorentz space*, in preparation.

[22] N.G. Meyers, *A theory of capacities for potentials of functions in Lebesgue Classes*, Math. Scand. **26** (1970) 255–292.

[23] L. Orsina, A. Prignet, *Nonexistence of solutions for some nonlinear elliptic equations involving measures*, Proc. Royal Soc. Edinburgh **130A** (2000), 167–187.

[24] G. Stampacchia, *Équations elliptiques du second ordre á coefficients discontinus*, Les Presses de l'Université de Montréal, Montréal, 1966.

Alberto Fiorenza
Universitá di Napoli
Dipartimento di Costruzioni e Metodi Matematici in Architettura
via Monteoliveto, 3
I-80134 Napoli, Italy

and

Consiglio Nazionale delle Ricerche
Istituto per le Applicazioni del Calcolo "Mauro Picone"
Sezione di Napoli
via Pietro Castellino, 111
I-80131 Napoli, Italy
e-mail: `fiorenza@unina.it`

Progress in Nonlinear Differential Equations
and Their Applications, Vol. 63, 267–277

Energy Forms on Non Self-similar Fractals

Uta Renata Freiberg and Maria Rosaria Lancia

Abstract. Some recent results on the construction of energy forms on certain classes of non self-similar fractal sets are presented. In order to overcome the lack of self-similarity, the energy for these sets is obtained by integrating a Lagrangian.

Mathematics Subject Classification (2000). Primary 31C25, 28A80; Secondary 58.

Keywords. Dirichlet form, Lagrangian, fractal manifold, non self-similar fractal, conformal mapping.

1. Introduction

In the last decades there was an increasing interest in the study of many physical phenomena such as percolation, diffusion through porous media (see, e.g., [8]), diffusion across highly conductive layers (see, e.g., [17], [22]). In this kind of problems, the media is often an "irregular and wild object" and, in the applications, it is often modelled by a fractal set. Due to this fact, there were many efforts to develop some tools of analysis on fractals (see also [14] and the references listed in). Fractals are non differentiable objects with Hausdorff dimension less than the topological dimension. Several approaches have been developed in order to build a potential theory on certain classes of fractals such as nested fractals (see [15], [18] and Section 2 in this paper) or p.c.f. fractals (see [13]), on which it is possible to construct an energy form (and hence a Laplacian) as a limit of approximating energies which are defined by suitable difference schemes. Note that sets from both these families are in particular self-similar fractals and the construction of the energy deeply relies on this property. Unfortunately, a lot of applications require to deal with "wilder objects" which for instance are no longer self-similar. Aim of this paper is to illustrate how to overcome the lack of self-similarity, in some special cases, by using a Lagrangian approach, i.e., the energy is obtained by integrating a local energy measure, the so-called *Lagrangian* (see [7], [19]–[22]).

More precisely, we will consider fractal sets G where the self-similarity is destroyed by matching or deforming some given self-similar sets. The energy form $\mathcal{E}_G$ on G is obtained by integrating a local energy measure $\mathcal{L}_G$ on G. In Section 2, we briefly recall the definition of nested fractals and the construction of energy forms as well as Lagrangians on them. In Section 3, we apply the Lagrangian approach to define energy forms on self-similar sets obtained by respectively matching (see Subsection 3.1) or conformal deforming (see Subsection 3.2) self-similar fractals. In Subsection 3.3, we combine matching and deforming techniques in order to extend our results to a wider class of non self-similar sets.

2. Energy and Lagrangian on self-similar fractals

In this section, we recall the construction of the energy form and the Lagrangian on a nested fractal in $\mathbb{R}^2$.

For any $B \subseteq \mathbb{R}^2$ we denote by $\mathcal{C}(B)$ the space of real-valued continuous functions on B, and by $\mathcal{C}(B)'$ its dual. $\mathcal{C}^{0,\beta}(B)$ denotes the space of all Hölder continuous functions on B with Hölder exponent β.

Assume that we are given a finite family of similitudes $\Psi = \{\psi_1, \ldots, \psi_N\}$ acting on $\mathbb{R}^2$ with the same ratio $L^{-1} \in (0,1)$. A set K is called *self-similar with respect to the family* Ψ if $K = \bigcup_{i=1}^{N} \psi_i(K)$. The existence and uniqueness of such a set K is proved in [10]. Note that – if, in addition, the well-known *open set condition* is satisfied – the Hausdorff dimension of K is given by $D_f = \frac{\ln N}{\ln L}$, and the D_f-dimensional Hausdorff measure of K is positive and finite, i.e., $0 < \mathcal{H}^{D_f}(K) < \infty$ (see [10]). We denote by μ the normalized D_f-dimensional Hausdorff measure $\mathcal{H}^{D_f}$, restricted to K, which turns out to be in addition self-similar with respect to the family Ψ, i.e., $\mu(A) = L^{-D_f} \sum_{i=1}^{N} \mu\left(\psi_i^{-1}A\right)$ for any Borel set $A \subseteq \mathbb{R}^2$ (see [10]). Note that μ is a so-called D_f-measure, i.e., there exist positive constants C_1, C_2 and r_0, such that

$$C_1 r^{D_f} \le \mu(B(x,r)) \le C_2 r^{D_f}, \qquad x \in K, r \in (0, r_0), \tag{2.1}$$

(see [11] for details).

In the following, we restrict ourselves to the particular case of so-called *nested fractals* (introduced by Lindstrøm, see [18]) which – in addition – satisfy certain symmetry and ramification properties crucial in constructing an energy form on K (see Kusuoka, [15]).

Let V be the set of fixed points of the maps $\psi_1, \ldots, \psi_N$, i.e., $\sharp V = N$. A point $P \in V$ is called *essential fixed point of the family* Ψ, if there exist $i, j \in \{1, \ldots, N\}$, $i \ne j$ and a point $Q \in V$ such that $\psi_i(P) = \psi_j(Q)$. Denote $V_0 := \{P_1, \ldots, P_M\}$, $M \le N$, the set of the essential fixed points of the family Ψ. They form a regular polygon P_F in the plane. We assume that it has unit side length. Every pair of endpoints of such a unit edge of P_F is called a pair of *nearest zero-neighbors* (see [18]). A self-similar fractal K is called *nested fractal* , if it is connected and if for any pair of different n-tuples $(i_1, \ldots, i_n), (j_1, \ldots, j_n) \in$

$\{1, \ldots, N\}^n$ the following *nesting condition* is satisfied:

$$\psi_{i_1} \circ \ldots \psi_{i_n}(K) \cap \psi_{j_1} \circ \ldots \psi_{j_n}(K) = \psi_{i_1} \circ \ldots \psi_{i_n}(V_0) \cap \psi_{j_1} \circ \ldots \psi_{j_n}(V_0).$$

Note, that in addition nested fractals have to fulfill a symmetry condition, see [18] for details. From the nesting axiom it follows that nested fractals are finitely ramified, well-known examples are Sierpinski gasket and the von Koch snowflake.

Define the increasing sequence of points $(V_n)_{n \geq 1}$ by $V_n := \bigcup_{i=1}^{N} \psi_i(V_{n-1}), n \geq 1$. Setting $V_* := \bigcup_{n \geq 0} V_n = \lim_{n \to \infty} V_n$ it holds that $K = \overline{V_*}$.

Two points p and q in V_n are called *nearest n-neighbors* – denoted in the following by $p \sim_n q$ – if there exists a n-tuple of indices $(i_1, \ldots, i_n) \in \{1, \ldots, N\}^n$ and two nearest zero-neighbors P_i and P_j in V_0 such that $p = \psi_{i_1} \circ \ldots \psi_{i_n}(P_i)$ and $q = \psi_{i_1} \circ \ldots \psi_{i_n}(P_j)$. Further, for any $n \geq 0$, we define a discrete measure on V_n by

$$\mu_n := \frac{a}{N^n} \sum_{p \in V_n} \delta_{\{p\}}, \tag{2.2}$$

where $\delta_{\{p\}}$ denotes the Dirac measure at the point p and a is the constant given by $a := \lim_{n \to \infty} \frac{\sharp V_n}{N^n}$.

Note that the sequence $(\mu_n)_{n \geq 1}$ is weakly convergent (i.e., in $\mathcal{C}(K)'$) to the measure μ (for example, the proof for the particular case that K is the von Koch curve can be found in [16]).

The construction of the energy form on K is based on finite difference schemes defined on the approximating sets $(V_n)_{n \geq 1}$. For any function $u : V_* \longrightarrow \mathbb{R}$, we define

$$\mathcal{E}_n[u] := \frac{1}{2} \, \varrho^n \sum_{p \in V_n} \sum_{q \sim_n p} (u(p) - u(q))^2, \tag{2.3}$$

where ϱ is a real number which can be determined by the so-called *Gaussian principle* (see for example Mosco [21], and see also the comments after Formula (2.7)). It can be shown (see [15]) that the sequence $(\mathcal{E}_n[u])_{n \geq 0}$ is non decreasing, the limit of the right-hand side of (2.3) exists and the limit form $\mathcal{E}_K[u] := \lim_{n \to \infty} \mathcal{E}_n[u]$ is non trivial (i.e., $\mathcal{E}_K \not\equiv \infty$) with domain

$$\mathcal{D}_*(\mathcal{E}_K) := \{u : V_* \longrightarrow \mathbb{R} \quad : \quad \mathcal{E}_K[u] < \infty\}.$$

Every function $u \in \mathcal{D}_*(\mathcal{E}_K)$ can be uniquely extended to be an element of $\mathcal{C}(K)$. We denote this extension still by u and we set $\mathcal{D} := \{u \in \mathcal{C}(K) : \mathcal{E}_K[u] < \infty\}$, where $\mathcal{E}_K[u] := \mathcal{E}_K[u_{|V_*}]$. Hence $\mathcal{D} \subseteq \mathcal{C}(K) \subseteq L^2(K, \mu)$. We now define the space $\mathcal{D}(\mathcal{E}_K)$ to be the completion of $\mathcal{D}$ in the norm

$$||u||_{\mathcal{E}_K} := \left(||u||^2_{L^2(K,\mu)} + \mathcal{E}_K[u] \right)^{1/2}. \tag{2.4}$$

$\mathcal{D}(\mathcal{E}_K)$ is injected in $L^2(K, \mu)$ and is a Hilbert space with the scalar product associated to the norm (2.4). Then we extend $\mathcal{E}_K$ as usual on the completed space

$\mathcal{D}(\mathcal{E}_K)$. By $\mathcal{E}_K(\cdot,\cdot)$ we denote the bilinear form defined on $\mathcal{D}(\mathcal{E}_K) \times \mathcal{D}(\mathcal{E}_K)$ by polarization, i.e.,

$$\mathcal{E}_K(u,v) := \frac{1}{2}\left(\mathcal{E}_K[u+v] - \mathcal{E}_K[u] - \mathcal{E}_K[v]\right), \quad u,v \in \mathcal{D}(\mathcal{E}_K).$$

It is easy to see that for any pair $u,v \in \mathcal{D}(\mathcal{E}_K)$, the form $\mathcal{E}_K(\cdot,\cdot)$ is the limit of the sequence $\mathcal{E}_n(\cdot,\cdot)$ given by

$$\mathcal{E}_n(u,v) := \frac{1}{2}\,\varrho^n \sum_{p \in V_n} \sum_{q \sim_n p} [u(p) - u(q)]\,[v(p) - v(q)]. \tag{2.5}$$

$\mathcal{E}_K(\cdot,\cdot)$ with domain $\mathcal{D}(\mathcal{E}_K)$ is a Dirichlet form on the Hilbert space $L^2(K,\mu)$. The form $\mathcal{E}_K$ is regular and strongly local. Moreover, the functions in $\mathcal{D}(\mathcal{E}_K)$ posses a continuous representative, which is actually Hölder continuous on K, the Hölder exponent is depending on the Hausdorff dimension D_f of K as well as on the number ϱ; more precisely, it is given by $\beta = \frac{\ln \varrho}{2\ln N}$ (see [20]).

Now we construct the Lagrangian on K. For the concept of Lagrangians on fractals, i.e., the notion of a measure-valued local energy, we refer to [7], [19] and [20] (see also [2] and [21]).

It is easy to check that the approximating energy forms $\mathcal{E}_n$ on V_n, $n \geq 0$, defined in (2.5), can be written as

$$\mathcal{E}_n(u,v) = \int_{V_n} \nabla_n u \cdot \nabla_n v \, d\mu_n, \tag{2.6}$$

where μ_n is the discrete measure on K supported on V_n, given in (2.2); and for any $p \in V_n$ the "discrete gradient" is given by

$$\nabla_n u \cdot \nabla_n v(p) = \frac{1}{2} \sum_{q \sim_n p} \frac{u(p) - u(q)}{|p - q|^\delta} \frac{v(p) - v(q)}{|p - q|^\delta}, \quad u,v \in \mathcal{D}(\mathcal{E}), \tag{2.7}$$

where δ is – as explained in [21] – the unique positive number which yields (in view of formulae (2.6) and (2.7)) a non trivial limit of the sequence $(\mathcal{E}_n)_{n \geq 0}$. Note that δ is not only determined by the Hausdorff dimension of the fractal K, but also by the ramification properties of the underlying "pre-fractal networks". This means that, from the viewpoint of the energy, the "effective distance" on the fractals is no longer given by the Euclidean metric, but by a certain power δ of it, i.e., by a quasi-metric. For any nested fractal, the "structural constants" N, L, ϱ and δ are in the following relationship

$$N\varrho = L^{2\delta} \tag{2.8}$$

(see [20]); thus the Hölder exponent can be also expressed by $\beta = \delta - \frac{D_f}{2}$ (see also [16]).

For the particular case that K is the von Koch curve, the proof of the following proposition can be found in [4], for a general fractal it is sketched in [21].

Proposition 2.1. *Let A be any subset of K. For every $u, v \in \mathcal{D}(\mathcal{E}_K)$, the sequence of measures given by*

$$\mathcal{L}_K^{(n)}(u,v)(A) := \int_{A \cap V_n} \nabla_n u \cdot \nabla_n v \, d\mu_n, \qquad n \geq 0,$$

weakly converges in $\mathcal{C}(K)'$ to a signed finite Radon measure $\mathcal{L}_K(u,v)$ on K as $n \to \infty$, the so-called Lagrangian measure on K. Moreover, it holds that $\mathcal{E}_K(u,v) = \int_K d\mathcal{L}_K(u,v), \quad u, v \in \mathcal{D}(\mathcal{E}_K)$.

The measure-valued map $\mathcal{L}_K$ on $\mathcal{D}(\mathcal{E}_K) \times \mathcal{D}(\mathcal{E}_K)$ is bilinear, symmetric and positive (i.e., $\mathcal{L}_K[u] := \mathcal{L}_K(u,u) \geq 0$ is a positive measure). This measure-valued Lagrangian takes on the fractal K the role of the Euclidean Lagrangian $d\mathcal{L}(u,v) = \nabla u \cdot \nabla v \, dx$. Note, that in the case of the Koch curve the Lagrangian $\mathcal{L}_K$ is absolutely continuous with respect to the volume measure μ (see [2]); on the contrary this is not true on most nested fractals (see [15]).

3. Energy form on non self-similar fractals

Aim of this section is to apply the Lagrangian approach to the construction of energy forms on non self-similar fractals which are obtained by deforming and matching nested fractals.

3.1. Matching

In [4], a simple example of a non self-similar fractal has been considered, namely the closed fractal curve F obtained as $F = \bigcup_{i=1}^3 K_i = \bigcup_{i=4}^6 K_i$, where the sets $K_1, \ldots, K_6$ are von Koch curves (see, e.g., [3]).

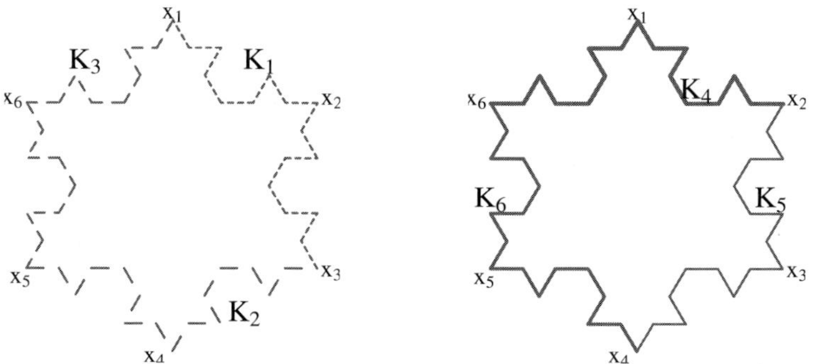

FIGURE 1. a: first decomposition; b: second decomposition.

Due to the special feature of F (see Figure 1.a and 1.b), and combining tools from differential and fractal geometry, it is possible to regard F as a "fractal manifold", which can be described by an atlas $\mathcal{A} = \{(U_i, \varphi_i)\}_{i=1}^6$, where the charts are just given by $U_i = \overset{\circ}{K_i}$, where $\overset{\circ}{K_i}$ denotes the set K_i without its endpoints;

and φ_i is the unique orthogonal mapping from K_i to a fixed reference Koch curve K, $i = 1, \ldots, 6$. Set $\mu_i := \varphi_i^{-1}\mu$, $i = 1, \ldots, 6$, where μ is the normalized D_f-dimensional Hausdorff measure restricted to K (see Section 2). Equip F with the finite Borel measure $\mu_F := \mu_1 + \mu_2 + \mu_3 = \mu_4 + \mu_5 + \mu_6$.

In this case, the Lagrangian $\mathcal{L}_F$ is locally defined on F as the image measure of the Lagrangian $\mathcal{L}_K$ on K with respect to the corresponding map φ_i^{-1}, i.e.,

$$\mathcal{L}_F(w, z)(A) := \mathcal{L}_K(w \circ \varphi_i^{-1}, z \circ \varphi_i^{-1})(\varphi_i(A)), \quad A \subseteq K_i, \quad w, z \in \mathcal{D}_F, \quad (3.1)$$

with

$$\mathcal{D}_F := \{w : F \longrightarrow \mathbb{R} \quad : \quad w \circ \varphi_i^{-1} \in \mathcal{D}(\mathcal{E}_K), \quad i = 1, \ldots, 6\},$$

where $\mathcal{D}(\mathcal{E}_K)$ is the space of all functions of finite energy on the reference set K. In [4], Subsection 4.1, it is shown, that definition in (3.1) is independent of the choice of the chart. Moreover, $\mathcal{L}_F(w, z)$ is uniquely extendible to any Borel subset of F, hence to a finite Borel measure supported on F, by using the additivity property of measures.

We now define the energy form on the fractal F by integrating its local energy measure, i.e.,

$$\mathcal{E}_F(u, v) := \int_F d\mathcal{L}_F(u, v), \qquad u, v \in \mathcal{D}_F.$$

It turns out that $(\mathcal{E}_F, \mathcal{D}_F)$ is a strongly local, closed, regular Dirichlet form on $L^2(F, \mu_F)$, i.e., there exists (see, e.g., Chap. 6, Theorem 2.1 in [12]) a unique self-adjoint, non positive operator Δ_F on $L^2(F, \mu_F)$ – with domain $\mathcal{D}(\Delta_F) \subseteq \mathcal{D}_F$, dense in $L^2(F, \mu_F)$ – such that

$$\mathcal{E}_F(u, v) = - \int_F (\Delta_F u)\, v d\mu_F, \qquad u \in \mathcal{D}(\Delta_F), v \in \mathcal{D}_F. \quad (3.2)$$

This Laplacian on F is locally given by the localized Laplacians on the Koch curves building F (see Subsection 5 in [4]). The latter fact has a nice stochastic interpretation in terms of a strong reflection principle (see Subsection 6 in [4]). The analogue of this in the language of the Dirichlet forms is given by the "natural" fact that the energy of a function u on F can be obtained as the sum of the energy of the restrictions of u to the Koch curves K_1, K_2 and K_3, or K_4, K_5 and K_6 respectively (see Theorem 4.6. in [4]). It is worth to be pointed out that no matching condition at the junction points is needed.

3.2. Deforming

Other examples of non self-similar fractals obtained by suitably deforming a nested fractal have been considered in [5]. Let K be a nested fractal as in Section 2, let $g : U \subset \mathbb{R}^2 \longrightarrow \mathbb{R}^2$ be a conformal $\mathcal{C}^1$-diffeomorphism, where U is an open set in $\mathbb{R}^2$ containing the set K. This yields that the differential Dg is given by $(Dg)(x) = f(x)O(x), x \in U$, where $f(\cdot)$ is a real-valued, positive, continuous function on U, and $O(x)$ is an orthogonal 2×2-matrix for any $x \in U$. Let $G := g(K)$ denote the deformed fractal.

The Hausdorff dimension $\dim_{\mathrm{H}}$ of G remains unchanged because g is in particular a bi-Lipschitz mapping. From [3], Proposition 2.2, it follows that $0 < \mathcal{H}^{D_f}(G) < \infty$, hence $\dim_{\mathrm{H}} G = \frac{\ln N}{\ln L}$.

As it was done in Section 2, we approximate G by an increasing sequence of finite sets. Set $W_n := g(V_n)$ and $W_* := \bigcup_{n \geq 0} W_n = g(V_*)$. It holds that $G = \overline{W_*}$. For any $n \geq 0$, two points p and q in W_n are nearest n-neighbors – denoted in the following also by $p \sim_n q$ – if and only if $g^{-1}(p)$ and $g^{-1}(q)$ are nearest n-neighbors in V_n.

We equip G with the image measure $\tilde{\mu} := g\mu$ of μ under g, i.e., $\tilde{\mu}(A) := \mu(g^{-1}A)$ for any Borel subset A of $g(U)$. Of course, $\operatorname{supp} \tilde{\mu} = G$ and $\tilde{\mu}(G) = 1$. On the other hand, $\tilde{\mu}$ can be described as the weak limit of a sequence of discrete measures which are supported on the approximating sets $(W_n)_{n \geq 0}$. Define $\tilde{\mu}_n := g\mu_n$, then it holds that $\operatorname{supp} \tilde{\mu}_n = W_n$, $\tilde{\mu}_n = \frac{a}{N^n} \sum_{p \in W_n} \delta_{\{p\}}$ (see Section 2). From the weak convergence of the sequence μ_n it follows that $\tilde{\mu}_n \rightharpoonup \tilde{\mu}$. Moreover, $\tilde{\mu}$ is a D_f-measure on G (see(2.1)), and therefore it is equivalent to $\mathcal{H}^{D_f}$ (see [11], Chapter III).

We introduce the Lagrangian $\mathcal{L}_G$ on the deformed fractal G which is obtained as the weak limit of a sequence of suitable defined discrete Lagrangians $\mathcal{L}_G^{(n)}$ supported on W_n, $n \geq 0$.

Let $\mathcal{L}_K$ and $(\mathcal{E}_K, \mathcal{D}(\mathcal{E}_K))$ be as in Section 2. We introduce the linear space

$$\mathcal{D}_G := \{u : G \longrightarrow \mathbb{R} \quad : \quad u \circ g \in \mathcal{D}(\mathcal{E}_K)\} = g^{-1}[\mathcal{D}(\mathcal{E}_K)].$$

Note that these are the functions of finite energy on G, because g is a $\mathcal{C}^1$-diffeomorphism; and they are still Hölder continuous with the same Hölder exponent.

For any $u \in \mathcal{D}_G$, we define a sequence of measures $\left(\mathcal{L}_G^{(n)}[u]\right)_{n \geq 0}$ by

$$\mathcal{L}_G^{(n)}[u](A) := \int_{A \cap W_n} |\tilde{\nabla}_n u|^2 d\tilde{\mu}_n,$$

where A is a Borel subset of G.

The crucial point is that in the definition of the discrete gradients $\tilde{\nabla}_n u$ the "effective distance" is now given by a suitable power of the arc length, instead of a suitable power of the Euclidean distance. This exponent turns out to be the same as for the undeformed fractal (see Section 3 in [5]).

Fix $n \geq 0$, $p \in W_n$ and $u \in \mathcal{D}_G$. Proceeding as in Section 2, we define the square of the discrete gradient of u in $p \in W_n$ by

$$|\tilde{\nabla}_n u|^2(p) = \frac{1}{2} \sum_{q \sim_n p} \left(\frac{u(p) - u(q)}{l_{pq}^{\delta}}\right)^2,$$

where l_{pq} denotes the arc length of the curve Γ defined as the image (under g) of the line segment joining $g^{-1}(p)$ and $g^{-1}(q)$.

Then it holds (see [5]):

Proposition 3.1. *For any $u \in \mathcal{D}_G$ there exists a unique finite Radon measure $\mathcal{L}_G[u]$ supported on G, which we call* Lagrangian measure *on G, such that $\mathcal{L}_G^{(n)}[u] \rightharpoonup \mathcal{L}_G[u]$ as $n \to \infty$. Moreover, this limit measure is given by*

$$d\mathcal{L}_G[u](x) = [f(g^{-1}(x))]^{-2\delta} d\mathcal{L}_K[u \circ g](g^{-1}(x)),$$

i.e.,

$$\mathcal{L}_G[u](A) = \int_A d\mathcal{L}_G[u](x) = \int_{g^{-1}(A)} [f(y)]^{-2\delta} d\mathcal{L}_K[u \circ g](y)$$

for any Borel set A in G.

The energy form on G is defined by integrating the Lagrangian given in (3.3), i.e., for any $u \in D_G$ we set $\mathcal{E}_G[u] := \int_G d\mathcal{L}_G[u]$, and the energy norm is given by $\|\cdot\|_{\mathcal{E}_G} := \left(\mathcal{E}_G[\cdot] + \|\cdot\|^2_{L^2(G,\tilde{\mu})}\right)^{1/2}$. The form $(\mathcal{E}_G, \mathcal{D}_G)$ is a regular, strongly local Dirichlet form on $L^2(G, \tilde{\mu})$ (as well as on $L^2(G, \hat{\mu})$, where $\hat{\mu} := \frac{1}{\mathcal{H}^{D_f}(G)} \mathcal{H}^{D_f} \llcorner_G$) (see Theorem 4.3 and Corollary 4.4 in [5]).

The above result has a probabilistic counterpart that is there exists a strong Markovian process $(X_t)_{t \geq 0}$ with continuous paths on G, which can be regarded as the "natural Brownian motion" on G. Proceeding analogously as in Subsection 3.1, it follows that there exists a unique self-adjoint, non positive operator $\widetilde{\Delta}_G$ on $L^2(G, \tilde{\mu})$ (or, on $L^2(G, \hat{\mu})$) – with domain $\mathcal{D}(\widetilde{\Delta}_G) \subseteq \mathcal{D}_G$, dense in $L^2(G, \tilde{\mu})$ (or, in $L^2(G, \hat{\mu})$ resp.) – which is the "natural" Laplacian on the "curved" fractal G, hence a "fractal Laplace-Beltrami-operator". Spectral asymptotics of $\widetilde{\Delta}_G$ will be determined in the forthcoming paper [6].

3.3. Matching and deforming

Let $K_1, \ldots, K_m$, $m \geq 2$, be nested fractals with (possibly different) Hausdorff dimensions $D_1, \ldots, D_m$ and equipped with their normalized D_i-dimensional Hausdorff measures, denoted by $\mu_1, \ldots, \mu_m$, $i = 1, \ldots, m$. As pointed out in Section 2, we can define on $K_1, \ldots, K_m$ energy forms $\mathcal{E}_1, \ldots, \mathcal{E}_m$ with domains $\mathcal{D}_1, \ldots, \mathcal{D}_m$ as well as Lagrangians $\mathcal{L}_{K_1}, \ldots, \mathcal{L}_{K_m}$ with corresponding structural constants ϱ_i and δ_i (see (2.8)). We set $F := \bigcup_{i=1}^m K_i$ and we assume that F is connected, but in such a way that the parts $K_1, \ldots, K_m$ are "just-touching", i.e., they match in only a finite number of points:

$$\sharp \bigcup_{i \neq j} (K_i \cap K_j) < \infty. \tag{3.3}$$

Condition (3.3) ensures that the intersections $K_i \cap K_j, i \neq j, i, j = 1, \ldots, m$, do not charge energy (see Theorem 5.2 in [1]).

Set $\mu := \mu_1 + \cdots + \mu_m$. Then μ is a finite Borel measure on F. We introduce the following linear subspace of $L^2(F, \mu)$:

$$\mathcal{D} := \{u \in \mathcal{C}(F) \quad : \quad u_{|K_i} \in \mathcal{D}_i, i = 1, \ldots, m\}.$$

As $\mathcal{D}_i \subseteq \mathcal{C}^{0,\beta_i}(K_i)$, $i = 1,\ldots,m$, it follows that $\mathcal{D} \subseteq \mathcal{C}^{0,\beta}(F)$, where $\beta := \min_{i=1,\ldots,m} \beta_i$.

We define the Lagrangian on F by

$$\mathcal{L}_F[u] := \sum_{i=1}^m \mathbf{1}_{K_i}\mathcal{L}_{K_i}[u_{|K_i}], \qquad u \in \mathcal{D},$$

and the energy on F by

$$\mathcal{E}_F[u] := \int_F d\mathcal{L}_F[u], \qquad u \in \mathcal{D}.$$

In a similar way as done in [4], one can show that

$$\mathcal{E}_F[u] = \sum_{i=1}^m \mathcal{E}_i[u_{|K_i}], \qquad u \in \mathcal{D}.$$

In [9], the latter formula is used in order to define the energy form on a suitable union of nested fractals. We remark that the matching condition in [9] slightly differs from (3.3).

Moreover, the Lagrangian approach allows us, to consider additionally a conformal deformation g of the set F defined on an open set U containing F. Following the ideas of Subsection 3.2, the Lagrangian on $G := g(F)$ in this case would be given by

$$\begin{aligned}
\mathcal{L}_G[u] &:= \sum_{i=1}^m \mathbf{1}_{G_i}(x)\mathcal{L}_{G_i}[u_{|G_i}], && u \in \mathcal{D}_G, \\
&= \sum_{i=1}^m \mathbf{1}_{G_i}(x)[f(x)]^{-2\delta_i}\mathcal{L}_{K_i}[u_{|G_i} \circ g_{|G_i}^{-1}], && u \in \mathcal{D}_G,
\end{aligned}$$

where $\mathcal{D}_G := \{u : G \longrightarrow \mathbb{R} \;\; : \;\; u \circ g \in \mathcal{D}\}$ and $G_i := g(K_i)$, $i = 1,\ldots,m$.

Even more generally, the sets $K_1,\ldots,K_m$ can be deformed before matching. Let $g_i : U_i \longrightarrow \mathbb{R}^2$, $K_i \subseteq U_i$ be conformal mappings as in Subsection 3.2 with differentials $Dg_i(x) = f_i(x)O_i(x)$, $i = 1,\ldots,m$. Denote $G_i := g_i(K_i)$ and match G_i to a connected set G, such that condition (3.3) is fulfilled. In this case, the Lagrangian is given by

$$\mathcal{L}_G[u] = \sum_{i=1}^m \mathbf{1}_{G_i}(x)[f_i(x)]^{-2\delta_i}\mathcal{L}_{K_i}[u_{|G_i} \circ g_{|G_i}^{-1}], \qquad u \in \mathcal{D}_G,$$

where $\mathcal{D}_G := \{u \in \mathcal{C}(G) \;\; : \;\; u_{|G_i} \circ g_i \in \mathcal{D}_i,\ i = 1,\ldots,m\}$. In both cases, the energy form on G is given by $\mathcal{E}_G[u] := \int_G d\mathcal{L}_G[u], u \in \mathcal{D}_G$; and it turns out that $\mathcal{E}_G$ is a closed, regular, strongly local Dirichlet form on $L^2(G, \tilde{\mu})$, where $\tilde{\mu}$ is given by $\tilde{\mu} := g\mu = \sum_{i=1}^m g\mu_i$, or $\tilde{\mu} := \sum_{i=1}^m g_i\mu_i$ respectively.

With the same arguments as in Subsection 3.1 (see (3.2)), a notion of fractal Laplace-Beltrami-operator on such a "wild" set G can be given; moreover, the notion of a strong Markovian diffusion process (hence, of a "natural Brownian motion") is provided (see [7], and also Section 6 in [4]).

References

[1] Bassat, B., Strichartz R. and Teplyaev, *What is not in the domain of the Laplacian on a Sierpinski gasket type fractal.* J. Funct. Anal.,**166**, 192–217, (1999)

[2] Capitanelli, R., *Lagrangians on homogeneous spaces.* PhD Thesis Univ. di Roma "La Sapienza", 2001

[3] Falconer, K. J., *The geometry of fractal sets.* Cambridge Univ. Press., Cambridge, 1985

[4] Freiberg, U.R. and Lancia, M.R., *Energy form on a closed fractal curve.* Z. Anal. Anwendungen, **23** no. 1, 115–137, (2004)

[5] Freiberg, U.R. and Lancia, M.R., *Energy forms on conformal images of nested fractals.* preprint MeMoMat, **15**, 2004

[6] Freiberg, U.R. and Lancia, M.R., *Can one hear the curvature of a fractal? Spectral asymptotics of fractal Laplace-Beltrami-operators.* in preparation

[7] Fukushima, M., Oshima, Y. and Takeda, M., *Dirichlet forms and symmetric Markov processes*, de Gruyter Studies in Mathematics, vol. 19, Berlin Eds. Bauer Kazdan, Zehnder 1994

[8] Goldstein, S., *Random walks and diffusions on fractals.* in "Percolation theory and ergodic theory of infinite particle systems", Minneapolis, Minn. 1984/85, 121–129; IMA Vol. Math. Appl. 8, Springer, New York, Berlin, 1987

[9] Hambly, B. and Kumagai, T., *Diffusion processes on fractal fields: heat kernel estimates and large deviations.* Probab. Theory Relat. Fields, **127**,(3), 305–352, (2003)

[10] Hutchinson, J.E., *Fractals and self-similarity.* Indiana Univ. Math. J. **30**, 713–747, (1981)

[11] Jonnson, A. and Wallin, H., *Function spaces on subsets of $\mathbb{R}^n$.* Math. Rep. Ser. 2 **1**, (1984)

[12] Kato, T., *Pertubation theory for linear operators.* 2nd edit., Springer, 1977

[13] Kigami, J., *Harmonic calculus on p.c.f. self-similar sets.* Trans. Am. Math. Soc. **335**, 721–755, (1993)

[14] Kigami, J., *Analysis on fractals.* Cambridge Univ. Press., Cambridge, 2001

[15] Kusuoka, S., *Diffusion processes on nested fractals.* Lecture Notes in Math. 1567, Springer, 1993

[16] Lancia, M.R. and Vivaldi, M.A., *Lipschitz spaces and Besov traces on self-similar fractals.* Rend. Accad. Naz. Sci. XL Mem. Mat. Appl.(5) **23**, 101–116, (1999)

[17] Lancia, M.R., *Second-order transmission problems across a fractal surface.* Rend. Accad. Naz. Sci. XL Mem. Mat. Appl.(1) **27**, 191–213, (2003)

[18] Lindstrøm, T., *Brownian Motion on Nested Fractals.* Memoirs Amer. Math. Soc. **420**, (1990)

[19] Mosco, U., *Composite media and asymptotic Dirichlet forms.* J. Funct. Anal. **123** no. 2, 368–421, (1994)

[20] Mosco, U., *Lagrangian metrics on fractals.* Proc. Symp. Appl. Math, 54, Amer. Math. Soc., R.Spigler and S. Venakides eds., 301–323, (1998)

[21] Mosco, U., *Energy functionals on certain fractal structures.* J. Convex Anal. **9**, 581–600, (2002)

[22] Mosco, U., *Highly conductive fractal layers*. Proc. Conf. "Whence the boundary conditions in modern physics?" Acad. Lincei, Rome, (2002)

Uta Renata Freiberg
Mathematisches Institut
Friedrich-Schiller-Universität Jena
Ernst-Abbé-Platz 1–4
D-07740 Jena, Germany
e-mail: `uta@mathematik.uni-jena.de`

Maria Rosaria Lancia
Dipartimento di Metodi e Modelli
      Matematici per le Scienze Applicate
Università degli Studi di Roma "La Sapienza"
Via A. Scarpa 16
I-00161 Roma, Italy
e-mail: `lancia@dmmm.uniroma1.it`

Progress in Nonlinear Differential Equations
and Their Applications, Vol. 63, 279–290

Measure Data and Numerical Schemes for Elliptic Problems

Thierry Gallouët

Dedicated to H. Brezis in the occasion of his 60th birthday

Abstract. In order to show existence of solutions for linear elliptic problems with measure data, a first classical method, due to Stampacchia, is to use a duality argument (and a regularity result for elliptic problems). Another classical method is to pass to the limit on approximate solutions obtained with regular data (converging towards the measure data). A third method is presented. It consists to pass to the limit on approximate solutions obtained with numerical schemes such that Finite Element schemes or Finite Volume schemes. This method also works for convection-diffusion problems which lead to non coercive elliptic problems with measure data. Thanks to a uniqueness result, the convergence of the approximate solutions as the mesh size vanishes is also achieved.

Mathematics Subject Classification (2000). Primary 35J25; Secondary 65N30.

Keywords. Elliptic equation, measure data, numerical schemes.

1. Introduction

The first result of existence and uniqueness of solutions for the Dirichlet problem for a linear elliptic equation (with possibly discontinuous coefficients and) with measure data is probably due to G. Stampacchia in his paper of 1965, see [1]. In this paper, G. Stampacchia use a duality method. A regularity result on a primal problem leads to an existence and uniqueness result on the dual problem. It is interesting to notice that the solution obtained by this method satisfies the equation with a stronger sense than the classical weak sense (such as (2.10) below) as it is shown by the counterexample given in Prignet [2], which is an adaptation of Serrin [3].

In the seventies, H. Brezis studied some semilinear elliptic equations such as:

$$-\Delta u + g(u) = \mu \quad \text{in } \Omega,$$
$$u = 0 \quad \text{on } \partial\Omega, \tag{1.1}$$

with a nondecreasing function $g \in C(\mathbb{R}, \mathbb{R})$. The case $\mu \in L^1(\Omega)$ is solved in the well-known papers of Brezis-Strauss [4], for the case where Ω is a bounded open subset of $\mathbb{R}^N$ with a smooth boundary, and of Bénilan-Brezis-Crandall [5] for the case $\Omega = \mathbb{R}^N$ (in this latter case, one assumes $g(0) = 0$ and the boundary condition "$u = 0$" has to be changed in a convenient condition). A well-known result of Bénilan-Brezis is devoted to the case of the Thomas-Fermi equation where μ is a measure on Ω, see [6] and the recent paper [7]. In fact, in the case of (1.1), the function g makes very different the cases "$\mu \in L^1(\Omega)$" and "μ measure on Ω". Indeed, if Ω is a bounded open subset of $\mathbb{R}^N$ with a smooth boundary and if $g \in C(\mathbb{R}, \mathbb{R})$ is such that $g(s)s \geq 0$ for all $s \in \mathbb{R}$, then, the problem (1.1) has a unique solution for all $\mu \in L^1(\Omega)$. But, the existence part of this result is not always true if μ is a measure on Ω. For instance, let $p \in]1, \infty[$, $g(s) = |s|^{p-1}s$ and μ be a measure on Ω. Then, (1.1) has a solution if and only if $\mu \in L^1(\Omega) + W^{-2,p}(\Omega)$. This latter condition is equivalent to say that $|\mu|(A) = 0$ for for every borelian subset of Ω whose $W^{2,p'}$-capacity is zero, see Gallouët-Morel [8] and Baras-Pierre [9].

Following the works of H. Brezis, the case of quasilinear equations with the classical Leray-Lions conditions may be studied:

$$-\operatorname{div}(a(\cdot, u, \nabla u)) = \mu \text{ in } \Omega, \qquad u = 0 \text{ on } \partial\Omega. \tag{1.2}$$

Here also, one obtains, for all measure μ on Ω, the existence of a solution to (1.2), see Boccardo-Gallouët [10] and [11].

In order to obtain these existence results (for (1.1) or (1.2)), a classical method is to consider approximate solutions obtained with a sequence of regular functions $(\mu_n)_{n \in \mathbb{N}}$, bounded in $L^1(\Omega)$ and $\star$-weakly converging to μ (with also some approximations of the function g in the case of (1.1)) and then to obtain some estimates on this sequence of approximate solutions and to pass to the limit as $n \to \infty$ (it is for this last step that some difference occurs between "L^1" and "measure" in the case of (1.1)).

In this paper, we will present a third method to obtain existence of solutions for elliptic problems with measure data. It consists to pass to the limit on the solution obtained with a discretization of the equation by a numerical scheme (such as a Finite Element scheme). This method has a double interest since it gives the existence of a solution for the problem considered and it gives a way to compute an approximation of this solution (especially if one has also a uniqueness result). In some cases, it is also possible to have some error estimates. This question of computation of the solution of an elliptic problem with measure data is crucial for some engineering problems. An example is given by the reservoir simulation in petroleum engineering. In this example, measure data have to be considered since the diameter of a well (about 10 cm) is very small with respect to a typical mesh size (about 100 m). It leads to source terms in the equations which are measures supported on some points (for some 2d models) or some lines (for 3d models), see Fabrie-Gallouët [12] for instance.

In Section 2, a model example is considered which is generalized in Section 3.

2. A model example

This section presents a result given in Gallouët-Herbin [13].

Let Ω be a polygonal open subset of $\mathbb{R}^2$ and $\mu \in M_b(\Omega)$, where $M_b(\Omega)$ denotes the set of bounded measures on Ω, that is the set of σ-additives applications from the borelian subsets of Ω to $\mathbb{R}$. An element $\mu \in M_b(\Omega)$ may be considered as an element of $(C(\overline{\Omega}))'$, setting $\mu(\varphi) = \int_\Omega \varphi d\mu$ if $\varphi \in C(\overline{\Omega})$. In the sequel, $C(\overline{\Omega})$ is endowed with its usual "sup-norm" and $\|\mu\|_{M_b}$ denotes the norm of μ in the dual space $(C(\overline{\Omega}))'$. One considers the Dirichlet problem with μ as datum:

$$
\begin{aligned}
-\Delta u &= \mu \quad \text{in } \Omega, \\
u &= 0 \quad \text{on } \partial\Omega.
\end{aligned}
\tag{2.1}
$$

In order to prove the existence of a (weak) solution to (2.1), the method developed in [10] considers a sequence $(\mu_n)_{n\in\mathbb{N}}$ of regular functions such that $\mu_n \to \mu$ for the $\star$-weak topology of $C(\overline{\Omega})'$ and the sequence $(u_n)_{n\in\mathbb{N}} \subset H_0^1(\Omega)$ of (weak) solutions of (2.1) with μ_n instead of μ, that is

$$
\begin{aligned}
-\Delta u_n &= \mu_n \quad \text{in } \Omega, \\
u_n &= 0 \quad \text{on } \partial\Omega.
\end{aligned}
\tag{2.2}
$$

The method developed in this paper is to consider a sequence of solutions of a numerical scheme as the mesh size goes to 0. Roughly speaking, it consists to "regularize the operator" (the discretized problem is a linear system in a finite-dimensional space) instead of "regularize the datum".

Let $\mathcal{M}$ be a Finite Element triangular mesh of Ω (see, e.g., Ciarlet [14]). One chooses the piecewise Finite Element approximation of (2.1). One sets $H = \{u \in C(\overline{\Omega}); u_{|K} \in P^1 \text{ for all } K \in \mathcal{M}\}$, where P^1 denotes the set of affine functions, and $H_0 = \{u \in H; u = 0 \text{ on } \partial\Omega\}$. The Finite Element approximation of (2.1) leads to the following problem:

$$
\begin{aligned}
&u_{\mathcal{M}} \in H_0, \\
&\int_\Omega \nabla u_{\mathcal{M}} \cdot \nabla v \, dx = \int_\Omega v \, d\mu, \quad \forall v \in H_0.
\end{aligned}
\tag{2.3}
$$

It is classical that (2.3) has a unique solution. The aim is to proves the convergence of $u_{\mathcal{M}}$ to some u, as the mesh size goes to zero, and that u is the unique solution of (2.1) in a convenient sense. The main difficulty is to obtain some estimates on $u_{\mathcal{M}}$.

In order to obtain these estimates, one recalls the way to obtain some estimates on the solution u_n of (2.2) (the method of [10]). Since $(\mu_n)_{n\in\mathbb{N}} \subset L^1(\Omega)$ and $\mu_n \to \mu$ for the $\star$-weak topology of $C(\overline{\Omega})'$, the sequence $(\mu_n)_{n\in\mathbb{N}}$ is bounded in $L^1(\Omega)$. Indeed, in order to simplify, one may assume that $\|\mu_n\|_{L^1} \le \|\mu\|_{M_b}$ for all n. Then, let $\theta > 1$ and define:

$$
\varphi(s) = \int_0^s \frac{1}{(1+|t|)^\theta} dt; \ s \in \mathbb{R}.
$$

Taking $\varphi(u_n)$ as test function in the weak formulation of (2.2) (note that $\varphi(u_n) \in H_0^1(\Omega)$) leads to:

$$\int_\Omega \frac{|\nabla u_n|^2}{(1+|u_n|)^\theta} dx \le C_\theta \|\mu\|_{M_b}, \tag{2.4}$$

where $C_\theta = \int_0^\infty \frac{1}{(1+|t|)^\theta} dt < \infty$ (and $|\cdot|$ denotes the Euclidean norm in $\mathbb{R}^d$, for any $d \ge 1$).

Using Hölder Inequality, Sobolev embedding and the fact that θ can be chosen arbitrarily close to 1, one deduces from (2.4) the existence, for all $q < 2$ (if Ω is a bounded open of $\mathbb{R}^d$, $d \ge 2$, the bound on q is $q < \frac{d}{d-1}$), of C_q, only depending on Ω, q and $\|\mu\|_{M_b}$ such that:

$$\int_\Omega |\nabla u_n|^q dx \le C_q.$$

A quite similar method can be used in order to obtain some estimates on the solution $u_\mathcal{M}$ of (2.3). The first difficulty is that $\varphi(u_\mathcal{M})$ does not belong to H_0, then it is not possible to take $v = \varphi(u_\mathcal{M})$ in (2.3). But, we can take for v the interpolate of $\varphi(u_\mathcal{M})$. Indeed, let $\mathcal{V}$ the set of vertices of $\mathcal{M}$ and ϕ_K the Finite Element basis function associated to $K \in \mathcal{V}$ (that is $\phi_K \in H$, $\phi_K(K) = 1$ and $\phi_K(L) = 0$ if $L \in \mathcal{V}$, $L \ne K$). One has, with $u_K = u_\mathcal{M}(K)$ for all $K \in \mathcal{V}$:

$$u_\mathcal{M} = \sum_{K \in \mathcal{V}} u_K \phi_K.$$

Taking $v = \sum_{K \in \mathcal{V}} \varphi(u_K)\phi_K$ in (2.3) leads to:

$$\sum_{(K,L) \in (\mathcal{V})^2} T_{K,L}(u_K - u_L)(\varphi(u_K) - \varphi(u_L)) \le C_\theta \|\mu\|_{M_b}, \tag{2.5}$$

where $T_{K,L} = -\int_\Omega \nabla\phi_K \cdot \nabla\phi_L dx$ and noting that $\sum_{L \in \mathcal{V}} T_{K,L} = 0$, for all $K \in \mathcal{V}$ since $\sum_{L \in \mathcal{V}} \phi_L(x) = 1$ for all $x \in \Omega$.

In order to deduce from (2.5) a $W_0^{1,q}$-estimate on $u_\mathcal{M}$ (for $1 \le q < 2$), an additional hypothesis is assumed. It is supposed that, the mesh $\mathcal{M}$ satisfies, for some positive ζ, the following Delaunay and non degeneracy conditions:

(i) For any interior edge of $\mathcal{M}$, the sum of the angles facing that edge is less or equal to $\pi - \zeta$,

(ii) For any edge lying on the boundary, the facing angle is less or equal to $\frac{\pi}{2} - \zeta$, $\qquad$ (2.6)

(iii) For any angle θ of any triangle T of the mesh $\mathcal{M}$, $\theta \ge \zeta$.

Under this hypothesis, it follows from (2.5) the existence, for all $q < 2$, of C_q, only depending on Ω, q, $\|\mu\|_{M_b}$ and ζ such that:

$$\|u_\mathcal{M}\|_{W_0^{1,q}(\Omega)} \le C_q. \tag{2.7}$$

A way to prove (2.7), using (2.5), can be done with similar results using Finite Volume schemes, see Gallouët-Herbin [15] or Droniou-Gallouët-Herbin [16]. Indeed,

$u_\mathcal{M} = \sum_{K \in \mathcal{V}} u_K \phi_K$ is solution of (2.3) if and only if the family $(u_K)_{K \in \mathcal{V}}$ is solution of

$$\sum_{L \in \mathcal{V}} T_{K,L}(u_K - u_L) = \int_\Omega \phi_K d\mu, \ \forall K \in \mathcal{V},$$
$$u_K = 0, \ \forall K \in \mathcal{V} \cap \partial\Omega. \tag{2.8}$$

The left-hand side of the first equation of (2.8) is the same than the left-hand side obtained with the classical Finite Volume scheme on the Voronoï mesh associated to the set $\mathcal{V}$. The control volume (of this Voronoï mesh) associated to $K \in \mathcal{V}$ is the set of points of Ω whose distance to K is less than its distance to any other element of $\mathcal{V}$. Thanks to Condition (2.6), the control volumes of the Voronoï mesh are also defined by the orthogonal bisectors of the edges of $\mathcal{M}$, see Figure 1. The fact that the schemes (Finite Element on $\mathcal{M}$ and Finite Volume on the Voronoï mesh associated to $\mathcal{V}$) differ only by the right-hand sides is due to the following computation for any $T \in \mathcal{M}$:

$$-\int_T \nabla\phi_K \cdot \nabla\phi_L dx = \frac{1}{2}\cotan(\theta_{K,L}),$$

where $\theta_{K,L}$ is the angle of T facing the edge with vertices K and L. Hence, for the edge of $\mathcal{M}$ whose vertices are K, L, denoted by $K|L$:

$$T_{K,L} = \frac{m_{K,L}}{d_{K,L}},$$

where $m_{K,L}$ denotes the distance between the points intersecting the orthogonal bisectors in each of the triangles with vertices K and L (except for the case $K \in \mathcal{V} \cap \partial\Omega$ and $L \in \mathcal{V} \cap \partial\Omega$ which has no importance), and $d_{K,L}$ denotes the distance between K and L.

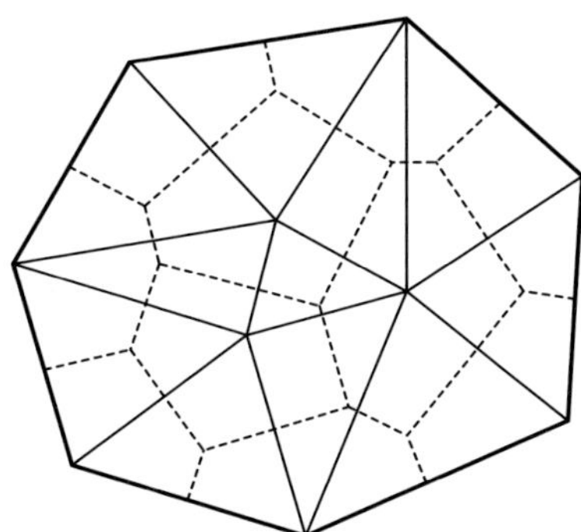

FIGURE 1. Continuous line: Finite Element mesh. Dashed line: Voronoï mesh associated to the vertices of the Finite Element mesh.

It is now possible possible to use the results of [15] (or [16]) which use the Hölder Inequality and a discrete version of the Sobolev embedding. It gives, for

$1 \leq q < 2$, the existence for $\overline{C}_q$, only depending on Ω, q and $\|\mu\|_{M_b}$ such that:

$$\sum_{K|L \in \{\text{edges of } \mathcal{M}\}} m_{K,L} d_{K,L} \left(\frac{u_K - u_L}{d_{K,L}} \right)^q \leq \overline{C}_q,$$

from which follows (2.7) for some C_q, only depending on Ω, q, $\|\mu\|_{M_b}$ and ζ.

Thanks to these $W_0^{1,q}$-estimates on $u_{\mathcal{M}}$, it is now possible to pass to the limit as size($\mathcal{M}$) goes to zero, where size($\mathcal{M}$) is the supremum of the diameters of the elements of $\mathcal{M}$.

Assuming $u_{\mathcal{M}} \to u$ for the weak topology of $W_0^{1,q}$, for all $1 \leq q < 2$, as size($\mathcal{M}$)$\to$ 0 (indeed, it is not possible, up to now, to assume such a convergence, one has to consider subsequences of sequences of meshes satisfying (2.6)), let $\psi \in C_c^\infty(\Omega)$ (a regular function with compact support). Taking $v = \psi_{\mathcal{M}} = \sum_{K \in \mathcal{V}} \psi(K)\phi_K$ in (2.3) (this is possible since $\psi_{\mathcal{M}} \in H_0$) gives:

$$\int_\Omega \nabla u_{\mathcal{M}} \cdot \nabla \psi_{\mathcal{M}} dx = \int_\Omega \psi_{\mathcal{M}} d\mu. \tag{2.9}$$

Since $\psi_{\mathcal{M}} \to \psi$, $\nabla \psi_{\mathcal{M}} \to \nabla \psi$ uniformly on Ω and $u_{\mathcal{M}} \to u$ for the weak topology of $W_0^{1,q}$, as size($\mathcal{M}$) $\to 0$, (2.9) gives that u satisfies:

$$\int_\Omega \nabla u \cdot \nabla \psi dx = \int_\Omega \psi d\mu.$$

Then, since $u \in W_0^{1,q}(\Omega)$ for all $1 \leq q < 2$ and since $W_0^{1,r}(\Omega) \subset C(\overline{\Omega})$ for all $r > 2$, a density argument gives that u is solution of:

$$\begin{aligned} &u \in \cap_{1 \leq q < 2} W_0^{1,q}(\Omega), \\ &\int_\Omega \nabla u \cdot \nabla \psi dx = \int_\Omega \psi d\mu, \ \forall \psi \in \cup_{r > 2} W_0^{1,r}(\Omega). \end{aligned} \tag{2.10}$$

The solution of (2.10) is unique (this is also true for a more general elliptic operator in dimension 2, but not for a general elliptic operator with discontinuous coefficients, in dimension $d \geq 3$, replacing 2 by $\frac{d}{d-1}$ and 2 by d in the two assertions of (2.10), a counterexample is in [2]).

Finally, thanks to this uniqueness result, it is proven that $u_{\mathcal{M}} \to u$ for the weak topology of $W_0^{1,q}$, for all $1 \leq q < 2$, as size($\mathcal{M}$) $\to 0$, $\mathcal{M}$ satisfying (2.6) (with a fixed $\zeta > 0$). This gives the following theorem:

Theorem 2.1. *Let Ω be a polygonal open subset of $\mathbb{R}^2$, $\mu \in M_b(\Omega)$ and $\zeta > 0$. For a Finite Element mesh $\mathcal{M}$ of Ω satisfying Condition (2.6), let $u_{\mathcal{M}}$ be the solution of (2.3). Then, $u_{\mathcal{M}} \to u$, unique solution of (2.10), for the weak topology of $W_0^{1,q}(\Omega)$, for all $1 \leq q < 2$, as size($\mathcal{M}$) $\to 0$.*

The convergence which is proven in Theorem 2.1 is only a weak convergence in $W_0^{1,q}(\Omega)$ for all $q < 2$. Then, it gives the (strong) convergence in $L^q(\Omega)$ for all $q < \infty$. It is perhaps also possible to prove a strong convergence in $W_0^{1,q}(\Omega)$ for any $q < 2$. In some cases, such that a Dirac measure for μ, it is possible to obtain some error estimates, see Scott [17].

The generalization of this proof of convergence for a Finite Element method in dimension $d = 3$ is not clear. It needs some additional work. In the following section, a generalization is given for a convection-diffusion operator, in dimension $d = 2$ or 3, using a Finite Volume method.

3. Convection-diffusion equations

This section presents a result given in Droniou-Gallouët-Herbin [16] (where more general problems are considered).

Let Ω be a polygonal (for $d = 2$) or polyhedral (for $d = 3$) open subset of $\mathbb{R}^d$ ($d = 2$ or 3). Let $v \in C(\overline{\Omega})^d$ and $\mu \in M_b(\Omega)$, the problem under consideration is:

$$-\Delta u + \mathrm{div}(vu) = \mu \quad \text{in } \Omega,$$
$$u = 0 \quad \text{on } \partial\Omega. \tag{3.1}$$

Such a problem is studied, for instance, in Droniou [18], where an existence and uniqueness result is given using the method of Stampacchia (see [1]), that is a regularity result and a duality argument. The objective, here, is to obtain an existence result, passing to the limit on numerical schemes (and this gives also the convergence of numerical schemes).

Remark 3.1. For some $v \in (C(\overline{\Omega}))^d$, the problem 3.1 appears to be associated to a noncoercive operator. Let $A : H_0^1(\Omega) \to H^{-1}(\Omega)$ be defined by $Au = -\Delta u + \mathrm{div}(vu)$ for $u \in H_0^1(\Omega)$. Then, it may exist some $u \in H_0^1(\Omega)$, $u \neq 0$, such that $\langle Au, u \rangle_{H^{-1}, H_0^1} = 0$, which leads to the noncoercivity of A.

A solution of (3.1) is a function u satisfying (using the fact that $W_0^{1,r}(\Omega) \subset C(\overline{\Omega})$ for $r > d$):

$$u \in \cap_{1 \leq q < \frac{d}{d-1}} W_0^{1,q}(\Omega),$$
$$\int_\Omega \nabla u \cdot \nabla \psi dx - \int vu \cdot \nabla \psi = \int_\Omega \psi d\mu, \ \forall \psi \in \cup_{r > d} W_0^{1,r}(\Omega). \tag{3.2}$$

The uniqueness of the solution of (3.2) is quite simple, using a regularity result on the dual problem to (3.2) (see [16] or [18]). In order to prove an existence result, a discretization of (3.1) by a Finite Volume scheme is used.

In [16] a large class of "admissibles" meshes of Ω is considered. Here, in order to simplify, one considers only some particular meshes. Let $\mathcal{T}$ be a mesh of Ω. One assumes that $\mathcal{T}$ is the Voronoï mesh associated to a family $\mathcal{V}$ of points of $\overline{\Omega}$ with the assumption that any point of $\partial\Omega$ belongs to a control volume (or its boundary) associated to an element of $\mathcal{V}$ which is also belonging to $\partial\Omega$ (this is always possible, adding to $\mathcal{V}$ some points on $\partial\Omega$ if necessary). In the sequel, a Voronoï mesh satisfying this property on the points of $\partial\Omega$ will be called a "genuine Voronoï mesh". An example is given in the preceding section. Indeed, the Voronoï mesh associated to the vertices of a Finite Element mesh $\mathcal{M}$ satisfying Condition (2.6) is a genuine Voronoï mesh, see Figure 2. The definition of a Voronoï mesh gives that the element of $\mathcal{T}$ are some open sets. In order to take into account the fact that the measure μ may charge some parts of the edges of $\mathcal{T}$, the elements of $\mathcal{T}$ are slightly modified such that $\mathcal{T}$ is now a (borelian) partition of Ω.

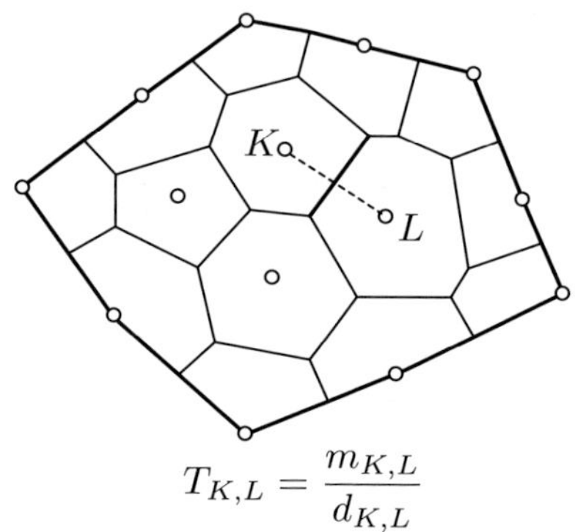

$$T_{K,L} = \frac{m_{K,L}}{d_{K,L}}$$

FIGURE 2. A genuine Voronoï mesh.

Let $K \in \mathcal{V}$. The control volume associated to K is denoted by V_K and $\mu_K = \mu(V_K)$. For $K \in \mathcal{V}$, the set of elements L of $\mathcal{V}$ such that V_K and V_L have a common edge is denoted by $\mathcal{N}_K$. If $L \in \mathcal{N}_K$, the common edge to V_K and V_L is denoted by $\sigma_{K,L}$ and its $(d-1)$-Lebesgue measure is denoted by $m_{K,L}$. The normal unit vector on $\sigma_{K,L}$, outward K, is denoted by $n_{K,L}$ (so that $n_{L,K} = -n_{K,L}$). Furthermore $d_{K,L}$ is the distance between K and L and:

$$T_{K,L} = \frac{m_{K,L}}{d_{K,L}}.$$

The discretization of (3.1) is performed with the classical Finite Volume scheme for the diffusion term and an upwind Finite Volume scheme for the convection term:

$$\sum_{L \in \mathcal{N}_L} T_{K,L}(u_K - u_L) + m_{K,L}v_{K,L}u_{K,L} = \mu_K, \ \forall K \in \mathcal{V} \cap \Omega,$$
$$u_K = 0, \ \forall K \in \mathcal{V} \cap \partial\Omega, \tag{3.3}$$

where $v_{K,L}$ is the mean value of $v \cdot n_{K,L}$ on $\sigma_{K,L}$ and $u_{K,L}$ is equal to u_K or u_L depending on the sign of $v_{K,L}$:

$$\begin{aligned} u_{K,L} &= u_K \quad \text{if } v_{K,L} > 0, \\ u_{K,L} &= u_L \quad \text{if } v_{K,L} < 0. \end{aligned} \tag{3.4}$$

The system (3.3)–(3.4) appears to be a linear system of N unknowns, namely $\{u_K, K \in \mathcal{V} \cap \Omega\}$, and N equations, where N is the number of elements of $\{K \in \mathcal{V} \cap \Omega\}$. Existence and uniqueness of the solution of this system is an easy consequence of the following property of positivity (interesting for its own sake), which is due to the upwind choice of $u_{K,L}$ (that is (3.4)):

$$\left. \begin{array}{l} \{u_K, K \in \mathcal{V}\} \text{ solution of } (3.3)\text{–}(3.4) \\ \mu_K \geq 0 \text{ for all } K \in \mathcal{V} \end{array} \right\} \Rightarrow u_K \geq 0 \text{ for all } K \in \mathcal{V}. \tag{3.5}$$

The proof of (3.5) is classical. If M is the matrix which determines the linear system (3.3)–(3.4), after an ordering of the unknowns, the property (3.5) is: $X \in \mathbb{R}^N$, $MX \geq 0 \Rightarrow X \geq 0$, which a consequence of the same property on $M^\star$, namely $X \in \mathbb{R}^N$, $M^\star X \geq 0 \Rightarrow X \geq 0$.

The solution $\{u_K,\ K \in \mathcal{V}\}$ of (3.3)–(3.4) gives an approximate solution of (3.1) $u_\mathcal{V}$ defined by:

$$u_\mathcal{V}(x) = u_K \text{ if } x \in V_K,\ K \in \mathcal{V}. \tag{3.6}$$

The proof that $u_\mathcal{V}$ converges to u, solution of (3.2), as the mesh size goes to 0, is now divided in four steps:

1. Estimates on $u_\mathcal{V}$ for a so-called discrete $W_0^{1,q}$-norm, for $1 \leq q < \frac{d}{d-1}$ (note that $u_\mathcal{V} \notin W_0^{1,q}(\Omega)$ except for some very particular cases !).
2. Relative compactness in $L^q(\Omega)$, for $1 \leq q < \frac{d}{d-2}$, of the family of approximate solutions.
3. Any possible limit of the approximate solutions as the mesh size goes to 0 is belonging to $W_0^{1,q}(\Omega)$ for $1 \leq q < \frac{d}{d-1}$.
4. Any possible limit of the approximate solutions as the mesh size goes to 0 is solution of (3.2).

With this four steps, the uniqueness of the solution of (3.2) gives that $u_\mathcal{V}$ converges to u, solution of (3.2), as the mesh size goes to 0, in $L^q(\Omega)$, for $1 \leq q < \frac{d}{d-2}$.

The main arguments of these four steps are now described.

Step 1. Estimates on $u_\mathcal{V}$. Using the method of Section 2, it is quite easy to obtain some estimates on $u_\mathcal{V}$ in the case where $\operatorname{div}(v) \geq 0$ (which gives some coercivity). But, it is not so easy without this assumption. Indeed, a first step is to control $\operatorname{meas}(\{u_\mathcal{V} \geq k\})$, as $k \to \infty$, uniformly with respect to $\mathcal{V}$. This is possible thanks to an estimate on $\ln(1 + |u_\mathcal{V}|)$. The way to obtain this estimate on $\ln(1 + |u_\mathcal{V}|)$ is described below in the continuous case that for the weak solution $u \in H_0^1(\Omega)$ of (3.1) when $\mu \in H^{-1}(\Omega) \cap L^1(\Omega)$.

Let $\varphi \in C^1(\mathbb{R}, \mathbb{R})$ be the function defined in Section 2 for $\theta = 2$, that is $\varphi(s) = \int_0^s \frac{1}{(1+|s|)^2}$ for $s \in \mathbb{R}$. Taking $\varphi(u)$ as test function in the weak formulation of (3.1) leads to:

$$\begin{aligned}
\int_\Omega \frac{|\nabla u|^2}{(1+|u|)^2}\,dx &\leq C_2 \|\mu\|_{M_b} + \int_\Omega \frac{|v||u||\nabla u|}{(1+|u|)^2}\,dx \\
&\leq C_2 \|\mu\|_{M_b} + \|v\|_\infty \int_\Omega \frac{|\nabla u|}{1+|u|}\,dx,
\end{aligned} \tag{3.7}$$

with $C_2 = \int_0^\infty \varphi(s)ds = 1$ and $\|v\|_\infty = \sup_{x \in \Omega} |v(x)| < \infty$.

Using Cauchy-Schwarz Inequality, Inequality (3.7) gives a bound on $\nabla \ln(1 + |u|)$ in $L^2(\Omega)$, only depending on v, $\|\mu\|_{M_b}$ and Ω. Then, since $\ln(1 + |u|) \in H_0^1(\Omega)$, Poincaré Inequality gives a bound on $\ln(1 + |u|)$ in $L^2(\Omega)$ only depending on v, $\|\mu\|_{M_b}$ and Ω.

A similar estimate holds for $u_\mathcal{V}$, solution of the discretized problem, namely (3.3)–(3.4) and (3.6). The bound on $\ln(1 + |u_\mathcal{V}|)$ in $L^2(\Omega)$ is also only depending on v, $\|\mu\|_{M_b}$ and Ω. The proof of this bound uses the same arguments, with some technical difficulties, and uses the upwind choice of $u_{K,L}$ in (3.4).

The bound on $\ln(1 + |u_\mathcal{V}|)$ in $L^2(\Omega)$ gives a bound on $\mathrm{meas}\{|u_\mathcal{V}| \geq k\})$, namely:

$$\mathrm{meas}\{|u_\mathcal{V}| \geq k\}) \leq \frac{C}{(\ln(1+k))^2}, \tag{3.8}$$

where C is only depending on v, $\|\mu\|_{M_b}$ and Ω. Using this bound, it is now possible to obtain estimates on the so-called discrete $W_0^{1,q}$-norm of $u_\mathcal{V}$ (recall that, generally, $u_\mathcal{V} \notin W_0^{1,q}(\Omega)$), for $1 \leq q < \frac{d}{d-1}$. This discrete $W_0^{1,q}$-norm is defined, for $u_\mathcal{V}$ satisfying (3.6) and such that $u_K = 0$ if $K \in \mathcal{V} \cap \partial\Omega$, by:

$$\|u_\mathcal{V}\|_{1,q,\mathcal{V}}^q = \sum_{(K,L);\ L \in \mathcal{N}_K} m_{K,L} d_{K,L} \left(\frac{u_K - u_L}{d_{K,L}} \right)^q.$$

A bound on $\|u_\mathcal{V}\|_{1,q,\mathcal{V}}$ is obtained, for $1 \leq q < \frac{d}{d-1}$, when $u_\mathcal{V}$ is solution of (3.3)–(3.4) and (3.6), using (3.8), the function φ of Section 2, with $\theta > 1$ (close to 1), and the functions T_k and S_k defined by $T_k(s) = \max(-k, \min(s, k))$, $S_k(s) = s - T_k(s)$, for $s \in \mathbb{R}$. It is also used, for proving this estimate on $\|u_\mathcal{V}\|_{1,q,\mathcal{V}}$, that, if $L \in \mathcal{N}_K$, the distance from K to $\sigma_{K,L}$ is equal to the distance from L to $\sigma_{K,L}$. The conclusion of this step is that, for $1 \leq q < \frac{d}{d-1}$, there exists C_q, only depending on v, $\|\mu\|_{M_b}$ and Ω, such that:

$$\|u_\mathcal{V}\|_{1,q,\mathcal{V}} \leq C_q. \tag{3.9}$$

Step 2. Relative compactness in $L^q(\Omega)$, for $1 \leq q < \frac{d}{d-2}$, of the family of approximate solutions. With the discrete $W_0^{1,q}$-norm and $q < d$, a discrete version of the Sobolev embedding holds. Here also, the fact that, if $L \in \mathcal{N}_K$, the distance from K to $\sigma_{K,L}$ is equal to the distance from L to $\sigma_{K,L}$ is used. There exists S_q, only depending on q, such that, if $u_\mathcal{V}$ is defined by (3.6) and $u_K = 0$ for $K \in \mathcal{V} \cap \partial\Omega$:

$$\|u_\mathcal{V}\|_{L^{q^\star}(\Omega)} \leq S_q \|u_\mathcal{V}\|_{1,q,\mathcal{V}}, \tag{3.10}$$

where $q^\star = \frac{qd}{d-q}$.

Then, if $u_\mathcal{V}$ is the solution of (3.3)–(3.4) and (3.6), Estimate (3.9) (where $1 \leq q < \frac{d}{d-1}$) leads, with (3.10), to an estimate on $u_\mathcal{V}$ in $L^r(\Omega)$ for $1 \leq r < \frac{d}{d-2}$. This estimate gives the relative weak-compactness in $L^r(\Omega)$ of the family of approximate solutions (that is the family of $u_\mathcal{V}$, solution of (3.3)–(3.4) and (3.6), as $\mathcal{V}$ describes all the possible sets of points of Ω leading to a genuine Voronoï mesh). In order to obtain the relative (strong-)compactness of the family of approximate solutions, an equivalent to the Rellich theorem, using the norm $\|\cdot\|_{1,q,\mathcal{V}}$ instead of the $W_0^{1,q}$-norm, is needed. This compactness theorem is, thanks to the Kolmogorov compactness theorem, a consequence of the following inequality, which holds for $q \leq 2$, $h \in \mathbb{R}^d$ and any $u_\mathcal{V}$ defined by (3.6) and such that $u_K = 0$ for $K \in \mathcal{V} \cap \partial\Omega$:

$$\int_{\mathbb{R}^d} |u_\mathcal{V}(x+h) - u_\mathcal{V}(x)|^q \leq |h|(|h| + C\,\mathrm{size}(\mathcal{V}))^{q-1} \|u_\mathcal{V}\|_{1,q,\mathcal{V}}, \tag{3.11}$$

where C is only depending on Ω, size($\mathcal{V}$) is the supremum of the diameters of the elements of the Voronoï mesh associated to $\mathcal{V}$, and $u_{\mathcal{V}}$ is defined outside Ω by setting $u_{\mathcal{V}}(x) = 0$ if $x \notin \Omega$.

Estimate (3.9) (where $1 \le q < \frac{d}{d-1}$) gives, with (3.10) and (3.11), the relative compactness of the family of approximate solutions in $L^q(\Omega)$ for $1 \le q < 2$, thanks to the Kolmogorov compactness theorem. Then, using the estimate on $u_{\mathcal{V}}$ in $L^r(\Omega)$ for $1 \le r < \frac{d}{d-2}$, the relative compactness of the family of approximate solutions is obtained in $L^q(\Omega)$ for $1 \le q < \frac{d}{d-2}$.

Step 3. Let $u_{\mathcal{V}}$ be the solution of (3.3)–(3.4) and (3.6). Assuming that $u_{\mathcal{V}}$ converges to some u in $L^q(\Omega)$, for all $1 \le q < \frac{d}{d-2}$, as size($\mathcal{V}$) $\to 0$, the fact that $u \in W_0^{1,q}(\Omega)$ for all $1 \le q < \frac{d}{d-1}$ is a consequence of Estimate (3.9) and (3.11). Indeed, for $h \in \mathbb{R}^d$, $h \ne 0$, (3.11) gives with (3.9) (recall that $u_{\mathcal{V}}$ is defined outside Ω by setting $u_{\mathcal{V}}(x) = 0$ if $x \notin \Omega$):

$$\int_{\mathbb{R}^d} \frac{|u_{\mathcal{V}}(x+h) - u_{\mathcal{V}}(x)|^q}{|h|^q} \le \frac{|h|(|h| + C\,\text{size}(\mathcal{V}))^{q-1}}{|h|^q} C_q,$$

which leads, for $1 \le q < \frac{d}{d-1}$, passing to the limit as size($\mathcal{V}$) $\to 0$:

$$\int_{\mathbb{R}^d} \frac{|u(x+h) - u(x)|^q}{|h|^q} \le C_q, \tag{3.12}$$

where, here also, u is defined outside Ω by setting $u(x) = 0$ if $x \notin \Omega$. Inequality (3.12) gives $\nabla u \in L^q(\mathbb{R}^d)$ and therefore, since $u = 0$ outside Ω, $u \in W_0^{1,q}(\Omega)$.

Step 4. The proof of this step is easier (at least for a regular v). Indeed, let $u_{\mathcal{V}}$ be the solution of (3.3)–(3.4) and (3.6). Assuming that $u_{\mathcal{V}}$ converges to some u in $L^q(\Omega)$, for all $1 \le q < \frac{d}{d-2}$, as size($\mathcal{V}$) $\to 0$, the preceding step gives that $u \in W_0^{1,q}(\Omega)$ for all $1 \le q < \frac{d}{d-1}$. Taking $\psi \in C_0^\infty(\Omega)$, (3.2) is proven, passing to the limit on the numerical scheme (3.3)–(3.4). Then, a density argument gives (3.2) for all $\psi \in \cup_{r>d} W_0^{1,r}(\Omega)$ and this concludes Step 4.

As usual, the steps 3 and 4 hold for "subsequences of sequences of approximate solutions" and it is the uniqueness of the solution of (3.2) which gives, finally, the convergence of all the family, that is the convergence of $u_{\mathcal{V}}$ to u, unique solution of (3.2), in $L^q(\Omega)$, for all $1 \le q < \frac{d}{d-2}$, as size($\mathcal{V}$) $\to 0$. Then, the conclusion of this proof is the following theorem:

Theorem 3.2. *Let Ω be a polygonal (for $d = 2$) or polyhedral (for $d = 3$) open subset of $\mathbb{R}^d$ ($d = 2$ or 3). Let $v \in C(\overline{\Omega})^d$ and $\mu \in M_b(\Omega)$. For a genuine Voronoï mesh associated to a set $\mathcal{V}$ of points of Ω, let $u_{\mathcal{V}}$ be the solution of (3.3)–(3.4) and (3.6). Then, $u_{\mathcal{V}}$ converges to u, unique solution of (3.2), in $L^q(\Omega)$, for all $1 \le q < \frac{d}{d-2}$, as size($\mathcal{V}$) $\to 0$.*

References

[1] G. Stampacchia, *Le problème de Dirichlet pour les équations elliptiques du second ordre à coefficients discontinus*. Ann. Inst. Fourier **15** (1965), 189–258.

[2] A. Prignet, *Remarks on existence and uniqueness of solutions of elliptic problems with right-hand side measures*. Rend. Mat. Appl. **15** (1995), 321–337.

[3] J. Serrin, *Pathological solutions of elliptic differential equations*. Ann. Scuola Norm. Pisa (1964), 385–387.

[4] H. Brezis and W. Strauss, *Semilinear elliptic equations in L^1*. J. Math. Soc. Japan **25** (1973), 565–590.

[5] P. Bénilan, H. Brezis and M. Crandall, *A semilinear elliptic equations in L^1*. Ann. Scuola Norm. Sup. Pisa **2** (1975), 523–555.

[6] H. Brezis, *Some variational problems of the Thomas-Fermi type*. In *Variational Inequalities* (Ed. Cottle, Gianessi-Lions) (Wiley, New York) (1980), 53–73.

[7] P. Bénilan and H. Brezis, *Nonlinear problems related to the Thomas-Fermi equation*. Journal of Evolution Equations **3** n°4 (2003), 673–770.

[8] T. Gallouët and J.-M. Morel, *Resolution of a semilinear equation in L^1*. Proceedings of the Royal Society of Edinburgh **96A** (1984), 275–288.

[9] P. Baras and M. Pierre, *Singularités éliminables pour des équations semi-linéaires*. Ann. Inst. Fourier **34** n°1 (1984), 185–206.

[10] L. Boccardo and T. Gallouët, *Nonlinear Elliptic and Parabolic Equations involving Measures Data*. J. of Functional Analysis **87** n°1 (1989), 149–169.

[11] L. Boccardo and T. Gallouët, *Nonlinear elliptic equations with right-hand side measures*. Comm. PDE **17** n°3 and 4 (1992), 641–655.

[12] P. Fabrie and T. Gallouët, *Modeling wells in porous media flows*. Mathematical Models and Methods in Applied Sciences **10** n°5 (2000), 673–709.

[13] T. Gallouët and R. Herbin, *Convergence of linear finite elements for diffusion equations with measure data*. C. R. Math. Acad. Sci. Mathématiques **338** issue 1 (2004), 81–84.

[14] P.G. Ciarlet, *Basic error estimates for elliptic problems*. In *Handbook of Numerical Analysis II* (North-Holland, Amsterdam) (1991), 17–352.

[15] T. Gallouët and R. Herbin, *Finite volume methods for diffusion problems and irregular data*. In *Finite volumes for complex applications, Problems and Perspectives, II* (Hermes) (1999), 155–162.

[16] J. Droniou, T. Gallouët and R. Herbin, *A finite volume scheme for noncoercive elliptic equation with measure data*. SIAM J. Numer. Anal. **41** n°6 (2003), 1997–2031.

[17] R. Scott, *Finite Element Convergence for Singular Data*. Numer. Math. **21** (1973), 317–327.

[18] J. Droniou, *Solving convection-diffusion equations with mixed, Neumann and Fourier boundary conditions and measures as data, by a duality method*. Adv. Differential Equations **5** n° 10–12 (2000), 1341–1396.

Thierry Gallouët
LATP, CMI
F-13453 Marseille cedex 13, France
e-mail: `gallouet@latp.univ-mrs.fr`

Progress in Nonlinear Differential Equations
and Their Applications, Vol. 63, 291–297
© 2005 Birkhäuser Verlag Basel/Switzerland

Brezis-Nirenberg Problem and Coron Problem for Polyharmonic Operators

Yuxin Ge

1. Introduction

Let $K \in \mathbb{N}$ and $\Omega \subset \mathbb{R}^N (N \geq 2K + 1)$ be a regular bounded domain in $\mathbb{R}^N$. We consider the semilinear polyharmonic problem

$$(-\Delta)^K u = |u|^{s-2} u + f(x, u) \qquad \text{in } \Omega \tag{1}$$

$$u > 0 \qquad \text{in } \Omega \tag{2}$$

$$u = (-\Delta)u = \cdots = (-\Delta)^{K-1} u = 0 \qquad \text{on } \partial\Omega \tag{3}$$

where

$$s := \frac{2N}{N - 2K},$$

denotes the critical Sobolev exponent and $f(x, u)$ is a lower-order perturbation of u^{s-1} in the sense that $\lim\limits_{u \to +\infty} \dfrac{f(x, u)}{u^{s-1}} = 0$ uniformly in $x \in \Omega$. The equation (1) is of variational type. Let

$$H_\theta^K(\Omega) = \left\{ v \in H^K(\Omega) \ \Big| \ (-\Delta)^i v = 0 \text{ on } \partial\Omega \quad \forall 0 \leq i < \left[\frac{K+1}{2}\right] \right\},$$

where $[\frac{K+1}{2}] = M + 1$ if $K = 2M + 1$ is odd and $[\frac{K+1}{2}] = M + 1$ when $K = 2M + 2$ is even. We endow the Hilbert space $H_\theta^K(\Omega)$ with the scalar product

$$(u, v)_\Omega = \begin{cases} \displaystyle\int_\Omega ((-\Delta)^M u)((-\Delta)^M v) & \text{if } K = 2M \\[2mm] \displaystyle\int_\Omega (\nabla(-\Delta)^M u)(\nabla(-\Delta)^M v) & \text{if } K = 2M + 1, \end{cases} \tag{4}$$

and denote by $\| \cdot \|_{K,2,\Omega}$ the corresponding norm. Thus solutions of (1) correspond to critical points of the energy functional

$$E(u) = \frac{1}{2} \|u\|_{K,2,\Omega}^2 - \frac{1}{s} \int_\Omega |u|^s - \int_\Omega F(x, u), \tag{5}$$

where $F(x, u) = \int_0^u f(x, t)dt$. Our motivation for the problem (1) to (3) comes from the fact that it resembles some variational problems in geometry and physics where lack of compactness occurs. For example, when $K = 1$, it arises from the famous Yamabe's problem and when $K = 2$, it is similar to a conformally covariant operator studied by Paneitz. For related problems, we infer [2], [4], [5], [15], [18] and the references therein.

When $K = 1$, Brezis and Nirenberg have studied the existence of positive solutions of (1) to (3). In particular, when $f(x, u) = \lambda u$, where $\lambda \in \mathbb{R}$ is a constant, they have discovered the following remarkable phenomenon: the qualitative behavior of the set of solutions of (1) to (3) is highly sensitive to N the dimension of the space. To state their result precisely, let us denote by $\lambda_1 > 0$ the first eigenvalue of $-\Delta$ in Ω. When $K = 1$, Brezis and Nirenberg have shown that, in dimension $N \geq 4$, there exists a positive solution of (1) to (3), if and only if $\lambda \in (0, \lambda_1)$; while, in dimension $N = 3$ and when $\Omega = B_1$ is the unit ball, there exists a positive solution of (1) to (3), if and only if $\lambda \in (\frac{\lambda_1}{4}, \lambda_1)$. Since the embedding $H_0^1(\Omega) \hookrightarrow L^6(\Omega)$ is not compact, the functional E does not satisfy the (P.-S.) condition. But it satisfies the (P.-S.) condition at certain energy levels small than $\frac{1}{N} S^{N/2}(\mathbb{R}^N)$, where $S(\mathbb{R}^N)$ is the best Sobolev constant for the embedding $H_0^1(\Omega) \hookrightarrow L^{2N/(N-2)}(\Omega)$. The energy of critical points found by Brezis and Nirenberg is essentially small than $\frac{1}{N} S^{N/2}(\mathbb{R}^N)$. Later on, many authors have considered the general polyharmonic problem (1) with $K \geq 1$, under the boundary conditions (3) or with homogenous Dirichlet boundary conditions given by

$$D^k u = 0 \qquad \text{on } \partial\Omega, \qquad \text{for } k = 0, \ldots, K - 1. \tag{6}$$

Here the $D^k u$ denotes any derivative of order k of the function u. The energy of solutions found by them is under certain energy level on which the (P.-S.) condition satisfies, see, e.g., [7], [12], [16].

On the other hand, using a Pohozaev identity, it is well known that if Ω is star sharped, there is no solution of the problem (1) to (3) (see [4]) when $K = 1$ and $f \equiv 0$. In this case, the concentration phenomenon occurs when we minimize the energy functional E on the manifold $\{u \in H_0^1(\Omega) \quad | \int_\Omega |u|^{2N/(N-2)} = 1\}$. This fact permits Coron in [6] to find a critical point for a perforated domain with the small holes in the higher energy level. Very recently, Coron's strategy is exploited again by several people for polyharmonic problem, see, e.g., [1], [3], [8].

In this paper, we will study the existence of positive solutions for the polyharmonic problem (1) to (3) (see also [11]). As in [4], we will fill out the sufficient conditions to find positive solutions for general domains. In the second part, when Brezis and Nirenberg's strategy does not work, we will see the concentration phenomenon occurs. So this fact leads us to search for positive solutions in the higher energy level by Coron's strategy. As a consequence, we will show the problem (1) to (3) admits always a non trivial solution for perforated domains with the small holes. For simplicity, we summarize our main result on the following simple example.

Theorem 1. *Assume $f(x,u) = \mu u^q$ for some $\mu > 0$ and some $q \in (1, s-1)$. Let Ω be a bounded annular domain satisfying*

(A) $\exists\ \epsilon_1 \in (0,1)$ *and* $\epsilon_2 > 0$ *s.t.* $A(\epsilon_1, 1) = \{x \in \mathbb{R}^N \mid \epsilon_1 < |x| < 1\} \subset \Omega$ *and*
$\quad\ B(0, \epsilon_2) = \{x \in \mathbb{R}^N \mid |x| < \epsilon_2\} \subset \Omega^c.$

Then, there exists $\eta > 0$ such that if $\epsilon_1 < \eta$, the problem (1) to (3) admits a non trivial solution in Ω.

2. Existence of positive solutions for general domains

In this section, we will search for positive solutions for the problem (1) to (3) for general domains. Our analysis is an adaptation of Brezis and Nirenberg's paper [4].

We assume that

(H1) $f(x,u) : \Omega \times [0, +\infty) \to [0, +\infty)$ is measurable in x, continuous in u and
$\quad\ $ that $\displaystyle\sup_{x \in \Omega, 0 \leq u \leq M} |f(x,u)| < \infty$ for every $M > 0$.

Moreover, we assume that $f(x,u)$ can be written as
(H2) $f(x,u) = a(x)u + g(x,u)$

with

(H3) $a(x) \in L^\infty(\Omega)$;

(H4) $g(x,u) = o(u)$ as $u \to 0^+$ uniformly in x;

(H5) $g(x,u) = o(u^{s-1})$ as $u \to +\infty$ uniformly in x.

Furthermore, we suppose that the operator $(-\triangle)^K - a(x)$ has its least eigenvalue positive in $H_\theta^K(\Omega)$, that is, $\exists \alpha > 0$ such that

(H6) $\|u\|_{K,2,\Omega}^2 - \displaystyle\int_\Omega a(x)u^2 \geq \alpha\|u\|_{K,2,\Omega}^2, \qquad \forall u \in H_\theta^K(\Omega).$

(H7) $\frac{\partial f}{\partial u}(x,u)$ is continuous on $\Omega \times \mathbb{R}^+$;

(H8) $|\frac{\partial f}{\partial u}(x,u)| \leq Cu^{s-2}, \forall u > 0$ uniformly in $x \in \Omega$;

(H9) $f_1(x,u) = \frac{f(x,u)}{u}$ is non decreasing in $u > 0$ for a.e. $x \in \Omega$.

(H10) $\frac{\partial^2 f}{\partial u^2}(x,u)$ is continuous on $\Omega \times \mathbb{R}^+$;

(H11) $|\frac{\partial^2}{\partial u^2}(f(x,u)u)| \leq Cu^{s-2}, \forall u > 0$ uniformly in $x \in \Omega$.

From (H1) to (H5), it follows that

$$f(x,0) = 0 \quad \forall x \in \Omega \quad \text{and} \quad \lim_{u \to +\infty} \frac{f(x,u)}{u^{s-1}} = 0 \quad \text{uniformly in } x.$$

Hence, f is a lower-order perturbation of u^{s-1}. As we look for positives solutions, we define $f(x,u) = 0, \forall x \in \Omega, \forall u \leq 0$. Set

$$F(x,u) = \int_0^u f(x,t)dt \qquad \forall x \in \Omega \text{ and } u \in \mathbb{R}.$$

We consider the following energy functional

$$E_1(u) = \frac{1}{2}\|u\|_{K,2,\Omega}^2 - \frac{1}{s}\int_\Omega (u^+)^s - \int_\Omega F(x,u), \qquad \forall u \in H_\theta^K(\Omega), \qquad (7)$$

where $u^+ = \max(|u|,0)$ designates the positive part of u. Clearly, E_1 is a C^1 functional on the Hilbert space $H_\theta^K(\Omega)$. Moreover, it follows from the Maximum principle that critical points of E_1 satisfy the equations (1)–(3). We define the best constant for the embedding $H_\theta^K(\Omega) \hookrightarrow L^s(\Omega)$

$$S_{K,\theta}(\Omega) := \inf_{v \in H_\theta^K(\Omega)\setminus\{0\}} \frac{\|v\|_{K,2,\Omega}^2}{\|v\|_{L^s(\Omega)}^2}$$

and Minimax value

$$\kappa_1 := \inf_{v \in H_\theta^K(\Omega)\setminus\{0\}} \sup_{t \geq 0} E_1(tv). \qquad (8)$$

In [10], we prove that $S_{K,\theta}(\Omega)$ is independent of Ω and $S_{K,\theta}(\Omega) = S_K(\mathbb{R}^N) := \inf_{v \in H^K(\mathbb{R}^N)\setminus\{0\}} \frac{\|v\|_{K,2,\mathbb{R}^N}^2}{\|v\|_{L^s(\mathbb{R}^N)}^2}$. Using Brezis and Nirenberg's strategy, we prove the following result.

Theorem 2. *Under Assumptions* (H1) *to* (H6), *we have*

$$\kappa_1 \leq \frac{K}{N}(S_K(\mathbb{R}^N))^{\frac{N}{2K}}. \qquad (9)$$

In addition, suppose

$$\kappa_1 < \frac{K}{N}(S_K(\mathbb{R}^N))^{\frac{N}{2K}}. \qquad (10)$$

Then, the problem (1) *to* (3) *admits a non trivial solution.*

    As in [4], a direct calculation shows (10) holds under one of the following hypotheses

  (i)   when $N > 4K, \exists \alpha, \beta, \mu \in (0, +\infty)$ s.t.
        $f(x,u) \geq \mu,$ for a.e. $x \in \Omega_0$ and $\forall u \in (\alpha, \beta)$
  (ii)   when $N = 4K, \exists \mu, A \in (0, +\infty)$ s.t.
        either $f(x,u) \geq \mu u,$ for a.e. $x \in \Omega_0$ and $\forall u \in [0, A]$
        or $f(x,u) \geq \mu u,$ for a.e. $x \in \Omega_0$ and $\forall u \in [A, +\infty)$
  (iii)   when $2K < N < 4K,$ $\displaystyle\lim_{u \to +\infty} \frac{f(x,u)u}{u^{\frac{4K}{N-2K}}} = +\infty,$ uniformly in $x \in \Omega_0,$

$$(11)$$

where Ω_0 is some non empty open subset of Ω.

3. Existence of positive solutions for some perforated domains

We define $\mathcal{M} = \{v \in H_\theta^K(\Omega) \setminus \{0\} \mid \|v\|_{K,2,\Omega}^2 = \|v^+\|_{L^s(\Omega)}^s + \int_\Omega f(x, v^+)v\}$. It is clear that $\mathcal{M}$ is a complete C^1 (resp. $C^{1,1}$) Finsler manifold under Assumptions

(H1) to (H9) (resp. (H1) to (H11)). We can show the mini-max value κ_1 is just the minimum of E_1 on $\mathcal{M}$. Thanks to Theorem 2, if $\kappa_1 < \frac{K}{N}(S_K(\mathbb{R}^N))^{\frac{N}{2K}}$, we can find a non trivial solution to the problem (1) to (3). In this section, we will study the remainder cases. When $\kappa_1 = \frac{K}{N}(S_K(\mathbb{R}^N))^{\frac{N}{2K}}$, two cases are possible:

(i) either there exists some $u \in \mathcal{M}$ satisfying

$$E_1(u) = \kappa_1, \tag{12}$$

(ii) or

$$E_1(v) > \kappa_1, \qquad \forall v \in \mathcal{M}. \tag{13}$$

In the first matter, u is a solution for the problem (1) to (3); in the latter one, the concentration phenomenon occurs. More precisely, we have the following result.

Theorem 3. *Suppose the Assumptions (H1) to (H9) are satisfied. Moreover, assume $\kappa_1 = \frac{K}{N}(S_K(\mathbb{R}^N))^{\frac{N}{2K}}$ and (13). Let $(u_n) \subset \mathcal{M}$ be a minimizing sequence for E_1, that is,*

$$\lim_{n \to \infty} E_1(u_n) = \kappa_1. \tag{14}$$

Then there exists $x_0 \in \overline{\Omega}$ such that

$$\mu_n := \zeta_\Omega \, F_K(u_n) \, dx \rightharpoonup S_K(\mathbb{R}^N)\delta_{x_0} \text{ weakly in } \mathcal{M}(\mathbb{R}^N)$$

and

$$\nu_n := \zeta_\Omega \, |u_n|^s \, dx \rightharpoonup S_K(\mathbb{R}^N)\delta_{x_0} \text{ weakly in } \mathcal{M}(\mathbb{R}^N),$$

where $\mathcal{M}(\mathbb{R}^N)$ denotes the space of non-negative Radon measures on $\mathbb{R}^N$ with finite mass, δ_{x_0} denotes Dirac measure concentrated at x_0 with mass equal to 1, ζ_Ω designates the indicatrix function of the set Ω and

$$F_K(v) := \begin{cases} ((-\Delta)^M v)^2 & \text{if } K = 2M \text{ is even} \\[2mm] |\nabla(-\Delta)^M v|^2 & \text{if } K = 2M + 1 \text{ is odd} . \end{cases}$$

We see the case in Theorem 3 occurs when $f(x, u) = \mu u^q$ for some small $\mu > 0$ and $q \in [1, \frac{6K-N}{N-2K}]$ (see [11]). On the other hand, if the assumptions in Theorem 3 are verified, we imply the level sets of E_1 on $\mathcal{M}$ near the minimum have non trivial topology provided Ω has non trivial topology. This fact permits us to apply Coron's strategy to search for the critical points for the problem (1) to (3) in the higher level sets for some perforated domains with small holes. For this aim, we show first a compactness citeron from Pohozaev identity: any $(P.S.)_\beta$ sequence $(u_n) \subset \mathcal{M}$ for $\beta \in (\frac{K}{N}(S_K(\mathbb{R}^N))^{\frac{N}{2K}}, \frac{2K}{N}(S_K(\mathbb{R}^N))^{\frac{N}{2K}})$ is precompact. Therefore, we can establish our main result in this section.

Theorem 4. *Let Ω be a bounded domain satisfying (A). Assume (H1) to (H11) hold. Then, there exists $\eta > 0$ such that if $\epsilon_1 < \eta$, the problem (1) to (3) admits a non trivial solution in Ω.*

References

[1] M.O. Ahmedou and F. Ebobisse, *On a nonlinear fourth-order elliptic equation involving the critical Sobolev exponent*, Nonlinear Anal., Theory Methods Appl. **52A**, 1535–1552 (2003).

[2] T. Aubin, *Equations différentielles non linéaires et problème de Yamabe concernant la courbure scalaire*, J. Math. pur. appl. 55, 269–296 (1976).

[3] T. Bartsch, T. Weth and M. Willem, *A Sobolev inequality with remainder term and critical equations on domains with topology for the polyharmonic operator*, Calc. Var. Partial Differ. Equ. **18**, 253–268 (2003).

[4] H. Brezis and L. Nirenberg, *Positive solutions of nonlinear elliptic equations involving critical Sobolev exponents*, Comm. Pure Appl. Math. 36, 437–477 (1983).

[5] S.-Y.A. Chang and P.C. Yang, *Extremal metrics of zeta function determinants on 4-manifolds*, Ann. Math. (2) 142, No.1, 171–212 (1995).

[6] J.M. Coron, *Topologie et cas limite des injections de Sobolev*, C. R. Acad. Sc. Paris, **299**, Ser. I (1984) 209–212.

[7] F. Gazzola, *Critical growth problems for polyharmonic operators*, Proc. R. Soc. Edinb., Sect. A, Math. 128, 251–263 (1998).

[8] F. Gazzola, H.C. Grunau and M. Squassina, *Existence and nonexistence results for critical growth biharmonic elliptic equations*, Calc. Var. Partial Differ. Equ. **18**, 117–143 (2003).

[9] Y. Ge, *Estimations of the best constant involving the L^2 norm in Wente's inequality and compact H-surfaces in Euclidean space*, Control, Optimisation and Calculus of Variations, Vol. **3**, (1998) 263–300.

[10] Y. Ge, *Sharp Sobolev inequalities in critical dimensions*, Mich. Math. J. **51**, 27–45 (2003).

[11] Y. Ge, *Positive solutions in semilinear critical problems for polyharmonic operators*, J. Math. Pures Appl. **84**, 199–245 (2005).

[12] H.C. Grunau, *Positive solutions to semilinear polyharmonic Dirichlet problems involving critical Sobolev exponents*, Calc. Var. and PDE 3, 243–252 (1995).

[13] P.L. Lions, *The concentration-compactness principle in the calculus of variations: The limit case. Part I and Part II*, Rev. Mat. Ibero. **1**(1) 145–201 (1985) and **1**(2) 45–121 (1985).

[14] R.S. Palais, *Lusternik-Schnirelman theory on Banach manifolds*, Topology 5, 115–132 (1966).

[15] S. Paneitz, *A quadratic conformally covariant differential operator for arbitrary pseudo-Riemannian manifolds*, preprint, (1983).

[16] P. Pucci and J. Serrin, *Critical exponents and critical dimensions for polyharmonic operators*, J. Math. Pures Appl. 69, 55–83, (1990).

[17] M. Struwe, *Variational Methods*, Springer, Berlin – Heidelberg – New York – Tokyo (1990).

[18] J. Wei and X. Xu, *Classification of solutions of higher-order conformally invariant equations*, Math. Ann. 313, No. 2, 207–228 (1999).

Yuxin Ge
Département de Mathématiques
Faculté de Sciences et Technologie
Université Paris XII – Val de Marne
61 avenue du Général de Gaulle
F-94010 Créteil Cedex, France
e-mail: `ge@univ-paris12.fr`

Progress in Nonlinear Differential Equations
and Their Applications, Vol. 63, 299–307
© 2005 Birkhäuser Verlag Basel/Switzerland

Local and Global Properties of Solutions of a Nonlinear Boundary Layer Equation

Mohammed Guedda

Abstract. We give a short survey of some results concerning solutions of the equation $f''' + \frac{1+\alpha}{2} f f'' - \alpha f'^2 = 0$, where $-\frac{1}{3} < \alpha < 0$. This equation arises in modeling the free convection, along a vertical flat plate embedded in a porous medium.

The analysis deals with existence, non-uniqueness and large t behavior of solutions to the above equation under certain conditions. We also consider the case where the solutions are singular and give the asymptotic behavior at the singular point, for $-1 \le \alpha < 0$.

1. Introduction

In this talk we are concerned with some results for solutions of the autonomous third order nonlinear differential equation

$$f''' + \frac{\alpha + 1}{2} f f'' - \alpha f'^2 = 0 \quad \text{on } (0, T), \tag{1.1}$$

where $0 < T \le \infty$ and $\alpha < 0$.

Equation (1.1) appears in the study of similarity solutions to problems of boundary-layer theory in some contexts of fluid mechanics [3], [5], [10], [11], [15], [17], [19].

Such equation with the boundary conditions

$$f(0) = 0, \quad f'(0) = 1, \quad \lim_{t \to +\infty} f'(t) = 0, \tag{1.2}$$

arises in the study of the free convection, along a vertical flat plate embedded in a porous medium. Here, the plate is impermeable and its temperature is assumed to be a power function with exponent equal to α:

$$T(x, y)_{|y=0} = T_\infty + Ax^\alpha, \tag{1.3}$$

where $A > 0$ and α are prescribed constants and T_∞ is the temperature far from the plate (see [11]). Coordinates (x, y) are measured along the plate and normal to it, with the origin at the leading edge. The x-axis being parallel to the direction of gravity but directed upwards.

The system

$$\frac{\partial u}{\partial x} + \frac{\partial v}{\partial y} = 0,\, u\frac{\partial T}{\partial x} + v\frac{\partial T}{\partial y} = \lambda\left(\frac{\partial^2 T}{\partial x^2} + \frac{\partial^2 T}{\partial y^2}\right),$$

where u and v are the velocity components, describes the 2D stationary heat convection. In porous media, u and v obey Darcy's law:

$$u = -k\mu^{-1}\left(\frac{\partial p}{\partial x} + \rho g\right), \quad v = -k\mu^{-1}\frac{\partial p}{\partial y},$$

with $\rho = \rho_\infty(1 - \beta T + \beta T_\infty)$. Here ρ is the $T-$dependent density, $\mu, \beta, k, \lambda, g$ are constants (viscosity, thermal expansion coefficient, permeability, thermal diffusivity, gravitational acceleration), p is the pressure and ρ_∞ is the value of ρ far from the plate. We suppose $u = 0$ for large y.

Introducing the stream function ψ by $u = \dfrac{\partial \psi}{\partial y}, v = -\dfrac{\partial \psi}{\partial x}$, using the boundary layer approximation $(\dfrac{\partial^2 T}{\partial x^2} = \dfrac{\partial^2 \psi}{\partial x^2} = 0)$, we obtain the system

$$\begin{cases} \dfrac{\partial^2 \psi}{\partial y^2} &= b^2\lambda A^{-1}\dfrac{\partial T}{\partial y}, \\ \lambda\dfrac{\partial^2 T}{\partial y^2} &= \dfrac{\partial T}{\partial x}\dfrac{\partial \psi}{\partial y} - \dfrac{\partial T}{\partial y}\dfrac{\partial \psi}{\partial x}, \end{cases} \tag{1.4}$$

with $\dfrac{\partial\psi(x,0)}{\partial x} = \dfrac{\partial\psi(x,\infty)}{\partial y} = 0$, where $b^2 = \rho_\infty\beta gk\mu^{-1}\lambda^{-1}A$.

We are looking for similarity solutions of (1.4) in standard form

$$\psi(x,y) = \lambda bx^{\frac{\alpha+1}{2}} f(t), \quad T = Ax^\alpha\theta(t) + T_\infty,$$

where $t = by/x^{\frac{1-\alpha}{2}}$ denotes the similarity variable. It can be checked that the shape functions f and θ satisfy the ODE system

$$\begin{cases} f'' &= \theta', \\ \theta'' &= \alpha\theta f' - m\theta' f, \end{cases} \tag{1.5}$$

where the primes denote differentiation with respect to t. From the boundary conditions, and (1.5), we have $f' = \theta$ and we obtain (1.1)–(1.2). Equation (1.1), with suitable boundary conditions, also arises in industrial manufacturing processes [3], in the excitation of liquid metals when placed in a high-frequency magnetic field [19] and in the context of boundary layer flow on permeable stretching surfaces with mass transfer parameter $a \neq 0$ [10], [17]. In the last situation initial conditions (1.2) take the form

$$f(0) = a, \quad f'(0) = 1. \tag{1.6}$$

The real a is also referred to as the suction/injection parameter. The case $a > 0$ corresponds to suction and $a < 0$ to injection of the fluid.

Note that in the case where $\alpha = 0$, equation (1.1) becomes

$$f''' + \frac{1}{2}ff'' = 0, \tag{1.7}$$

which is called the Blasius equation [6]. This equation with the boundary condition

$$f(0) = 0, \quad f'(0) = 0, \quad \lim_{t \to +\infty} f'(t) = 1, \tag{1.8}$$

was first solved numerically by Blasius. In [21] Weyl established the existence and uniqueness of solution to (1.7), (1.8) using functional analytical methods. In the same vein Callegary and Frieddman [8] Callegary and Nachman [9] proved the existence and uniqueness of solution to (1.7), (1.8) with the condition $f'(0) = \lambda, \lambda \geq 0$ instead of $f'(0) = 0$. In the case where $\lambda < 0$, it is proved by Hussaini, Lakin and Nachman [14] that the problem has a solution only for λ larger than a critical value λ_c.

Results concerning Problem (1.1), (1.2) can be found in [11] in which the numerical solution has been performed in the case where $-\frac{1}{3} < \alpha < 0$. For the case $\alpha > -\frac{1}{2}$ numerical investigations are in [2] and [15]. The mathematical analysis is also considered in [2]. The authors showed the non existence of solutions to (1.1), (1.2), where $\alpha < -\frac{1}{2}$, satisfying

$$\lim_{t \to \infty} f' f^2(t) = 0. \tag{1.9}$$

Recently Belhachmi, Brighi and Taous [5] showed the non existence of solutions to (1.1),(1.2) for $\alpha \leq -\frac{1}{2}$ without condition (1.9). Among other results they proved that this problem has an infinite number of solutions when $\alpha = -\frac{1}{3}$ whereas uniqueness holds for $0 \leq \alpha \leq \frac{1}{3}$.

We have two main goals in this paper. First, we investigate, in Section 2, the existence, non-uniqueness and large t behavior of solutions of (1.1), (1.2) for $-\frac{1}{3} \leq \alpha \leq 0$. Secondly, we address, in Section 3, the non-existence of global solutions and give the behavior of the blowing-up solutions for $-1 \leq \alpha \leq 0$.

2. Multiple solutions and large t behavior

Let $-\frac{1}{3} < \alpha < 0$ and $a > 0$. We consider the initial value problem,

$$\begin{cases} f''' + \dfrac{\alpha + 1}{2} f f'' - \alpha f'^2 = 0, \\ f(0) = a, \ f'(0) = 1, \ f''(0) = \gamma. \end{cases} \tag{2.1}$$

The real γ is regarded as the shooting parameter. For every $\gamma \in \mathbb{R}$ Problem (2.1) has a unique local solution f_γ defined on $(0, T_\gamma), T_\gamma \leq +\infty$. This solution is of class C^3 on $[0, T_\gamma)$, in fact $f_\gamma \in C^\infty$ and satisfies

$$f_\gamma''(t) + \frac{1+\alpha}{2} f_\gamma'(t) f_\gamma(t) = \gamma + \frac{1+\alpha}{2} a + \frac{3\alpha + 1}{2} \int_0^t f_\gamma'^{\,2}(s)\,ds, \tag{2.2}$$

and

$$(f_\gamma'' e^{mF})' = \alpha e^{mF} f_\gamma'^{\,2}, \quad \forall\, t < T_\gamma, \tag{2.3}$$

where $F(t) = \int_0^t f_\gamma(s)ds$. Property (2.2) indicates that f_γ cannot have a local maximum, for $\gamma > -\frac{1+\alpha}{2}a$. Let us note that if $T_\gamma < +\infty$, then $\lim_{t\uparrow T_\gamma} |f_\gamma(t)| + |f'_\gamma(t)| + |f''_\gamma(t)| = +\infty$. In fact the existence time T_γ is characterized by

Lemma 2.1. *If $T_\gamma < +\infty$, then $\lim_{t\uparrow T_\gamma} |f_\gamma(t)| = +\infty$.*

The proof is similar as in [12]. The first theorem we prove is the following.

Theorem 2.1. *Let $a > 0$ and $-\frac{1}{3} < \alpha < 0$. For any $\gamma > -\frac{1+\alpha}{2}a$ f_γ is global and goes to infinity with t. Moreover*

$$\lim_{t\to\infty} f'_\gamma(t) = \lim_{t\to\infty} f''_\gamma(t) = 0.$$

Remark 2.1. Since γ is arbitrary we deduce that Problem (1.1), (1.2) has an infinite number of solutions. These solutions are unbounded. This gives an answer to the open questions of [5].

To establish Theorem 2.1 we use Lemmas 2.2 and 2.3 below.

Lemma 2.2. *$f'_\gamma > 0, f_\gamma > 0$ on $(0, T_\gamma)$ and $T_\gamma = +\infty$; that is f_γ is global. Moreover f'_γ and f''_γ are bounded.*

Proof. It is not difficult to see that $f'_\gamma > 0, f_\gamma > 0$ on $(0, T_\gamma)$. To demonstrate that $T_\gamma = +\infty$ we consider a Lyapunov function for f_γ

$$E(t) = \frac{1}{2}(f''_\gamma(t))^2 - \frac{\alpha}{3}(f'_\gamma(t))^3,$$

which satisfies

$$E'(t) = -\frac{1+\alpha}{2}f_\gamma(f''_\gamma)^2 \leq 0,$$

thanks to $(2.1)_1$. Therefore E is bounded and then f''_γ and f'_γ are bounded, since $\alpha < 0$. This in turn implies that if $T_\gamma < \infty$ the function f_γ is bounded which is absurd. $\square$

Lemma 2.3. *$f_\gamma(t)$ tends to infinity with t, $f'_\gamma(t)$ and $f''_\gamma(t)$ tend to 0 as $t \to \infty$.*

Proof. From (2.3) it follows that f'_γ is monotone on $(t_1, +\infty)$, t_1 large enough. Since f'_γ is bounded there exists $l \in \mathbb{R}_+$ such that $\lim_{t\to+\infty} f'_\gamma(t) = l$. This implies in particular the existence of a sequence (t_n) tending to $+\infty$ with n such that $\lim_{n\to+\infty} f''_\gamma(t_n) = 0$ and then $\lim_{t\to+\infty} f''_\gamma(t) = 0$, by using the function E.

Next we suppose that f_γ is bounded, therefore $l = 0$. Subsequently

$$0 = \gamma + \frac{1+\alpha}{2}a\epsilon + \frac{3\alpha+1}{2}\int_0^{+\infty} f'_\gamma(t)^2 dt.$$

This is impossible if $\alpha > -\frac{1}{3}$. Therefore f_γ is unbounded and then $\lim_{t\to+\infty} f_\gamma(t) = +\infty$. It remains to prove that $l = 0$. Suppose that $l > 0$. Together with (2.2) we get, as t approaches infinity

$$f''_\gamma(t) = -\tfrac{1+\alpha}{2}l^2 t + \tfrac{3\alpha+1}{2}l^2 t + o(t), \qquad f''_\gamma(t) = \alpha l^2 t + o(t).$$

This is only possible if $\alpha = 0$. Then $l = 0$. $\square$

The following theorem shows that Problem (1.1), (1.2) has a solution for any $\alpha \in (-\frac{1}{2}, -\frac{1}{3})$ provided that $a \geq \sqrt{\frac{1}{1+\alpha}}$ [13].

Theorem 2.2. *For any $-\frac{1}{2} < \alpha < 0$ and any $a \geq \sqrt{\frac{1}{1+\alpha}}$, the problem*

$$\begin{cases} f''' + \dfrac{\alpha+1}{2} f f'' - \alpha f'^2 = 0, \\ f(0) = a, \ f'(0) = 1, \ f'(+\infty) = 0, \\ 0 \leq f'(t) \leq 1, \end{cases}$$

has at least one unbounded solution satisfying

$$\lim_{t \to +\infty} f_\gamma f''_\gamma(t) = 0, \ \lim_{t \to \infty} f_\gamma^2 f'_\gamma(t) = +\infty.$$

The next result deals with the large$-t$ behavior of any possible global solution such that

$$\lim_{t \to \infty} f(t) = +\infty, \tag{2.4}$$

where $-\frac{1}{2} < \alpha < 0$.

Theorem 2.3. *Suppose $-\frac{1}{2} < \alpha < 0$ Let f be a solution of (1.1), (2.4). Then there exists a constant, $A > 0$, such that*

$$f(t) = t^{\frac{1+\alpha}{1-\alpha}} (A + o(1)), \tag{2.5}$$

as $t \to +\infty$.

Proof. Let f be a global solution of (1.1), (2.4).

We claim that $\lim_{t \to +\infty} f'(t) f(t)^2 = +\infty$. In view of equation (1.1) f satisfies

$$\begin{cases} f(t)f''(t) - \dfrac{1}{2}f'(t)^2 + \dfrac{1+\alpha}{2}f'(t)f(t)^2 = \\ f(0)f''(0) - \dfrac{1}{2}f'(0)^2 + \dfrac{1+\alpha}{2}f'(0)f(0)^2 + (1+2\alpha)\displaystyle\int_0^t f(s)f'(s)^2 ds. \end{cases} \tag{2.6}$$

Hence, one sees that $f'(t)f^2(t) \to +\infty$ as $t \to +\infty$. Next we get, by differentiating (1.1) twice,

$$f^{(v)} + \frac{1+\alpha}{2} f f^{(iv)} + \left(\frac{1-3\alpha}{2}\right) f''^2 + (1-\alpha) f' f''' = 0. \tag{2.7}$$

Equation (2.7) asserts, in particular, that $f^{(iv)}$ has at most one zero. Therefore $f'''(t) > 0$ and then $f^{(iv)}(t) < 0$, for all $t \geq t_0, t_0$ large. Using again (1.1) and (2.7) one sees

$$\int_{t_0}^t \frac{f^{(iv)}}{f' f^2} ds + \frac{1-3\alpha}{2} \int_{t_0}^t \frac{f''}{f^2} ds$$

$$= \frac{1+\alpha}{2} \log\left(f^{-\alpha}(t_0) f'^{\frac{1+\alpha}{2}}(t_0)\right) + \frac{1+\alpha}{2} \log\left(f^\alpha(t) f'(t)^{-\frac{1+\alpha}{2}}\right),$$

for any $t \geq t_0$.

Because the functions $\dfrac{f^{(iv)}}{f'f^2}$ and $\dfrac{f''}{f^2}$ are integrable we deduce that the function $\log\left(f^\alpha f'^{-\frac{1+\alpha}{2}}\right)$ has a finite limit as t tends to infinity. Hence, there exists a constant, $C > 0$, such that

$$\lim_{t \to +\infty} f^\alpha(t) f'(t)^{-\frac{1+\alpha}{2}} = C,$$

which immediately leads to (2.5). $\qquad\qquad\qquad\qquad\qquad\qquad\qquad\qquad\square$

Note that if $\alpha = -\frac{1}{3}$ it follows from (2.2) that any solution of (1.1) satisfies the Riccati equation

$$f' + \frac{1}{6}f^2 = \lambda t + \beta,$$

where λ and β are reals. Therefore, if $\lambda > 0$ f is global, tends to infinity and satisfies (2.5). The case $\lambda = 0$ and $\beta \geq 0$ is easy to solve.

3. Behavior of singular solutions at the blowing-up point

In this section we present a result concerning the blowing-up solutions of (1.1). The result is mostly due to Alaa, Benlahsen and Guedda [1]. The problem of the blowing-up solutions to a boundary layer equation was first mentioned by Coppel [12]. The author classified all solutions of the differential equation

$$f''' + ff'' + \lambda(1 - f'^2) = 0, \tag{3.1}$$

where $0 \leq \lambda < 2$. In particular it is shown that for $0 \leq \lambda < 1/2$, any blowing-up solution satisfies $f'(t) \sim -(2 - \lambda)f(t)^2/6$ as $t \to T$, where $0 < T < \infty$ is the blow-up point of f.

Recently, the initial value problem

$$\begin{cases} f''' + \frac{1}{2}ff'' = 0 \quad \text{on } (0,T), \\ f(0) = a, f'(0) = b, f''(0) = \gamma, \end{cases} \tag{3.2}$$

where $a \in \mathbb{R}, b > 0$ and $\gamma \leq 0$, has been considered by Belhachmi, Brighi and Taous [4]. Among other results, it is shown with the help of the Comparison Principle, that there exists a $\gamma_\star \leq 0$ such that, for any $\gamma < \gamma_\star$ the unique local solution to (3.2) is not global.

Very recently it is indicated in [18] that the problem ($\alpha = -1$)

$$\begin{cases} f''' + f'^2 = 0 \quad \text{on } (0,\infty), \\ f(0) = a, f'(0) = 1, f'(\infty) = 0, \end{cases} \tag{3.3}$$

has no solution for any $a \in \mathbb{R}$. In fact, we can see by an easy argument that any local solution to (3.3) blows up at a finite point.

The absence of global solutions of the problem

$$\begin{cases} f''' + \frac{1}{2}ff'' = 0 \quad \text{on } (0,T), \\ f(0) = 1, f'(0) = 0, f''(0) = \gamma, \end{cases} \tag{3.4}$$

where $\gamma < 0$, has also been considered by Ishimura and Matsui [16]. By using the blow-up coordinate f'/f^2 the authors proved that for any $\gamma < 0$, the solution f to (3.4) blows up at a some point T_γ, and satisfies

$$\lim_{t\uparrow T_\gamma}(T_\gamma - t)f(t) = -6.$$

Let us note that $(3.4)_1$ has an explicit solution, g, given by $g(t) = -\frac{6}{T-t}$ and this solution satisfies also (1.1) for any α.

In this work we extend the results of [4], [16] to the problem

$$\begin{cases} f''' + \frac{1+\alpha}{2}ff'' = \alpha f'^2 & \text{on } (0,T), \\ f(0) = a, f'(0) = b, f''(0) = \gamma, \end{cases} \tag{3.5}$$

where $-1 \le \alpha \le 0, a \in \mathbb{R}, b \le 0, \gamma < 0$ and $T > 0$. As in Section 2 f_γ denotes the local solution of (3.5) defined on $(0, T_\gamma)$. We shall see that for any $\gamma < 0$ f_γ is not global, i.e., $\lim_{t\uparrow T_\gamma}|f_\gamma(t)| = +\infty$.

Theorem 3.1. *Let $b \le 0, a \in \mathbb{R}$. Assume that $-1 \le \alpha \le 0$. For any $\gamma < 0$ T_γ is finite and the function f_γ satisfies*

$$\lim_{t\uparrow T_\gamma} f_\gamma(t) = -\infty.$$

Having showed that f_γ blows up at a finite point, we determine its precise asymptotic behavior. The case $\alpha = -1$ is easy to solve. By using the property $E(t) = \frac{1}{2}f_\gamma''(t)^2 - \frac{\alpha}{3}f_\gamma'(t)^3 = E(0)$, for $\alpha = -1$ we deduce that $\lim_{t\to T_\gamma} f''^2 f'^3(t) = -\frac{2}{3}$, and then $\lim_{t\uparrow T_\gamma}(T_\gamma - t)f_\gamma(t) = -6$. So, in the remainder of this paper we assume that $-1 < \alpha \le 0$.

Theorem 3.2. *Let $b < 0, a \in \mathbb{R}$. Assume that $-1 < \alpha \le 0$. For any $\gamma < 0$ the solution f_γ satisfies*

$$\lim_{t\uparrow T_\gamma}(T_\gamma - t)f_\gamma(t) = -6. \tag{3.6}$$

To prove Theorem 2.2 we exploit an idea used in [16] for the Blasius equation and introduced by Toland [20]. First we reduce equation $(3.5)_1$ to a second order equation. To this end we regard f_γ as an independent variable. Since $\gamma < 0$ f_γ and f_γ' are monotone decreasing and tends $-\infty$ as t approaches T_γ. Therefore f_γ', f_γ are negative on some (T_0, T_γ). Without loss of generality we may assume that $f_\gamma(T_0) = 0$. In what follow we set

$$x = -f_\gamma, v(x) = f_\gamma'(t(x))^2.$$

Using $(3.5)_1$ we arrive at the second-order differential equation

$$v''(x) = -2\alpha\sqrt{v(x)} + \frac{1+\alpha}{2}x\frac{v'(x)}{\sqrt{v(x)}}, \quad x > 0. \tag{3.7}$$

The initial condition is given by

$$v(0) = f_\gamma'(T_0)^2 > 0, v'(0) = -2f_\gamma''(T_0) > 0. \tag{3.8}$$

From (3.7) the function

$$w(s) = \frac{v(x)}{x^4}, \quad x = e^s, x \geq x_0,$$

(x_0 large) satisfies

$$w'' + 7w' + 12w - 2\sqrt{w} - \frac{1+\alpha}{2}\frac{w'}{\sqrt{w}} = 0. \tag{3.9}$$

Therefore, the proof the Theorem 3.2 is a simple consequence of the following lemma.

Lemma 3.1. *There holds*

$$\lim_{s\to\infty} w(s) = \frac{1}{36}.$$

Acknowledgments

The author would like to thank B. Brighi and R. Kersner for stimulating discussions. This work was partially supported by Direction des Affaires Internationales (UPJV) Amiens, France.

References

[1] Alaa N., Benlahsen M. & Guedda M., *On blowing-up solutions of a similarity boundary layer equation,* Preprint, LAMFA, Université de Picardie Jules Verne 2003.

[2] Banks W.H.H., *Similarity solutions of the boundary layer equations for a stretching wall,* J. de Mécan. Théo. et Appl. 2 (1983) 375–392.

[3] Banks W.H.H. & Zaturska M. B., *Eigensolutions in boundary layer flow adjacent to a stretching wall,* IMA Journal of Appl. Math. 36 (1986) 375–392.

[4] Belhachmi, Z., Brighi, B. & Taous K., *On the concave solutions of the Blasius equation,* Acta Math. Univ. Comenian, 69, (2) (2000), 199–214.

[5] Belhachmi Z., Brighi B. & Taous K., *On a family of differential equation for boundary layer approximations in porous media,* Euro. Jnl. Appl. Math. 12 (2001) 513–528.

[6] Blasius H., Grenzschichten in Flüssigkeiten mit kleiner Reibung, Z. math. Phys. 56 (1908) 1–37.

[7] Brighi B., *On a similarity boundary layer equation,* Z. Anal. Anwendungen 21 (2002), no. 4, 931–948.

[8] Callegari A.J. & Frieddman M.B., *An analytical solution of a nonlinear, singular boundary value problem in the theory of viscous fluids,* J. Math. Analy. Appl. 21 (1968) 510–529.

[9] Callegari A.J. & Nachman A., *Some singular nonlinear differential equations arising in boundary layer theory,* J. Math. Analy. Appl. 64 (1978) 96–105.

[10] Chaudhary M. A. Merkin J. H. & Pop, I., *Similarity solutions in free-convection boundary layer flows adjacent to vertical permeable surfaces in porous media. I. Prescribed surface temperature,* Eur. J. Mech. B Fluids 14 no. 2 (1995) 217–237.

[11] CHENG, P. & MINKOWYCZ, W.J., *Free-convection about a vertical flat plate embedded in a porous medium with application to heat transfer from a dike*, J. Geophys. Res. 2 (14) (1977) 2040–2044.

[12] COPPEL W.A., *On a differential equation of boundary layer theory*, Phil. Trans. Roy. Soc. London, Ser. A 253 (1960) 101–136.

[13] GUEDDA M., *Similarity solutions of differential equations for boundary layer approximations in porous media*, ZAMP, J. of Appl. Math. Phy. to appear.

[14] HUSSAINI M.Y., LAKIN W.D. & NACHMAN A., *On similarity solutions of a boundary layer problem with upstream moving wall*, SIAM J. Appl. Math., 7, (4), (1987) 699–709.

[15] INGHAM, D.B. & BROWN, S.N., *Flow past a suddenly heated vertical plate in a porous medium*, J. Proc. R. Soc. Lond. A 403 (1986) 51–80.

[16] ISHIMURA N. & MATSUI S., *On Blowing-up solutions of the Blasius equation*, Disc. cont. Dyn. Syst., 9, (4), (2003), 985–992.

[17] MAGYARI E. & KELLER B., *Exact solutions for self-similar boundary layer flows induced by permeable stretching walls*, Eur. J. Mech. B Fluids 19 no. 1 (2000) 109–122.

[18] MAGYARI E., POP, I. & KELLER B., *The "missing" self-similar free convection boundary-layer flow over a vertical permeable surface in a porous medium*, Transp. Porous Media 46, no. 1 (2002), 91–102.

[19] MOFFATT H.K., *High-frequency excitation of liquid metal systems*, IUTAM Symposium: Metallurgical Application of Magnetohydrodynamics, Cambridge, 1982.

[20] TOLAND, J.F., *Existence and uniqueness of heteroclinic orbits for the equation* $\lambda u''' + u' = f(u)$, Proc. Roy. Soc. Edinburgh Sect. A, 109, 1-2, (1988), 23–36.

[21] WEYL H., *On the differential equations of the simplest boundary-layer problems*, Ann. Math. 253 (1942) 381–407.

Mohammed Guedda
LAMFA, CNRS UMR 6140
Université de Picardie Jules Verne
Faculté de Mathématiques et d'Informatique
33, rue Saint-Leu
F-80039 Amiens, France
e-mail: guedda@u-picardie.fr

Progress in Nonlinear Differential Equations
and Their Applications, Vol. 63, 309–318

Mathematical Models of Aggregation: The Role of Explicit Solutions

M.A. Herrero

To Professor Haïm Brezis, with affection and gratitude.

Abstract. We shortly review some classical models of aggregate formation from their elementary monomeric components. Particular attention is paid to the role played by explicit solutions in the overall evolution of the theory, for which some relevant results and open questions are stressed.

1. Introduction

A question that has attracted the attention of mankind since the beginning of recorded scientific thought is that of understanding the way in which complex structures can be formed out of a rather limited in choice (but large in number) elementary components. For instance, the views of Greek philosopher Democritus (ca. 470–ca. 400 BC) are summarized by Diogenes Laertius as follows:

> ... His opinions are these. The first principles of the universe are atoms and empty space ... The worlds are unlimited; they come into being and perish ... Further, the atoms are unlimited in size and number, and they are borne along in the whole universe in a vortex, and thereby generate all composite things – fire, water, air and earth; for even these are conglomerations of given atoms ... The sun and the moon have been composed of such smooth and spherical masses (i.e., atoms), and so also the soul, which is identical with reason ... ,

(cf. [7]). While many of the ideas in the previous paragraph have kept their appeal over the centuries, it has taken a long time to develop a theory that could quantitatively account for even some of the simplest cases of aggregation of individual units into larger condensates.

Perhaps the earliest mathematical model derived to deal with any such situation was that proposed by Smoluchowski (cf. [18] and also [5] for an illuminating survey) in the context of the theory of aggregation of colloids. Smoluchowski considered the case in which colloidal particles are dissolved in a solvent whose molecules are kept in motion by the ambient thermal energy. The impacts of solvent molecules into the (larger) colloid particles are the reason for the irregular, Brownian movement of the last. Whenever two of these colloidal units come sufficiently close to each other, they stick together, thus giving raise to a two-component aggregate (a dimer), which is assumed to undergo the same type of Brownian movement as individual units follow. Upon coming sufficiently close to wandering monomers, trimers, and in general larger aggregates, can also be formed. A question naturally arises, namely that of determining the distribution in time of k-mers, for any chain length $k \geq 1$.

In order to tackle this problem, Smoluchowski made a number of (rather strict) assumptions. To begin with, he postulated that, under the addition of a small amount of electrolyte, any monomeric particle was endowed with a "sphere of influence" of radius $R > 0$, so that aggregation will happen whenever two of these spheres of influence overlap. Furthermore, dimers, trimers, etc. will have their own sphere of influence with same radius $R > 0$. Under the additional assumptions that aggregation is irreversible (so that no breakage of k-mers is allowed) and homogeneous in space, he eventually derived the following set of equations for the concentrations $c_k(t)$ of k-mers at time $t > 0$:

$$\frac{dc_k}{dt} = \sum_{i+j=k} c_i c_j - 2c_k \sum_{j=1}^{\infty} c_j \qquad (k \geq 1). \tag{1.1}$$

In order to solve (1.1), Smoluchowski first assumed that only monomers are present initially, that is:

$$c_1(0) = c_0 > 0, \qquad c_j(0) = 0 \quad \text{for} \quad j \geq 2. \tag{1.2}$$

He then noticed that, upon adding equations (1.1), for $k \geq 1$ one has that:

$$\frac{d}{dt}\left(\sum_{k=1}^{\infty} c_k\right) = -\left(\sum_{k=1}^{\infty} c_k\right)^2$$

whence, using (1.2):

$$\sum_{k=1}^{\infty} c_k = \frac{c_0}{1 + c_0 t}. \tag{1.3}$$

Plugging (1.3) into (1.1), it is possible to recursively solve such system to obtain:

$$\begin{cases} c_1(t) = \frac{c_0}{(1+c_0 t)^2}, \\ c_k(t) = c_0 \left(\frac{(c_0 t)^{k-1}}{(1+c_0 t)^{k+1}}\right) \quad \text{for} \quad k \geq 2. \end{cases} \tag{1.4}$$

Notice that this implies that, while $c_1(t)$ decreases monotonically to zero as time increases, for $k \geq 2$ $c_k(t)$ first achieves a maximum to eventually decay to zero as $t \to \infty$.

While Smoluchowski theory was able to explain a number of experimental facts concerning colloid theory, it was soon realized that modifications had to be made to deal with ongoing technological developments, of which polymer theory was of paramount importance. We shall consider some of these extensions in our next section.

2. The polymerization equations

In the years around 1940, considerable effort was devoted to understanding the mechanisms of polymeric reactions, by which branched and unbranched linear chains were obtained by means of reactive processes starting with relatively simple molecules. From the very beginning, particular attention was payed to unravelling the so-called sol-gel transition. This is determined by the onset of a gel, described as an insoluble product exhibiting rheological properties quite different to those of the initially present reactants (the sol phase, characterised by its solubility). In the words of one of the leading figures in polymer chemistry at that time, P.J. Flory, one has that:

> ... since gelation occurs only when there is the possibility of unlimited growth in three dimensions, the conclusion that it is the result of the formation of infinitely large molecules (that is, molecules of an order of magnitude approaching that of the containing vessel) has been irresistible ...

(cf. [8]). A first attempt to develop a mathematical model of polymer formation was provided by the so-called Flory-Stockmayer theory ([8], [19], [20],...) which will be briefly recalled below.

Consider a system composed of N identical monomeric units, each carrying $f \geq 2$ functional groups capable of reaction with each other. Let c_n be the number of polymeric molecules composed of n units (n-mers). The total number of units is then:

$$\sum_{n \geq 1} n\, c_n = N \,, \tag{2.1}$$

and the number of molecules is:

$$\sum_{n \geq 1} c_n = M \,. \tag{2.2}$$

Notice that the upper limit of the index n changes in a finite system as the reaction proceeds, but it can never exceed N. Assume now that the following assumptions are satisfied:

(a.1)   Intramolecular reactions, leading to cyclic structures do not occur.

(a.2)   At any stage during the reaction, all unreacted functional groups are considered to be equally reactive, regardless of the size of the molecule to which they are attached.

In 1943, W.H. Stockmayer proposed in [19] a method for computing c_n for any given extent of reaction, thus extending previous results by P.J. Flory. It consisted in maximizing the function:

$$\Omega\left(c_n\right) = N! \prod_n \left(\frac{\omega_n}{n!}\right)^{c_n} \frac{1}{c_n!} \tag{2.3}$$

subject to the conditions that N and M (cf. (2.1), (2.2)) should remain constant. Notice that $\Omega\left(c_n\right)$ represents the total number of ways in which the N given units may be formed into c_1 monomers, c_2 dimers, $\ldots$ c_n n-mers, etc. On the other hand, ω_n is the number of ways in which n units may form an n-mer, assuming that assumption (a.1) holds. As shown in [19], one has that:

$$\omega_n = \frac{f^n \left(n\left(f-1\right)\right)!}{\left(\left(f-2\right)n+2\right)!} \tag{2.4}$$

The most probable distribution of molecular sizes for a chosen extent of reaction (that is, the solution of the maximization problem stated above) is given by:

$$c_n = \frac{A\,\omega_n \xi^n}{n!} \tag{2.5}$$

(cf. [19], formula (5)), where A and ξ are Lagrange's multipliers that are determined (and given a physical significance) by means of (2.1) and (2.2). More precisely, Stockmayer considered the question of how to estimate sums of the type:

$$\sum_{n\geq1} c_n\,, \quad \sum_{n\geq1} n\,c_n\,, \quad \sum_{n\geq1} n^2 c_n\,,\ldots \tag{2.6}$$

and to this end he replaced the finite sums in (2.6) by infinite series:

$$S_i = \sum_{n=1}^{\infty} n^i c_n \quad \text{for} \quad i = 0, 1, 2, \ldots \tag{2.7}$$

On introducing the auxiliary variables x and α given by

$$f\,\xi = x = \alpha\left(1-\alpha\right)^{f-2}, \tag{2.8}$$

classical summation techniques yield the following result:

Let $f \geq 3$. Then for $0 \leq x < x_c = \frac{(f-2)^{f-2}}{(f-1)^{f-1}}$ one has that:

$$S_0 = \frac{\alpha\left(1-\alpha f/2\right)}{\left(1-\alpha\right)^2 f}\,, \quad S_1 = \frac{\alpha}{\left(1-\alpha\right)^2 f}\,. \tag{2.9}$$

Moreover, S_2 diverges at $x = x_c$.

In view of (2.9) and (2.8), and since $S_0 = \frac{M}{A}$, $S_1 = \frac{N}{A}$ where A is as in (2.5), one may check that the series S_2 in (2.7) diverges at

$$\alpha = \alpha_c = \frac{1}{f-1}. \tag{2.10}$$

Furthermore, since

$$\alpha = \frac{2N - 2M}{fN} \tag{2.11}$$

it follows that α represents the reacted fraction of functional groups. Stockmayer then suggested that when $f \geq 3$ and α reaches the value α_c in (2.10), a gel fraction will appear, that can be mathematically characterised by the divergence of the second-moment series S_2. As he pointed out, "in a system of finite N this sum never diverges, but undergoes a sudden increase as soon as x exceeds x_c, which is due almost entirely to an increase in the terms of high n" (cf. [19], p. 48). Incidentally, no such threshold value appears when $f = 2$, in which case no sol-gel transition is possible.

The arguments recalled before are of a static nature, in that computations are always made for a fixed extent of reaction α. However, in Appendix C at the end of [19], Stockmayer proposed that the distribution formula for c_n, that in view of (2.5)–(2.11) can be written in the form:

$$c_n(\alpha) = \frac{fN(1-\alpha)^2}{\alpha} \frac{(fn-n)!}{n!\,(fn - 2n + 2)!} \left(\alpha(1-\alpha)^{f-2} \right)^n \tag{2.12}$$

can be given a dynamic meaning. Namely, on setting

$$\alpha(t) = \frac{fNt}{1 + fNt} \tag{2.13}$$

he claimed that $c_n(t) \equiv c_n(\alpha(t))$ happen to solve the kinetic system

$$\frac{dc_n}{dt} = \frac{1}{2} \sum_{i+j=k} a_{j,k} c_i c_j - c_n \sum_1^\infty a_{j,n} c_j \qquad (n \geq 1) \tag{2.14}$$

where

$$a_{j,k} = ((f-2)j + 2)((f-2)k + 2). \tag{2.15}$$

Note that (2.14), (2.15) represent a generalization of the original Smoluchowski's equation (1.1), and the coagulation coefficients $\{a_{j,k}\}$ in (2.15) are proportional to the number of possible reactions of free groups in a j-mer with those of a k-mer. Concerning the proposed solution (2.12), (2.13) to equations (2.14)–(2.15) some remarks are in order. To begin with, since N is assumed to be constant, it is expected to be valid only up to the gelation time t_g (defined as the root of $\alpha(t) = \alpha_c$, α_c as in (2.10)). Moreover, in view of (2.11), (2.13) the sum M cannot stay constant in the interval $0 < t < t_g$.

3. Kinetic coagulation equations: mathematical theory

Since the sixties of the last century, mathematicians grew increasingly interested in equations of type (2.14). A key role in that trend was played by J.B. Mc Leod's seminal work [15]. In this article, he considered equation (2.14) with monodisperse initial conditions:

$$c_1(0) = 1 \quad , \quad c_j(0) = 0 \quad \text{for} \quad j > 1, \tag{3.1}$$

with the choice of coagulation coefficients

$$a_{j,k} = jk, \tag{3.2}$$

which can be considered as a limit case of (2.15). In article [15], the author proved that, for $0 < t < 1$, the problem under consideration (i.e., that consisting of (2.14), (3.1), (3.2)) has a unique solution $\{c_j(t)\}$ such that

$$\sum_1^\infty nc_n(t) = 1. \tag{3.3}$$

Actually, as shown in [15], as long as (3.3) holds, (2.14) can be sufficiently simplified so as to allow for explicit integration which yields

$$c_j(t) = \frac{j^{j-3}t^{j-1}}{(j-1)!}e^{-jt} \quad \text{for} \quad j \geq 1 \quad \text{and} \quad 0 < t < 1. \tag{3.4}$$

A striking property of the problem is that (3.3) is proved to be no longer valid for $t > 1$, so that (3.4) cannot remain true after $t = 1$. However in 1981 Leyvraz and Tschudi [13] were able to extend Mc Leod's solution in a suitable way. More precisely, they proved that the function $c_j(t)$ given by:

$$c_j(t) = \begin{cases} \frac{j^{j-3}t^{j-1}}{(j-1)!}e^{-jt} & \text{for} \quad 0 < t < 1, \\ \frac{j^{j-3}e^{-j}}{(j-1)!t} & \text{for} \quad t > 1 \end{cases} \tag{3.5}$$

solves (2.14), (3.1) and (3.2), and is such that

$$\sum_1^\infty jc_j(t) = \begin{cases} 1 & \text{for} \quad 0 < t < 1, \\ \frac{1}{t} & \text{for} \quad t > 1. \end{cases} \tag{3.6}$$

Moreover, $\sum j^2 c_j(t)$ diverges at $t = 1$ (while remaining finite for $0 < t < 1$). This was considered to be a mathematical illustration of the sol-gel transition, in which the mass of the sol was continuously diminishing due to the onset (and subsequent growth) of the non-reacting gel phase.

While the choice of coagulation coefficients made in (3.2) is rather academic, the sharpness of the results derived in [15] and [13] contributed to trigger a keen interest in dealing with equations of type (2.14) involving physically-motivated aggregation coefficients $\{a_{j,k}\}$ (of which the simplest choice is perhaps the Flory-Stockmayer case (2.15)). This in turn has led to the development of a mathematical theory for equations (2.14) whose coefficients $\{a_{j,k}\}$ satisfy rather general growth conditions in j and k. While various outstanding questions remain to be ascertained as yet (some of them will be recalled in a concluding Section at the end of this note), a good deal is already known concerning the asymptotics of solutions for various choices of the coefficients $\{a_{j,k}\}$. We refer to da Costa [4] and Leyvraz [12] for details on such results.

4. Reversible coagulation: the impact of fragmentation

In aggregation processes, it is generally observed that as clusters grow in size, fragmentation effects become more important, so that irreversible models need to be replaced by reversible ones. As a matter of fact, the balance between coagulation and fragmentation is the main driving force leading to nontrivial equilibrium distributions. Mathematically, if we start from equations of type (2.14), one is then naturally led to consider systems of the type:

$$\frac{dc_j}{dt} = \frac{1}{2} \sum_{k=1}^{j-1} (a_{j-k,k}\, c_{j-k}c_k - b_{j-k,k}c_j) - \sum_{k=1}^{\infty} (a_{j,k}c_jc_k - b_{j,k}c_{j+k}) \tag{4.1}$$

for $j \geq 1$, where $\{b_{j,k}\}$ denote the fragmentation rates in the situation being examined. As it turns out, physical considerations impose compatibility conditions on the coefficients $\{a_{j,k}\}$ and $\{b_{j,k}\}$. As recalled for instance in [6], it is natural to require:

(A) A detailed balance condition: at any stationary state, the number of i-mers lost to j-mers and k-mers ($j + k = i$) through fragmentation is exactly balanced by the number of i-mers formed by coagulation of j-mers and k-mers, that is:

$$b_{i,j}c_{i+j}(\infty) = a_{i,j}c_i(\infty)\, c_j(\infty) \quad (i \geq 1, j \geq 1), \tag{4.2}$$

where for any $j \geq 1$, $c_j(\infty) = \lim_{t\to\infty} c_j(t)$.

(B) A normalization condition: the total fragmentation rate should be proportional to the number of bounds, i.e.:

$$\frac{1}{2} \sum_{i+j=k} b_{i,j} = \lambda(k-1) \quad \text{for} \quad k \geq 1, \tag{4.3}$$

for some positive constant λ called the fragmentation strength of the process.

As noticed in [6], p.306, one has that $\lambda = \exp(\varepsilon/\kappa_B T)$, where ε is the Gibbs free energy of a single chemical bond, T the absolute temperature and κ_B is Boltzmann's constant. The kinetic theory of irreversible polymerization (corresponding to $b_{i,j} \equiv 0$) is then recovered in the limit $\varepsilon \to -\infty$ (so that $\lambda \to 0$), corresponding to infinitely strong bonds.

One of the most striking results derived in [6] consists in extending Stockmayer's solution (2.12), (2.13) to the reversible process (4.1) with $\{a_{j,k}\}$ satisfying Flory-Stockmayer assumption (2.15) under the hypotheses that (4.2), (4.3) hold true. Recently, a similar result has been established for the limit case $a_{i,j} = ij$

316   M.A. Herrero

under assumptions (4.2), (4.3) by means of a different argument (cf. [10]). Both solutions display a sol-gel transition even in the presence of fragmentation effects.

5. Related problems and open questions

Despite of their simplicity (or perhaps due to it), kinetic equations as (2.14) and (4.1) are still being used to test experimental results. The behavior predicted by these equations for appropriate choices of their coefficients describes what is sometimes referred to as mathematical gelation (cf. for instance [17]). A careful comparison between theory and experiments reveals that, when using simple models as (2.14), gelation is often predicted to occur much faster than actually observed ([17]).

One of the reasons for that disagreement seems to stem from the fact that coagulation coefficients of the type (2.14) have only a limited degree of approximation. In some cases, these coefficients may vary in time; in other situations, their functional form may be quite inadequate. This last situation is particularly relevant when one tries to incorporate into the model some information on the geometry of the gel structure that will eventually unfold.

Even if we restrict our attention to homogeneous (i.e., only time-dependent) equations as (2.14), there may be reasons to analyse continuous models instead of discrete ones. A typical example might be:

$$\frac{dc_n}{dt}(x,t) = \frac{1}{2}\int_0^x K(y, x-y)\, c(y,t)\, c(x-y,t)\, dy$$

$$- c(x,t)\int_0^\infty K(x,y)\, c(y,t)\, dy\,, \quad (5.1)$$

for a suitable choice of the coagulation kernel $K(x,y)$. Roughly speaking, continuous equations might be suitable to incorporate screening effects in the coagulation kinetics. A few explicit solutions are also available in this case; some of them are recalled in [1], where the reader will find an interesting discussion on the relation between continuous and discrete deterministic models and the underlying stochastic phenomena which are macroscopically represented by the former.

An intriguing question in both discrete and continuous models consists in understanding how to detect observed phenomena (as the sol-gel transition) in finite systems, obtained for instance by suitable truncation in equations (2.14) or (5.1). As it has been observed previously in Section 2 infinite systems are usually introduced as a way to facilitate computations, but the models under consideration are necessarily finite. The reduction to finite-size systems has been explored in some particular situations (cf. for instance [9], [4]) but some of the main goals (as for instance efficiently tracking gelation) remain still elusive.

Once a number of explicit solutions are available in the literature, it is natural to wonder what role these solutions play in the asymptotics of the processes under consideration. Oddly enough, only preliminary studies on their stability seem to have been done so far. A question related to this is the identification of stable asymptotic regimes. An old conjecture in this field is that self-similarity is expected to play a key role in the large-time dynamics. A recent result in this direction is provided in [16].

Finally, it is natural to wonder what kind of models would arise when inhomogeneities are taken into account, and the space structure is explicitly incorporated into the equations. A mathematically simple way of doing so consists in transforming ODE equations as (2.14) into PDE systems by inserting Laplacian operators there. Leaving aside a number of questions on the time and space scales thereby involved, this approach has been explored by a number of authors (cf. for instance [3], [14], [11], [2]) and is being currently subject to active investigation.

References

[1] D.J. Aldous: Deterministic and stochastic models for coalescence (aggregation and coagulation): a review of the mean-field theory for probabilists. Bernouilli 5, 1 (1999), 3–48.

[2] H. Amann: Coagulation-fragmentation processes. Arch. Rat. Mech. Anal. 151 (2000), 339–366.

[3] Ph. Benilan and D. Wrzosek: On an infinite system of reaction-diffusion equations. Adv. Math. Sci. Appl. 7 (1997), 351–366.

[4] F.P. da Costa: A finite-dimensional dynamical model for gelation in coagulation processes. J. Nonlinear Sci. 8 (1998), 619–653.

[5] S. Chandrasekhar: Stochastic problems in physics and astronomy. Rev. Mod. Phys. 15 (1943), 1–91.

[6] P. van Dongen and M.H. Ernst: Kinetics of reversible polymerization. J. Stat. Phys. 37 (1984), 301–329.

[7] Diogenes Laertius: Lives of eminent philosophers II. Loeb Classical Library, Harvard University Press (1979).

[8] P.J. Flory: Molecular size distribution in three dimensional polymers I. Gelation. J. Am. Chem. Soc. 63 (1941), 3038–3090.

[9] A. Fasano and F. Rosso: Dynamics of droplets in an agitated dispersion with multiple breakage and unbounded fragmentation rate. University of Florence preprint series nr. 10 (2004).

[10] M.A. Herrero and M. Rodrigo: A discrete kinetic system related to coagulation-fragmentation problems. Preprint (2004).

[11] M.A. Herrero, J.J.L. Velázquez and D. Wrzosek: Sol-gel transition in a coagulation-diffusion model. Physica D 141 (2000), 221–247.

[12] F. Leyvraz: Scaling theory and exactly solved models in the kinetics of irreversible aggregation. Preprint (2004).

[13] F. Leyvraz and H.R. Tschudi: Singularities in the kinetics of coagulation processes. J. Phys. A. 14 (1981), 3389–3405.

[14] P. Laurençot and D. Wrzosek: The Becker-Döring model with diffusion II. The long time behavior. J. Diff. Equations 148 (1998), 268–291.

[15] J.B. Mc Leod: On an infinite set of non-linear differential equations. Quart. J. Math. Oxford 2 (1962), 119–128.

[16] G. Menon and R.L. Pego: Approach to self-similarity in Smoluchowski's coagulation equations. Max Planck Institut, Leipzig, Preprint nr. 82 (2003).

[17] P. Sandkühler, J. Sefcik and M. Morbidelli: Kinetics of gel formation in dilute dispersions with strong attractive particle interactions. Adv. in Colloid and Interface Science 108–109 (2004), 133–143.

[18] M. von Smoluchowski: Drei Vorträge über Diffusion, Brownsche Bewegung und Koagulation von Kolloiden. Physik Z. 17 (1916), 557–585.

[19] W.H. Stockmayer: Theory of molecular size distribution and gel formation in branched-chain polymers. J. Chem. Phys. 11 (1943), 45–55.

[20] R.M. Ziff: Kinetics of polymerization. J. Stat. Phys. 23, 2 (1980), 241–263.

M.A. Herrero
Departamento de Matemática Aplicada
Facultad de Matemáticas
Universidad Complutense
E-28040 Madrid, Spain
e-mail: Miguel_Herrero@mat.ucm.es

Progress in Nonlinear Differential Equations
and Their Applications, Vol. 63, 319–328

Metastable Behavior of Premixed Gas Flames

S. Kamin, H. Berestycki, L. Kagan and G. Sivashinsky

To Haïm Brezis, inspiring scholar, good friend, and great human being.

1. Introduction

In this paper we discuss the behavior of solutions of the initial-boundary value
problem

$$\Phi_t - \frac{1}{2}|\nabla\Phi|^2 = \varepsilon\Delta\Phi + \Phi - \langle\Phi\rangle$$

$$\text{in } S = \{(x, y, t) : (x, y) \in D, \ t > 0\}$$

(1.1)

$$\Phi(0, x, y) = \Phi_0(x, y), \ (x, y) \in D \tag{1.2}$$

$$\frac{\partial\Phi}{\partial n} = 0 \quad \text{on} \quad \partial D \times (0, \infty) \tag{1.3}$$

where

$$\langle\Phi\rangle = \frac{1}{|D|} \int\!\!\int_D \Phi(t, x, y)dxdy \ .$$

is the space average over the domain D.

The question under investigation is the long-time dynamics of the solution
for small positive ε. The expression 'metastable behavior' appearing in the title
means that the solution slowly evolves over an exponentially long time, provided
$\varepsilon \ll 1$. The plan of the paper is as follows:

  a) The physical model and experimental results (Section 2),
  b) The numerical simulations for rectangular and elliptic domains (Section 3),
  c) Rigorous mathematical results for one-dimensional and rectangular domains
    (Section 4).

The case of the general domain remains an open problem.

There are several papers dealing with metastable behavior for various physical
problems. In the earlier works by Carr and Pego [10], Fusco and Hale [11] the
Allen-Cahn equation is studied and the exponentially slow motion of the solution is
proved. Several papers have dealt with the Allen-Cahn and Cahn-Hillard equations
(see [1], [2], [4], [8], [12], [14], [24]–[27]). The flame front model was considered by
the present authors in [5]–[7] and also by Sun and Ward in [23]. In [23] the formal

asymptotic expansion for the movement of the solution is presented for a one-dimensional case.

For general surveys on the metastability the reader may consult [25] and [27]. In particular, dynamics of the solution for the constrained Allen-Cahn equation is studied in [25], employing formal asymptotic approaches.

2. Physical model

The problem (1.1)–(1.3) appears in the description of the premixed gas flames in vertical tubes. The premixed flame is a self-sustained wave of an exothermic chemical reaction propagating through a reactive gaseous mixture. This is a classical case of a free interface system. Indeed, in the flame, the bulk of the heat release normally occurs in a narrow layer, the *reaction zone*. This zone separates the cold combustible mixture from hot combustion products. The width of the reaction zone is often much smaller than the typical length scale of the underlying flow field, thereby allowing one to consider the flame as a geometric interface. The dynamics and geometry of this surface are strongly coupled with those of the background gas flow.

The main motivation of the present study is a specific dynamic phenomenon occurring in premixed gas flames in vertical tubes subjected to the buoyancy effect.

The thermal expansion of the gas accompanying flame propagation makes the latter sensitive to the external acceleration. In upward propagating flames, the cold (denser) mixture is superimposed over the hot (less dense) combustion products. Hence, the planar front separating the cold and hot gases is subjected to the classical effect of Rayleigh-Taylor instability. As a result, the flame front becomes convex towards the cold gas [16], [22]. In combustion, in contrast to the Rayleigh-Tailor problem, the interface is permeable, since here the gas has a nonzero normal velocity relative to the flame front.

As is known from many experimental observations, upward propagating flames often assume a characteristic paraboloidal shape with the tip of the paraboloid located somewhere near the channel's centerline. Flames where the tip slides along the channel's wall have also been observed [21], however, this type of flame configuration has received less attention. Upward flame propagation may thus occur through different but seemingly stable geometrical realizations. The present study is intended to give a better understanding of the pertinent nonlinear phenomenology, which, it transpires, is rather interesting.

As a mathematical model we shall employ the weakly nonlinear flame interface evolution equation similar to that proposed by Rakib and Sivashinsky [17].

In suitably chosen units the flame evolution is described by the model (1.1)–(1.3) involving only one equation and one parameter [15], [17]. Here Φ is the perturbation of the planar flame.

The requirement $\varepsilon \ll 1$ means that the flame width is small compared to the characteristic length-scale of the tube cross-section.

3. Numerical simulations and results

Numerical simulations are conducted for the rectangular and elliptic domains. The paraboloid

$$\Phi = -0.5\left((x - x_0)^2 + (y - y_0)^2\right) \tag{3.1}$$

is utilized as the initial condition. For the rectangular domain $D = (0,1) \times (0,1)$ the corresponding numerical solution is discussed in [7]. The problem (1.1)–(1.3) is solved for $\varepsilon = 0.01$ at the two different initial conditions: (a) the symmetry case where the flame-tip is at $x_0 = y_0 = 0.51$ and (b) the asymmetry case with $x_0 = 0.51$ and $y_0 = 0.55$. Figure 1 depicts the temporal evolution of the flame speed, $V = <\Phi>_t$. The trajectories of the flame-tip are shown on Figure 2.

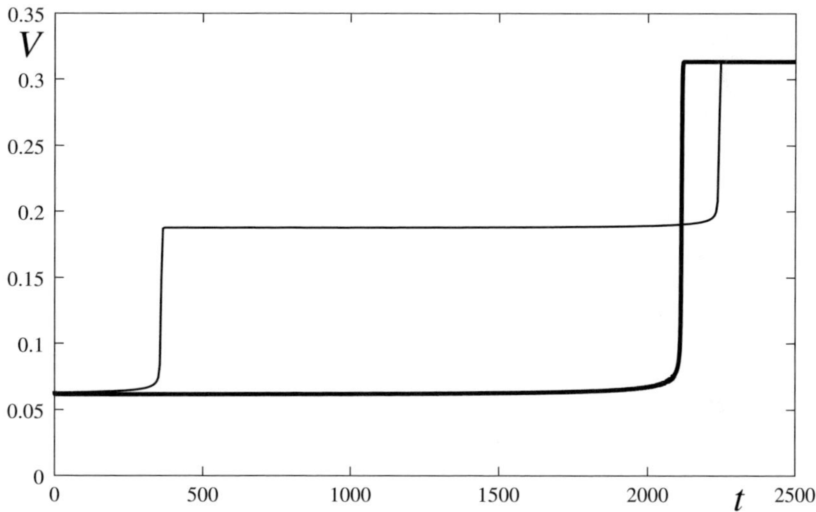

FIGURE 1. Temporal evolution of the flame speed V for the symmetry (bold line) and asymmetry (thin line) cases. Rectangular domain.

The current work is concerned also with the elliptic domain

$$\left(\frac{x}{A}\right)^2 + \left(\frac{y}{B}\right)^2 \leq 1, \tag{3.2}$$

with $A = 2$ and $B = 1$. In the initial condition (3.1) the flame-tip is set at $x_0 = 1$, $y_0 = 0.05$ to make the initial configuration slightly asymmetric. In line with the rectangular case [7] the flame maintains an almost paraboloidal shape over the whole area (3.2) except for the thin boundary layer. Figure 3 depicts the flame speed V vs. t. Figure 4 shows the flame-tip trajectory.

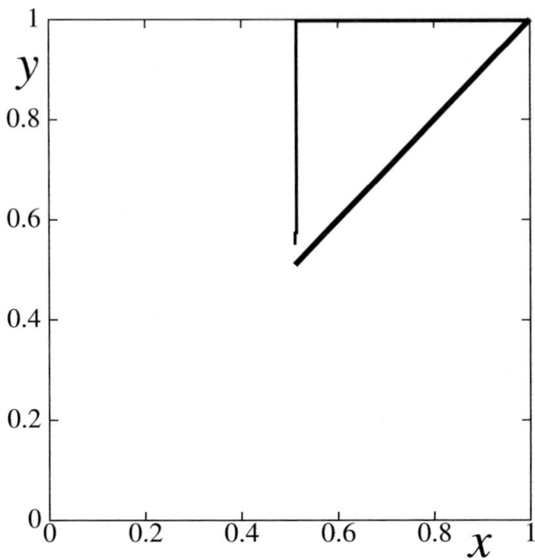

FIGURE 2. Trajectories of the maxima for the symmetry (bold line) and asymmetry (thin line) cases. Rectangular domain.

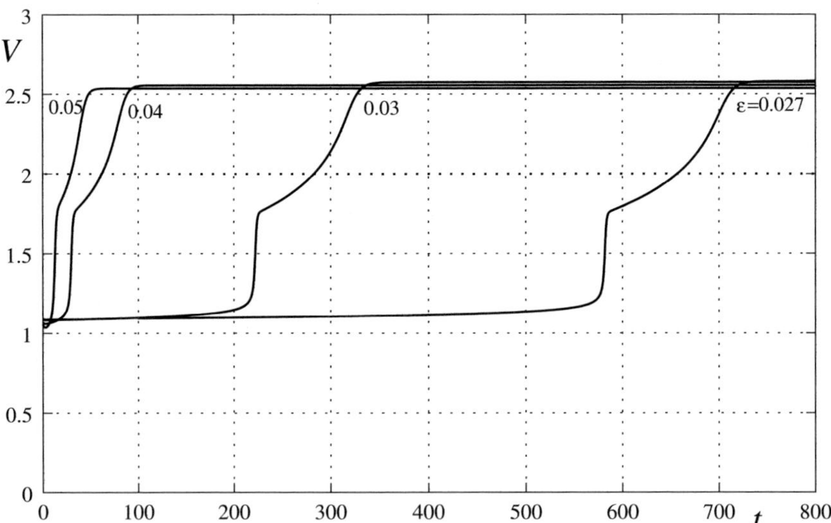

FIGURE 3. Temporal evolution of the flame speed V for several values of ε. Elliptic domain.

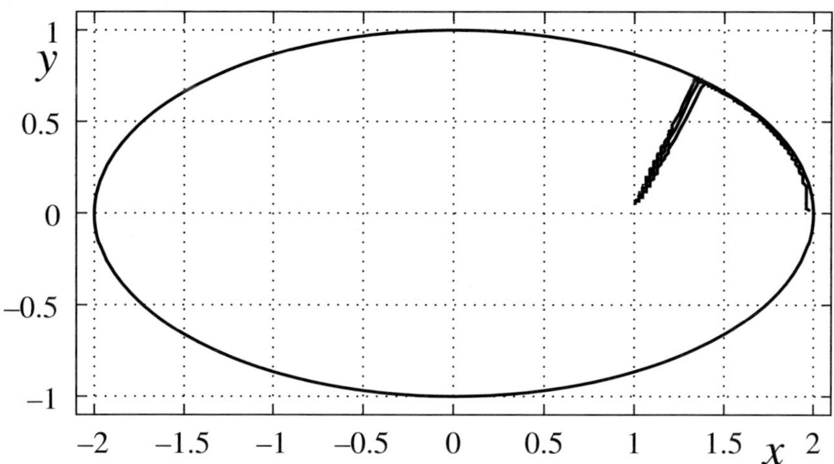

FIGURE 4. Trajectories of the maxima for several values of ε. Elliptic domain.

Since the initial condition (3.1) does not satisfy the boundary conditions (1.3) there is a short transient period of readjustment, upon which the flame displays a metastable behavior, i.e., the flame moves upward, preserving its shape (Figures 1, 3).

After a certain relatively long time period the flame-tip jumps to the nearest point on the boundary (Figures 2, 4). This event is accompanied by a marked increase in the flame speed. Upon reaching the boundary, the flame-tip drifts slowly along the boundary until it reaches the final (stable) position, which coincides with the point of maximum curvature ($x = 2$, $y = 0$).

Note that in the rectangular case this appears to be one of the corner points. In the final state the flame speed reaches its highest value (Figures 1, 3). As shown in the next section, for the rectangular domain, the problem may also be tackled analytically.

4. One-dimensional case and rectangular domains

In our papers [6], [7] we proved several results concerning the metastable behavior of solutions. For the convenience of the reader we present them here.

In the one-dimensional version of (1.1) we set

$$u = -\Phi_x$$

and obtain the equation

$$u_t - \varepsilon u_{xx} + uu_x - u = 0 \quad \text{for} \quad x \in (0, 1), \quad t > 0 \tag{4.1}$$

$$u(t, 0) = u(t, 1) = 0 \tag{4.2}$$

together with initial data

$$u(0, x) = u_0(x) \ . \tag{4.3}$$

This is a nonlinear equation of Burgers type which is well studied. Note that when $u(t, \cdot)$ is close to linear, then $\Phi(t, \cdot)$ is close to parabola. The tip of this parabola corresponds to the point where u vanishes. We consider initial data $u_0(x)$ which change sign at most once, and analyze the dynamics of solutions when ε is a small positive number.

Let us first consider the stationary problem

$$\varepsilon f'' - f f' + f = 0 \quad \text{in} \quad (0, 1) \tag{4.4}$$

$$f(0) = f(1) = 0 \ . \tag{4.5}$$

where $\varepsilon > 0$ is a parameter.

Theorem 4.1. [6] *There exists no nontrivial solution (i.e., $f \not\equiv 0$) of* (4.4), (4.5) *when $\varepsilon \geq \pi^{-2}$. For every ε, $0 < \varepsilon < \pi^{-2}$, there exists a unique positive solution f_ε^+. Likewise, there exists a unique negative solution f_ε^-.*

Moreover, $f_\varepsilon^+(x) \to x$ uniformly on compact sets of $[0, 1)$ and

$$\frac{df_\varepsilon^+(x)}{dx}(0) = 1 - O(e^{-\beta/\varepsilon}), \quad \beta > 0$$

We denote by $a_\varepsilon(t)$, $0 < t$, the curve of zeros of $u_\varepsilon(t, \cdot)$ in the interval $(0,1)$.

Theorem 4.2. [6] *Suppose that $u_0 < 0$ in $(0, a_0)$, $u_0 > 0$ in $(a_0, 1)$, for some $a_0 \in (0, 1)$ and $a_0 \neq \frac{1}{2}$. Then for any δ, η, there exist a time $T > 0$ and constants $\alpha > 0$, $\varepsilon_0 > 0$ such that for $\varepsilon \leq \varepsilon_0$*

$$|a_\varepsilon(t) - a_0| < \delta \ \text{ for all } t \ \ 0 \leq t \leq T_\varepsilon = e^{\frac{\alpha}{\varepsilon}}$$

$$|u_\varepsilon(t, x) - (x - a_0)| \leq \delta \ \text{ for all } x \in [\eta, 1 - \eta] \ \text{ and for all } t \ T \leq t \leq T_\varepsilon.$$

The proof of this theorem is based on (i) analysis of the equation without viscosity ($\varepsilon = 0$); (ii) the use of Theorem 4.1 for construction of sub- and super-solutions.

Next we present the results for the rectangular domain. Let $D = (0, 1) \times (0, 1)$.

Assumption A1. There exist constants $a > 0$, $b > 0$ such that $a < b$ and

$$\frac{\partial \Phi_0}{\partial x} > 0 \quad \text{for} \quad (x, y) \in \{(0, a) \times [0, 1]\}$$

$$\frac{\partial \Phi_0}{\partial x} < 0 \quad \text{for} \quad (x, y) \in \{(b, 1) \times [0, 1]\} \ .$$

Assumption A2. There exist constants $c > 0$, $d > 0$ such that $c < d$ and

$$\frac{\partial \Phi_0}{\partial y} > 0 \quad \text{for} \quad \{[0, 1] \times (0, c)\}$$

$$\frac{\partial \Phi_0}{\partial y} < 0 \quad \text{for} \quad \{[0, 1] \times (d, 1)\} \ .$$

A paraboloid with its tip at the point $(x_0, y_0) \in (0,1) \times (0,1)$ is an example of a function which satisfies (A1)–(A2).

Without loss of generality we also assume that $\Phi_0(x,y) \in C^2(\overline{D})$ and $\frac{\partial \Phi_0}{\partial n} = 0$ on ∂D.

Theorem 4.3. [7] *Suppose $\Phi_0(x,y)$ satisfies (A1)–(A2) and $\Phi = \Phi_\varepsilon(t,x,y)$ is the solution of (1.1)–(1.3). Let δ be an arbitrary small number less than $\min\{a, 1 - b, c, 1 - d\}$. Then there are constants $\alpha > 0$ and $\varepsilon_0 > 0$ such that for all $\varepsilon < \varepsilon_0$*

$$-\frac{\partial \Phi_\varepsilon}{\partial x} < 0 \quad for \quad (x,y) \in (0, a - \delta) \times (0,1) \tag{4.6}$$

$$-\frac{\partial \Phi_\varepsilon}{\partial x} > 0 \quad for \quad (x,y) \in (b + \delta, 1) \times (0,1) \tag{4.7}$$

$$-\frac{\partial \Phi_\varepsilon}{\partial y} < 0 \quad for \quad (x,y) \in (0,1) \times (0, c - \delta) \tag{4.8}$$

$$-\frac{\partial \Phi_\varepsilon}{\partial y} > 0 \quad for \quad (x,y) \in (0,1) \times (d + \delta, 1) \tag{4.9}$$

for all t, $0 \leq t \leq T_\varepsilon := e^{\alpha/\varepsilon}$.

Remark 1. Let D_1 be a rectangle $(a,b) \times (c,d)$. It follows from (4.6)–(4.9) that the points of maximum of $\Phi_\varepsilon(t,x,y)$ for any fixed t remain inside D_1 for an exponentially long time. This fact is an evidence of the metastable behavior of the solution.

Remark 2. The values of α and ε_0 in Theorem 4.3 depend only on the values of a, b, c, d and δ.

Theorem 4.4. *Let $f_\varepsilon^+(x)$ be the unique positive solution of the ODE boundary value problem (4.4), (4.5) Suppose $\Phi_0(x,y)$ satisfies (A1)–(A2) and $b < \frac{1}{2}$, $d < \frac{1}{2}$. Then for all ε sufficiently small*

$$\lim_{t \to \infty} \left[\Phi_\varepsilon(t,x,y) - \Phi_\varepsilon(t,0,0) \right] = -\left(\int_0^x f_\varepsilon^+(\sigma)d\sigma + \int_0^y f_\varepsilon^+(\sigma)d\sigma \right). \tag{4.10}$$

The convergence in (4.10) is uniform in D.

Remark 3. The geometrical meaning of Theorem 4.4 is that as $t \to \infty$ the shape of $\Phi(t, \cdot)$ tends to a bell with its tip at the corner $(0,0)$. This corner is the nearest one to the point of maximum of initial conditions.

Exactly the same behavior demonstrates the flame in numerical simulation of Section 3.

Appendix. Numerical scheme

The numerical solution of the problem (1.1)–(1.3) is obtained by means of a finite-difference code with a uniform spatio-temporal grid. The spatio- temporal steps are specified by the resolution tests.

The first-order spatial derivatives of (1.1) are approximated by the central differences, whilst the second-order derivatives by the 3-point stencils. The temporal approximations for the rectangular and elliptic domains are different. The rectangular domain is tackled by the implicit Alternative Directions technique [13], while for the elliptic domain an explicit scheme is employed. Both approaches are of the first-order accuracy.

For the elliptic domain the rectangular grid employed is one step wider than the domain itself. When approximating the boundary conditions (1.3) the points outside the ellipse (3.2) are treated as ghost-points. Each ghost-point has the associated mirror-point inside the ellipse. By virtue of (1.3) the value of Φ at the ghost-points and at the mirror-points are identical. At the mirror-points Φ is evaluated by the bilinear approximation based on four neighboring points. This leads to the system of linear algebraic equations for Φ at the ghost-points. At each time-step the system is solved by an interactive procedure. For more details the reader may consult [19].

Acknowledgments

These studies were supported in part by the German-Israeli Foundation under Grant No. 695-15.10.01, the United States-Israel Binational Science Foundation under Grant No. 2002008, the Israel Science Foundation under Grant Nos. 67-01, 278-03, and the European Community Program RTN-HPRN-CT-2002-00274.

References

[1] Alikakos, N.D., Bates, P.W., Fusco, G.: Slow motion for the Cahn-Hilliard equation in one space dimension. *J. Diff. Eq.* **90** (1991), 81–135.

[2] Alikakos, N., Fusco, G.: Slow dynamics for the Cahn-Hilliard equation in higher spatial dimensions: the motion of bubbles. *Arch. Rational Mech. Anal.* **141** (1998), 1–61.

[3] Aronson, D.G., Crandall, M.G., Peletier, L.A.: Stabilization of solutions of a degenerate nonlinear diffusion problem. *Nonlinear Anal. TMA* **6** (1982), 1001–1022.

[4] Bates P.W., Xun J.: Metastable patterns for the Cahn-Hilliard equation: Parts I and II. *J. Diff. Equat.* **111** (1994), 421–457; *J. Diff. Equat.* **117** (1995), 165–216.

[5] Berestycki, H., Kamin, S., Sivashinsky, G.: Nonlinear dynamics and metastability in a Burgers type equation (for upward propagating flames). *C.R. Acad. Sci. Paris*, Ser I, **321** (1995), 185–190.

[6] Berestycki, H., Kamin, S., Sivashinsky, G.: Metastability in a flame front evolution equation. *Interfaces and Free Boundaries*, **3** (2001), 361–392.

[7] Berestycki, H., Kagan, L., Kamin, S., Sivashinsky, G.: Metastable behavior of premixed gas flames in rectangular channels. *Interfaces and Free Boundaries*, **6** (2004), 423–438.

[8] Bronsard, L., Hilhorst, D.: On the slow dynamics for the Cahn-Hilliard equation in one-space dimensions. *Proc. R. Soc. London A* **439** (1992), 669–682.

[9] Bronsard, L., Kohn, R.V.: On the slowness of phase boundary motion in one space dimension. *Comm. Pure Appl. Math.* **43** (1990), 983–998.

[10] Carr, J., Pego, R.: Metastable patterns in solutions of $u_t = \varepsilon^2 u_{xx} - f(u)$. *Comm. Pure Appl. Math* **42** (1989), 523–576.

[11] Fusco G., Hale J.K.,: Slow motion manifolds, dormant instability and singular perturbations. *J. Dyn. Diff. Equat.* **1** (1989), 75–94.

[12] Grant, Ch.P.: Slow motion in one-dimensional Cahn-Morral systems. *SIAM J. Appl. Math* **26** (1995), 21–34.

[13] Godunov, C.K., Ryabenki, V.S. *The Theory of Difference Schemes.* North-Holland, 1964

[14] Laforgue, J.G., O'Malley, R.E.: Shock layer movement of Burgers equation. *SIAM J. Appl. Math.* **55** (1995), 332–348.

[15] Mikishev, A.B., Sivashinsky, G.I.: Quasi-equilibrium in upward propagating flames. *Physics Letters A* **175** (1993), 409–414.

[16] Pelce-Savornin, C., Quinard, J., Searby, G.: The flow field of a curved flame propagating freely upwards. *Combustion Science and Technology* **58**, (1988), 337.

[17] Rakib, Z., Sivashinsky, G.I.: Instabilities in upward propagating flames. *Combust. Sci. Technol.* **54** (1987), 69–84.

[18] Sattinger, D.H.: Monotone methods in nonlinear elliptic and parabolic boundary value problems. *Ind. Univ. Math. J.* **21** (1972), 979–1000.

[19] Shaojie, X., Aslam, T., Stewart, D.S. High resolution numerical simulation of ideal and non-ideal compressible reacting flows with embedded internal boundaries. *Combustion Theory and Modelling* **1** (1997), 113–142.

[20] Smoller, J.: *Shock Waves and Reaction-Diffusion Equations.* Springer-Verlag, 1983.

[21] Sohrab, S.H. Private communication.

[22] Strehlow, R.A. *Combustion Fundamentals.* McGraw-Hill, New York (1985) pp. 349.

[23] Sun, X., Ward M.J., Metastability for a generalized Burgers equation with applications to propagating flame-fronts. *European J. Appl. Math.* **10** (1999), 27–53.

[24] Sun, X., Ward M., Dynamics and coarsening of interfaces for the viscous Cahn-Hilliard Equation in one spatial dimension. *Studies Appl. Math.* **105** (2000), 203–234.

[25] Ward, M.J., Reyna, L.G.: Internal layers, small eigenvalues and the sensitivity of metastable motion. *SIAM J. Appl. Math.* **55** (1995), 425–445.

[26] Ward, M.: Exponential Asymptotics and convection-diffusion-reaction models, in *Analyzing Multiscale Phenomena Using Singular Perturbation Methods.* (J. Cronin, R. O'Mally ed.) *Proc. Symp. Appl. Math.*, **56** (1998), 151–184.

[27] Ward, M.J.: Metastable dynamics and exponential asymptotics in multi-dimensional domains, in *Multiple Time-Scale Dynamical Systems*, IMA Volumes in Mathematics and its Applications (eds. C.K.R.T. Jones, A. Khibnik), **122** (2000), 233–260.

S. Kamin
School of Mathematical Sciences
Tel Aviv University
Ramat Aviv
Tel-Aviv 69978, Israel
e-mail: `kamin@post.tau.ac.il`

H. Berestycki
Ecole des Hautes Etudes
   en Sciences Sociales
CAMS
54, Boulevard Raspail
F-75006 Paris, France

L. Kagan
School of Mathematical Sciences
Tel Aviv University
Ramat Aviv
Tel Aviv 69978, 69978, Israel

G. Sivashinsky
School of Mathematical Sciences
Tel Aviv University
Ramat Aviv
Tel Aviv 69978, 69978, Israel

Progress in Nonlinear Differential Equations
and Their Applications, Vol. 63, 329–341

Recent Progress on Boundary Blow-up

Satyanad Kichenassamy

To Haïm Brezis on the occasion of his sixtieth birthday celebration

Abstract. We report on the solution of two long-standing conjectures on the boundary behavior of maximal solutions of semilinear elliptic equations, focusing on the proof of the boundary regularity of the hyperbolic radius in higher dimensions. The main tool is the reduction of the problem to a degenerate equation of Fuchsian type, for which new Schauder-type estimates are proved. We also sketch an algorithm suitable for large classes of applications.

Haïm Brezis' work has always been characterized by a particular combination of clarity and depth. His thorough understanding of the most difficult aspects of Functional Analysis, combined with a consistent emphasis on specific problems raised by applications, has led him to identify fundamental mathematical difficulties in the simplest models, and to see, earlier than others, difficulties with linear paradigms. He was therefore led to give Nonlinear Analysis its proper place: not limited anymore to being the theoretical counterpart of computing, he established it as an independent subject, at the forefront of Mathematics, both pure and applied. Two examples come to mind:

(a) his early recognition of the importance of compactly supported solutions of nonlinear PDEs, at a time where the search for better and better criteria for unique continuation in linear PDEs was quite active;

(b) his understanding of the need to study strongly singular solutions beyond existing paradigms: distributional framework, nonlinear Fourier analysis, or variational methods, for all of which he provided rather subtle reasons why they are inadequate [6, 10]. As for his lucid style of exposition, a good example is the early work [9].

It was my good fortune to start my thesis under his direction a little over twenty years ago, on quasilinear problems with singularities [16]. The results reported here also pertain to the issue of understanding strongly singular solutions of nonlinear PDEs. In fact, the impetus for this work was given by Haïm Brezis in a conversation in 1998, on what the elliptic counterpart of my results on blow-up for nonlinear wave equations should be. This latter work also turned out, more recently, to be relevant to a question he had raised in the graduate course he taught in 1984 [11].

1. Introduction

We report on the recent proof of the following results.

Consider the Liouville equation

$$-\Delta u + 4e^{2u} = 0 \tag{1}$$

on a bounded domain in $\mathbb{R}^2$ of class $C^{2+\alpha}$, for some $\alpha \in (0,1)$. It has a maximal solution $u_\Omega(x)$, which tends to infinity as $x \in \Omega$ approaches the boundary. Our first result is

Theorem 1.1. $v_\Omega := \exp(-u_\Omega)$ *is of class* $C^{2+\alpha}$ *up to the boundary of* Ω.

Next, consider the Loewner-Nirenberg equation

$$-\Delta u + n(n-2)u^{\frac{n+2}{n-2}} = 0 \tag{2}$$

on a bounded domain in $\mathbb{R}^n$ of class $C^{2+\alpha}$, for some $\alpha \in (0,1)$, with $n \geq 3$. It has a maximal solution $u_\Omega(x)$, which tends to infinity as $x \in \Omega$ approaches the boundary. Our second result is

Theorem 1.2. $v_\Omega := u_\Omega^{-2/(n-2)}$ *is of class* $C^{2+\alpha}$ *up to the boundary of* Ω.

In both cases, the boundary behavior reflects the geometry of the domain:

Theorem 1.3.
$$v_\Omega(x) = 2d(x) - d(x)^2[H(x) + o(1)]$$
as $d(x) \to 0$, *where* $H(x)$ *is the mean curvature of* $\partial\Omega$ *at the point of* $\partial\Omega$ *closest to* x. *In addition,* v_Ω *is a classical solution of*

$$v_\Omega \Delta v_\Omega = \frac{1}{2}(|\nabla v_\Omega|^2 - 4). \tag{3}$$

This detailed information on the boundary behavior is relevant to the numerical computation of u_Ω.

Thus, via a nonlinear change of unknown, one can reduce the search for a singular solution of a problem with boundary blow-up to the search for a classical solution of the transformed problem.

Theorem 1.1 is proved in [17, 18], and Theorem 1.2 in [19]. They had been conjectured by Bandle and Flucher [2], to which we refer for an overview of the applications of the function u_Ω. In a nutshell,

- (i) u_Ω provides a universal bound on all classical solutions irrespective of boundary conditions (Keller [15], Osserman [29]);
- (ii) u_Ω enables one to construct an intrinsic metric on Ω which makes it complete (Loewner and Nirenberg [27]);
- (iii) the minima of u_Ω, known as hyperbolic centers, are close to the points of concentration arising in several variational problems of recent interest.

The two-dimensional problem has a long history, starting with Bieberbach [8]. In fact, v_Ω, known as the hyperbolic radius, reduces to the conformal radius in two dimensions if Ω is simply connected. Earlier contributions to the problem include [1, 3, 4, 7, 12, 25, 28].

Remark 1. The number n plays a double role: it determines the space dimension as well as the nonlinearity. It is in fact the latter which is essential for the methods of this paper. Consider for instance the problem

$$-\Delta u + 24u^2 = 0, \tag{4}$$

in three dimensions to fix ideas. It should be handled by the methods of this paper with $n = 6$. The details will be considered elsewhere. Since this problem has an interpretation in terms of super-diffusions, it would be interesting to know whether our results, at least in this special case, admit of a probabilistic interpretation.

The method of proof relies on the method of Fuchsian Reduction (FR), introduced in the early nineties to describe asymptotics of solutions of nonlinear wave equations near blow-up singularities. The following points may be stressed:

- It is necessary to consider second-order asymptotics of the solution.
- The reduced equation makes the discovery of comparison functions natural and systematic.
- One needs to go beyond scaled interior Schauder estimates.

Quite generally, existing proofs of the Schauder estimates [13] work by perturbing a constant-coefficient model (the Laplacian); in the present situation, the appropriate local model is not the Laplacian, but a model Fuchsian operator with quadratic degeneracy, for which we prove directly the appropriate estimates. Note that weighted Sobolev spaces are not convenient here, not only because the data of the problem are Hölder-continuous, but also because we do not have sufficient estimates on derivatives to start a bootstrap in these spaces.

An overview of the method and its main applications are given next, in a form suitable for a wide range of applications. General results on Fuchsian elliptic PDEs are collected in Section 3, and the main steps of the proof of Theorem 1.2 are given in Section 4.

2. Fuchsian Reduction

2.1. Principle of the method

Consider a PDE, of any type, in any number of variables, which we write symbolically

$$F[u] = 0.$$

We are interested in constructing solutions which become singular precisely on a hypersurface of equation $\Phi = 0$. The solution should be the sum of a singular part, generally given in closed form, and a remainder which vanishes as fast as possible as Φ tends to zero. As a result, one obtains not only existence, but an actual asymptotic formula for singular solutions. Advantages include: (i) a substitute for numerical computations precisely near the places where they fail; (ii) a parametrization of singular solutions by data characterizing the singular behavior, thereby enabling one to settle the question of stability of blow-up; (iii) in many cases, a geometric interpretation of several terms of the expansion.

The method proceeds by reducing the PDE to an equation which degenerates at the singularity. We review in this section the main steps of this algorithm, and treat a few simple examples in the next. We focus on single equations: systems may be reduced along similar lines. For further details, one may refer to the surveys [22, 23], and the introductions of the papers [5, 14, 20, 21].

In general terms, the method succeeds because it identifies a *scale-invariant structure* in the directions normal to the singular set. It also shows that the solution is determined by the coefficients of its asymptotic expansion near the singularity, and the equation of the singularity locus. As a result, one can study (i) how the solution depends on the singularity locus; (ii) which types of singularities may be deformed without change in character (stability of the blow-up pattern).

2.1.1. Leading-order analysis. Let us change coordinates so that $T := \Phi$ becomes the first coordinate.

The first step is to seek a pair (u_0, ν) such that, if the solution satisfies $u \sim u_0 T^\nu$ as $T \to 0$, then the most singular terms in the equation cancel each other; in practice, this often happens because two such terms precisely balance each other, producing the required cancellation. Hence the term "leading balance," which was introduced in soliton theory to refer to any such pair (u_0, ν).

More precisely, insert in the equation $u_0 T^\nu$ for u, where u_0 does not blow-up for $T = 0$; assume, to fix ideas, that

$$F[u_0 T^\nu] = f[u_0, \nu] T^\rho (1 + o(T)).$$

One then chooses u_0 and ν so that

$$f[u_0, \nu] \equiv 0 \text{ and } u_0 \not\equiv 0. \tag{5}$$

There may be more than one pair (u_0, ν) satisfying this requirement. Also, it may be necessary to consider more general expressions than T^ν, including logarithms of T, exponentials, inverse logarithms, variable powers, etc. The power case will suffice for the rest of the paper.

2.1.2. Renormalized unknown and first reduction. Once the leading balance has been identified, one focuses on solutions such that

$$u \sim u_0 T^\nu \text{ as } T \to 0 + .$$

Define a *renormalized unknown v* by

$$u = u_0 T^\nu (1 + Tv). \tag{6}$$

One then finds that the resulting equation has, under very general assumptions on F, the form

$$P(T\partial_T + 1)v = Tg[v] \tag{7}$$

where P is a polynomial. We are not interested in all solutions of this equation: only in those for which v remains bounded as $T \to 0+$. Indeed, if v is not controlled, we may not have $u \sim u_0 T^\nu$.

The nonlinearity g may involve derivatives of v, as well as other variables. An equation such as (7) is said to be *Fuchsian*, because equation $P(T\partial_T)v = 0$ is the simplest type of equation with a regular singular point at $T = 0$, that is, an equation to which Fuchs-Frobenius theory applies. The roots of the polynomial P are known as *resonances* or *indices*.

Note however that Fuchs-Frobenius theory, which yields a basis of branched solutions for holomorphic solutions of linear PDE with holomorphic coefficients, generally does not apply to the nonlinear PDE (7), not only because the coefficients may not be analytic, but also because it does not consider sufficiently general series expansions. Nevertheless, the basic observation, which goes back to Euler, and was developed in connection with special functions in the late nineteenth century, that the solutions should involve non-integral powers, and possibly logarithms, even for PDEs with analytic coefficients, remains valid. Fuchsian Reduction represents the systematic application of "Weierstrass' view-point" to nonlinear PDEs.

Remark 2. In some applications, it is appropriate to perform a more complicated reduction than the one described in equation (6); in particular, one may need to define T to be $\Phi^{1/s}$, for s large enough, and the r.h.s. of (7) may have the form $T^\varepsilon g[v]$, where $0 < \varepsilon < 1$. Also, it is equivalent, but sometimes more convenient, to define v by

$$u = u_0 T^\nu (1 + Tv)^\nu. \tag{8}$$

2.1.3. Formal solution. At this stage, one may appeal to general theorems on spaces of formal series, such as are given in [5, 20, 22], to derive a formal expansion of u, and hence v. Of course, if the solution or the coefficients of the equation have limited regularity, only the first few terms of the series are meaningful.

The most common type of formal series has the form

$$u = T^\nu \sum_{j \geq 0} p_j(\ln T) T^j, \tag{9}$$

where p_j is a polynomial in $\ln T$, with coefficients depending on other variables, and such that

$$\deg p_j \leq lj,$$

where the number l can be predicted from the knowledge of P (see [22, 20] and their references).

In addition, if j is a root of P of multiplicity m_j, one can prove that the coefficients of $(\ln T)^k$ in p_j are arbitrary if $k < m_j$.

The formal series is shown next to be associated to actual solutions. After this has been accomplished, one is left with an asymptotic representation of solutions near the singularity, which renders the same services as a closed-form solution.

2.1.4. Second reduction. Once a formal solution to high order has been obtained, it is occasionally useful to perform a *second reduction*

$$v = \tilde{v} + T^m w,$$

where $\tilde{v}$ is obtained by truncating the formal solution to sufficiently high order, and m is large. It turns out that the resulting equation is again Fuchsian, and the polynomial $P(X)$ is simply replaced by $P(X + m)$. If m is large enough, P will have no positive root, and the expansion of w will be entirely determined by $\tilde{v}$.

2.1.5. Justification of the series. We may now appeal to a number of existence-regularity theorems for Fuchsian PDEs which make the justification of the series more or less routine (see [24, 20, 23]). In the analytic case, there are two generalizations of the Cauchy-Kowalewska theorem which apply to Fuchsian PDEs: the first is adapted to series containing only powers of T and $\ln T$, and the second applies to general continuous dependence on T and analytic dependence on the "space" variables. In the Sobolev case, there is a generalization of the theory of symmetric-hyperbolic systems [24].

The elliptic case is the subject of this paper.

2.1.6. Computing the blow-up set. The reader will have noticed that the process begins with the knowledge of the singular set. In fact, it is possible in many cases to recover the equation of the blow-up set from the Cauchy data in the following way: first define a map sending the set of arbitrary functions which determine the expansion (9) – including in it the suitably normalized equation of the blow-up set (the "singularity data") – to the set of Cauchy data on a nearby hypersurface on which the solution is regular; then, prove that this map is invertible. This program may be worked out in many cases of interest; the most difficult example to date, which requires the Nash-Moser inverse function theorem, may be found in [21]. The inverse map thus gives the equation of the blow-up set in terms of the Cauchy data, as desired.

Even for semi-linear wave equations in 1D, this seemingly roundabout approach is the only one, to date, which can relate the higher regularity of the blow-up set to the regularity of the Cauchy data.

2.2. Examples and applications

The main applications of reduction techniques to date, apart from Theorems 1.1 and 1.2, are

- The proof of a conjecture of Fefferman and Graham, on the local embedding of a Riemannian manifold as a null hypersurface of codimension two in a Lorentz manifold [20].
- The proof of the convergence of ARS-WTC expansions for integrable and non-integrable equations [22].
- Construction of inhomogeneous solutions of Einstein's equations with asymptotically velocity-dominated behavior in the sense of Eardley, Liang and Sachs [23].

- Singular expansions for solutions of nonlinear wave equations near blow-up, continuation beyond blow-up etc. [11, 23].
- Behavior of the temperature near the center in the point-source stellar model studied by von Neumann and Chandrasekhar [14].

We give the elements of Fuchsian reduction in four simple examples, identifying the leading term, renormalized unknown, and resonances. Examples 1 and 3 refer in fact to regular solutions, and show that reduction techniques generalize the Cauchy and Dirichlet problems.

2.3. Example 1: Cauchy problem

Consider, to fix ideas, the wave equation

$$u_{tt} - \Delta u = 0.$$

Taking $T = t$ and u_0 an arbitrary function of x, we let

$$u = u_0(x) + tv(x, t).$$

We find $tv_{tt} + 2v_t - t\Delta u = 0$. This is a special case of the *Euler-Poisson-Darboux equation*. Multiply through by t. Since $t^2\partial_{tt} + 2tv_t = (t\partial_t)(t\partial_t + 1)$, we obtain the Fuchsian equation

$$P(t\partial_t + 1)v = t^2\Delta v,$$

with $P(X) = X(X - 1)$. The resonances are therefore 0 and 1.

2.4. Example 2: Korteweg-de Vries equation

Consider the equation

$$F[u] := u_t - 6uu_x + u_{xxx} = 0,$$

and take $T = \Phi = x$, to fix ideas. One finds, as $x \to 0$, that if $u_0(x, t)$ is any smooth function,

$$F[u_0 x^\nu] = u_{0t}x^\nu - 6\nu u_0^2 x^{2\nu-1}(1 + o(1)) + \nu(\nu - 1)(\nu - 2)u_0^3 x^{\nu-3}(1 + o(1)).$$

There are two types of leading balances:

1. If $\nu \geq 0$, then $\rho = \nu - 3 < \min(2\nu - 1, \nu)$. In that case,

$$f[u_0, \nu] = \nu(\nu - 1)(\nu - 2)u_0^3,$$

   and we must take $\nu = 0$, 1, or 2. This means we are dealing with solutions such that $u \sim u_0$, $u \sim xu_0$ or $u \sim x^2 u_0$. Now, such solutions are given by the analytic Cauchy problem with data given for $x = 0$; we therefore do not further discuss this case.

2. If $\nu < 0$, then ρ equals $\nu - 3$ or $2\nu - 1$, depending on which one is smaller. We therefore distinguish three sub-cases:
   - If $\rho = \nu - 3 < 2\nu - 1$, one finds that $f[u_0, \nu]$ is proportional to u_0^3, so that it is not possible to satisfy the requirements (5).
   - If $\rho = 2\nu - 1 < \nu - 3$, one finds that $f[u_0, \nu]$ is proportional to u_0^2, so that again, no non-trivial solution may be found.

- If $\rho = \nu - 3 = 2\nu - 1$, in which case the second and third terms in the equation precisely balance each other, we find $\nu = -2$, $\rho = -5$,

$$f[u_0, \nu] = \nu(\nu - 1)(\nu - 2)u_0 - 6\nu u_0^2 = 12u_0(u_0 - 2),$$

  so that we must take $u_0 = 2$. We are therefore interested in solutions such that

$$u \sim 2x^{-2}.$$

Let us therefore focus on solutions which behave like $2/x^2$, and define v by $u = x^{-2}(2 + xv)$. It is convenient to let $D = x\partial_x$ and to re-write F in the form

$$F[u] = u_t - 6x^{-1}uDu + x^{-3}D(D - 1)(D - 2)u.$$

Using the property $D(t^m u) = t^m(D + m)u$, one finds

$$F[x^{-2}(2 + xv)] = x^{-4}[P(D + 1)v + xg(x, v, v_t, Dv)],$$

where $P(X) = (X + 1)(X - 4)(X - 6)$, so that the resonances are -1, 4 and 6. The theory now gives that the Korteweg-de Vries equation has convergent series solutions of the form

$$\sum_{j \geq 0} u_j x^{-2+j},$$

in which the coefficients u_4 and u_6 are arbitrary functions, and all the others may be computed inductively.

Remark 3. More generally, one may construct such series solutions whenever $\Phi = x - \psi(t)$, where ψ is arbitrary. For "non-integrable" equations, a similar result holds, with the difference that the u_j are polynomials in $\ln x$, with coefficients which depend on t; their degrees increase linearly with j, and the constant term is arbitrary for $j = 4$ or 6. In all cases, there are three arbitrary functions of t, if the normalized equation of the singular set $(x = \psi(t))$ is included in the list (u_4, u_6, ψ) of "singularity data". Thus, the integrable cases differ from the general case by the relative simplicity of the form of the expansion.

2.5. Example 3: Dirichlet problem

Consider the Dirichlet problem for the Laplace equation on the domain Ω, with $u = g$ on $\partial\Omega$. Let d denote the distance to the boundary. We may then define the renormalized unknown v by $u = g + dv$, so that, multiplying the equation $\Delta u = 0$ by d, v solves the Fuchsian elliptic equation

$$d^2 \Delta v + 2d\nabla d \cdot \nabla v + d(v\Delta d + \Delta g) = 0.$$

In a system of coordinates in which $T = d$, one finds that the equation has the form

$$(T^2 \partial_{TT} + 2T\partial_T)v = O(T).$$

Since $T^2 \partial_{TT} + 2T\partial_T = D(D + 1) = P(D + 1)$, with $P(X) = X(X - 1)$; the resonances are therefore 0 and 1, corresponding to the fact that the Dirichlet and Neumann data may be prescribed arbitrarily in the analytic case. In the non-analytic case, the Dirichlet data determine the Neumann data, so that the appropriate question is whether the *regularity* of the solution reflects the regularity of the Dirichlet data, since, by Taylor's theorem, regularity implies that the first few terms of an expansion in powers of d describe correctly the asymptotic behavior at the boundary.

2.6. Example 4: equation (2)

It is helpful to consider first the ODE

$$-u_{TT} + n(n-2)u^{(n+2)/(n-2)} = 0.$$

Leading-order analysis leads to the behavior $u^{-2/(n-2)} \sim 2T$. Furthermore, if we let $u^{-2/(n-2)} = 2T + T^2 w$, we find an equation of the form

$$(D+2)(D+1-n)w = O(T).$$

The resonances are therefore -1 and n, which suggests that logarithmic terms will not arise until terms involving T^n in the expansion of $u^{-2/(n-2)}$.

Motivated by this computation, we define v, and a renormalized unknown w, by the relations $u^{-2/(n-2)} = v = 2d + d^2 w$, where d is the distance to the boundary. One first finds

$$v\Delta v = \frac{n}{2}[|\nabla v|^2 - 4],$$

and then

$$Lw + 2\Delta d = M_w(w), \tag{10}$$

where

$$L := d^2 \Delta + (4-n)d\nabla d \cdot \nabla + (2-2n),$$

and M_w is a linear operator with w-dependent coefficients, defined by

$$M_w(f) := \frac{nd^2}{2(2+dw)}[2f\nabla d \cdot \nabla w + d\nabla w \cdot \nabla f] - 2df\Delta d.$$

The analysis of this equation is the object of the last two sections of this paper.

2.7. Example 4: equation (1)

In the case of equation $-\Delta u + 4\exp(2u) = 0$, leading-order analysis leads to the formal behavior

$$u \sim \ln\frac{1}{2d}.$$

We therefore consider the equation satisfied by $v = \exp(-u)$, which reads

$$v\Delta v = |\nabla v|^2 - 4.$$

We expect $v \sim 2d$. We therefore define the renormalized unknown w by $v = 2d + d^2 w$. One finds that w solves precisely equation (10), where L is defined by the same formula as in the previous example, but with $n = 2$.

3. Two types of Fuchsian elliptic operators

Let $d(x)$ denote the distance of x to $\partial\Omega$ and $\Omega_\delta = \{0 < d < \delta\}$, where δ is chosen small enough for d to be of class $C^{2+\alpha}(\overline{\Omega}_\delta)$. We say that $u \in C^{k+\alpha}_\sharp(\overline{\Omega}_\delta)$ if $d^j u \in C^{j+\alpha}_\sharp(\overline{\Omega}_\delta)$ for $0 \le j \le k$.

Let $(a^{ij}) \in C^\alpha(\overline{\Omega}_\delta)$ be uniformly elliptic. An operator A is said to be of type (I) if it has the form

$$A := \partial_i(d^2 a^{ij}(x)\partial_j) + db^i(x)\partial_i + c(x),$$

where b^i and c belong to $L^\infty(\Omega_\delta)$.

It is said to be of type (II) if it has the form

$$A := d^2 a^{ij}(x)\partial_{ij} + db^i(x)\partial_i + c(x),$$

where b^i and c belong to $C^\alpha(\overline{\Omega}_\delta)$.

Operator L, which may be written $\mathrm{div}(d^2\nabla) + (2-n)d\nabla d \cdot \nabla + (2-2n)$, is of type (I) as well as (II).

The results we will need are the following.

Theorem 3.1. *If*

1. *A is of type (I), and*
2. *Af and f are in $L^\infty(\Omega_\delta)$,*

then df and $d^2\nabla f$ belong to $C^\alpha(\overline{\Omega}_{\delta'})$ for $\delta' < \delta$, and $d\nabla f$ is bounded near $\partial\Omega$.

This is proved in two dimensions in Theorem 5.1 of [18]; the proof applies without modification in n dimensions.

Theorem 3.2. *If*

1. *A is of type (I), and*
2. *Af and f are $O(d^\alpha)$ as $d \to 0$,*

then $f \in C^{1+\alpha}_\sharp(\overline{\Omega}_{\delta'})$ for $\delta' < \delta$.

This corresponds to Theorem 5.2 in [18]: in the latter paper, it is assumed that $Af = O(d)$, and that $n = 2$, but the proof proceeds *verbatim* for any n, if one only knows that $Af = O(d^\alpha)$.

Theorem 3.3. *If*

1. *A is of type (II),*
2. *$Af \in C^\alpha(\overline{\Omega}_\delta)$,*
3. *$f \in C^{1+\alpha}_\sharp(\overline{\Omega}_\delta)$,*

then $f \in C^{2+\alpha}_\sharp(\overline{\Omega}_{\delta'})$ for $\delta' < \delta$.

The (short) proof of this result may be found in [19].

4. Outline of proof of Theorem 1.2

Recall that we have reduced the problem to the degenerate elliptic equation of Fuchsian type (10) for the renormalized unknown w defined by $u = (2d + d^2w)^{-(n-2)/2}$. Note also that the assumptions on $\partial\Omega$ ensure that d is of class $C^{2+\alpha}$ near the boundary. We wish to prove that $2d + d^2w$ is of class $C^{2+\alpha}$ near (and up to) the boundary.

Step I. One first proves, by a comparison argument combined with regularity estimates, that w and $d^2\nabla w$ are bounded near $\partial\Omega$; it follows that operator $L - M_w$ is of type (I).

Step II. Since w and $(L - M_w)w$ are both bounded near $\partial\Omega$, Theorem 3.1 shows that $d\nabla w$ is bounded near $\partial\Omega$, so that $M_w(w) = O(d)$ as $d \to 0$.

Step III. One finds w_0, defined near the boundary, such that

$$Lw_0 + 2\Delta d = 0, \tag{11}$$

and

$$d^k w_0 \in C^{k+\alpha} \text{ for } k = 0, 1, 2,$$

and proves that one can formally set $d = 0$ in equation (11), so that

$$w_0 = \Delta d/(n - 1) = -H$$

on the boundary. This accounts for the role of the mean curvature H of the boundary.

Step IV. Let $Z = w - w_0$. One proves, using comparison functions involving d, that $Z = O(d)$.

Step V. Since Z and LZ are both $O(d)$ one first gets, by the "type (I)" Theorem 3.2, that Z and $d\nabla Z$ are of class C^α. It follows, by inspection of the definition of $M_w(w)$, that LZ is in fact of class C^α near and up to the boundary.

Step VI. Since LZ, Z and $d\nabla Z$ are C^α near and up to the boundary, Theorem 3.3 gives that d^2Z is of class $C^{2+\alpha}$. Since $w = w_0 + Z$, we find that $2d + d^2w$ is of class $C^{2+\alpha}$ near the boundary, QED.

References

[1] Bandle C., Essén M., On the solution of quasilinear elliptic problems with boundary blow-up, Symposia Math. 35 (1994) 93–111.

[2] Bandle C., Flucher M., Harmonic radius and concentration of energy; hyperbolic radius and Liouville's equations $\Delta U = e^U$ and $\Delta U = U^{\frac{n+2}{n-2}}$, SIAM Review 38 (1996) 191–238.

[3] Bandle C., Marcus M., On second-order effects in the boundary behavior of large solutions of semilinear elliptic problems, Differ. and Integral Equations 11 (1998) 23–34.

[4] Bandle C., Marcus M., Asymptotic behavior of solutions and their derivatives, for semilinear elliptic problems with blow-up on the boundary, Ann. IHP (Analyse Non Linéaire) 12 (2) (1995) 155–171.

[5] Bentrad, A., Kichenassamy, S., A linear Fuchsian equation with variable indices, J. of Differential Equations, 190 (1) (2003) 64–80.

[6] Bénilan, B., Brezis, H., Nonlinear problems related to the Thomas-Fermi equation, J. of Evolution Equations 3 (4) (2003) 673–770.

[7] Berhanu S., Porru G., Qualitative and quantitative estimates for large solutions to semilinear equations, Commun. Applied Analysis 4 (1) (2000) 121–131.

[8] Bieberbach, L., $\Delta u = e^u$ und die automorphen Funktionen, Math. Ann. 77 (1916) 173–212.

[9] Brezis, H., *Opérateurs maximaux monotones et semi-groupes de contraction dans les espaces de Hilbert*, North-Holland, Amsterdam, 1973.

[10] Brezis H., Vázquez, J. L., Blow-up of solutions of some nonlinear elliptic problems, Rev. Mat. Univ. Complutense Madrid 10 (2) (1997) 443–469

[11] Cabart, C., Kichenassamy, S., Explosion et normes L^p pour l'équation des ondes non linéaire cubique, C. R. Acad. Sci. Paris, sér. 1, 355 (11) (2002) 903–908.

[12] Caffarelli L.A., Friedman A., Convexity of solutions of semilinear elliptic equations, Duke Math. J. 52 (1985) 431–457.

[13] Gilbarg D., Trudinger N., *Elliptic Partial Differential Equations of Elliptic Type*, Springer, 1983.

[14] Jager, L., Kichenassamy, S., Stellar models and irregular singularities, Communications in Contemporary Mathematics 5 (5) (2003) 719–735.

[15] Keller J.B., On solutions of $\Delta u = f(u)$, Comm. Pure Appl. Math. 10 (1957) 503–510.

[16] Kichenassamy S., Quasilinear problems with singularities, Manuscripta Math. 57 (1987) 281–313.

[17] Kichenassamy S., Régularité du rayon hyperbolique, C. R. Acad. Sci. Paris, sér. 1, 338 (1) (2004) 13–18.

[18] Kichenassamy S., Boundary blow-up and degenerate equations, J. Functional Analysis 215 (2) (2004) 271–289.

[19] Kichenassamy S., Boundary behavior in the Loewner-Nirenberg problem, in J. Functional Analysis 222 (1) (2005) 98–113.

[20] Kichenassamy S., On a conjecture of Fefferman and Graham, Advances in Math. 184 (2004) 268–288.

[21] Kichenassamy S., The blow-up problem for exponential nonlinearities, Communications in PDE, 21 (1&2) (1996) 125–162.

[22] Kichenassamy S., WTC expansions and non-integrable equations, Studies in Applied Mathematics 102 (1999) 1–26.

[23] Kichenassamy S., Stability of blow-up patterns for nonlinear wave equations, *in:* Nonlinear PDEs, Dynamics and Continuum Physics, (J. Bona, K. Saxton and R. Saxton eds.), Contemporary Mathematics 255 (2000) 139–162.

[24] Kichenassamy, S., Fuchsian equations in Sobolev spaces and blow-up, Journal of Differential Equations, 125 (1996) 299–327.

[25] Kondrat'ev V.A., Nikishkin V.A., Asymptotics, near the boundary, of a solution of a singular boundary-value problem for a semilinear elliptic equation, Differ. Eqs. 26 (1990) 345–348.

[26] Lazer A., McKenna P.J., Asymptotic behavior of boundary blow-up problems, Differ. and Integral Eqs. 7 (1994) 1001–1019.

[27] Loewner C., Nirenberg L., Partial differential equations invariant under conformal or projective transformations, in: *Contributions to Analysis*, Ahlfors L. et al. (Eds.), Acad. Press, 1974, pp. 245–272.

[28] Marcus M., Véron L., Uniqueness and asymptotic behavior of solutions with boundary blow-up for a class of nonlinear elliptic equations, Ann. IHP (Analyse Non Linéaire) 14 (1997) 237–274.

[29] Osserman R., On the inequality $\Delta u \geq f(u)$, Pacific J. Math. 7 (1957) 1641–1647.

Satyanad Kichenassamy
Laboratoire de Mathématiques (UMR 6056)
CNRS & Université de Reims Champagne-Ardenne
Moulin de la Housse, B.P. 1039
F-51687 Reims Cedex 2, France
e-mail: `satyanad.kichenassamy@univ-reims.fr`

Progress in Nonlinear Differential Equations
and Their Applications, Vol. 63, 343–351

Maximum Principle for Bounded Solutions of the Telegraph Equation: The Case of High Dimensions

Jean Mawhin

To Haïm Brezis, for many years of inspiration and friendship

1. Introduction

The equation for the transmission of an electrical signal through a telegraph line
is given by

$$LV_{tt} + \left(R + \frac{GL}{C} \right) V_t - \frac{1}{C} V_{xx} + \frac{GR}{C} V = 0, \tag{1}$$

where V is the electrical potential, L the self-inductance per unit of length, R
the resistance per unit of length, C the capacity per unit of length, G the conductance per unit of length. In 1855, motivated by the problem of establishing
a telegraphic line between Great Britain and United States, Thomson [10] modelled the propagation of signals through an immersed cable by equation (1) with
$L = G = 0$, which is equivalent to Fourier's heat equation. Three years later,
Kirchhoff [2] deduced equation (1) with no conductance ($G = 0$) from Weber's
electromagnetic theory. Equation (1) was obtained in 1876 by Heaviside [1] from
Maxwell theory. All those authors only computed special solutions of (1) through
separation of variables. The first general solution of equation (1) was first given
by Poincaré [8] in 1893, using Fourier transforms, and one year later by Picard [7],
using Riemann's method.

Maximum principles, in their various forms, play an important role in the
study of linear and nonlinear second order partial differential equations of elliptic
and parabolic type [9]. For the linear elliptic equation

$$-\Delta u(x) + \lambda u(x) = f(x), \quad x \in \Omega \subset \mathbb{R}^N \tag{2}$$

with Dirichlet, Neumann or Robin boundary conditions

$$Bu = 0, \quad x \in \partial\Omega \tag{3}$$

or with periodic boundary conditions

$$x \in \mathbb{T}^N = (\mathbb{R}/2\pi\mathbb{Z})^N \tag{4}$$

one (weak) form of the *maximum principle* is :

$$(\forall\, \lambda > 0)(\forall\, N \geq 1) : f \geq 0 \;\Rightarrow\; u \geq 0.$$

This result is basic in justifying the *method of upper and lower solutions* for semi-linear elliptic equations

$$-\Delta u(x) = F(x, u(x)) \tag{5}$$

with the boundary conditions (3) or (4). It allows to show the convergence of the iterations

$$-\Delta u_{k+1}(x) + \lambda u_{k+1}(x) = F(x, u_k(x)) + \lambda u_k(x) \quad (k \in \mathbb{N}),$$

$$u_0 = \alpha \text{ or } u_0 = \beta,$$

where $\alpha \leq \beta$ are respectively a lower and an upper solution of equation (5), i.e., verify the differential inequalities

$$-\Delta\alpha(x) \leq F(x, \alpha(x)), \quad -\Delta\beta(x) \geq F(x, \beta(x)),$$

and suitable inequalities on the boundary of Ω, depending on the boundary conditions. Similar results hold for linear and semilinear parabolic equations

$$u_t(t, x) - \Delta u(t, x) + \lambda u(t, x) = 0$$

$$u_t(t, x) - \Delta u(t, x) = F(t, x, u(t, x)),$$

wit suitable initial and boundary conditions.

Equation (1) is a hyperbolic equation, and it is well known that, in contrast to elliptic and parabolic ones, this class of partial differential equations does not admit similar maximum principles (see, e.g., [9]). In recent papers [5, 3, 4, 6], Ortega, Robles-Pérez and the author have shown that maximum principles in the sense of the positivity of the inverse operator exist for time-periodic or time-bounded solutions of the telegraph equation

$$\mathcal{L}u + \lambda u = f(t, x) \tag{6}$$

with periodic boundary conditions in the space variable, where $\mathcal{L}$ is the telegraph operator defined by

$$\mathcal{L}u := u_{tt} + cu_t - \Delta u, \tag{7}$$

when $c > 0$,

$$0 < \lambda < \frac{c^2}{4}, \tag{8}$$

and the space dimension is smaller or equal to three. For $N = 1$, $\mathcal{L}$ can be seen as the sum of the two ordinary differential operators $L_1 u := u_{tt} + cu_t$ and $L_2 u = -u_{xx}$.

Applied to functions depending only upon t and bounded over $\mathbb{R}$, one can show that $(L_1 + \lambda I)^{-1} f(t) \geq 0$ for $f(t) \geq 0$ if and only if inequality (8) holds; on the other hand, applied to functions depending only upon x and T-periodic, one can show that $(L_2 + \lambda I)^{-1} f(x) \geq 0$ for $f(x) \geq 0$ if and only if $\lambda > 0$. Consequently, a maximum principle for the solutions of (6) bounded in t and periodic in x can only hold if inequality (8) is satisfied. The results of the four above-mentioned papers show that this necessary condition is also sufficient when $1 \leq N \leq 3$, and that no maximum principle exist for $N \geq 4$. So, this type of maximum principle for the telegraph equation differs from the corresponding one for the elliptic or parabolic equations in two respects:

1. The maximum principle does not hold for all $\lambda > 0$ but only for λ verifying (8).
2. The maximum principle only holds in space dimensions $N \leq 3$.

We describe those results here, and refer to [3, 4, 6]) for the complete proofs.

2. Maximum principles for the telegraph equation when $N \leq 3$

We first define the concept of *bounded solution* we are interested in. Let us define the operator $\mathcal{L}^*$ by

$$\mathcal{L}^* u := u_{tt}(t, x) - c u_t(t, x) - \Delta u(t, x).$$

Definition 1. *If $f \in L^\infty(\mathbb{R} \times \mathbb{T}^3)$, a weak bounded solution of (6) is a function $u \in L^\infty(\mathbb{R} \times \mathbb{T}^3)$, such that, for all $\varphi \in \mathcal{D}(\mathbb{R} \times \mathbb{T}^3)$, one has*

$$\int_{\mathbb{R} \times \mathbb{T}^3} [u \mathcal{L}^* \varphi + \lambda u \varphi] = \int_{\mathbb{R} \times \mathbb{T}^3} f \varphi,$$

in other words such that

$$\mathcal{L} u + \lambda u = f \quad in \quad \mathcal{D}'(\mathbb{R} \times \mathbb{T}^3).$$

The following lemma is useful.

Lemma 1. *If there exists $\Lambda > 0$ such that, for each $f \in L^\infty(\mathbb{R} \times \mathbb{T}^3)$, equation*

$$\mathcal{L} u + \Lambda u = f \quad in \quad \mathcal{D}'(\mathbb{R} \times \mathbb{T}^3) \tag{9}$$

has a unique solution $u \in L^\infty(\mathbb{R} \times \mathbb{T}^3)$, and that $u \geq 0$ a.e. whenever $f \geq 0$ a.e., then the same is true for equation (6) for all $\lambda \in (0, \Lambda]$.

Proof. Let us define the linear operator $\mathcal{R}$ by

$$\mathcal{R} : L^\infty(\mathbb{R} \times \mathbb{T}^3) \to L^\infty(\mathbb{R} \times \mathbb{T}^3), f \mapsto u$$

where $u \in L^\infty(\mathbb{R} \times \mathbb{T}^3)$ is the unique weak solution of equation (9). Notice first that if $M = \|f\|_{L^\infty(\mathbb{R} \times \mathbb{T}^3)}$, then $U_+ = \frac{M}{\Lambda}$ (resp. $U_- = -\frac{M}{\Lambda}$) is the unique bounded solution of

$$\mathcal{L} u + \Lambda u = M \quad in \quad \mathcal{D}'(\mathbb{R} \times \mathbb{T}^3)$$
$$(resp. \quad \mathcal{L} u + \Lambda u = -M \quad in \quad \mathcal{D}'(\mathbb{R} \times \mathbb{T}^3)),$$

and hence

$$\mathcal{L}(U_+ - u) + \Lambda(U_+ - u) = M - f \quad \text{in} \quad \mathcal{D}'(\mathbb{R} \times \mathbb{T}^3),$$
$$\mathcal{L}(u - U_-) + \Lambda(u - U_-) = f + M \quad \text{in} \quad \mathcal{D}'(\mathbb{R} \times \mathbb{T}^3).$$

The maximum principle implies that $U_- \leq u(t, x) \leq U_+$ a.e. and hence

$$\|\mathcal{R}f\|_{L^\infty(\mathbb{R}\times\mathbb{T}^3)} \leq \frac{1}{\Lambda}\|f\|_{L^\infty(\mathbb{R}\times\mathbb{T}^3)}. \tag{10}$$

Now, if $\lambda \in \,]0, \Lambda]$, we can write equation (6) as

$$\mathcal{L}u + \Lambda u = (\Lambda - \lambda)u + f \quad \text{in} \quad \mathcal{D}'(\mathbb{R} \times \mathbb{T}^3)$$

or, equivalently

$$u = \mathcal{R}[(\Lambda - \lambda)u + f]. \tag{11}$$

As, using (10),

$$\|(\Lambda - \lambda)\mathcal{R}\|_{L^\infty(\mathbb{R}\times\mathbb{T}^3)} \leq \frac{\Lambda - \lambda}{\Lambda} < 1,$$

the corresponding Neumann series is convergent and

$$u = [I - (\Lambda - \lambda)\mathcal{R}]^{-1}\mathcal{R}f = \sum_{k=0}^{\infty}(\Lambda - \lambda)^k \mathcal{R}^{k+1} f. \tag{12}$$

As $f \geq 0$ implies that $\mathcal{R}^k f \geq 0$ for all $k \geq 1$, the result follows. $\qquad\square$

We now state the main result of this section.

Theorem 1. *For each $\lambda \in \left(0, \frac{c^2}{4}\right]$ and each $f \in L^\infty(\mathbb{R} \times \mathbb{T}^3)$, there exists a unique $u \in L^\infty(\mathbb{R} \times \mathbb{T}^3)$ such that*

$$\mathcal{L}u + \lambda u = f \quad \text{in} \quad \mathcal{D}'(\mathbb{R} \times \mathbb{T}^3).$$

Furthermore if $f \geq 0$ a.e., then $u \geq 0$ a.e.

Sketch of the proof. It is divided in three steps. *1. Uniqueness.* It is proved by contradiction for $f = 0$. First reduce the problem to the case of a smooth solution by convolution, then take the Fourier coefficient $v(t)$ of an eigenfunction of Laplace-Beltrami operator on $\mathbb{T}^3$ such that $v(t) \not\equiv 0$, and deduce that the ordinary differential equation

$$\ddot{v}(t) + c\dot{v}(t) + \eta v(t) = 0$$

has a nontrivial bounded solution for some $\eta > 0$, a contradiction.

2. Existence and maximum principle for $\lambda = \frac{c^2}{4}$. Introduce the function

$$\chi(x) := \frac{1}{4\pi|x|}e^{-\frac{c|x|}{2}} \in L^1(\mathbb{R}^3),$$

so that

$$\|\chi\|_{L^1(\mathbb{R}^3)} = \frac{4}{c^2}.$$

Then define the Radon measure $\mathcal{U}_3$ by

$$\langle \mathcal{U}_3, \varphi \rangle = \frac{1}{4\pi} \int_{\mathbb{R}^3} e^{-\frac{c|x|}{2}} \frac{\varphi(|x|, x)}{|x|} \, dx,$$

so that

$$\|\mathcal{U}_3\|_{M(\mathbb{R} \times \mathbb{R}^3)} = \frac{4}{c^2}.$$

Using Kirchhoff's formula

$$\varphi(0, 0) = \frac{1}{4\pi} \int_{\mathbb{R}^3} \frac{(\Box \varphi)(|x|, x)}{|x|} \, dx,$$

one sees that $\mathcal{U}_3$ solves the equation

$$\mathcal{L}u + \frac{c^2}{4} u = \delta \quad \text{in} \quad \mathcal{D}'(\mathbb{R} \times \mathbb{R}^3).$$

Let

$$\psi(t, x) := e^{-\frac{ct}{2}} \varphi(t, x).$$

Then,

$$\mathcal{L}^* \varphi + \frac{c^2}{4} \varphi = e^{\frac{ct}{2}} \Box \psi$$

$$\langle (\mathcal{L} + \frac{c^2}{4}) \mathcal{U}_3, \varphi \rangle = \langle \mathcal{U}_3, e^{\frac{ct}{2}} \Box \psi \rangle = \psi(0, 0) = \langle \delta, \varphi \rangle.$$

Therefore

$$u(t, x) := (\mathcal{U}_3 * f)(t, x) = \int_{\mathbb{R} \times \mathbb{R}^3} f(t - \tau, x - \xi) d\mathcal{U}_3(\tau, \xi).$$

Consequently, $u \in L^\infty(\mathbb{R} \times \mathbb{R}^3)$,

$$\|u\|_{L^\infty} \leq \|\mathcal{U}_3\|_M \|f\|_{L^\infty} = \frac{4}{c^2} \|f\|_{L^\infty}$$

and

$$\mathcal{L}u + \frac{c^2}{4} u = f \quad \text{in} \quad \mathcal{D}'(\mathbb{R} \times \mathbb{R}^3).$$

Now, as $f \in L^\infty(\mathbb{R} \times \mathbb{T}^3)$, we have $u \in L^\infty(\mathbb{R} \times \mathbb{T}^3)$. Finally, use a partition of unity periodic in x to show that one can replace $\mathcal{D}'(\mathbb{R} \times \mathbb{R}^3)$ by $\mathcal{D}'(\mathbb{R} \times \mathbb{T}^3)$.

3. *Existence and maximum principle for* $\lambda \in \left(0, \frac{c^2}{4}\right)$. Use Part 2 of the proof and Lemma 9) with $\Lambda = \frac{c^2}{4}$. $\qquad\square$

Remark 1. It can be shown by examples that $\frac{c^2}{4}$ is optimal and that the maximum principle is not strong.

Remark 2. It follows from Lemma 9) that

$$\|u\|_{L^\infty} \leq \frac{1}{\lambda} \|f\|_{L^\infty}.$$

Remark 3. In the physical notations of equation (1), the threshold value $\lambda = \frac{c^2}{4}$ corresponds to the case where

$$RC = GL$$

already emphasized by Heaviside [1], as being the one under which a signal is not dispersed through the correspond telegraph equation.

Remark 4. It is easy to show that if $f \in L^\infty((T/2\pi)\mathbb{T} \times \mathbb{T}^3)$, the unique weak solution $u \in L^\infty(\mathbb{R} \times \mathbb{T}^3)$ of

$$\mathcal{L}u + \lambda u = f$$

is such that $u \in L^\infty((T/2\pi)\mathbb{T} \times \mathbb{T}^3)$.

Similar existence, uniqueness and maximum principle results can be deduced from Theorem 1 by a method of descent. For example, if $N = 2$, and $f = f(t, x_1, x_2) : \mathbb{R} \times \mathbb{T}^2 \to \mathbb{R}$, one defines $\widetilde{f} : \mathbb{R} \times \mathbb{T}^3 \to \mathbb{R}$, by

$$\widetilde{f}(t, x_1, x_2, x_3) = f(t, x_1, x_2).$$

Now, $L^\infty(\mathbb{R} \times \mathbb{T}^2) \hookrightarrow L^\infty(\mathbb{R} \times \mathbb{T}^3)$, $\mathcal{D}'(\mathbb{R} \times \mathbb{T}^2) \hookrightarrow \mathcal{D}'(\mathbb{R} \times \mathbb{T}^3)$, and Theorem 1 implies the existence of $w \in L^\infty(\mathbb{R} \times \mathbb{T}^3)$ such that

$$\mathcal{L}w + \lambda w = \widetilde{f} \quad \text{in} \quad \mathcal{D}'(\mathbb{R} \times \mathbb{T}^3).$$

Then $u := \frac{1}{2\pi} \int_{\mathbb{T}} w(t, x_1, x_2, x_3) \, dx_3 \in L^\infty(\mathbb{R} \times \mathbb{T}^2)$ and

$$\mathcal{L}u + \lambda u = f \quad \text{in} \quad \mathcal{D}'(\mathbb{R} \times \mathbb{T}^2).$$

Better regularity results can be proved in lower dimensions, as indicated in the following statements (see [3, 6]).

Theorem 2. *For each* $\lambda \in \left(0, \frac{c^2}{4}\right]$, *and each* $f \in L^\infty(\mathbb{R} \times \mathbb{T}^2)$, *there exists a unique* $u \in L^\infty(\mathbb{R} \times \mathbb{T}^2)$ *such that*

$$\mathcal{L}u + \lambda u = f \quad in \quad \mathcal{D}'(\mathbb{R} \times \mathbb{T}^2).$$

Furthermore if $f \geq 0$ *a.e. then* $u \geq 0$ *a.e., and* u *is continuous.*

Theorem 3. *For each* $\lambda \in \left(0, \frac{c^2}{4}\right]$, *and each* $f \in L^\infty(\mathbb{R} \times \mathbb{T})$, *there exists a unique* $u \in L^\infty(\mathbb{R} \times \mathbb{T}))$ *such that*

$$\mathcal{L}u + \lambda u = f \quad in \quad \mathcal{D}'(\mathbb{R} \times \mathbb{T}).$$

Furthermore if $f \geq 0$ *a.e. then* $u \geq 0$ *a.e., and* $u \in W^{1,\infty}(\mathbb{R} \times \mathbb{T})$.

Remark 5. Various applications of the above results to a method of upper and lower solutions for semilinear telegraph equations and to the existence of bounded or almost periodic solutions of dissipative sine-Gordon or Duffing equations can be found in [3, 6].

3. Obstruction to a maximum principle for $N = 4$

The fundamental solution of the wave operator can be computed explicitly also in dimension $N \geq 4$, but is not a measure anymore and so cannot be positive. In this section we exploit this fact to show that an analogous of Theorem 1 is not valid for $N = 4$ and $\lambda = \frac{c^2}{4}$.

Consider the problem of finding $u \in L^\infty(\mathbb{R} \times \mathbb{T}^4)$ such that

$$\mathcal{L}u + \frac{c^2}{4}u = f(t,x) \quad \text{in } \mathcal{D}'(\mathbb{R} \times \mathbb{T}^4), \tag{13}$$

where f is a given function in $L^\infty(\mathbb{R} \times \mathbb{T}^4)$. This problem has at most one solution because the proof of uniqueness of Theorem 1 is easily adapted. The existence of solution is more delicate and we introduce a specific class of smooth functions. A function f is said to belong to $BC^\infty(\mathbb{R} \times \mathbb{T}^4)$ if it belongs to $C^\infty(\mathbb{R} \times \mathbb{T}^4)$ and all its successive derivatives $\partial^\alpha f$, $0 \leq |\alpha| < \infty$, are bounded.

Proposition 1. *For each $f \in BC^\infty(\mathbb{R} \times \mathbb{T}^4)$, there exists a unique $u \in BC^\infty(\mathbb{R} \times \mathbb{T}^4)$ solving*

$$\mathcal{L}u + \frac{c^2}{4}u = f(t,x) \quad \text{in } \mathcal{D}'(\mathbb{R} \times \mathbb{T}^4)$$

Sketch of the proof. Let

$$\mathcal{N}(s,\xi) = \frac{1}{2\sqrt{s + |\xi|^2}} e^{-\frac{c}{2}\sqrt{s + |\xi|^2}}, \quad s > 0,\ \xi \in \mathbb{R}^4.$$

Define $\mathcal{U}_4 \in \mathcal{D}'(\mathbb{R} \times \mathbb{R}^4)$ by

$$\langle \mathcal{U}_4, \phi \rangle = -\frac{1}{2\pi^2} \int_0^\infty \frac{1}{\sqrt{s}} \frac{d}{ds} \left(\int_{\mathbb{R}^4} \mathcal{N}(s,\xi)\phi(\sqrt{s + |\xi|^2},\xi)\, d\xi \right) ds.$$

Then

$$|\langle \mathcal{U}_4, \phi \rangle| \leq k_1 \left[\|\phi\|_{L^\infty} + \|\frac{\partial \phi}{\partial t}\|_{L^\infty} \right]$$

and $\mathcal{U}_4$ solves

$$\mathcal{L}u + \frac{c^2}{4}u = \delta \quad \text{in } \mathcal{D}'(\mathbb{R} \times \mathbb{R}^4).$$

Indeed, if we define $\mathcal{E}_4 \in \mathcal{D}'(\mathbb{R} \times \mathbb{R}^4)$ by

$$\langle \mathcal{E}_4, \phi \rangle = -\frac{1}{2\pi^2} \int_0^\infty \frac{1}{\sqrt{s}} \frac{d}{ds} \left(\int_{\mathbb{R}^4} \frac{\phi(\sqrt{s + |\xi|^2},\xi)}{2\sqrt{s + |\xi|^2}}\, d\xi \right) ds,$$

then $\mathcal{E}_4$ solves

$$u_{tt} - \Delta_x u = \delta \quad \text{in } \mathcal{D}'(\mathbb{R} \times \mathbb{R}^4),$$
$$\langle \mathcal{U}_4, e^{\frac{c}{2}t}\Box\psi \rangle = \langle \mathcal{E}_4, \Box\psi \rangle$$

for all $\psi \in \mathcal{D}(\mathbb{R} \times \mathbb{R}^4)$, and the rest goes as in the proof for $\mathcal{U}_3$ in Theorem 1. For $f \in \mathcal{D}(\mathbb{R} \times \mathbb{R}^4)$, define $u = \mathcal{U}_4 * f$. Then $u \in BC^\infty(\mathbb{R} \times \mathbb{R}^4)$,

$$\mathcal{L}u + \frac{c^2}{4}u = f \quad \text{in } \mathcal{D}'(\mathbb{R} \times \mathbb{R}^4), \qquad \|u\|_{L^\infty} \leq k_1 \left[\|f\|_{L^\infty} + \|\frac{\partial f}{\partial t}\|_{L^\infty} \right].$$

Notice that the formula $\mathcal{U}_4 \star f$ defining u still makes sense for $f \in BC^\infty(\mathbb{R} \times \mathbb{R}^4)$. For $f \in BC^\infty(\mathbb{R} \times \mathbb{R}^4)$ define now

$$u(t,x) = \langle \mathcal{U}_4, T_{(t,x)}\check{f}\rangle, \quad (t,x) \in \mathbb{R} \times \mathbb{R}^4$$

$$T_{(t,x)}f(\tau,\xi) = f(t+\tau, x+\xi), \quad \check{f}(\tau,\xi) = f(-\tau,-\xi).$$

One can show that $u \in BC^\infty(\mathbb{R} \times \mathbb{T}^4)$ is the required solution. Furthermore,

$$u(0,0) = -\frac{1}{2\pi^2}\int_0^\infty \frac{1}{\sqrt{s}}\frac{d}{ds}\left(\int_{\mathbb{R}^4} \mathcal{N}(s,\xi)\check{f}(\sqrt{s+|\xi|^2},\xi)d\xi\right) ds. \qquad \square$$

We also need the following lemma, whose proof is straightforward.

Lemma 2. *Let $\chi \in C^\infty(\mathbb{R})$ be such that*

$$\chi' \geq 0, \ \chi' \geq \frac{1}{2} \ on \ [\frac{5}{4}, \frac{7}{4}],$$

$$\chi(s) = 1 \ if \ s \geq 2, \ \chi(s) = 0 \ if \ s \leq 1.$$

Then, for each $G \in C^1(0,\infty) \cap C[0,\infty)$ such that

$$G(0) > 0, \ \int_0^\infty s^{-1/2}|G'(s)|ds < +\infty,$$

one has

$$\lim_{\varepsilon \to 0^+} \int_0^\infty \frac{1}{\sqrt{s}}\frac{d}{ds}\left(G(s)\chi(s/\varepsilon)\right) ds = +\infty.$$

Finally, the following Proposition implies the claimed non-existence of the maximum principle.

Proposition 2. *There exist a sequence (f_n) in $BC^\infty(\mathbb{R} \times \mathbb{T}^4)$ with $0 \leq f_n \leq 1$ such that, if*

$$\mathcal{L}u_n + \frac{c^2}{4}u_n = f_n \quad in \quad \mathcal{D}'(\mathbb{R} \times \mathbb{T}^4),$$

one has

$$u_n(0,0) \to -\infty \quad as \quad n \to \infty.$$

Sketch of the proof. Define $\psi \in C^\infty(\mathbb{R})$ even and such that

$$\psi = 1 \ if \ 0 \leq t \leq \frac{\pi}{2}, \ \psi = 0 \ if \ t \geq \pi, \ 0 \leq \psi \leq 1.$$

For $\|x\|_\infty \leq \pi$ and $t \in \mathbb{R}$ define

$$f_\varepsilon(t,x) = \chi\left(\frac{t^2 - |x|^2}{\varepsilon}\right)\psi(t),$$

with χ like in Lemma 2. Then, f_ε vanishes in a neighborhood of $\|x\|_\infty = \pi$. Extend f_ε by periodicity to

$$f_\varepsilon \in C^\infty(\mathbb{R} \times \mathbb{T}^4), \ 0 \leq f_\varepsilon \leq 1, \ f_\varepsilon \in BC^\infty(\mathbb{R} \times \mathbb{T}^4).$$

One can show that

$$\int_{\mathbb{R}^4} \mathcal{N}(s,\xi)\check{f}_\varepsilon(\sqrt{s+|\xi|^2},\xi)\,d\xi = \chi(s/\varepsilon)G(s).$$

Finally, use Lemma 2 to check that, for $\varepsilon \searrow 0$,

$$u_\varepsilon(0,0) = -\frac{1}{2\pi^2}\int_0^\infty \frac{1}{\sqrt{s}}\frac{d}{ds}\left(G(s)\chi(s/\varepsilon)\right)ds \to -\infty. \qquad \square$$

References

[1] O. Heaviside, On the extra current, *Phil. Mag.* (1876) (5) **2**, 135.

[2] G. Kirchhoff, Über die Bewegung der Elektrizität in Drähten, *Pogg. Ann. der Phys.* 100 (1857), p. 193, 251.

[3] J. Mawhin, R. Ortega and A.M. Robles-Pérez, A maximum principle for bounded solutions of the telegraph equations and applications to nonlinear forcings, *J. Math. Anal. Appl.* **251** (2000), 695–709.

[4] J. Mawhin, R. Ortega and A.M. Robles-Pérez, A maximum principle for bounded solutions of the telegraph equation in space dimension three, *C.R. Acad. Sci. Paris, Ser. I* **334** (2002), 1089–1094.

[5] R. Ortega and A.M. Robles-Pérez, A maximum principle for periodic solutions of the telegraph equation, *J. Math. Anal. Appl.* **221** (1998), 625–651.

[6] J. Mawhin, R. Ortega and A.M. Robles-Pérez, Maximum principles for bounded solutions of the telegraph equation in space dimensions two and three and applications, *J. Differential Equations*, 208 (2005), 42–63.

[7] E. Picard, Sur une équation aux dérivées partielles de la théorie de la propagation de l'électricité, *Bull. Sci. Math.* 22 (1894), 2–8.

[8] H. Poincaré, Sur la propagation de l'électricité, *CRAS Paris* 117 (1893), 1027–1032.

[9] M.H. Protter and H. Weinberger, Maximum Principles in Differential Equations, Prentice Hall, Englewoods Cliffs, N.J., 1967.

[10] W. Thomson, On the theory of the electric telegraph, *Proc. Roy. Soc.* 7 (1855).

Jean Mawhin
Département de mathématique
Université Catholique de Louvain
B-1348 Louvain-la-Neuve, Belgium
e-mail: `mawhin@math.ucl.ac.be`

Progress in Nonlinear Differential Equations
and Their Applications, Vol. 63, 353–364
© 2005 Birkhäuser Verlag Basel/Switzerland

Kolmogorov Equations
in Physics and in Finance

Andrea Pascucci

Abstract. This paper contains a survey of results about linear and nonlinear partial differential equations of Kolmogorov type arising in physics and in mathematical finance. Some recent pointwise estimates proved in collaboration with S. Polidoro are also presented.

Mathematics Subject Classification (2000). AMS Subject Classification: 35K57, 35K65, 35K70.

1. Introduction

We consider a class of the differential equations of Kolmogorov type of the form

$$Lu \equiv \sum_{i,j=1}^{p_0} a_{ij}(z)\partial_{x_i x_j}u + \sum_{i=1}^{p_0} a_i(z)\partial_{x_i}u + \sum_{i,j=1}^{N} b_{ij}x_i\partial_{x_j}u + c(z)u - \partial_t u = 0, \quad (1.1)$$

where $z = (x,t) \in \mathbb{R}^N \times \mathbb{R}$ and $1 \leq p_0 \leq N$. By convenience, hereafter the term "Kolmogorov equation" will be shortened to KE. We assume the following hypotheses:

H.1 the matrix $A_0 = (a_{ij})_{i,j=1,\ldots,p_0}$ is symmetric and uniformly positive definite in $\mathbb{R}^{p_0}$: there exists a positive constant μ such that

$$\frac{|\eta|^2}{\mu} \leq \sum_{i,j=1}^{p_0} a_{ij}(z)\eta_i\eta_j \leq \mu|\eta|^2, \qquad \forall \eta \in \mathbb{R}^{p_0}, \ z \in \mathbb{R}^{N+1}; \qquad (1.2)$$

Investigation supported by the University of Bologna. Funds for selected research topics.

H.2 the matrix $B \equiv (b_{ij})$ has constant real entries and takes the following block from:

$$\begin{pmatrix} * & B_1 & 0 & \cdots & 0 \\ * & * & B_2 & \cdots & 0 \\ \vdots & \vdots & \vdots & \ddots & \vdots \\ * & * & * & \cdots & B_r \\ * & * & * & \cdots & * \end{pmatrix} \tag{1.3}$$

where B_j is a $p_{j-1} \times p_j$ matrix of rank p_j, with

$$p_0 \geq p_1 \geq \cdots \geq p_r \geq 1, \qquad p_0 + p_1 + \cdots + p_r = N,$$

and the $*$-blocks are arbitrary.

The prototype of (1.1) is the following equation

$$\partial_{x_1 x_1} u + x_1 \partial_{x_2} u - \partial_t u = 0, \qquad (x_1, x_2, t) \in \mathbb{R}^3, \tag{1.4}$$

whose fundamental solution was explicitly constructed by Kolmogorov [25]. In his celebrated paper [23], Hörmander generalized this result to *constant coefficients KEs*, i.e., equations of the form (1.1), with constant a_{ij} and $a_i = c \equiv 0$ for $i = 1, \ldots, p_0$, satisfying the following condition:

$\mathrm{Ker}(A)$ *does not contain non-trivial subspaces which are invariant for B.* (1.5)

In (1.5), A denotes the $N \times N$ matrix

$$A = \begin{pmatrix} A_0 & 0 \\ 0 & 0 \end{pmatrix}. \tag{1.6}$$

Let us recall that, for constant coefficients equations, condition (1.5) is equivalent to the structural assumptions H.1–H.2 which in turn are equivalent to the classical Hörmander condition:

$$\mathrm{rank\ Lie}\,(X_1, \ldots, X_{p_0}, Y) = N + 1, \tag{1.7}$$

at any point of $\mathbb{R}^{N+1}$. In (1.7), $\mathrm{Lie}\,(X_1, \ldots, X_{p_0}, Y)$ denotes the Lie algebra generated by the vector fields

$$X_i = \sum_{j=1}^{p_0} a_{ij} \partial_{x_j}, \ i = 1, \ldots, p_0, \qquad \text{and} \qquad Y = \langle x, BD \rangle - \partial_t, \tag{1.8}$$

where $\langle \cdot, \cdot \rangle$ and D respectively denote the inner product and the gradient in $\mathbb{R}^N$. A proof of the equivalence of these conditions is given by Kupcov in [26], Theorem 3 and by Lanconelli and Polidoro in [30], Proposition A.1.

Equation (1.4) is the lowest dimension version of the following ultraparabolic equation in $\mathbb{R}^{N+1}$ with $N = 2n$:

$$\sum_{j=1}^{n} \partial_{x_j}^2 + \sum_{j=1}^{n} x_j \partial_{x_{n+j}} - \partial_t = 0. \tag{1.9}$$

Kolmogorov introduced (1.9) in 1934 in order to describe the probability density of a system with $2n$ degree of freedom. The $2n$-dimensional space is the phase space, $(x_1, \ldots, x_n)$ is the velocity and $(x_{n+1}, \ldots, x_{2n})$ the position of the system. We also recall that (1.9) is a prototype for a family of evolution equations arising in the kinetic theory of gases that take the following general form

$$Yu = \mathcal{J}(u). \tag{1.10}$$

Here $\mathbb{R}^{2n} \ni x \longmapsto u(x,t) \in \mathbb{R}$ is the density of particles which have velocity $(x_1, \ldots, x_n)$ and position $(x_{n+1}, \ldots, x_{2n})$ at time t,

$$Yu \equiv \sum_{j=1}^{n} x_j \partial_{x_{n+j}} u + \partial_t u$$

is the so-called *total derivative of u* and $\mathcal{J}(u)$ describes some kind of collision. This last term can take different form, it may also occur in non-divergence form and its coefficients may depend on $z \in \mathbb{R}^{2n+1}$ as well as on the solution u. For instance, in the usual Fokker-Planck equation, we have

$$\mathcal{J}(u) = -\sum_{i,j=1}^{n} \partial_{x_i} \left(a_{ij} \partial_{x_j} u + b_i u \right) + \sum_{i=1}^{n} a_i \partial_{x_i} u + cu \tag{1.11}$$

where a_{ij}, a_i, b_i and c are functions of z. In the Boltzmann-Landau equation (see [9], [31] and [32])

$$\mathcal{J}(u) = \sum_{i,j=1}^{n} \partial_{x_i} \left(a_{ij}(\cdot, u) \partial_{x_j} u \right),$$

and the coefficients depend on the unknown function through some integral expressions. This kind of operator is studied as a simplified version of the Boltzmann collision operator. A description of wide classes of stochastic processes and kinetic models leading to equations of the previous type can be found in the classical monographs [10], [16] and [11].

Linear KEs also arise in mathematical finance in some generalization of the celebrated Black & Scholes model [7]. Consider a "stock" whose price S_t is given by the stochastic differential equation

$$d S_t = \mu_0 S_t \, dt + \sigma S_t \, dW_t, \tag{1.12}$$

where μ_0 and σ are positive constants and W_t is a Wiener process. Also consider a "bond" whose price B_t only depends on a constant interest rate r:

$$B_t = B_0 e^{t r}.$$

Finally, consider an "European option" which is a contract which gives the *right* (but not the *obligation*) to buy the stock at a given "exercise price" E and at a given "expiry time" T. The problem studied in [7] is to find a fair price of the option contract. Under some assumptions on the financial market, Black & Scholes show that the price of the option, as a function of the time and of the stock price $V(t, S_t)$, is the solution of the following partial differential equation

$$-rV + \frac{\partial V}{\partial t} + rS\frac{\partial V}{\partial S} + \frac{1}{2}\sigma^2\frac{\partial^2 V}{\partial S^2} = 0$$

in the domain $(S,t) \in \mathbb{R}^+ \times]0,T[$, with the *final condition*

$$V(T, S_T) = \max(S_T - E, 0).$$

In the last decades the Black & Scholes theory has been developed by many authors and mathematical models involving KEs have appeared in the study of the so-called path-dependent contingent claims (see, for instance, [1], [4], [5] and [48]). *Asian options* are options whose exercise price is not fixed as a given constant E, but depends on some average of the history of the stock price. In this case, the value of the option at the expiry time T is (for a a geometric average option):

$$V(S_T, M_T) = \max\left(S_T - e^{\frac{M_T}{T}}, 0\right), \quad M_t = \int_0^t \log(S_\tau)d\tau.$$

If we suppose by simplicity that the interest rate is $r = 0$, the Black & Scholes method leads to the following degenerate equation

$$S^2 \partial_S^2 V + (\log S)\partial_M V + \partial_t V = 0, \qquad S, t > 0, \ M \in \mathbb{R} \tag{1.13}$$

which can be reduced to the KE (1.4) by means of an elementary change of variables (see [6], page 479). A numerical study of the solution of the Cauchy problem related to (1.13) is also proposed in [6].

A recent motivation in finance comes from the model by Hobson & Rogers [22]. In the Black & Scholes theory the hypothesis that the volatility σ in the stochastic differential equation (1.12) is constant contrasts with the empirical observations. Aiming to overcome this problem, many authors proposed different models based on a stochastic volatility (see [18] for a survey). However the presence of a second Wiener process leads some difficulties in the arbitrage argument underlying the Black & Scholes theory. The model proposed by Hobson and Rogers for European options assumes that the volatility only depends on the difference between the present stock price and the past price. This simple model seems to capture the features observed in the market and avoid the problems related to the use of many sources of randomness.

As in the study of Asian options, in the Hobson & Rogers model for European options the value of the option $V(t, S_t, M_t)$ is supposed to depend on the time t, on the price of the stock S_t, on some average M_t and must satisfy the following differential equation

$$\frac{1}{2}\sigma^2(S, M)\left(\partial_S^2 V - \partial_S V\right) + (S - M)\partial_M V + \partial_t V = 0, \tag{1.14}$$

that is a KE with Hölder continuous coefficients. In the recent paper [15] the Cauchy problem related to (1.14) has been studied numerically. In [13] the stability and the rate of convergence of different numerical methods for solving (1.14) are tested. The numerical schemes proposed in these papers rely on the approximation of the directional derivative Y by the finite difference $-\frac{u(x,y,t)-u(x,y+\delta x,t-\delta)}{\delta}$: hence this method, which is respectful of the non-Euclidean geometry of the Lie group, seems to provide a good approximation of the solution.

Finally we recall that KEs with *non linear total derivative term* of the form

$$\Delta_x u + \partial_y g(u) - \partial_t u = f, \qquad x = (x_1, \ldots, x_n) \in \mathbb{R}^n, \ y, t \in \mathbb{R}, \tag{1.15}$$

have been considered for convection-diffusion models (cf. [19] and [36]), for pricing models of options with memory feedback (cf. [40]) and for mathematical models for utility functional and decision making (cf. [2], [3], [12] and [38]). The linearized equation of (1.15)

$$g'(u)\partial_y v - \partial_t v = -\Delta_x v,$$

if $g'(u)$ is different from zero and smooth enough, can be reduced to the form (1.1) with $N = n + 2$,

$$A = \begin{pmatrix} 1 & \cdots & 0 & 0 \\ \vdots & \ddots & \vdots & \vdots \\ 0 & \cdots & 1 & 0 \\ 0 & \cdots & 0 & 0 \end{pmatrix} \qquad \text{and} \qquad B = \begin{pmatrix} 0 & \cdots & 0 & 1 \\ \vdots & \ddots & \vdots & \vdots \\ 0 & \cdots & 0 & 0 \\ 0 & \cdots & 0 & 0 \end{pmatrix}.$$

2. Constant coefficients Kolmogorov equations

We call *constant coefficients KE* any equation of the form

$$Ku \equiv \sum_{i,j=1}^{p_0} a_{ij}\partial_{x_i x_j} u + \langle x, BDu \rangle - \partial_t = 0, \tag{2.1}$$

with constant a_{ij}'s and satisfying hypotheses H.1–H.2. We set

$$\mathcal{C}(t) = \int_0^t E(s)AE^T(s)ds, \qquad t \in \mathbb{R},$$

where

$$E(t) = e^{-tB^T}. \tag{2.2}$$

It is known (see, for instance, [30]) that H.1–H.2 are equivalent to condition

$$\mathcal{C}(t) > 0, \qquad \forall t > 0. \tag{2.3}$$

If (2.3) holds then a fundamental solution to (2.1) is given by

$$\Gamma(x, t, \xi, \tau) = \Gamma(x - E(t - \tau)\xi, t - \tau), \tag{2.4}$$

where $\Gamma(x, t) = 0$ if $t \leq 0$ and

$$\Gamma(x, t) = \frac{(4\pi)^{-\frac{N}{2}}}{\sqrt{\det \mathcal{C}(t)}} \exp\left(-\frac{1}{4}\langle \mathcal{C}^{-1}(t)x, x \rangle - t \operatorname{tr}(B)\right), \qquad \text{if } t > 0. \tag{2.5}$$

Let us remark that $\Gamma(\cdot, \cdot)$ is a C^∞ function outside the diagonal of $\mathbb{R}^{N+1} \times \mathbb{R}^{N+1}$.

The denomination "constant coefficients KE" stems from the theory of parabolic PDEs. Indeed a constant coefficients parabolic equation is nothing more that a translation invariant equation on the Euclidean space. Similarly, a constant coefficients KE has the remarkable property of being invariant with respect to the *non-Euclidean* left translations in the Lie group law

$$(x,t) \circ (\xi,\tau) = (\xi + E(\tau)x, t + \tau), \qquad (x,t),(\xi,\tau) \in \mathbb{R}^N \times \mathbb{R},$$

with $E(\cdot)$ as in (2.2). The class of constant coefficients KEs contains a significant subclass of equations which are also invariant with respect to a suitable dilation group. Indeed, given B in the form (1.3), let us consider the family of dilations in $\mathbb{R}^{N+1}$:

$$\delta_\lambda = \mathrm{diag}(\lambda I_{p_0}, \lambda^3 I_{p_1}, \ldots, \lambda^{2r+1} I_{p_r}, \lambda^2),$$

where I_{p_j} denotes the $p_j \times p_j$ identity matrix. Then K is δ_λ-homogeneous of degree two, i.e.,

$$K \circ \delta_\lambda = \lambda^2 \left(\delta_\lambda \circ K\right), \qquad \forall \lambda > 0,$$

if and only if all the $*$-blocks in (1.3) are zero matrices. The proofs of these statements are contained in [27] and [30]. When the $*$-blocks in B are zero, the dilations $(\delta_\lambda)_{\lambda>0}$ are a group of automorphisms of the Lie group $\mathcal{G} = (\mathbb{R}^{N+1}, \circ)$. Equipped with them, $\mathcal{G}$ becomes a homogeneous group with homogeneous dimension $Q + 2$, where

$$Q = p_0 + 3p_1 + \cdots + (2r+1)p_r, \tag{2.6}$$

(see [26], page 288, and [30], Remark 2.1).

As in classical theory, constant coefficients KEs serve as an essential class of prototypes and many results can be extended to the general situation of variable coefficients by perturbation arguments: in the next sections we present a survey of the main results for KEs with variable coefficients.

3. Kolmogorov equations with regular coefficients

In view of the invariance properties of constant coefficients KEs with respect to $\mathcal{G}$, it is natural to expect that the intrinsic geometry underlying L is that one determined by $\mathcal{G}$. Let $\alpha_1, \ldots, \alpha_N$ be the strictly positive integers such that

$$\delta_\lambda = \mathrm{diag}\left(\lambda^{\alpha_1}, \ldots, \lambda^{\alpha_N}, \lambda^2\right)$$

and define, for every $z \in \mathbb{R}^{N+1} \setminus \{0\}$, $\|z\|_\mathcal{G} = \rho$ where ρ is the unique positive solution to the equation

$$\frac{t^2}{\rho^4} + \sum_{j=1}^{N} \frac{x_j^2}{\rho^{2\alpha_j}} = 1, \qquad z = (x_1, \ldots, x_N, t).$$

We agree to let $\|z\|_\mathcal{G} = 0$ if $z = 0$. Then $z \longmapsto \|z\|_\mathcal{G}$ is a δ_λ-homogeneous function of degree one, continuous on $\mathbb{R}^{N+1}$, strictly positive and of class C^∞ in $\mathbb{R}^{N+1} \setminus \{0\}$. If we define

$$d_\mathcal{G}(z,\zeta) = \|\zeta^{-1} \circ z\|_\mathcal{G}, \qquad z, \zeta \in \mathbb{R}^{N+1},$$

then $(\mathbb{R}^{N+1}, d_\mathcal{G})$ is a (pseudo-)metric space. We say that a function f is B-Hölder continuous of order $\alpha \in]0,1]$ on a domain Ω of $\mathbb{R}^{N+1}$, and we write $f \in C_B^\alpha(\Omega)$, if there exists a constant C such that

$$|f(z) - f(\zeta)| \leq C d_{\mathcal{G}}(z, \zeta)^{\alpha}, \qquad \forall z, \zeta \in \Omega.$$

Assuming that the coefficients a_{ij}, $a_i, c \in C_B^{\alpha}(\mathbb{R}^{N+1})$, for $i, j = 1, \ldots, p_0$, are bounded functions, a fundamental solution Γ for the operator L in (1.1) can be constructed by adapting the Levi's parametrix method to the Lie group and metric structures related to the matrix B (see [14] and [41] which improve and generalize the previous results by Weber [47], Il'in [24] and Sonin [46]).

The Levi's parametrix method also provides a global upper bound for Γ. Indeed let Γ^{ε} denote the fundamental solution to the constant coefficients KE

$$L^{\varepsilon} = (\mu + \varepsilon)\Delta_{\mathbb{R}^{p_0}} + Y \tag{3.1}$$

where $\varepsilon > 0$, μ is as in (1.2), $\Delta_{\mathbb{R}^{p_0}}$ denotes the Laplacian in the variables $x_1, \ldots, x_{p_0}$ and Y is the vector fields in (1.8). Then for every positive ε and T, there exists a constant C, only dependent on μ, B, ε and T, such that

$$\Gamma(z, \zeta) \leq C \, \Gamma^{\varepsilon}(z, \zeta) \tag{3.2}$$

for any $z, \zeta \in \mathbb{R}^{N+1}$ with $0 < t - \tau < T$. Similar estimates also hold for the derivatives of Γ (see [14] and [41]).

For operators in divergence form

$$L = \sum_{i,j=1}^{p_0} \partial_{x_i}\left(a_{ij}(z)\partial_{x_j}\right) + Y \tag{3.3}$$

with null $*$-blocks in (1.3), a lower bound for Γ analogous to (3.2) also holds. This result relies on a Harnack inequality which is invariant with respect to the translations and dilations in $\mathcal{G}$ (see [41], Theorem 1.3 which extends some Harnack inequalities for constant coefficients Kolmogorov operators first appeared in [28], [20] and [30]).

Theorem 3.1. (Polidoro [42]) *Let Γ be the fundamental solution of the divergence form operator (3.3). There exists a positive constant m such that, if Γ^- denotes the fundamental solution of*

$$L^- = m^{-1}\Delta_{p_0} + \langle x, BD \rangle - \partial_t,$$

then, for every $T > 0$, there exists a positive constant C^- such that

$$C^-\Gamma^-(z, \zeta) \leq \Gamma(z, \zeta) \tag{3.4}$$

for every $z = (x, t)$, $\zeta = (\xi, \tau) \in \mathbb{R}^{N+1}$, $0 < t - \tau < T$.

We would like to emphasize that the functions Γ^- and Γ^{ε} appearing in (3.2) and (3.4) have the explicit form (2.4)–(2.5), with the matrix A in (1.6) replaced by $m^{-1}\mathrm{diag}(I_{p_0}, 0, \ldots, 0)$ and $(\mu + \varepsilon)\,\mathrm{diag}(I_{p_0}, 0, \ldots, 0)$ respectively. Theorem 3.1 was proved in [42] by using a technique which is inspired by a method of Aronson and Serrin for classical parabolic operators. The core of the method used in [42] is a kind of discretization of the connectivity Theorem of Carathéodory-Razewski-Chow.

We also recall some interior regularity results. The following Schauder type estimates proved in [34] (see also [33] and [36]) improve and generalize the previous ones contained in [21], [45] and [17]: for every bounded open set Ω_1 such that $\overline{\Omega}_1 \subseteq \Omega$ where Ω is a subset of $\mathbb{R}^{N+1}$, there exists a constant $C > 0$ such that

$$|u|_{2+\alpha,\Omega_1} \leq C \left(\sup_{\Omega} |u| + |Lu|_{\alpha,\Omega} \right),$$

for any u smooth real function defined on Ω. Here $|\cdot|_{\alpha,\Omega}$ and $|\cdot|_{2+\alpha,\Omega_1}$ denote suitable Hölder norms defined in terms of $d_{\mathcal{G}}$. In [34], the interior Schauder estimates are also used to study a first boundary value problem for L. We also quote the paper [29] in which a boundary value problem for the non-linear equation

$$\sum_{i,j=1}^{p_0} \partial_{x_i} \left(a_{ij}(z, u)\partial_{x_j} \right) + Yu = 0. \tag{3.5}$$

was studied. In [29] the a priori estimates of [34] are used as crucial tools.

The L^p regularity theory for weak solutions to equations in divergence or non-divergence form has been studied in [8], [35], [43] and [44]. In [8] and [43], interior regularity properties of strong solutions to the non-divergence form equation $Lu = f$ were studied. The main results are some L^p_{loc} estimates of the derivatives of the solution u and its Hölder continuity in terms of some L^q_{loc} norm of f. The key tools are some deep continuity results for singular integrals. The same techniques, suitably adapted, were used in [35] and in [44] in order to prove interior regularity results for weak solutions to the equation $Lu = \sum_{i=1}^{p_0} \partial_{x_i} F_i$ with L as in (3.3).

4. Kolmogorov equations with measurable coefficients

As said in the previous section, the Hölder estimates for weak solutions to (3.3) have been used for the study of nonlinear KEs. However the dependence of the Hölder constant on the regularity of the coefficients forces quite restrictive hypotheses on the nonlinearity. In order to remove such restrictions, regularity results for solutions to linear equations with merely measurable coefficients are needed. A first result in such a direction has been recently proved by the author in collaboration with S. Polidoro. In [39], the local boundedness of the weak solutions to L is proved only assuming the uniform positivity condition (1.2). The main result in [39] is the following theorem.

Theorem 4.1. *Let u be a non-negative weak solution to*

$$\sum_{i,j=1}^{p_0} \partial_{x_i} \left(a_{ij}(z)\partial_{x_j} \right) + \langle x, B\nabla u \rangle - \partial_t u = 0 \tag{4.1}$$

in a domain Ω. Let r, ρ, $0 < \frac{r}{2} \leq \rho < r$, be such that $\overline{H_r} \subseteq \Omega$ where H_r denotes a suitable cylindrical domain of radius r. Then there exists a positive constant C, only dependent on μ and on the homogeneous dimension Q (cf. (2.6)) such that, *for every $p > 0$, it holds*

$$\sup_{H_\rho} u^p \leq \frac{C}{(r-\rho)^{Q+2}} \int_{H_r} u^p. \tag{4.2}$$

Estimate (4.2) also holds for every $p < 0$ such that $u^p \in L^1(H_r)$.

This theorem is proved in [39] by using an iterative procedure analogous to the one introduced by Moser in the classical elliptic and parabolic cases. As it is well known, the Moser's technique is based on a combination of Caccioppoli type estimates with the classical Sobolev inequality. Actually the weak solutions to (4.1) satisfy a Caccioppoli type estimate, however this estimate only gives a L^2_{loc} bound of the first order derivatives $\partial_{x_j} u$ for $j = 1, \ldots, p_0$ and does not give any information on the others $(N - p_0)$ spatial derivatives. Thus, if $p_0 < N$, this lack of information cannot be restored by the usual Sobolev embedding theorem.

The key idea in [39] *is to prove a Sobolev type inequality for non negative sub- and super-solutions to* (4.1), good enough to be successfully combined with the previous "weak" Caccioppoli inequality. To be more specific, let us first recall the definition of weak sub- and super-solution to. We say that a function $u \in L^2_{\mathrm{loc}}(\Omega)$, Ω open subset of $\mathbb{R}^{N+1}$, is a *weak sub-solution* to (4.1) if the weak derivatives $\partial_{x_1} u, \ldots, \partial_{x_{p_0}} u$ and Yu exist, belong to $L^2_{\mathrm{loc}}(\Omega)$ and

$$\int_\Omega -\langle ADu, D\varphi \rangle + \varphi Yu \geq 0, \qquad \forall \varphi \in C_0^\infty(\Omega), \ \varphi \geq 0.$$

If $-u$ is a weak sub-solution, we say that u is a weak super-solution. Then, the following Caccioppoli type estimate holds (cf. [39], Proposition 3.2)

Proposition 4.2. *Let u be a non-negative weak sub-solution of (4.1) in Ω. Let $\rho, r > 0$, $\frac{r}{2} \leq \rho < r$, and $\overline{H_r} \subseteq \Omega$. Then, there exists a constant C, only dependent on μ in (1.2) and on the homogeneous dimension Q, such that*

$$\|\partial_{x_j} u^p\|_{L^2(H_\rho)} \leq \frac{C\sqrt{1+\varepsilon}}{\varepsilon} \|u^p\|_{L^2(H_r)}, \quad where \quad \varepsilon = \frac{|2p-1|}{4p}, \tag{4.3}$$

for every $j = 1, \ldots, p_0$ and $p < 0$ or $p \geq 1$. The same inequality holds for non-negative weak super-solutions and $p \in]0, 1/2[$.

The key Sobolev type inequality for weak sub- and super-solutions proved in [39] is the following.

Proposition 4.3. *Let u be a non-negative weak sub-solution to (4.1) and let r, ρ be as in the previous Proposition 4.2. Then $u \in L^{2\kappa}_{\mathrm{loc}}(H_\rho)$, $\kappa = 1 + \frac{2}{Q}$, and there exists a constant C, only dependent on μ and Q, such that*

$$\|u\|_{L^{2\kappa}(H_\rho)} \leq \frac{c}{r-\rho} \left(\|u\|_{L^2(H_r)} + \sum_{j=1}^{p_0} \|\partial_{x_j} u\|_{L^2(B_r)} \right). \tag{4.4}$$

The same inequality holds for non-negative super-solutions.

Inequalities (4.3)–(4.4) allow to start up an iterative procedure analogous to the classical Moser's one and to prove Theorem 4.1. We also recall that Theorem 4.1 has been used in [37] to obtain a pointwise global upper bound for the fundamental solution of (4.1).

Theorem 4.4. *There exists two positive constants C and ε, only dependent on μ in (1.2) and on B, such that*

$$\Gamma(x,t,\xi,\tau) \leq C\,\Gamma^{\varepsilon}(x,t,\xi,\tau), \qquad \forall x,\xi \in \mathbb{R}^{N},\ t > \tau,$$

where Γ^{ε} is the fundamental solution to (3.1).

We remark explicitly that Theorem 4.4 improves inequality (3.2) in that C is independent of the modulus of continuity of the coefficients.

References

[1] B. ALZIARY, J.P. DÉCAMPS, AND P.F. KOEHL, *A P.D.E. approach to Asian options: analytical and numerical evidence*, J. Banking Finance, 21 (1997), pp. 613–640.

[2] F. ANTONELLI, E. BARUCCI, AND M.E. MANCINO, *Asset pricing with a forward-backward stochastic differential utility*, Econom. Lett., 72 (2001), pp. 151–157.

[3] F. ANTONELLI AND A. PASCUCCI, *On the viscosity solutions of a stochastic differential utility problem*, J. Differential Equations, 186 (2002), pp. 69–87.

[4] G. BARLES, *Convergence of numerical schemes for degenerate parabolic equations arising in finance theory*, in Numerical methods in finance, Cambridge Univ. Press, Cambridge, 1997, pp. 1–21.

[5] J. BARRAQUAND AND T. PUDET, *Pricing of American path-dependent contingent claims*, Math. Finance, 6 (1996), pp. 17–51.

[6] E. BARUCCI, S. POLIDORO, AND V. VESPRI, *Some results on partial differential equations and Asian options*, Math. Models Methods Appl. Sci., 11 (2001), pp. 475–497.

[7] F. BLACK AND M. SCHOLES, *The pricing of options and corporate liabilities*, J. Political Economy, 81 (1973), pp. 637–654.

[8] M. BRAMANTI, M.C. CERUTTI, AND M. MANFREDINI, L^P *estimates for some ultraparabolic operators with discontinuous coefficients*, J. Math. Anal. Appl., 200 (1996), pp. 332–354.

[9] C. CERCIGNANI, *The Boltzmann equation and its applications*, Springer-Verlag, New York, 1988.

[10] S. CHANDRESEKHAR, *Stochastic problems in physics and astronomy*, Rev. Modern Phys., 15 (1943), pp. 1–89.

[11] S. CHAPMAN AND T.G. COWLING, *The mathematical theory of nonuniform gases*, Cambridge University Press, Cambridge, third ed., 1990.

[12] G. CITTI, A. PASCUCCI, AND S. POLIDORO, *Regularity properties of viscosity solutions of a non-Hörmander degenerate equation*, J. Math. Pures Appl. (9), 80 (2001), pp. 901–918.

[13] M. DI FRANCESCO, P. FOSCHI, AND A. PASCUCCI, *Analysis of an uncertain volatility model*, preprint, (2004).

[14] M. DI FRANCESCO AND A. PASCUCCI, *On a class of degenerate parabolic equations of Kolmogorov type*, preprint available on-line at http://www.dm.unibo.it/~pascucci/, (2004).

[15] ——, *On the complete model with stochastic volatility by Hobson and Rogers*, to appear in R. Soc. Lond. Proc. Ser. A Math. Phys. Eng. Sci., (2004).

[16] J.J. DUDERSTADT AND W.R. MARTIN, *Transport theory*, John Wiley & Sons, New York-Chichester-Brisbane, 1979. A Wiley-Interscience Publication.

[17] S.D. EIDELMAN, S.D. IVASYSHEN, AND H.P. MALYTSKA, *A modified Levi method: development and application*, Dopov. Nats. Akad. Nauk Ukr. Mat. Prirodozn. Tekh. Nauki, 5 (1998), pp. 14–19.

[18] T.W. EPPS, *Pricing derivative securities*, World Scientific, Singapore, 2000.

[19] M. ESCOBEDO, J.L. VÁZQUEZ, AND E. ZUAZUA, *Entropy solutions for diffusion-convection equations with partial diffusivity*, Trans. Amer. Math. Soc., 343 (1994), pp. 829–842.

[20] N. GAROFALO AND E. LANCONELLI, *Level sets of the fundamental solution and Harnack inequality for degenerate equations of Kolmogorov type*, Trans. Amer. Math. Soc., 321 (1990), pp. 775–792.

[21] T.G. GENČEV, *On ultraparabolic equations*, Dokl. Akad. Nauk SSSR, 151 (1963), pp. 265–268.

[22] D.G. HOBSON AND L.C.G. ROGERS, *Complete models with stochastic volatility*, Math. Finance, 8 (1998), pp. 27–48.

[23] L. HÖRMANDER, *Hypoelliptic second order differential equations*, Acta Math., 119 (1967), pp. 147–171.

[24] A.M. IL'IN, *On a class of ultraparabolic equations*, Dokl. Akad. Nauk SSSR, 159 (1964), pp. 1214–1217.

[25] A. KOLMOGOROV, *Zufllige Bewegungen. (Zur Theorie der Brownschen Bewegung.).*, Ann. of Math., II. Ser., 35 (1934), pp. 116–117.

[26] L.P. KUPCOV, *The fundamental solutions of a certain class of elliptic-parabolic second order equations*, Differencial'nye Uravnenija, 8 (1972), pp. 1649–1660, 1716.

[27] ——, *The mean value property and the maximum principle for second order parabolic equations*, Dokl. Akad. Nauk SSSR, 242 (1978), pp. 529–532.

[28] ——, *On parabolic means*, Dokl. Akad. Nauk SSSR, 252 (1980), pp. 296–301.

[29] E. LANCONELLI AND F. LASCIALFARI, *A boundary value problem for a class of quasilinear operators of Fokker-Planck type*, in Proceedings of the Conference "Differential Equations", Ann. Univ. Ferrara Sez. VII (N.S.), vol. 41 suppl., 1996, pp. 65–84.

[30] E. LANCONELLI AND S. POLIDORO, *On a class of hypoelliptic evolution operators*, Rend. Sem. Mat. Univ. Politec. Torino, 52 (1994), pp. 29–63.

[31] E.M. LIFSCHITZ AND L.P. PITAEVSKII, *Teoreticheskaya fizika ("Landau-Lifshits"). Tom 10*, "Nauka", Moscow, 1979. Fizicheskaya kinetika. [Physical kinetics].

[32] P.-L. LIONS, *On Boltzmann and Landau equations*, Philos. Trans. Roy. Soc. London Ser. A, 346 (1994), pp. 191–204.

[33] A. LUNARDI, *Schauder estimates for a class of degenerate elliptic and parabolic operators with unbounded coefficients in* $\mathbb{R}^N$, Ann. Scuola Norm. Sup. Pisa Cl. Sci. (4), 24 (1997), pp. 133–164.

[34] M. MANFREDINI, *The Dirichlet problem for a class of ultraparabolic equations*, Adv. Differential Equations, 2 (1997), pp. 831–866.

[35] M. MANFREDINI AND S. POLIDORO, *Interior regularity for weak solutions of ultraparabolic equations in divergence form with discontinuous coefficients*, Boll. Unione Mat. Ital. Sez. B Artic. Ric. Mat. (8), 1 (1998), pp. 651–675.

[36] A. PASCUCCI, *Hölder regularity for a Kolmogorov equation*, Trans. Amer. Math. Soc., 355 (2003), pp. 901–924.

[37] A. PASCUCCI AND S. POLIDORO, *A Gaussian upper bound for the fundamental solutions of a class of ultraparabolic equations*, J. Math. Anal. Appl., 282 (2003), pp. 396–409.

[38] ———, *On the Cauchy problem for a nonlinear Kolmogorov equation*, SIAM J. Math. Anal., 35 (2003), pp. 579–595.

[39] ———, *The Moser's iterative method for a class of ultraparabolic equations*, Commun. Contemp. Math., Vol.6 n.2 (2004), pp. 1–23.

[40] R. PESZEK, *PDE models for pricing stocks and options with memory feedback*, Appl. Math. Finance, 2 (1995), pp. 211–223.

[41] S. POLIDORO, *On a class of ultraparabolic operators of Kolmogorov-Fokker-Planck type*, Matematiche (Catania), 49 (1994), pp. 53–105.

[42] ———, *A global lower bound for the fundamental solution of Kolmogorov-Fokker-Planck equations*, Arch. Rational Mech. Anal., 137 (1997), pp. 321–340.

[43] S. POLIDORO AND M.A. RAGUSA, *Sobolev-Morrey spaces related to an ultraparabolic equation*, Manuscripta Math., 96 (1998), pp. 371–392.

[44] ———, *Hölder regularity for solutions of ultraparabolic equations in divergence form*, Potential Anal., 14 (2001), pp. 341–350.

[45] J.I. ŠATYRO, *The smoothness of the solutions of certain degenerate second order equations*, Mat. Zametki, 10 (1971), pp. 101–111.

[46] I.M. SONIN, *A class of degenerate diffusion processes*, Teor. Verojatnost. i Primenen, 12 (1967), pp. 540–547.

[47] M. WEBER, *The fundamental solution of a degenerate partial differential equation of parabolic type*, Trans. Amer. Math. Soc., 71 (1951), pp. 24–37.

[48] P. WILMOTT, S. HOWISON, AND J. DEWYNNE, *Option pricing*, Oxford Financial Press, Oxford, 1993.

Andrea Pascucci
Dipartimento di Matematica
Università di Bologna
Piazza di Porta S. Donato 5
I-40126 Bologna, Italy
e-mail: `pascucci@dm.unibo.it`

Progress in Nonlinear Differential Equations
and Their Applications, Vol. 63, 365–374

Harnack Inequalities and Gaussian Estimates for a Class of Hypoelliptic Operators

Sergio Polidoro

Abstract. We announce some results obtained in a recent study [13], concerning a general class of hypoelliptic evolution operators in $\mathbb{R}^{N+1}$. A Gaussian lower bound for the fundamental solution and a global Harnack inequality are given.

Mathematics Subject Classification (2000). 35K57, 35K65, 35K70.

1. Introduction

In this note we will discuss a result obtained in a recent study by Andrea Pascucci and myself [13]. Let us consider the linear second order operator in $\mathbb{R}^{N+1}$ of the form

$$L = \sum_{p=1}^{m} X_p^2 + X_0 - \partial_t. \tag{1.1}$$

In (1.1) $z = (x, t)$ denotes the point in $\mathbb{R}^{N+1}$ and the X_p's are smooth vector fields on $\mathbb{R}^N$, i.e.,

$$X_p(x) = \sum_{j=1}^{N} a_j^p(x)\partial_{x_j}, \qquad p = 0, \ldots, m,$$

where any a_j^p is a C^∞ function. In the sequel we also consider the X_p's as vector fields in $\mathbb{R}^{N+1}$, moreover we denote

$$Y = X_0 - \partial_t, \qquad \text{and} \qquad \lambda \cdot X \equiv \lambda_1 X_1 + \cdots + \lambda_m X_m, \tag{1.2}$$

for $\lambda = (\lambda_1, \ldots, \lambda_m) \in \mathbb{R}^m$. We say that a curve $\gamma : [0, T] \to \mathbb{R}^{N+1}$ is L-admissible if it is absolutely continuous and satisfies

$$\gamma'(s) = \lambda(s) \cdot X(\gamma(s)) + Y(\gamma(s)), \qquad \text{a.e. in } [0, T],$$

for suitable piecewise constant real functions $\lambda_1, \ldots, \lambda_m$. We next state our main assumptions:

H.1 there exists a homogeneous Lie group $\mathbb{G} = \left(\mathbb{R}^{N+1}, \circ, \delta_\lambda\right)$ such that

    (i) $X_1, \ldots, X_m, Y$ are left translation invariant on $\mathbb{G}$;

    (ii) $X_1, \ldots, X_m$ are δ_λ-homogeneous of degree one and Y is δ_λ-homogeneous of degree two;

H.2 for every $(x, t), (\xi, \tau) \in \mathbb{R}^{N+1}$ with $t > \tau$, there exists an L-admissible path $\gamma : [0, T] \to \mathbb{R}^{N+1}$ such that $\gamma(0) = (x, t)$, $\gamma(T) = (\xi, \tau)$.

Let us recall that a Lie group $\mathbb{G} = \left(\mathbb{R}^{N+1}, \circ\right)$ is called *homogeneous* if there exists a family of dilations $(\delta_\lambda)_{\lambda>0}$ on $\mathbb{G}$. Hypotheses H.1–H.2 imply that $\mathbb{R}^N$ has a direct sum decomposition

$$\mathbb{R}^N = V_1 \oplus \cdots \oplus V_n$$

such that, if $x = x^{(1)} + \cdots + x^{(n)}$ with $x^{(k)} \in V_k$, then the dilations are

$$\delta_\lambda(x^{(1)} + \cdots + x^{(n)}, t) = (\lambda x^{(1)} + \cdots + \lambda^n x^{(n)}, \lambda^2 t),$$

for any $(x, t) \in \mathbb{R}^{N+1}$ and $\lambda > 0$. The natural number $Q = 2 + \sum_{k=1}^{n} k \dim V_k$ is usually called the *homogeneous dimension* of $\mathbb{G}$ with respect to $(\delta_\lambda)_{\lambda>0}$. We also introduce the following δ_λ-homogeneous norm on $\mathbb{R}^N$:

$$|x|_\mathbb{G} = \max \left\{ \left|x_i^{(k)}\right|^{\frac{1}{k}} \mid k = 1, \ldots, n, \ i = 1, \ldots, m_k \right\}.$$

Operators of the form (1.1), verifying assumptions H.1–H.2, have been introduced by Kogoj and Lanconelli in [5] and [6]. Under these hypotheses the Hörmander condition holds:

$$\text{rank Lie}\{X_1, \ldots, X_m, Y\}(z) = N + 1, \qquad \forall z \in \mathbb{R}^{N+1}; \tag{1.3}$$

hence L in (1.1) is hypoelliptic (i.e., every distributional solution to $Lu = 0$ is smooth; see, for instance, Proposition 10.1 in [5]) and has a fundamental solution Γ which is smooth out of the pole and δ_λ-homogeneous of degree $2 - Q$:

$$\Gamma(\delta_\lambda z) = \lambda^{2-Q} \Gamma(z), \qquad \lambda > 0. \tag{1.4}$$

Hence operator (1.1) belongs to the general class of hypoelliptic operators on homogeneous groups first studied by Folland [3], Rothschild and Stein [17], Nagel, Stein and Wainger [11].

The main result in [5] is an invariant (local) Harnack inequality for L; one-side Liouville theorems are given in [6]. In the paper [13] a *non-local* Harnack inequality has been proved (see Theorem 5.1 below). Moreover in [13] it is proved a lower bound for the fundamental solution Γ of the operator L, under the assumption that the group $\mathbb{G}$ has step three, i.e.,

$$\mathbb{R}^N = V_1 \oplus V_2 \oplus V_3.$$

Proposition 1.1. *Let L be the operator in (1.1) on a group of step three and Γ its fundamental solution. There exists a positive constant C such that*

$$\Gamma(x, t) \geq \frac{C}{t^{\frac{Q-2}{2}}} \exp\left(-C \frac{|x|_\mathbb{G}^6}{t^3}\right), \qquad \forall (x, t) \in \mathbb{R}^N \times \mathbb{R}^+. \tag{1.5}$$

In the above statement $\Gamma(\cdot)$ denotes the fundamental solution of L with pole at the origin. Due to the left $\circ$-invariance of Γ, we have that $\Gamma(z, \zeta) = \Gamma(\zeta^{-1} \circ z)$ and a lower bound analogous to (1.5) also holds for $\Gamma(\cdot, \zeta)$.

The above estimate looks rather rough, since it is natural to expect $\frac{|x|_{\mathbb{G}}^2}{t}$ in the exponent in (1.5). Indeed the following Gaussian upper bound has been proved by by Kogoj and Lanconelli in [5]:

$$\Gamma(x, t) \leq \frac{C}{t^{\frac{Q-2}{2}}} \exp\left(-\frac{|x|_{\mathbb{G}}^2}{Ct}\right), \qquad \forall x \in \mathbb{R}^N, \ t > 0, \tag{1.6}$$

being C a positive constant. However it is known that the fundamental solution of the (Kolmogorov) operator $\partial_{x_1}^2 + x_1 \partial_{x_2} - \partial_t$ is

$$\Gamma(x_1, x_2, t) = \frac{\sqrt{3}}{2\pi t^2} \exp\left(-\frac{x_1^2}{t} - 3\frac{x_1 x_2}{t^2} - 3\frac{x_2^2}{t^3}\right), \qquad x_1, x_2 \in \mathbb{R}, \ t > 0 \tag{1.7}$$

(see (3.4) below) so that, in particular,

$$\Gamma(0, x_2, t) = \frac{\sqrt{3}}{2\pi t^2} \exp\left(-3\frac{x_2^2}{t^3}\right) = \frac{\sqrt{3}}{2\pi t^2} \exp\left(-3\frac{|(0, x_2)|_{\mathbb{G}}^6}{t^3}\right). \tag{1.8}$$

On the other hand, we have

$$\Gamma(x_1, 0, t) = \frac{\sqrt{3}}{2\pi t^2} \exp\left(-\frac{x_1^2}{t}\right) = \frac{\sqrt{3}}{2\pi t^2} \exp\left(-\frac{|(x_1, 0)|_{\mathbb{G}}^2}{t}\right),$$

so that neither (1.5) nor (1.6) are sharp. However we can hope to sharpen (1.5) at least in some component of x. The following example shows that further hypotheses on the operator L are needed to obtain such a result. Consider the operator $\widetilde{L} = X^2 + Y$ in $\mathbb{R}^3$, where

$$X = \partial_{x_1} + 3x_1{}^2 \partial_{x_2}, \qquad \text{and} \qquad Y = x_1 \partial_{x_2} - \partial_t.$$

It is straightforward to verify H.1, H.2 for $\widetilde{L}$ (the dilations are $\delta_\lambda(x_1, x_2, t) = (\lambda x_1, \lambda^3 x_2, \lambda^2 t)$). The fundamental solution is $\widetilde{\Gamma}(x_1, x_2, t) = \Gamma(x_1, x_2 - x_1^3, t)$ with Γ in (1.7), then

$$\widetilde{\Gamma}(x_1, 0, t) = \frac{\sqrt{3}}{2\pi t^2} \exp\left(-\frac{|x_1|^2}{t} - 3\frac{|x_1|^4}{t^2} - 3\frac{|x_1|^6}{t^3}\right), \qquad \forall (x_1, t) \in \mathbb{R} \times \mathbb{R}^+. \tag{1.9}$$

In Sections 2, 3, and 4 we give some examples of operators that motivate our study. In these particular cases we will give sharp estimates for the case of a Lie algebra of step three (see Propositions 2.1, 3.1 and 4.1 below). The results stated in Sections 2 and 3 agree with the known results for the same kind of operators, the results stated in Section 4 are new. Last section contains an outline of the proof.

2. Heat operators on Carnot groups

Consider the operator L in (1.1) under assumptions H.1 and

$$\text{rank Lie}\{X_1, \ldots, X_m\}(x) = N, \qquad \forall x \in \mathbb{R}^N. \tag{2.1}$$

In this case $\mathbb{G} = \left(\mathbb{R}^N, \circ, \delta_\lambda\right)$ is a *Carnot (or stratified) group* (see, for instance, [3] and [19]). When $X_0 \equiv 0$ in (1.1), we have

$$L = \Delta_\mathbb{G} - \partial_t, \tag{2.2}$$

where as usual $\Delta_\mathbb{G}$ denotes the *canonical sub-Laplacian on* $\mathbb{G}$:

$$\Delta_\mathbb{G} = \sum_{p=1}^{m} X_p^2.$$

We recall the well-known Gaussian upper and lower bounds for heat kernels due to Jerison and Sánchez-Calle [4], Kusuoka and Stroock [8], Varopoulos, Saloff-Coste and Coulhon [19]. These results apply to Lie groups which are not necessarily homogeneous. We also quote the more recent and accurate estimates by Saloff-Coste and Stroock [18], Bonfiglioli, Lanconelli and Uguzzoni [2].

More generally condition (2.1) is satisfied by operators (1.1) of the form

$$L = \Delta_\mathbb{G} + X_0 - \partial_t, \tag{2.3}$$

with $X_0 \in \text{Lie}\{X_1, \ldots, X_m\}$. Operators of this kind have been considered by Alexopoulos in [1].

The following statement contains the global lower bound and the global Harnack inequality for a parabolic operator on a Carnot group of step three proved in [13]. The estimates are in accord with the classical ones given in [4], [8] and in [1].

Proposition 2.1. *Let L be a parabolic operator on a Carnot group of step three and let $z_0 = (x_0, t_0) \in \mathbb{R}^{N+1}$, $T > 0$. Then:*

- *There exists a positive constant C such that*

$$\Gamma(x, t) \geq \frac{C}{t^{\frac{Q-2}{2}}} \exp\left(-C\frac{|x|_\mathbb{G}^2}{t}\right), \qquad \forall(x, t) \in \mathbb{R}^N \times \mathbb{R}^+. \tag{2.4}$$

- *There exist two constants $c > 0$ and $C > 1$, only dependent on L, such that, if u is a non-negative solution to $Lu = 0$ in $\mathbb{R}^N \times \,]t_0 - cT, t_0 + T]$, then*

$$u(z_0) \leq \exp\left(C\left(1 + \frac{|x|_\mathbb{G}^2}{t}\right)\right) u(z_0 \circ z), \tag{2.5}$$

for every $z = (x, t) \in \mathbb{R}^N \times \,]0, T]$.

3. Kolmogorov type operators

Assume $X_p = \partial_p$, $p = 1, \ldots, m$, and the coefficients of X_0 are linear functions of $x \in \mathbb{R}^N$:

$$X_0 = \langle x, B\nabla \rangle$$

for a constant $N \times N$ matrix B. Then

$$L = \sum_{p=1}^{m} \partial_p^2 + X_0 - \partial_t. \tag{3.1}$$

This kind of operator has been extensively studied (see [10] and [9] for a comprehensive bibliography). It is known that H.1–H.2 for L are equivalent to the following hypothesis:

[H.3]: the matrix B takes the form

$$B = \begin{pmatrix} 0 & B_1 & 0 & \cdots & 0 \\ 0 & 0 & B_2 & \cdots & 0 \\ \vdots & \vdots & \vdots & \ddots & \vdots \\ 0 & 0 & 0 & \cdots & B_n \\ 0 & 0 & 0 & \cdots & 0 \end{pmatrix} \tag{3.2}$$

for some basis of $\mathbb{R}^N$, where B_k is a $d_k \times d_{k+1}$ matrix of rank d_k, $k = 1, 2, \ldots, n$ with $m = d_1 \geq d_2 \geq \cdots \geq d_{n+1} \geq 1$ and $d_1 + \cdots + d_{n+1} = N$.

The equivalence of [H.3] and the couple of hypotheses H.1-(1.3) has been proved in [10]. As said before H.1–H.2 yield H.1-(1.3), on the other hand in [16] it is proved the converse implication for Kolmogorov operators.

Under assumption [H.3], the dilations are

$$\delta_\lambda = \operatorname{diag}(\lambda I_{d_1}, \lambda^3 I_{d_2}, \ldots, \lambda^{2n+1} I_{d_{n+1}}, \lambda^2), \qquad \lambda > 0, \tag{3.3}$$

where I_{d_k} is the $d_k \times d_k$ identity matrix. Moreover the fundamental solution of L in (1.1) is explicitly known:

$$\Gamma(z) = \frac{1}{\sqrt{(4\pi)^N \det \mathcal{C}(t)}} \exp\left(-\frac{1}{4}\langle \mathcal{C}^{-1}(t)x, x\rangle\right), \tag{3.4}$$

for $t > 0$, and $\Gamma(z) = 0$ for $t \leq 0$. In (3.4), we denote

$$E(t) = \exp(-tB^T) \quad \text{and} \quad \mathcal{C}(t) = \int_0^t E(s)AE^T(s)ds, \tag{3.5}$$

where B^T is the transpose matrix of B. We remark that condition [H.3] ensures that $\mathcal{C}(t) > 0$ for any $t > 0$ (cf. Proposition A.1 in [10], see also [7]). In this case the group law is

$$(x,t) \circ (\xi, \tau) = (\xi + E(\tau)x, t + \tau), \qquad (x,t), (\xi, \tau) \in \mathbb{R}^{N+1}. \tag{3.6}$$

In the sequel we call $\mathbb{K} \equiv (\mathbb{R}^{N+1}, \circ)$ a *Kolmogorov group*.

Non-local Harnack inequalities for this kind of operator are proved in [14], moreover Gaussian estimates for the fundamental solution are given in [15], [16] and [12] in the case of non-constant coefficients of the second order derivatives.

The following statement contains the global lower bound and the global Harnack inequality for operators on a Kolmogorov group of step three proved in [13]. Also in this case the estimates are in accord with the ones given in [10].

Proposition 3.1. *Let L be a Kolmogorov type operator on a group $\mathbb{K}$ of step three and let $z_0 = (x_0, t_0) \in \mathbb{R}^{N+1}$, $T > 0$. Then:*

- *There exists a positive constant C such that*

$$\Gamma(x,t) \geq \frac{C}{t^{\frac{Q-2}{2}}} \exp\left(-C\left(\frac{\left|x^{(1)}\right|_{\mathbb{K}}^2}{t} + \frac{\left|x^{(3)}\right|_{\mathbb{K}}^6}{t^3}\right)\right), \qquad \forall (x,t) \in \mathbb{R}^N \times \mathbb{R}^+. \quad (3.7)$$

- *There exist two constants $c > 0$ and $C > 1$, only dependent on L, such that, if u is a non-negative solution to $Lu = 0$ in $\mathbb{R}^N \times \,]t_0 - cT, t_0 + T]$, then*

$$u(z_0) \leq \exp\left(C\left(1 + \frac{\left|x^{(1)}\right|_{\mathbb{K}}^2}{t} + \frac{\left|x^{(3)}\right|_{\mathbb{K}}^6}{t^3}\right)\right) u(z_0 \circ z), \quad (3.8)$$

for every $z = (x,t) \in \mathbb{R}^N \times \,]0,T]$.

4. Operators on linked groups

Let $\mathbb{L} = \mathbb{G} \triangle \mathbb{K}$ be the linked group of a Carnot group $\mathbb{G}$ on $\mathbb{R}^m \times \mathbb{R}^n$ and a Kolmogorov group $\mathbb{K}$ on $\mathbb{R}^m \times \mathbb{R}^r \times \mathbb{R}$, as defined by Kogoj and Lanconelli in [5] (Sect. 10). We consider the operator

$$L = \Delta_{\mathbb{G}} + Y. \quad (4.1)$$

For reader's convenience, we recall here the definition of link of Carnot and Kolmogorov groups. Consider a Carnot group

$$\mathbb{G} = \left(\mathbb{R}^m \times \mathbb{R}^n, \circ, \delta_\lambda^{\mathbb{G}}\right),$$

where (x, y) denotes the point in $\mathbb{R}^m \times \mathbb{R}^n$ and assume that

$$X_p = \partial_p + a^p(x, y)\nabla_y, \qquad p = 1, \dots, m. \quad (4.2)$$

Hence the dilations and the group law take the following form:

$$\delta_\lambda^{\mathbb{G}}(x, y) = \left(\lambda x, \rho_\lambda^{\mathbb{G}} y\right), \qquad (x, y) \circ (x', y') = (x + x', Q(x, y, x', y')).$$

Moreover the Kolmogorov group is[1]

$$\mathbb{K} = \left(\mathbb{R}^m \times \mathbb{R}^r \times \mathbb{R}, \circ, \delta_\lambda^{\mathbb{K}}\right),$$

where we denote (x, w, t) the point in $\mathbb{R}^m \times \mathbb{R}^r \times \mathbb{R}$. We assume that

$$Y = X_0(x, w) - \partial_t = \langle (x, w), B\nabla_{(x,w)} \rangle - \partial_t. \quad (4.3)$$

---

[1] We use the same notation "$\circ$" for the composition law in different groups; the context will avoid ambiguity.

The dilations (3.3) and the group law (3.6) will be denoted by:

$$\delta_\lambda^{\mathbb{K}}(x, w, t) = \left(\lambda x, \rho_\lambda^{\mathbb{K}} w, \lambda^2 t\right),$$

$$(x, w, t) \circ (x', w', t') = \left(x + x', R(x, w, t, x', w', t'), t + t'\right).$$

The link $\mathbb{L} = \mathbb{G}\triangle\mathbb{K}$ is defined as follows:

$$\mathbb{L} = \left(\mathbb{R}^m \times \mathbb{R}^n \times \mathbb{R}^r \times \mathbb{R}, \circ, \delta_\lambda^{\mathbb{L}}\right),$$

where

$$\delta_\lambda^{\mathbb{L}}(x, y, w, t) = \left(\lambda x, \rho_\lambda^{\mathbb{G}} y, \rho_\lambda^{\mathbb{K}} w, \lambda^2 t\right)$$

and

$$(x,y,w,t) \circ (x',y',w',t') = (x+x', Q(x,y,x',y'), R(x,w,t,x',w',t'), t+t'). \tag{4.4}$$

It turns out that $\mathbb{L}$ is a homogeneous group, the X_p's and Y (considered as vector fields on $\mathbb{R}^m \times \mathbb{R}^n \times \mathbb{R}^r \times \mathbb{R}$) satisfy H.1–H.2 (see Propositions 10.4 and 10.5 in [5]). Let explicitly note that the operations defined in $\mathbb{L}$ extend the ones in $\mathbb{G}$ and $\mathbb{K}$. In particular we have

$$(x, y, 0, 0) \circ (x', y', 0, 0) = ((x, y) \circ (x', y'), 0, 0). \tag{4.5}$$

The following statement contains the global lower bound and the global Harnack inequality for operators on a linked group of step three proved in [13].

Proposition 4.1. *Let L be the operator in* (4.1) *on a linked group* $\mathbb{L} = \mathbb{G}\triangle\mathbb{K}$ *of step three. Let* $z_0 = (\xi_0, \eta_0, \omega_0, t_0) \in \mathbb{R}^{N+1} \equiv \mathbb{R}^m \times \mathbb{R}^n \times \mathbb{R}^r \times \mathbb{R}$ *and* $T > 0$. *Then:*

- *There exists a positive constant* C *such that*

$$\Gamma(x, y, w, t) \geq \frac{C}{t^{\frac{Q-2}{2}}} \exp\left(-C\left(\frac{|(x, y)|_{\mathbb{L}}^2}{t} + \frac{|w|_{\mathbb{L}}^6}{t^3}\right)\right), \qquad \forall (x, y, w, t) \in \mathbb{R}^N \times \mathbb{R}^+.$$

$$\tag{4.6}$$

- *There exist two constants* $c > 0$ *and* $C > 1$, *only dependent on* L, *such that, if* u *is a non-negative solution to* $Lu = 0$ *in* $\mathbb{R}^N \times \,]t_0 - cT, t_0 + T]$, *then*

$$u(z_0) \leq \exp\left(C\left(1 + \frac{|(x, y)|_{\mathbb{L}}^2}{t} + \frac{|w|_{\mathbb{L}}^6}{t^3}\right)\right) u(z_0 \circ z), \tag{4.7}$$

for every $z = (x, y, w, t) \in \mathbb{R}^N \times \,]0, T]$.

5. Outline of the proof

The main tool in the proof of Propositions 2.1, 3.1 and 4.1 is the following *non-local* Harnack inequality given in [13], Theorem 1.1:

Theorem 5.1. *Let* $z_0 = (x_0, t_0) \in \mathbb{R}^{N+1}$ *and* $s > 0$. *There exist two constants* $c, C > 1$, *only dependent on* L, *such that*

$$u(\exp(s(\lambda \cdot X + Y))(z_0)) \leq C^{1+s|\lambda|^2} u(z_0), \tag{5.1}$$

for every non-negative solution u *to* $Lu = 0$ *in* $\mathbb{R}^N \times \,]t_0 - cs, t_0]$, $\lambda \in \mathbb{R}^m$.

In the above statement we denoted (as usual) $\exp(sX)(z) = \gamma(s)$, where γ is the (unique and globally defined) solution to the Cauchy problem $\gamma' = X(\gamma); \gamma(0) = z$. In the sequel we also use the following notation $e^X = \exp(X)(0)$ and recall that $\exp(X)(z) = z \circ e^X$.

The above Harnack inequality has been proved by using repeatedly the invariant local Harnack inequality by Kogoj and Lanconelli [5], Theorem 7.1. In order to state the Harnack inequality in [5], we set some notations. Given $r > 0$, $\varepsilon \in {]0, 1[}$ and $z_0 \in \mathbb{R}^{N+1}$, we put

$$\mathcal{C}_r(z_0) = z_0 \circ \delta_r(\mathcal{C}_1),$$
$$\mathcal{S}_r^{(\varepsilon)}(z_0) = z_0 \circ \delta_r(\mathcal{S}_1^{(\varepsilon)}),$$

where

$$\mathcal{C}_1 = \{z = (x, t) \in \mathbb{R}^{N+1} \mid \|z\|_{\mathbb{G}} \leq 1,\ t \leq 0\},$$
$$\mathcal{S}_1^{(\varepsilon)} = \{z = (x, -\varepsilon) \mid z \in \mathcal{C}_1\}.$$

Theorem 5.2. *Let O be an open set in $\mathbb{R}^{N+1}$ containing $\mathcal{C}_r(z_0)$ for some $z_0 \in \mathbb{R}^{N+1}$ and $r > 0$. Given $\varepsilon \in {]0, 1[}$, there exist two positive constants $\theta = \theta(L, \varepsilon)$ and $C = C(L, \varepsilon)$ such that*

$$\sup_{\mathcal{S}_{\theta r}^{(\varepsilon)}(z_0)} u \leq C u(z_0), \tag{5.2}$$

for every non-negative solution u of L in O.

The connectivity assumption H.2 and Theorem 5.1 directly yield a global Harnack inequality for positive solutions to $Lu = 0$ of the form:

$$u(x, t) \leq H(x, t, \xi, \tau)\, u(\xi, \tau), \qquad \forall (x, t), (\xi, \tau) \mathbb{R}^{N+1},\ t < \tau. \tag{5.3}$$

When we are able to find explicitly an L-admissible path γ connecting (x, t) to (ξ, τ), then we can express explicitly $H(x, t, \xi, \tau)$ and obtain a more useful estimate. Aiming to take into account of the homogeneous structure of the Lie group, we construct such a γ by considering separately the commutators of different homogeneity of $X_1, \ldots, X_m, Y$. We remark that these commutators can be conveniently approximated by L-admissible paths: for instance, the direction of the commutator $[X_p, X_q]$ can be obtained by using the integral curves of $X_p, X_q, -X_p, -X_q$. To be more specific, by using the Campbell-Hausdorff formula, we have

$$e^{X_p+Y} \circ e^{X_q+Y} \circ e^{-X_p+Y} \circ e^{-X_q+Y} = e^{4Y+[X_p,X_q]+R_2},$$

where the error term R_2 contains commutators δ_λ-homogeneous of order greater than two. This fact is well known and has been used by many authors in the study of the regularity of "elliptic" and "parabolic" operators of the form

$$\sum_{p=1}^{m} X_p^2 \qquad \text{and} \qquad \sum_{p=1}^{m} X_p^2 - \partial_t, \tag{5.4}$$

respectively. However the study of operator (1.1) involves commutators of the form $[X_p, Y]$ that do not occur in the examples (5.4). In this case, we have to use a different combination of vector fields, namely

$$e^{X_p+Y} \circ e^{-X_p+Y} = e^{2Y+[X_p,Y]+R_3},$$

where R_3 is an error term of order three. The above argument can be adapted to commutators of higher length and leads to explicit estimates of H in (5.3) which are stated in Propositions 2.1, 3.1 and 4.1.

We omit here the details of the proof, that are contained in the paper [13].

References

[1] G.K. ALEXOPOULOS, *Sub-Laplacians with drift on Lie groups of polynomial volume growth*, Mem. Amer. Math. Soc., 155 (2002), pp. x+101.

[2] A. BONFIGLIOLI, E. LANCONELLI, AND F. UGUZZONI, *Uniform Gaussian estimates of the fundamental solutions for heat operators on Carnot groups.*, Adv. Differ. Equ., 7 (2002), pp. 1153–1192.

[3] G.B. FOLLAND, *Subelliptic estimates and function spaces on nilpotent Lie groups*, Ark. Mat., 13 (1975), pp. 161–207.

[4] D.S. JERISON AND A. SÁNCHEZ-CALLE, *Estimates for the heat kernel for a sum of squares of vector fields*, Indiana Univ. Math. J., 35 (1986), pp. 835–854.

[5] A.E. KOGOJ AND E. LANCONELLI, *An invariant Harnack inequality for a class of hypoelliptic ultraparabolic equations*, Mediterr. J. Math., 1 (2004), pp. 51–80.

[6] ———, *One-side Liouville theorems for a class of hypoelliptic ultraparabolic equations*, "Geometric Analysis of PDE and Several Complex Variables" Contemporary Mathematics Proceedings, (2004).

[7] L.P. KUPCOV, *The fundamental solutions of a certain class of elliptic-parabolic second order equations*, Differencial'nye Uravnenija, 8 (1972), pp. 1649–1660, 1716.

[8] S. KUSUOKA AND D. STROOCK, *Applications of the Malliavin calculus. III*, J. Fac. Sci. Univ. Tokyo Sect. IA Math., 34 (1987), pp. 391–442.

[9] E. LANCONELLI, A. PASCUCCI, AND S. POLIDORO, *Linear and nonlinear ultraparabolic equations of Kolmogorov type arising in diffusion theory and in finance*, in Nonlinear problems in mathematical physics and related topics, II, vol. 2 of Int. Math. Ser. (N. Y.), Kluwer/Plenum, New York, 2002, pp. 243–265.

[10] E. LANCONELLI AND S. POLIDORO, *On a class of hypoelliptic evolution operators*, Rend. Sem. Mat. Univ. Politec. Torino, 52 (1994), pp. 29–63. Partial differential equations, II (Turin, 1993).

[11] A. NAGEL, E.M. STEIN, AND S. WAINGER, *Balls and metrics defined by vector fields. I. Basic properties*, Acta Math., 155 (1985), pp. 103–147.

[12] A. PASCUCCI AND S. POLIDORO, *A Gaussian upper bound for the fundamental solutions of a class of ultraparabolic equations*, J. Math. Anal. Appl., 282 (2003), pp. 396–409.

[13] ———, *Harnack inequalities and Gaussian estimates for a class of hypoelliptic operators*, to appear in Trans. Amer. Math. Soc.

[14] ———, *On the Harnack inequality for a class of hypoelliptic evolution equations*, Trans. Amer. Math. Soc., 356 (2004), pp. 4383–4394 (electronic).

[15] S. POLIDORO, *On a class of ultraparabolic operators of Kolmogorov-Fokker-Planck type*, Matematiche (Catania), 49 (1994), pp. 53–105 (1995).

[16] ———, *A global lower bound for the fundamental solution of Kolmogorov-Fokker-Planck equations*, Arch. Rational Mech. Anal., 137 (1997), pp. 321–340.

[17] L.P. ROTHSCHILD AND E.M. STEIN, *Hypoelliptic differential operators and nilpotent groups*, Acta Math., 137 (1976), pp. 247–320.

[18] L. SALOFF-COSTE AND D.W. STROOCK, *Opérateurs uniformément sous-elliptiques sur les groupes de Lie*, J. Funct. Anal., 98 (1991), pp. 97–121.

[19] N.T. VAROPOULOS, L. SALOFF-COSTE, AND T. COULHON, *Analysis and geometry on groups*, vol. 100 of Cambridge Tracts in Mathematics, Cambridge University Press, Cambridge, 1992.

Sergio Polidoro
Dipartimento di Matematica
Università di Bologna
Piazza di Porta S. Donato 5
I-40126 Bologna, Italy
e-mail: `polidoro@dm.unibo.it`

Progress in Nonlinear Differential Equations
and Their Applications, Vol. 63, 375–388
© 2005 Birkhäuser Verlag Basel/Switzerland

How to Construct Good Measures

Augusto C. Ponce

Dedicated to H. Brezis in the occasion of his 60th birthday

Abstract. Given any continuous nondecreasing function $g : \mathbb{R} \to \mathbb{R}$, with $g(t) = 0$, $\forall t \le 0$, we show that there always exists some positive measure μ, concentrated on a set of zero Newtonian capacity, for which the problem

$$\begin{cases} -\Delta u + g(u) = \mu & \text{in } \Omega, \\ \qquad\qquad\ u = 0 & \text{on } \partial\Omega, \end{cases} \qquad (0.1)$$

admits a solution. This provides an affirmative answer to Open problem 2 raised by Brezis-Marcus-Ponce [3]. When $N \ge 3$ and $g(t) = e^t - 1$, $\forall t \ge 0$, Bartolucci-Leoni-Orsina-Ponce [1] proved that any measure $\mu \le 4\pi \mathcal{H}^{N-2}$ is good for problem (0.1). We present examples of other good measures which are not $\le 4\pi \mathcal{H}^{N-2}$.

Mathematics Subject Classification (2000). 35J60, 35B05.

Keywords. Nonlinear elliptic equations, good measures, Cantor sets.

1. Introduction

Let $\Omega \subset \mathbb{R}^N$, $N \ge 2$, be a smooth bounded domain. Let $g : \mathbb{R} \to \mathbb{R}$ be a continuous nondecreasing function such that $g(t) = 0$, $\forall t \le 0$. Given a bounded measure $\mu \in \mathcal{M}(\Omega)$, then u is a solution of

$$\begin{cases} -\Delta u + g(u) = \mu & \text{in } \Omega, \\ \qquad\qquad\ u = 0 & \text{on } \partial\Omega, \end{cases} \qquad (1.1)$$

if $u \in L^1(\Omega)$, $g(u) \in L^1(\Omega)$, and

$$-\int_\Omega u \Delta\zeta + \int_\Omega g(u)\zeta = \int_\Omega \zeta \, d\mu \quad \forall \zeta \in C^2(\overline{\Omega}), \ \zeta = 0 \text{ on } \partial\Omega.$$

We say that μ is a *good measure* (relative to g) if (1.1) has a solution u. We observe that u, whenever it exists, is unique. The study of problem (1.1), when $\mu \in L^1(\Omega)$, was initiated by Brezis-Strauss [5]. They established that every measure in $L^1(\Omega)$ is good. Later, Bénilan-Brezis [2] (see also Brezis-Véron [6]) proved that (1.1) need not have a solution for a given measure μ. In fact, if $N \geq 3$ and $g(t) = t^p$, $\forall t \geq 0$, for some $p \geq \frac{N}{N-2}$, then there exists no u satisfying (1.1) for $\mu = \delta_a$, $a \in \Omega$.

Let $\mathcal{G}(g)$ denote the set of good measures associated to g. One can show (see [3]) that $\mathcal{G}(g)$ is convex and closed with respect to the strong topology in $\mathcal{M}(\Omega)$.

A measure μ is *diffuse* if $\mu(A) = 0$ for every Borel set $A \subset \Omega$ such that $\mathrm{cap}\,(A) = 0$, where "cap" denotes the Newtonian (H^1) capacity. If $\mu \in \mathcal{M}(\Omega)$ and μ^+ is diffuse, then μ is good for every nonlinearity g (see [3, Corollary 3]). The converse is also true. Namely, if μ is good for every g, then μ^+ is diffuse (see [3, Theorem 5]). We can summarize this as

$$\Big\{ \mu \in \mathcal{M}(\Omega) : \mu^+ \text{ is a diffuse measure} \Big\} = \bigcap_g \mathcal{G}(g),$$

where the intersection is taken over all continuous nondecreasing functions $g : \mathbb{R} \to \mathbb{R}$ such that $g(t) = 0$, $\forall t \leq 0$.

One of our main results is the following

Theorem 1. *Given any g, we have*

$$\Big\{ \mu \in \mathcal{M}(\Omega) : \mu^+ \text{ is a diffuse measure} \Big\} \subsetneq \mathcal{G}(g).$$

In other words, for any fixed g, there exists a measure $\mu \in \mathcal{M}(\Omega)$, $\mu \geq 0$, such that $\mu \in \mathcal{G}(g)$, but μ is not diffuse.

Theorem 1 gives a positive answer to Open problem 2 in [3]. As we shall see below, the proof of Theorem 1 is constructive. In fact, it gives a recipe for explicitly obtaining the measure μ. Of course, such μ will heavily depend on the function g.

In dimension $N \geq 3$, Theorem 1 can be improved. Recall that any Borel set $A \subset \Omega$ such that $\mathcal{H}^{N-2}(A) < \infty$ has zero capacity (but the converse is false; see [7]). When $N \geq 3$, it is always possible to find good measures μ of the form $\mu = \alpha \mathcal{H}^{N-2}\lfloor_K$ for some compact set $K \subset \Omega$ and $\alpha > 0$. More precisely, we have

Theorem 2. *Assume $N \geq 3$. Given any g, there exists a compact set $K \subset \Omega$, $\mathcal{H}^{N-2}(K) \in (0, \infty)$, such that $\mu = \alpha \mathcal{H}^{N-2}\lfloor_K$ is good (relative to g) for every $\alpha > 0$.*

Theorem 2 is no longer true in dimension $N = 2$. In fact, problem (1.1) has no solution when $g(t) = e^t - 1$, $\forall t \geq 0$, and $\mu = \alpha \delta_a$, $a \in \Omega$, for any $\alpha > 4\pi$ (see Vázquez [12]).

One can also construct good measures $\mu \geq 0$ concentrated on a set of zero $\mathcal{H}^{N-2}$-measure. In fact,

Theorem 3. *Assume $N \geq 3$. For any g, there exists a good measure $\mu \geq 0$ such that $\mathcal{H}^{N-2}(\mathrm{supp}\,\mu) = 0$.*

When $N \geq 3$ and $g(t) = e^t - 1$, $\forall t \geq 0$, it has been established in [1] that if $\mu \leq 4\pi \mathcal{H}^{N-2}$, then μ is good. According to Theorems 2 and 3 above, there are other good measures which are *not* $\leq 4\pi \mathcal{H}^{N-2}$. The existence of such measures was suggested by L. Véron in a personal communication.

The construction presented here has been applied in the study of other related problems; see [4] and [8]. An alternative approach for obtaining good measures which are not diffuse might be found in some recent work of Marcus-Véron [10].

This paper is organized as follows. In Section 2, we define a Cantor-type set F associated to a subsequence (ℓ_{k_j}); as we shall see later on, the proofs of Theorems 1–3 rely on suitable choices of (ℓ_k) and (k_j). We then introduce a positive measure μ_F supported on F. In Section 3, we estimate the potential generated by μ_F in terms of (ℓ_{k_j}). In Section 4, we present the proofs of Theorems 2 and 3; as a corollary, we obtain Theorem 1 when $N \geq 3$. Finally, in Section 5, we prove Theorem 1 in the case $N = 2$.

2. Construction of the Cantor set F associated to the subsequence (ℓ_{k_j})

We shall assume for simplicity that $\Omega = Q_1$, the unit cube centered at 0. One of the main ingredients in the proofs of Theorems 1–3 will be the construction of a (generalized) Cantor set $F \subset \Omega$; see, e.g., [11]. We begin by describing the building blocks used in the definition of F.

Let $n \geq 1$ be an integer and let $0 < s \ll t$. We shall associate to the triple (s, t, n) a compact set $E(s, t, n) \subset [-\frac{t}{2}, \frac{t}{2}]^N$ in the following way. Let

$$\alpha = \frac{t - ns}{n - 1}. \tag{2.1}$$

For $j = 1, \ldots, n$, set

$$a_j = (j - 1)(s + \alpha) - \frac{t}{2} \quad \text{and} \quad b_j = a_j + s.$$

In particular, $a_1 = -\frac{t}{2}$ and $b_n = \frac{t}{2}$. We then define

$$E(s, t, n) = \bigcup_{1 \leq i_1, \ldots, i_N \leq n} [a_{i_1}, b_{i_1}] \times \cdots \times [a_{i_N}, b_{i_N}].$$

Thus, the set $E(s, t, n)$ is the union of n^N cubes of side s, uniformly distributed in $[-\frac{t}{2}, \frac{t}{2}]^N$. The distance between two components of $E(s, t, n)$ is $\geq \alpha$.

We now turn to the construction of F.

Let (ℓ_k) be a decreasing sequence of positive numbers such that

$$\ell_1 \leq \frac{1}{4} \quad \text{and} \quad \ell_{k+1} \leq \theta \, \ell_k \quad \forall k \geq 1, \tag{2.2}$$

for some $\theta \in (0, \frac{1}{2})$. The Cantor set F associated to the subsequence (ℓ_{k_j}) is defined by induction as follows.

Let $F_0 = Q_1$, $k_0 = 0$ and $\ell_0 = 1$. Let F_j be the set obtained after the jth step; F_j is the disjoint union of 2^{Nk_j} cubes Q_i of side ℓ_{k_j}. Let $x_1, \ldots, x_{2^{Nk_j}}$ denote the centers of each component of F_j (although it is not indicated, such points do depend on j). We then set

$$F_{j+1} = \bigcup_{i=1}^{2^{Nk_j}} E\left(\ell_{k_{j+1}}, \gamma \ell_{k_j}, 2^{(k_{j+1}-k_j)}\right) + x_i, \tag{2.3}$$

where $\gamma = \frac{1}{2} + \theta \in (\frac{1}{2}, 1)$. In particular, F_{j+1} is the union of $2^{Nk_{j+1}}$ disjoint cubes of side $\ell_{k_{j+1}}$. Moreover, since we are taking $t = \gamma \ell_{k_j}$, we have

$$d(F_{j+1}, \partial F_j) = \frac{1-\gamma}{2} \ell_{k_j} = \frac{1-2\theta}{4} \ell_{k_j}. \tag{2.4}$$

We also point out that the distance between any two components of F_{j+1} inside the cube $[-\gamma \ell_{k_j}, \gamma \ell_{k_j}]^N + x_i$ is $\geq \alpha$, where α is given by (2.1). Since (2.2) holds with $\theta < \frac{1}{2}$, we have

$$\alpha \sim \frac{\ell_{k_j}}{2^{(k_{j+1}-k_j)}}.$$

We finally set

$$F = \bigcap_{j=0}^{\infty} F_j.$$

We would like to emphasize the main feature in the construction of F. In order to obtain a standard Cantor set, inside each component Q_i of F_j one would take 2^N small cubes. In our case, we select $2^{N(k_{j+1}-k_j)}$ small cubes inside Q_i. This possibility of choosing many more cubes turns out to be crucial in the proofs of some of our main results.

3. Potential generated by the uniform measure μ_F concentrated on F

In this section, we present some basic estimates which will be used throughout this paper.

For each $j \geq 1$, let $\mu_j = \frac{1}{|F_{j+1}|} \chi_{F_{j+1}}$, where F_{j+1} is given by (2.3). The *uniform measure concentrated on F*, μ_F, is the weak* limit of (μ_j) in $\mathcal{M}(\Omega)$ as $j \to \infty$. In particular, $\mu_F \geq 0$ and $\mu_F(\Omega) = 1$. A key property satisfied by μ_F is given by the next

Lemma 1. *For every $x \in F_{j+1}$, $j \geq 0$, we have*

$$\mu_F\left(B_r(x)\right) \sim \begin{cases} \dfrac{1}{2^{Nk_{j+1}}} & \text{if } \ell_{k_{j+1}} \lesssim r \lesssim \dfrac{\ell_{k_j}}{2^{(k_{j+1}-k_j)}}, \\[2ex] \dfrac{1}{2^{Nk_j}} \left(\dfrac{r}{\ell_{k_j}}\right)^N & \text{if } \dfrac{\ell_{k_j}}{2^{(k_{j+1}-k_j)}} \lesssim r \lesssim \ell_{k_j}. \end{cases} \tag{3.1}$$

Here, we implicitly assume that $k_0 = 0$. We say that $a \lesssim b$ if there exists $C > 0$, depending on N and θ, such that $a \leq C\, b$. By $a \sim b$, we mean that $a \lesssim b$ and $b \lesssim a$.

Proof. We shall use the same notation as in the construction of F. Note that if

$$\ell_{k_{j+1}} \lesssim r \lesssim \frac{\ell_{k_j}}{2^{(k_{j+1}-k_j)}},$$

then $B_r(x)$ contains a single component $Q_{i,n}$ of F_{j+1}. Since

$$\mu_F(Q_{i,n}) = \frac{1}{2^{Nk_{j+1}}},$$

the first estimate in (3.1) follows.

We now assume

$$\frac{\ell_{k_j}}{2^{(k_{j+1}-k_j)}} \lesssim r \lesssim \ell_{k_j}.$$

Let Q_i be the component of F_j containing x. Recall that there are $2^{N(k_{j+1}-k_j)}$ components $Q_{i,n}$ of F_{j+1} contained in Q_i. Thus, the number of cubes $Q_{i,n}$ contained in $B_r(x)$ is of the order of $2^{N(k_{j+1}-k_j)} \left(\frac{r}{\ell_{k_j}}\right)^N$. Since, for each $Q_{i,n}$, $\mu_F(Q_{i,n}) = \frac{1}{2^{Nk_{j+1}}}$, we then have

$$\mu_F\big(B_r(x)\big) \sim 2^{N(k_{j+1}-k_j)} \left(\frac{r}{\ell_{k_j}}\right)^N \mu_F(Q_{i,n}) = \frac{1}{2^{Nk_j}} \left(\frac{r}{\ell_{k_j}}\right)^N.$$

The proof of the lemma is complete.

Let $v \in L^1(Q_1)$ be the unique solution of

$$\begin{cases} -\Delta v = \mu_F & \text{in } Q_1, \\ \quad\; v = 0 & \text{on } \partial Q_1. \end{cases} \tag{3.2}$$

A basic estimate satisfied by v is given by the following

Proposition 1. *Assume $N \geq 3$. Let $F \subset Q_1$ be the Cantor set associated to the subsequence (ℓ_{k_j}) and let v be the solution of (3.2). Then, there exist constants $C_1, C_2 > 0$ (depending on N and θ) such that*

$$C_1 \left(\frac{1}{\ell_{k_1}^{N-2}} + \sum_{i=1}^{j} \frac{1}{2^{Nk_i}\ell_{k_i}^{N-2}}\right) \leq v(x) \leq C_2 \left(\frac{1}{\ell_{k_1}^{N-2}} + \sum_{i=1}^{j} \frac{1}{2^{Nk_i}\ell_{k_i}^{N-2}}\right), \tag{3.3}$$

for every $x \in \partial F_j$, $j \geq 1$.

Proof. Let

$$w(x) = \frac{1}{N\omega_N} \int_0^\infty \frac{\mu_F\big(B_r(x)\big)}{r^{N-1}}\, dr \quad \forall x \in Q_1,$$

where $\omega_N = |B_1|$. By (2.4), for every $x \in \partial F_j$ we have

$$\mu_F\big(B_r(x)\big) = 0 \quad \text{if } r \lesssim \ell_{k_j},$$

so that

$$w(x) \sim \int_{\ell_{k_j}}^{\infty} \frac{\mu_F\big(B_r(x)\big)}{r^{N-1}} \, dr \quad \forall x \in \partial F_j.$$

Thus,

$$w(x) \sim \sum_{i=1}^{j-1} \int_{\ell_{k_{i+1}}}^{\ell_{k_i}} \frac{\mu_F\big(B_r(x)\big)}{r^{N-1}} \, dr + \int_{\ell_{k_1}}^{\infty} \frac{\mu_F\big(B_r(x)\big)}{r^{N-1}} \, dr \sim \sum_{i=1}^{j-1} \big(A_i + B_i\big) + \frac{1}{\ell_{k_1}^{N-2}},$$

where, by Lemma 1 and (2.2),

$$A_i = \int_{\ell_{k_{i+1}}}^{\frac{\ell_{k_i}}{2^{(k_{i+1}-k_i)}}} \frac{\mu_F\big(B_r(x)\big)}{r^{N-1}} \, dr \sim \frac{1}{2^{Nk_{i+1}}} \int_{\ell_{k_{i+1}}}^{\frac{\ell_{k_i}}{2^{(k_{i+1}-k_i)}}} \frac{dr}{r^{N-1}} \sim \frac{1}{2^{Nk_{i+1}} \ell_{k_{i+1}}^{N-2}}$$

and

$$B_i = \int_{\frac{\ell_{k_i}}{2^{(k_{i+1}-k_i)}}}^{\ell_{k_i}} \frac{\mu_F\big(B_r(x)\big)}{r^{N-1}} \, dr \sim \frac{1}{2^{Nk_i} \ell_{k_i}^{N}} \int_{\frac{\ell_{k_i}}{2^{(k_{i+1}-k_i)}}}^{\ell_{k_i}} r \, dr \sim \frac{1}{2^{Nk_i} \ell_{k_i}^{N-2}}.$$

Therefore,

$$w(x) \sim \sum_{i=1}^{j-1} \left(\frac{1}{2^{Nk_{i+1}} \ell_{k_{i+1}}^{N-2}} + \frac{1}{2^{Nk_i} \ell_{k_i}^{N-2}} \right) + \frac{1}{\ell_{k_1}^{N-2}} \sim \sum_{i=1}^{j} \frac{1}{2^{Nk_i} \ell_{k_i}^{N-2}} + \frac{1}{\ell_{k_1}^{N-2}}. \quad (3.4)$$

In other words, w satisfies (3.3). On the other hand, we have

$$d(F_1, \partial Q_1) = \frac{1-\gamma}{2} = \frac{1-2\theta}{4} > 0.$$

Since $w \geq 0$ and $-\Delta w = \mu_F$ in Q_1 (see Lemma 2 below), there exist constants $\tilde{C}_1, \tilde{C}_2 > 0$ such that

$$\tilde{C}_1 w \leq v \leq \tilde{C}_2 w \quad \text{on } F_1. \tag{3.5}$$

Combining (3.4) and (3.5), we obtain (3.3). This concludes the proof of the proposition.

We now establish a well-known fact used in the proof of Proposition 1:

Lemma 2. *Given $\mu \in \mathcal{M}(\mathbb{R}^N)$, let*

$$w(x) = \frac{1}{N\omega_N} \int_0^{\infty} \frac{\mu\big(B_r(x)\big)}{r^{N-1}} \, dr \quad \forall x \in \mathbb{R}^N. \tag{3.6}$$

Then,

$$-\Delta w = \mu \quad \text{in } \mathcal{D}'(\mathbb{R}^N).$$

Proof. We shall prove the lemma for $N \geq 3$; the case $N = 2$ is similar. We make the change of variables $r = s^{-\frac{1}{N-2}}$ in (3.6). Since $\frac{dr}{r^{N-1}} = -\frac{ds}{N-2}$, we get

$$
\begin{aligned}
N(N-2)\omega_N\, w(x) &= (N-2) \int_0^\infty \mu\Big(\{y \in \mathbb{R}^N : |x-y| < r\}\Big) \frac{dr}{r^{N-1}} \\
&= \int_0^\infty \mu\Big(\{y \in \mathbb{R}^N : |x-y| < s^{-\frac{1}{N-2}}\}\Big)\, ds \\
&= \int_0^\infty \mu\Big(\{y \in \mathbb{R}^N : \frac{1}{|x-y|^{N-2}} > s\}\Big)\, ds = \int_{\mathbb{R}^N} \frac{d\mu(y)}{|x-y|^{N-2}},
\end{aligned}
$$

from which the result follows. $\quad\blacksquare$

The counterpart of Proposition 1 in dimension $N = 2$ is given by

Proposition 2. *Assume $N = 2$. Let $F \subset Q_1$ be the Cantor set associated to the subsequence (ℓ_{k_j}) and let v be the solution of (3.2). Then, for every $j \geq 1$, we have*

$$
v \sim \left(\log \frac{1}{\ell_{k_1}} + \sum_{i=1}^{j} \frac{1}{4^{k_i}} \log \frac{1}{\ell_{k_i}} \right) \qquad \text{on } \partial F_j. \tag{3.7}
$$

The proof of Proposition 2 follows along the same lines and shall be omitted.

4. Proofs of Theorems 2 and 3

We start by recalling the definition of the (spherical) Hausdorff measure $\mathcal{H}^s$ in $\mathbb{R}^N$, where $0 \leq s \leq N$. Let $A \subset \mathbb{R}^N$ be a Borel set. Given $\delta > 0$, let

$$
\mathcal{H}^s_\delta(A) = \inf \left\{ \sum_i \omega_s r_i^s : K \subset \bigcup_i B_{r_i} \text{ with } r_i < \delta, \forall i \right\},
$$

where the infimum is taken over all coverings of A with open balls B_{r_i} of radii $r_i < \delta$, and $\omega_s = \frac{\pi^{s/2}}{\Gamma(\frac{s}{2}+1)}$. When s is a positive integer, then ω_s is the measure of the unit ball in $\mathbb{R}^s$. We then set

$$
\mathcal{H}^s(A) = \lim_{\delta \downarrow 0} \mathcal{H}^s_\delta(A).
$$

We have the following

Lemma 3. *Let F be the Cantor set associated to the subsequence (ℓ_{k_j}). Then,*

$$
\mathcal{H}^s(F) \sim \liminf_{j \to \infty} 2^{Nk_j} \ell_{k_j}^s. \tag{4.1}
$$

Moreover, if $\mathcal{H}^s(F) \in (0, \infty)$, then

$$
\mu_F = \frac{1}{\mathcal{H}^s(F)} \mathcal{H}^s \lfloor F. \tag{4.2}
$$

Proof.

Proof of (4.1). For $j \geq 1$ fixed, let (B_i) be a covering of F with 2^{Nk_j} balls of radii ℓ_{k_j}, where each ball B_i is concentric to some component of F_j. Then,

$$\mathcal{H}_\delta^s(F) \leq \omega_s \, 2^{Nk_j} \ell_{k_j}^s,$$

for every $\delta > \ell_{k_j}$. Thus,

$$\mathcal{H}^s(F) \leq \omega_s \liminf_{j \to \infty} 2^{Nk_j} \ell_{k_j}^s, \tag{4.3}$$

which gives $\lesssim$ in (4.1).

Conversely, if $\liminf_{j \to \infty} 2^{Nk_j} \ell_{k_j}^s = 0$, then it follows from (4.3) that $\mathcal{H}^s(F) = 0$ and we are done. We now assume that

$$\liminf_{j \to \infty} 2^{Nk_j} \ell_{k_j}^s > 0$$

(the limit above possibly being infinite). Given $0 < a < \liminf_{j \to \infty} 2^{Nk_j} \ell_{k_j}^s$, let $j_0 \geq 1$ be sufficiently large so that

$$2^{Nk_j} \ell_{k_j}^s \geq a \quad \forall j \geq j_0. \tag{4.4}$$

It then follows from Lemma 1 and (4.4) that there exists $C > 0$ such that

$$\mu_F\big(B_r(x)\big) \leq \frac{Cr^s}{a} \quad \forall x \in F, \quad \forall r \in (0, \ell_{j_0}). \tag{4.5}$$

Let $\delta \in (0, \ell_{j_0})$ and let (B_{r_i}) be a covering of F with balls of radii $r_i < \delta$. Without loss of generality, we may assume that each B_{r_i} is centered at some point of F. Thus, in view of (4.5), we have

$$\sum_i r_i^s \geq \frac{a}{C} \sum_i \mu_F(B_{r_i}) \geq \frac{a}{C} \mu_F\Big(\bigcup_i B_{r_i}\Big) = \frac{a}{C} \mu_F(F) = \frac{a}{C}.$$

This lower bound holds for any covering (B_{r_i}) such that $r_i < \delta$, $\forall i$. Therefore,

$$\mathcal{H}^s(F) \geq \mathcal{H}_\delta^s(F) \geq \frac{\omega_s}{C} a.$$

Since $a < \liminf_{j \to \infty} 2^{Nk_j} \ell_{k_j}^s$ was arbitrary, we conclude that

$$\mathcal{H}^s(F) \geq \frac{\omega_s}{C} \liminf_{j \to \infty} 2^{Nk_j} \ell_{k_j}^s.$$

This establishes (4.1).

Proof of (4.2). Assume $\mathcal{H}^s(F) \in (0, \infty)$. Let Q_i be a component of F_j, $j \geq 1$. By symmetry, we have

$$\mathcal{H}^s(F) = 2^{Nk_j} \mathcal{H}^s(Q_i \cap F).$$

Since $\mu_F(Q_i) = 2^{-Nk_j}$, we get

$$\mu_F(Q_i) = \frac{1}{\mathcal{H}^s(F)} \mathcal{H}^s \llcorner_F (Q_i). \tag{4.6}$$

Given $A \subset \mathbb{R}^N$ open, we may write $A \cap F = \bigcup_i (Q_i \cap F)$, where (Q_i) is a family of disjoint connected components among all F_j, $j \geq 1$. It then follows from (4.6) that

$$\mu_F(A) = \frac{1}{\mathcal{H}^s(F)} \mathcal{H}^s \lfloor_F (A) \quad \text{for every open set } A \subset \mathbb{R}^N.$$

Since μ_F and $\mathcal{H}^s \lfloor_F$ are Radon measures, (4.2) follows. This concludes the proof of the lemma.

We recall the following result (see [3, Theorem 4]):

Proposition 3. *Suppose $\mu_1 \in \mathcal{M}(\Omega)$ is a good measure for problem* (1.1). *Then, any measure $\mu_2 \leq \mu_1$ is also good.*

We now establish the

Proposition 4. *Assume $N \geq 3$. Let F be the Cantor set associated to the subsequence (ℓ_{k_j}). There exists $C > 0$ (depending on N and θ) such that if*

$$\sum_{j=1}^{\infty} g\left(C\alpha_0 \sum_{i=1}^{j+1} \frac{1}{2^{Nk_i} \ell_{k_i}^{N-2}} \right) 2^{Nk_j} \ell_{k_j}^N < \infty \quad \text{for some } \alpha_0 > 0, \tag{4.7}$$

then $\alpha_0 \mu_F \in \mathcal{G}(g)$.

Proof. Let

$$a = \frac{1}{\ell_{k_1}^{N-2}} \quad \text{and} \quad b_j = \sum_{i=1}^{j} \frac{1}{2^{Nk_i} \ell_{k_i}^{N-2}} \quad \forall j \geq 1.$$

Let v be the solution of (3.2). By Proposition 1, there exists $C_2 > 0$ such that

$$v(x) \leq C_2(a + b_j) \quad \forall x \in \partial F_j.$$

Note that v is harmonic in $(\text{int } F_j) \backslash F_{j+1}$. Thus, by the maximum principle,

$$v(x) \leq C_2(a + b_{j+1}) \quad \forall x \in F_j \backslash F_{j+1}.$$

Assume that $\lim_{j \to \infty} b_j < \infty$. In this case, we have $v \in L^\infty(\Omega)$; hence, $g(\alpha_0 v) \in L^1(\Omega)$. We then conclude that $\alpha_0 \mu_F + g(\alpha_0 v)$ is good. By Proposition 3, $\alpha_0 \mu_F$ is also a good measure.

We now assume that

$$\lim_{j \to \infty} b_j = \infty. \tag{4.8}$$

Since

$$|F_j \backslash F_{j+1}| \leq |F_j| = 2^{Nk_j} \ell_{k_j}^N,$$

then, for every $\alpha > 0$, we have

$$\int_\Omega g(\alpha v) = \sum_{j=1}^{\infty} \int_{F_j \backslash F_{j+1}} g(\alpha v) + \int_{\Omega \backslash F_1} g(\alpha v)$$

$$\leq \sum_{j=1}^{\infty} g\big(C_2 \alpha (a + b_{j+1}) \big) 2^{Nk_j} \ell_{k_j}^N + O(1).$$

Using (4.8), we have $C_2\alpha(a + b_{j+1}) \leq 2C_2\alpha b_{j+1}$ for every $j \geq 1$ sufficiently large. Therefore, if (4.7) holds with $C = 2C_2$, then $g(\alpha_0 v) \in L^1(\Omega)$, so that $\alpha_0\mu_F + g(\alpha_0 v)$ is a good measure. Applying Proposition 3 above, we conclude that $\alpha_0\mu_F \in \mathcal{G}(g)$.

We now present the

Proof of Theorem 2. Set $\ell_k = 2^{-\frac{N}{N-2}k}$, $\forall k \geq 1$. We now fix an increasing sequence of positive integers (k_j) such that

$$\frac{g(j^2)}{2^{\frac{2N}{N-2}k_j}} \leq \frac{1}{2^j} \quad \forall j \geq 1. \tag{4.9}$$

Let F be the Cantor set associated to the subsequence (ℓ_{k_j}). We claim that $\alpha\mu_F$ is good for every $\alpha > 0$.
In fact, since $2^{Nk_i}\ell_{k_i}^{N-2} = 1$ for every $i \geq 1$, we have

$$\sum_{i=1}^{j+1} \frac{1}{2^{Nk_i}\ell_{k_i}^{N-2}} = j + 1 \leq 2j \quad \forall x \in F_j \backslash F_{j+1}.$$

Moreover,

$$2^{Nk_j}\ell_{k_j}^{N} = \frac{1}{2^{\frac{2N}{N-2}k_j}}.$$

Thus, for every $\beta > 0$, we have

$$\sum_{j=1}^{\infty} g\left(\beta \sum_{i=1}^{j+1} \frac{1}{2^{Nk_i}\ell_{k_i}^{N-2}}\right) 2^{Nk_j}\ell_{k_j}^{N} \leq \sum_{j=1}^{\infty} \frac{g(2\beta j)}{2^{\frac{2N}{N-2}k_j}}. \tag{4.10}$$

Since $2\beta j \leq j^2$ for $j \geq 1$ sufficiently large, it then follows from (4.9) that the right-hand side of (4.10) is finite for every $\beta > 0$. Applying Proposition 4, we conclude that $\alpha\mu_F$ is a good measure for every $\alpha > 0$.
On the other hand, since $2^{Nk_j}\ell_{k_j}^{N-2} = 1$, $\forall j \geq 1$, we deduce from Lemma 3 that $\mathcal{H}^{N-2}(F) \in (0, \infty)$. Thus, by (4.2), we have

$$\mu_F = \frac{1}{\mathcal{H}^{N-2}(F)}\mathcal{H}^{N-2}\lfloor_F.$$

Therefore, $\alpha\mathcal{H}^{N-2}\lfloor_F$ is good for every $\alpha > 0$.

Proof of Theorem 3. Let (k_j) be an increasing sequence of positive integers such that

$$\frac{g(j^3)}{2^{\frac{2N}{N-2}k_j}} \leq \frac{1}{2^j} \quad \forall j \geq 1. \tag{4.11}$$

Let

$$\ell_k = \frac{1}{j^{\frac{1}{N-2}} 2^{\frac{Nk}{N-2}}} \quad \text{if } k_{j-1} < k \leq k_j,$$

with the convention that $k_0 = 0$. Let F be the Cantor set associated to the subsequence (ℓ_{k_j}). By Lemma 3, we know that $\mathcal{H}^{N-2}(F) = 0$. We now show that μ_F is a good measure relative to g.

Since $2^{Nk_i}\ell_{k_i}^{N-2} = \frac{1}{i}$, we have

$$\sum_{i=1}^{j+1} \frac{1}{2^{Nk_i}\ell_{k_i}^{N-2}} = \frac{(j+1)(j+2)}{2} \le 3j^2.$$

Moreover,

$$2^{Nk_j}\ell_{k_j}^{N} = \frac{1}{j^{\frac{N}{N-2}}\, 2^{\frac{2N}{N-2}k_j}} \le \frac{1}{2^{\frac{2N}{N-2}k_j}}.$$

Thus, for every $\beta > 0$, we have

$$\sum_{j=1}^{\infty} g\left(\beta \sum_{i=1}^{j+1} \frac{1}{2^{Nk_i}\ell_{k_i}^{N-2}}\right) 2^{Nk_j}\ell_{k_j}^{N} \le \sum_{j=1}^{\infty} \frac{g(3\beta j^2)}{2^{\frac{2N}{N-2}k_j}}. \tag{4.12}$$

Since $3\beta j^2 \le j^3$ for $j \ge 1$ sufficiently large, it then follows from (4.11) that the right-hand side of (4.12) is finite for every $\beta > 0$. Applying Proposition 4, we conclude that μ_F is a good measure. The proof of Theorem 3 is complete.

5. Proof of Theorem 1

When $N \ge 3$, Theorem 1 follows from Theorem 2 (or Theorem 3) and the following well-known

Proposition 5. *Let $K \subset \Omega$ be a compact set. If $\mathcal{H}^{N-2}(K) < \infty$, then $\mathrm{cap}\,(K) = 0$.*

We refer the reader to, e.g., [7] for a proof of Proposition 5.

We now deal with the case $N = 2$. We shall need the following

Lemma 4. *Assume $N = 2$. Let $F \subset \Omega$ be the Cantor set associated to the subsequence (ℓ_{k_j}). Then,*

$$\mathrm{cap}\,(F) = 0 \quad \text{if and only if} \quad \sum_{j=1}^{\infty} \frac{1}{4^{k_j}} \log \frac{1}{\ell_{k_j}} = \infty. \tag{5.1}$$

When F is a standard Cantor set, (5.1) is Carleson's test (see [7, p. 31]) for determining whether F has zero capacity. The same proof as in [7] can be used to establish Lemma 4. We present a different argument based on Proposition 2 above.

Proof of Lemma 4. ($\Leftarrow$) Suppose

$$\sum_{j=1}^{\infty} \frac{1}{4^{k_j}} \log \frac{1}{\ell_{k_j}} = \infty.$$

It then follows from Proposition 2 that $v = +\infty$ on F, where v is the solution of (3.2). Since v is superharmonic, we can apply Theorem 7.33 in [9] to conclude that $\mathrm{cap}\,(F) = 0$.

($\Rightarrow$) Assume that

$$\sum_{j=1}^{\infty} \frac{1}{4^{k_j}} \log \frac{1}{\ell_{k_j}} < \infty. \tag{5.2}$$

Let v be the solution of (3.2). It follows from (5.2) and Proposition 2 that v is uniformly bounded in Ω. Thus, the measure μ_F is diffuse. Since μ_F is concentrated in F, we must have $\operatorname{cap}(F) > 0$. The proof of Lemma 4 is complete.

Remark 1. *Here is the counterpart of (5.1) in dimension $N \geq 3$:*

$$\operatorname{cap}(F) = 0 \quad \text{if and only if} \quad \sum_{j=1}^{\infty} \frac{1}{2^{Nk_j} \ell_{k_j}^{N-2}} = \infty. \tag{5.3}$$

The proof of (5.3) follows along the same lines.

The analog of Proposition 4 in dimension $N = 2$ is given by the next

Proposition 6. *Assume $N = 2$. Let F be the Cantor set associated to the subsequence (ℓ_{k_j}). There exists $C > 0$ (depending on θ) such that if*

$$\sum_{j=1}^{\infty} g\left(C\alpha_0 \sum_{i=1}^{j+1} \frac{1}{4^{k_i}} \log \frac{1}{\ell_{k_i}} \right) 4^{k_j} \ell_{k_j}^2 < \infty \quad \text{for some } \alpha_0 > 0, \tag{5.4}$$

then $\alpha_0 \mu_F \in \mathcal{G}(g)$.

The proof of Proposition 6 is based on Proposition 2 and shall be omitted.

We may now present the

Proof of Theorem 1 completed. Let $\ell_k = 4^{-4^k}$, $\forall k \geq 1$. We now fix an increasing sequence of positive integers (k_j) such that

$$\frac{g(j^2)}{4^{4^{k_j}}} \leq \frac{1}{2^j} \quad \forall j \geq 1. \tag{5.5}$$

Let F be the Cantor set associated to the subsequence (ℓ_{k_j}). Note that

$$\frac{1}{4^{k_i}} \log \frac{1}{\ell_{k_i}} = \log 4 \quad \forall i \geq 1.$$

In particular,

$$\sum_{i=1}^{j+1} \frac{1}{4^{k_i}} \log \frac{1}{\ell_{k_i}} = (j+1)\log 4 \leq 4j.$$

It then follows from Lemma 4 that $\operatorname{cap}(F) = 0$. We now show that μ_F is a good measure.
Since

$$|F_j \backslash F_{j+1}| \leq |F_j| = 4^{k_j} \ell_{k_j}^2 = \frac{1}{4^2 \, 4^{k_j} - k_j} \leq \frac{1}{4^{4^{k_j}}},$$

then, for every $\beta > 0$, we have

$$\sum_{j=1}^{\infty} g\left(\beta \sum_{i=1}^{j+1} \frac{1}{4^{k_i}} \log \frac{1}{\ell_{k_i}}\right) 4^{k_j} \ell_{k_j}^2 \le \sum_{j=1}^{\infty} \frac{g(4\beta j)}{4^{4^{k_j}}}. \tag{5.6}$$

In view of (5.5), we conclude that the right-hand side of (5.6) is finite for every $\beta > 0$. Thus, by Proposition 6 above, μ_F is good. The proof of Theorem 1 is complete.

Acknowledgments

The author is deeply grateful to H. Brezis for his encouragement; this work is dedicated to him with admiration and gratitude.

We warmly thank L. Orsina, I. Shafrir, and L. Véron for interesting discussions. This work was supported by the NSF grant DMS-0111298 and Sergio Serapioni, Honorary President of Società Trentina Lieviti – Trento (Italy).

References

[1] D. Bartolucci, F. Leoni, L. Orsina, and A.C. Ponce, *Semilinear equations with exponential nonlinearity and measure data*. To appear in Ann. Inst. H. Poincaré Anal. Non Linéaire.

[2] Ph. Bénilan and H. Brezis, *Nonlinear problems related to the Thomas-Fermi equation*. J. Evol. Equ. **3** (2004), 673–770. Dedicated to Ph. Bénilan.

[3] H. Brezis, M. Marcus, and A.C. Ponce, *Nonlinear elliptic equations with measures revisited*. To appear in Annals of Math. Studies, Princeton University Press. Part of the results were announced in a note by the same authors: *A new concept of reduced measure for nonlinear elliptic equations*, C. R. Acad. Sci. Paris, Ser. I **339** (2004), 169–174.

[4] H. Brezis and A.C. Ponce, *Reduced measures on the boundary*. To appear in J. Funct. Anal.

[5] H. Brezis and W.A. Strauss, *Semilinear second-order elliptic equations in L^1*. J. Math. Soc. Japan **25** (1973), 565–590.

[6] H. Brezis and L. Véron, *Removable singularities for some nonlinear elliptic equations*. Arch. Rational Mech. Anal. **75** (1980/81), 1–6.

[7] L. Carleson, Selected problems on exceptional sets. Van Nostrand Mathematical Studies, No. 13, Van Nostrand, Princeton, 1967.

[8] L. Dupaigne, A.C. Ponce, and A. Porretta, *Elliptic equations with vertical asymptotes in the nonlinear term*. In preparation.

[9] L.L. Helms, Introduction to potential theory. Pure and Applied Mathematics, vol. XXII, Wiley-Interscience, New York, 1969.

[10] M. Marcus and L. Véron. *Nonlinear capacities associated to semilinear elliptic equations*. In preparation.

[11] P. Mattila, Geometry of sets and measures in Euclidean spaces. Cambridge Studies in Advanced Mathematics, vol. 44, Cambridge University Press, Cambridge, 1995.

[12] J.L. Vázquez, *On a semilinear equation in $\mathbb{R}^2$ involving bounded measures*. Proc. Roy. Soc. Edinburgh Sect. A **95** (1983), 181–202.

Augusto C. Ponce
Institute for Advanced Study
Princeton, NJ 08540, USA
e-mail: `augponce@math.ias.edu`

Progress in Nonlinear Differential Equations
and Their Applications, Vol. 63, 389–401
© 2005 Birkhäuser Verlag Basel/Switzerland

Bifurcation and Asymptotics for Elliptic Problems with Singular Nonlinearity

Vicenţiu Rădulescu

A mon Maître, avec reconnaissance

Abstract. We report on some recent existence and uniqueness results for elliptic equations subject to Dirichlet boundary condition and involving a singular nonlinearity. We take into account the following types of problems: (i) singular problems with sublinear nonlinearity and two parameters; (ii) combined effects of asymptotically linear and singular nonlinearities in bifurcation problems; (iii) bifurcation for a class of singular elliptic problems with subquadratic convection term. In some concrete situations we also establish the asymptotic behavior of the solution around the bifurcation point. Our analysis relies on the maximum principle for elliptic equations combined with adequate estimates.

Mathematics Subject Classification (2000). Primary 35J60; Secondary 35B32, 35B40.

Keywords. Singular nonlinearity, bifurcation, asymptotic analysis, maximum principle.

1. Motivation and previous results

I will report on some results contained in our recent papers [3, 7, 8, 9, 10, 11] that are closely related to the study of some problems on blow-up boundary solutions. More precisely, consider the elementary example

$$\begin{cases} \Delta u = u^p & \text{in } \Omega, \\ u > 0 & \text{in } \Omega, \\ u = +\infty & \text{on } \partial\Omega, \end{cases}$$

Partially supported by a research grant with the Romanian Academy and by the CNCSIS grant No. 308 (Nonlinearities and Singularities in Mathematical Physics).

where $\Omega \subset \mathbb{R}^N$ is a smooth bounded domain and $p > 1$. Then the function $v = u^{-1}$ satisfies

$$\begin{cases} -\Delta v = v^{2-p} - \dfrac{2}{v}|\nabla v|^2 & \text{in } \Omega, \\ v > 0 & \text{in } \Omega, \\ v = 0 & \text{on } \partial\Omega. \end{cases} \tag{1.1}$$

The above equation contains both singular nonlinearities (like v^{-1} or v^{2-p}, if $p > 2$) and a convection term (denoted by $|\nabla v|^2$). These nonlinearities make more difficult to handle problems like (1.1). Our purpose in this paper is to give an overview on some old and new results in this direction. We recall the pioneering paper [5] that contains one of the first existence results for singular elliptic problems. In fact, it is proved in [5] that the boundary value problem

$$\begin{cases} -\Delta u - u^{-\alpha} = -u & \text{in } \Omega, \\ u > 0 & \text{in } \Omega, \\ u = 0 & \text{on } \partial\Omega \end{cases}$$

has a solution, for any $\alpha > 0$. Let us now consider the problem

$$\begin{cases} -\Delta u - u^{-\alpha} = \lambda u^p & \text{in } \Omega, \\ u > 0 & \text{in } \Omega, \\ u = 0 & \text{on } \partial\Omega, \end{cases} \tag{1.2}$$

where $\lambda \geq 0$ and $\alpha, p \in (0, 1)$. In [4] it is proved that problem (1.2) has at least one solution for all $\lambda \geq 0$ and $0 < p < 1$. Moreover, if $p \geq 1$, then there exists λ^* such that problem (1.2) has a solution for $\lambda \in [0, \lambda^*)$ and no solution for $\lambda > \lambda^*$. In [4] it is also proved a related non-existence result. More exactly, the problem

$$\begin{cases} -\Delta u + u^{-\alpha} = u & \text{in } \Omega, \\ u > 0 & \text{in } \Omega, \\ u = 0 & \text{on } \partial\Omega \end{cases}$$

has no solution, provided that $0 < \alpha < 1$ and $\lambda_1 \geq 1$ (that is, if Ω is "small"), where λ_1 denotes the first eigenvalue of $(-\Delta)$ in $H_0^1(\Omega)$.

Problems related to multiplicity and uniqueness become difficult even in simple cases. In [16] it is studied the existence of radial symmetric solutions to the problem

$$\begin{cases} \Delta u + \lambda(u^p - u^{-\alpha}) = 0 & \text{in } B_1, \\ u > 0 & \text{in } B_1, \\ u = 0 & \text{on } \partial B_1, \end{cases}$$

where $\alpha > 0$, $0 < p < 1$, $\lambda > 0$, and B_1 is the unit ball in $\mathbb{R}^N$. Using a bifurcation theorem of Crandall and Rabinowitz, it has been shown in [16] that there exists $\lambda_1 > \lambda_0 > 0$ such that the above problem has no solutions for $\lambda < \lambda_0$, exactly one solution for $\lambda = \lambda_0$ or $\lambda > \lambda_1$, and two solutions for $\lambda_0 < \lambda \leq \lambda_1$.

Our purpose in this survey paper is to present various existence, and nonexistence results for several classes of singular elliptic problems. We also take into account bifurcation nonlinear problems and establish the precise rate decay of the solution in some concrete situations. We intend to reflect the "competition" between different quantities, such as: sublinear or superlinear nonlinearities, singular nonlinear terms (like $u^{-\alpha}$, for $\alpha > 0$), convection nonlinearities (like $|\nabla u|^q$, with $0 < q \leq 2$), as well as sign-changing potentials.

2. A singular problem with sublinear nonlinearity

Consider the following boundary value problem with two parameters:

$$\begin{cases} -\Delta u + K(x)g(u) = \lambda f(x, u) + \mu h(x) & \text{in } \Omega, \\ u > 0 & \text{in } \Omega, \\ u = 0 & \text{on } \partial\Omega, \end{cases} \qquad (2.1)$$

where Ω is a smooth bounded domain in $\mathbb{R}^N$ ($N \geq 2$), $K, h \in C^{0,\gamma}(\overline{\Omega})$, with $h > 0$ on Ω, and λ, μ are positive real numbers. We suppose that $f : \overline{\Omega} \times [0, \infty) \to [0, \infty)$ is a Hölder continuous function which is positive on $\overline{\Omega} \times (0, \infty)$. We also assume that f is non-decreasing with respect to the second variable and is sublinear, that is,

($f1$) the mapping $0, \infty) \ni s \longmapsto \dfrac{f(x, s)}{s}$ is non-increasing for all $x \in \overline{\Omega}$;

($f2$) $\displaystyle\lim_{s\downarrow 0} \frac{f(x, s)}{s} = +\infty$ and $\displaystyle\lim_{s\to\infty} \frac{f(x, s)}{s} = 0$, uniformly for $x \in \overline{\Omega}$.

We assume that $g \in C^{0,\gamma}(0, \infty)$ is a non-negative and non-increasing function. A fundamental role in our analysis will be played by the numbers

$$K^* := \max_{x\in\overline{\Omega}} K(x), \qquad K_* = \min_{x\in\overline{\Omega}} K(x).$$

Our first theorem is a non-existence result and it concerns nonlinearities with strong blow-up rate at the origin (like $u^{-\alpha}$, with $\alpha \geq 1$).

Theorem 2.1. *Assume that $K_* > 0$ and f satisfies $(f1)$–$(f2)$. If $\int_0^1 g(s)ds = +\infty$, then problem (2.1) has no classical solution, for any λ, $\mu > 0$.*

Next, we assume that the growth of the nonlinearity is described by the following conditions:

($g1$) $\displaystyle\lim_{s\downarrow 0} g(s) = +\infty$;

($g2$) there exist C, $\delta_0 > 0$ and $\alpha \in (0, 1)$ such that $g(s) \leq Cs^{-\alpha}$, for all $s \in (0, \delta_0)$.

The above conditions $(g1)$ and $(g2)$ are fulfilled by singular nonlinearities like $g(u) = u^{-\alpha}$, with $\alpha \in (0, 1)$. Obviously, hypothesis $(g2)$ implies the following Keller-Osserman type condition around the origin:

($g3$) $\displaystyle\int_0^1 \left(\int_0^t g(s)ds \right)^{-1/2} dt < \infty.$

As proved by Bénilan, Brezis and Crandall [1], condition $(g3)$ is equivalent to the *property of compact support*, that is, for every $h \in L^1(\mathbb{R}^N)$ with compact support, there exists a unique $u \in W^{1,1}(\mathbb{R}^N)$ with compact support such that $\Delta u \in L^1(\mathbb{R}^N)$ and $-\Delta u + g(u) = h$, a.e. in $\mathbb{R}^N$. That it is why it is natural to try to find solutions in the class

$$\mathcal{E} = \{\, u \in C^2(\Omega) \cap C(\overline{\Omega}); \ \Delta u \in L^1(\Omega)\}.$$

In the case where the potential $K(x)$ has a constant sign, the following results hold.

Theorem 2.2. *Assume that $K_* > 0$, f satisfies $(f1)$–$(f2)$, and g satisfies $(g1)$–$(g2)$. Then there exists $\lambda_*, \mu_* > 0$ such that:*

- *problem (2.1) has at least one solution in $\mathcal{E}$ either if $\lambda > \lambda_*$ or if $\mu > \mu_*$.*
- *problem (2.1) has no solution in $\mathcal{E}$ if $\lambda < \lambda_*$ and $\mu < \mu_*$.*

Moreover, if either $\lambda > \lambda_*$ or if $\mu > \mu_*$, then problem (2.1) has a maximal solution in $\mathcal{E}$ which is increasing with respect to λ and μ.

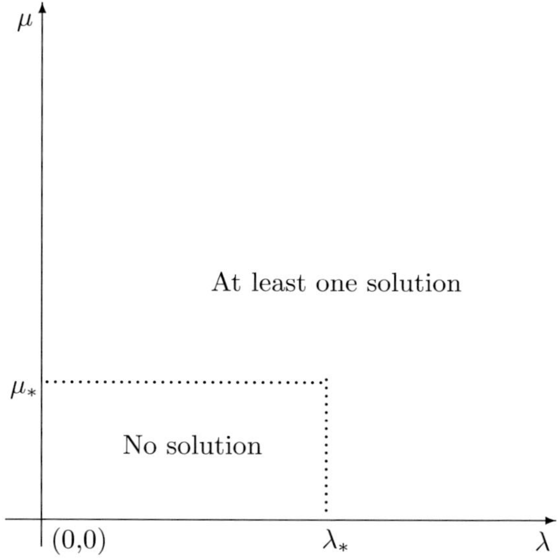

FIGURE 1. The dependence on λ and μ in Theorem 2.2

At this stage we are not able to describe the behavior in the following cases:

(i) $[\lambda = \lambda_*$ and $0 < \mu \leq \mu_*]$ and

(ii) $[0 < \lambda \leq \lambda_*$ and $\mu = \mu_*]$.

We conjecture that existence or non-existence results can be established in conjunction with a more precise description of the decay rate of the potential coefficients and nonlinearities.

Theorem 2.3. *Assume that $K^* \leq 0$, f satisfies conditions $(f1)$–$(f2)$ and g satisfies $(g1)$–$(g2)$. Then problem (2.1) has a unique solution $u_{\lambda,\mu}$ in $\mathcal{E}$, for any λ, $\mu > 0$. Moreover, $u_{\lambda,\mu}$ is increasing with respect to λ and μ.*

The following result give partial answers in the case where the potential $K(x)$ changes sign.

Theorem 2.4. *Assume that $K^* > 0 > K_*$, f satisfies $(f1)$–$(f2)$ and g verifies $(g1)$–$(g2)$. Then there exist λ_* and $\mu_* > 0$ such that problem (2.1) has at least one solution $u_{\lambda,\mu} \in \mathcal{E}$, provided that either $\lambda > \lambda_*$ or $\mu > \mu_*$. Moreover, for $\lambda > \lambda_*$ or $\mu > \mu_*$, $u_{\lambda,\mu}$ is increasing with respect to λ and μ.*

The proofs of the above results rely on the sub- and super-solution method for elliptic equations combined with adequate comparison principles. We refer to [8] for complete details and additional results.

A natural question is to see what happens if assumption $(f1)$ holds true, but if $\lim_{s\to\infty} f(x,s)/s$ is **not** zero. We give in what follows a precise description in the case where $K \leq 0$. More exactly, we consider the problem

$$\begin{cases} -\Delta u = \lambda f(u) + a(x)g(u) & \text{in } \Omega, \\ u > 0 & \text{in } \Omega, \\ u = 0 & \text{on } \partial\Omega, \end{cases} \qquad (2.2)$$

where $a \in C^{0,\gamma}(\overline{\Omega})$, $a \geq 0$, $a \not\equiv 0$ in $\overline{\Omega}$, and

$$(f3) \quad \lim_{s\to\infty} \frac{f(s)}{s} = m \in (0,\infty).$$

Let λ_1 be the first Dirichlet eigenvalue of $(-\Delta)$ in Ω and $\lambda^* := \lambda_1/m$. Set $a_* := \min_{x\in\overline{\Omega}} a(x)$ and $d(x) := \text{dist}\,(x, \partial\Omega)$.

Theorem 2.5. *Assume that conditions $(f1)$, $(f3)$, $(g1)$, and $(g2)$ are fulfilled. Then the following hold.*

  (i) *If $\lambda \geq \lambda^*$, then problem (2.2) has no solutions in $\mathcal{E}$.*
 (ii) *If $a_* > 0$ (resp. $a_* = 0$) then problem (2.2) has a unique solution $u_\lambda \in \mathcal{E}$ for all $-\infty < \lambda < \lambda^*$ (resp. $0 < \lambda < \lambda^*$) with the properties:*
    (ii1) *u_λ is strictly increasing with respect to λ;*
    (ii2) *there exist two positive constants c_1, $c_2 > 0$ depending on λ such that $c_1\, d(x) \leq u_\lambda(x) \leq c_2\, d(x)$, for all $x \in \Omega$;*
    (ii3) *$\lim_{\lambda\nearrow\lambda^*} u_\lambda = +\infty$, uniformly on compact subsets of Ω.*

Proof. The first part of the proof relies on standard arguments based on the maximum principle (see [3] for details). The most interesting part of the proof concerns (ii3) and, due to the special character of our problem, we will be able to show that, in this case, L^2-boundedness implies H_0^1-boundedness! We refer to [14] for a related problem and further results.

Let $u_\lambda \in \mathcal{E}$ be the unique solution of (2.2) for $0 < \lambda < \lambda^*$. We prove that $\lim_{\lambda\nearrow\lambda^*} u_\lambda = +\infty$, uniformly on compact subsets of Ω. Suppose the contrary. Since $(u_\lambda)_{0<\lambda<\lambda^*}$ is a sequence of nonnegative super-harmonic functions in Ω then, by Theorem 4.1.9 in [12], there exists a subsequence of $(u_\lambda)_{\lambda<\lambda^*}$ [still denoted by $(u_\lambda)_{\lambda<\lambda^*}$] which is convergent in $L^1_{\text{loc}}(\Omega)$.

We first prove that $(u_\lambda)_{\lambda < \lambda^*}$ is bounded in $L^2(\Omega)$. We argue by contradiction. Suppose that $(u_\lambda)_{\lambda < \lambda^*}$ is not bounded in $L^2(\Omega)$. Thus, passing eventually at a subsequence we have $u_\lambda = M(\lambda) w_\lambda$, where

$$M(\lambda) = \|u_\lambda\|_{L^2(\Omega)} \to \infty \quad \text{as } \lambda \nearrow \lambda^* \text{ and } w_\lambda \in L^2(\Omega), \ \|w_\lambda\|_{L^2(\Omega)} = 1. \quad (2.3)$$

Using $(f1)$, $(g2)$ and the monotonicity assumption on g, we deduce the existence of A, B, C, $D > 0$ $(A > m)$ such that

$$f(t) \le At + B, \quad g(t) \le Ct^{-\alpha} + D, \quad \text{for all } t > 0. \quad (2.4)$$

This implies

$$\frac{1}{M(\lambda)}\left(\lambda f(u_\lambda) + a(x)g(u_\lambda)\right) \to 0 \quad \text{in } L^1_{\text{loc}}(\Omega) \text{ as } \lambda \nearrow \lambda^*$$

that is,

$$-\Delta w_\lambda \to 0 \quad \text{in } L^1_{\text{loc}}(\Omega) \text{ as } \lambda \nearrow \lambda^*. \quad (2.5)$$

By Green's first identity, we have

$$\int_\Omega \nabla w_\lambda \cdot \nabla \phi \, dx = -\int_\Omega \phi \, \Delta w_\lambda \, dx = -\int_{\text{Supp}\,\phi} \phi \, \Delta w_\lambda \, dx \quad \forall \phi \in C_0^\infty(\Omega). \quad (2.6)$$

Using (2.5) we derive that

$$\left|\int_{\text{Supp}\,\phi} \phi \, \Delta w_\lambda \, dx\right| \le \int_{\text{Supp}\,\phi} |\phi||\Delta w_\lambda| \, dx$$
$$\le \|\phi\|_{L^\infty} \int_{\text{Supp}\,\phi} |\Delta w_\lambda| \, dx \to 0 \quad \text{as } \lambda \nearrow \lambda^*. \quad (2.7)$$

Combining (2.6) and (2.7), we arrive at

$$\int_\Omega \nabla w_\lambda \cdot \nabla \phi \, dx \to 0 \quad \text{as } \lambda \nearrow \lambda^*, \quad \forall \phi \in C_0^\infty(\Omega). \quad (2.8)$$

By definition, the sequence $(w_\lambda)_{0 < \lambda < \lambda^*}$ is bounded in $L^2(\Omega)$.

We claim that $(w_\lambda)_{\lambda < \lambda^*}$ is bounded in $H_0^1(\Omega)$. Indeed, using (2.4) and Hölder's inequality, we have

$$\int_\Omega |\nabla w_\lambda|^2 = -\int_\Omega w_\lambda \Delta w_\lambda = \frac{-1}{M(\lambda)} \int_\Omega w_\lambda \Delta u_\lambda$$
$$= \frac{1}{M(\lambda)} \int_\Omega [\lambda w_\lambda f(u_\lambda) + a(x)g(u_\lambda)w_\lambda]$$
$$\le \frac{\lambda}{M(\lambda)} \int_\Omega w_\lambda(Au_\lambda + B) + \frac{\|a\|_\infty}{M(\lambda)} \int_\Omega w_\lambda(Cu_\lambda^{-\alpha} + D)$$
$$= \lambda A \int_\Omega w_\lambda^2 + \frac{\|a\|_\infty C}{M(\lambda)^{1+\alpha}} \int_\Omega w_\lambda^{1-\alpha} + \frac{\lambda B + \|a\|_\infty D}{M(\lambda)} \int_\Omega w_\lambda$$
$$\le \lambda^* A + \frac{\|a\|_\infty C}{M(\lambda)^{1+\alpha}} |\Omega|^{(1+\alpha)/2} + \frac{\lambda B + \|a\|_\infty D}{M(\lambda)} |\Omega|^{1/2}.$$

From the above estimates, it is easy to see that $(w_\lambda)_{\lambda<\lambda^*}$ is bounded in $H_0^1(\Omega)$, so the claim is proved. Then, there exists $w \in H_0^1(\Omega)$ such that (up to a subsequence)

$$w_\lambda \rightharpoonup w \quad \text{weakly in} \quad H_0^1(\Omega) \quad \text{as } \lambda \nearrow \lambda^* \tag{2.9}$$

and, since $H_0^1(\Omega)$ is compactly embedded in $L^2(\Omega)$,

$$w_\lambda \to w \quad \text{strongly in} \quad L^2(\Omega) \quad \text{as } \lambda \nearrow \lambda^*. \tag{2.10}$$

On the one hand, by (2.3) and (2.10), we derive that $\|w\|_{L^2(\Omega)} = 1$. Furthermore, using (2.8) and (2.9), we infer that

$$\int_\Omega \nabla w \cdot \nabla \phi \, dx = 0, \qquad \text{for all } \phi \in C_0^\infty(\Omega).$$

Since $w \in H_0^1(\Omega)$, using the above relation and the definition of $H_0^1(\Omega)$, we get $w = 0$. This contradiction shows that $(u_\lambda)_{\lambda<\lambda^*}$ is bounded in $L^2(\Omega)$. As above for w_λ, we can derive that u_λ is bounded in $H_0^1(\Omega)$. So, there exists $u^* \in H_0^1(\Omega)$ such that, up to a subsequence,

$$\begin{cases} u_\lambda \rightharpoonup u^* & \text{weakly in } H_0^1(\Omega) \text{ as } \lambda \nearrow \lambda^*, \\ u_\lambda \to u^* & \text{strongly in } L^2(\Omega) \text{ as } \lambda \nearrow \lambda^*, \\ u_\lambda \to u^* & \text{a.e. in } \Omega \text{ as } \lambda \nearrow \lambda^*. \end{cases} \tag{2.11}$$

Now we can proceed to obtain a contradiction. Multiplying by φ_1 in (2.2) and integrating over Ω we have

$$-\int_\Omega \varphi_1 \Delta u_\lambda = \lambda \int_\Omega f(u_\lambda)\varphi_1 + \int_\Omega a(x)g(u_\lambda)\varphi_1, \quad \text{for all } 0 < \lambda < \lambda^*. \tag{2.12}$$

On the other hand, by $(f1)$ it follows that $f(u_\lambda) \geq mu_\lambda$ in Ω, for all $0 < \lambda < \lambda^*$. Combining this with (2.12) we obtain

$$\lambda_1 \int_\Omega u_\lambda \varphi_1 \geq \lambda m \int_\Omega u_\lambda \varphi_1 + \int_\Omega a(x)g(u_\lambda)\varphi_1, \quad \text{for all } 0 < \lambda < \lambda^*. \tag{2.13}$$

Notice that by $(g1)$, (2.11) and the monotonicity of u_λ with respect to λ we can apply the Lebesgue convergence theorem to find

$$\int_\Omega a(x)g(u_\lambda)\varphi_1 \, dx \to \int_\Omega a(x)g(u^*)\varphi_1 \, dx \text{ as } \lambda \nearrow \lambda_1.$$

Passing to the limit in (2.13) as $\lambda \nearrow \lambda^*$, and using (2.11), we obtain

$$\lambda_1 \int_\Omega u^* \varphi_1 \geq \lambda_1 \int_\Omega u^* \varphi_1 + \int_\Omega a(x)g(u^*)\varphi_1.$$

Hence $\int_\Omega a(x)g(u^*)\varphi_1 = 0$, which is a contradiction. Therefore $\displaystyle\lim_{\lambda \nearrow \lambda^*} u_\lambda = +\infty$, uniformly on compact subsets of Ω. This concludes the proof. $\qquad\square$

3. Bifurcation and asymptotics for a singular elliptic equation with convection term

Problems of this type arise in the study of non-Newtonian fluids, boundary layer phenomena for viscous fluids, chemical heterogeneous catalysts, cellular automata and interacting particle systems with self-organized criticality, as well as in the theory of Van der Waals interactions in thin films spreading on solid surfaces (see, e.g., [2, 6, 15]).

We are concerned in this section with singular elliptic problems of the following type

$$\begin{cases} -\Delta u = g(u) + \lambda |\nabla u|^p + \mu f(x, u) & \text{in } \Omega, \\ u > 0 & \text{in } \Omega, \\ u = 0 & \text{on } \partial\Omega, \end{cases} \qquad (3.1)$$

where $\Omega \subset \mathbb{R}^N$ $(N \geq 2)$ is a bounded domain with smooth boundary, $0 < p \leq 2$, and λ, $\mu \geq 0$. We suppose that $f : \overline{\Omega} \times [0, \infty) \to [0, \infty)$ is a Hölder continuous function which is non-decreasing with respect to the second variable and is positive on $\overline{\Omega} \times (0, \infty)$. We assume that $g : (0, \infty) \to (0, \infty)$ is a Hölder continuous function which is non-increasing and $\lim_{s \searrow 0} g(s) = +\infty$. As in the previous section, we denote by λ_1 the first eigenvalue of $(-\Delta)$ in $H_0^1(\Omega)$. By the monotony of g, there exists $a := \lim_{s \to \infty} g(s) \in [0, \infty)$.

The next result concerns the case $\lambda = 1$ and $1 < p \leq 2$.

Theorem 3.1. *Assume $\lambda = 1$ and $1 < p \leq 2$. Then the following properties hold true.*
  (i) *If $p = 2$ and $a \geq \lambda_1$, then problem (3.1) has no solutions.*
 (ii) *If either [$p = 2$ and $a < \lambda_1$] or if $1 < p < 2$, then there exists $\mu^* > 0$ such that problem (3.1) has at least one classical solution for $\mu < \mu^*$ and no solutions exist if $\mu > \mu^*$.*

In what follows the asymptotic behavior of the nonlinear smooth term $f(x, u)$ will play a decisive role. We impose the following assumptions:

$(f4)$ there exists $c > 0$ such that $f(x, s) \geq cs$ for all $(x, s) \in \overline{\Omega} \times [0, \infty)$;

$(f5)$ the mapping $(0, \infty) \ni s \longmapsto f(x, s)/s$ is non-decreasing for all $x \in \overline{\Omega}$;

$(f6)$ the mapping $(0, \infty) \ni s \longmapsto f(x, s)/s$ is non-increasing for all $x \in \overline{\Omega}$;

$(f7)$ $\lim\limits_{s \to \infty} f(x, s)/s = 0$, uniformly for $x \in \overline{\Omega}$.

We first consider the case $\lambda = 1$ and $0 < p \leq 1$.

Theorem 3.2. *Assume $\lambda = 1$ and $0 < p \leq 1$. Then the following properties hold true.*
  (i) *If f satisfies either $(f4)$ or $(f5)$, then there exists $\mu^* > 0$ such that problem (3.1) has at least one classical solution for $\mu < \mu^*$ and no solutions exist if $\mu > \mu^*$.*
 (ii) *If $0 < p < 1$ and f satisfies $(f6)$–$(f7)$, then problem (3.1) has at least one solution for all $\mu \geq 0$.*

We now analyze the case $\mu = 1$. Our framework is related to the sublinear case, described by assumptions $(f6)$ and $(f7)$.

Theorem 3.3. *Assume $\mu = 1$ and f satisfies assumptions $(f6)$ and $(f7)$. Then the following properties hold true.*

(i) *If $0 < p < 1$, then problem (3.1) has at least one classical solution for all $\lambda \geq 0$.*

(ii) *If $1 \leq p \leq 2$, then there exists $\lambda^* \in (0, \infty]$ such that problem (3.1) has at least one classical solution for $\lambda < \lambda^*$ and no solution exists if $\lambda > \lambda^*$. Moreover, if $1 < p \leq 2$, then λ^* is finite.*

Related to the above result we raise the following **open problem:** if $p = 1$ and $\mu = 1$, is λ^* a finite number?

Theorem 3.3 shows the importance of the convection term $\lambda|\nabla u|^p$ in the singular problem (3.1). Indeed, according to Theorem 2.3 and for any $\mu > 0$, the boundary value problem

$$\begin{cases} -\Delta u = u^{-\alpha} + \lambda|\nabla u|^p + \mu u^\beta & \text{in } \Omega, \\ u > 0 & \text{in } \Omega, \\ u = 0 & \text{on } \partial\Omega \end{cases} \tag{3.2}$$

has a unique solution, provided that $\lambda = 0$ and $\alpha, \beta \in (0, 1)$. Theorem 3.3 shows that if λ is not necessarily 0, then the following situations may occur : (i) problem (3.2) has solutions if $p \in (0, 1)$ and for all $\lambda \geq 0$; (ii) if $p \in (1, 2)$ then there exists $\lambda^* > 0$ such that problem (3.2) has a solution for any $\lambda < \lambda^*$ and no solution exists if $\lambda > \lambda^*$.

We give in what follows a complete description in the special case $f \equiv 1$ and $p = 2$. More precisely, we consider the problem

$$\begin{cases} -\Delta u = g(u) + \lambda|\nabla u|^2 + \mu & \text{in } \Omega, \\ u > 0 & \text{in } \Omega, \\ u = 0 & \text{on } \partial\Omega. \end{cases} \tag{3.3}$$

A key role in this case will be played by the asymptotic behavior of the singular term g. In the statement of the next result we remark some similarities with Theorem 2.5.

Theorem 3.4. *The following properties hold true.*

(i) *Problem (3.3) has solution if and only if $\lambda(a + \mu) < \lambda_1$.*

(ii) *Assume $\mu > 0$ is fixed, g is decreasing and let $\lambda^* := \lambda_1/(a+\mu)$. Then problem (3.3) has a unique solution u_λ for all $\lambda < \lambda^*$ and the sequence $(u_\lambda)_{\lambda<\lambda^*}$ is increasing with respect to λ. Moreover, if $\limsup_{s\searrow 0} s^\alpha g(s) < +\infty$, for some $\alpha \in (0, 1)$, then the sequence of solutions $(u_\lambda)_{0<\lambda<\lambda^*}$ has the following properties:*

(ii1) *for all $0 < \lambda < \lambda^*$ there exist two positive constants c_1, c_2 depending on λ such that $c_1\, d(x) \leq u_\lambda \leq c_2\, d(x)$ in Ω;*

(ii2) *$u_\lambda \in C^{1,1-\alpha}(\overline{\Omega}) \cap C^2(\Omega)$;*

(ii3) *$u_\lambda \longrightarrow +\infty$ as $\lambda \nearrow \lambda^*$, uniformly on compact subsets of Ω.*

We refer to [10] for complete proofs and further details.

4. An elliptic problem with strong singular nonlinearity and convection term

We study the boundary value problem

$$\begin{cases} -\Delta u = p(x)g(u) + q(x)|\nabla u|^a & \text{in } \Omega, \\ u > 0 & \text{in } \Omega, \\ u = 0 & \text{on } \partial\Omega, \end{cases} \tag{4.1}$$

where $\Omega \subset \mathbb{R}^N$ ($N \geq 2$) is a smooth bounded domain, $0 < a < 1$ and $q \in C^{0,\alpha}(\overline{\Omega})$, $q > 0$ in $\overline{\Omega}$. The potential $p \in C^1(\Omega)$ satisfies

$$c_1 \, d(x)^\beta \leq |p(x)| \leq c_2 \, d(x)^\beta, \qquad \text{for all } x \in \Omega, \tag{4.2}$$

where $c_1, c_2 > 0$, and β is a real number. This assumption shows that the potential $p(x)$ can admit a singular boundary behavior (corresponding to $\beta < 0$).

Throughout this section we suppose that $g \in C^1(0, \infty)$ is a positive decreasing function such that $\lim_{s \searrow 0} g(s) = +\infty$. The blow-up rate of g at the origin is described by the following assumption:

$(g4)$ there exists $\gamma > \max\{1, \beta + 1\}$ such that $\lim_{s \searrow 0} s^\gamma g(s) \in (0, \infty)$.

Observe that the stronger decay of the singular nonlinearity g around the origin [described by our assumption $(g4)$] implies that g does **not** obey the Keller-Osserman type condition $(g3)$.

From (4.2) we deduce that p does not vanish in Ω. Our first result concerns the case $p < 0$ in Ω.

Theorem 4.1. *Assume that g satisfies $(g4)$, p is negative in Ω, and condition (4.2) is fulfilled. Then problem (4.1) has no classical solutions.*

Proof. Let φ_1 be the normalized positive eigenfunction corresponding to the first eigenvalue λ_1 of $(-\Delta)$ in $H_0^1(\Omega)$. Then $\lambda_1 > 0$, $\varphi_1 \in C^2(\overline{\Omega})$, and

$$C_1 \, d(x) \leq \varphi_1(x) \leq C_2 \, d(x), \qquad x \in \Omega, \tag{4.3}$$

for some positive constants C_1 and C_2. From (4.2) and (4.3) it follows that there exist $\tau_1, \tau_2 > 0$ such that

$$\tau_1 \varphi_1(x)^\beta \leq |p(x)| \leq \tau_2 \varphi_1(x)^\beta, \qquad \text{for all } x \in \Omega. \tag{4.4}$$

Fix $C > 0$ such that $\|q\|_\infty^2 C^{a-1} < \lambda_1$ and define $\psi : [0, \infty) \to [0, \infty)$ by $\psi(s) = s^a/(s^2 + C)$. Then ψ attains its maximum at $\bar{s} = [Ca/(2 - a)]^{1/2}$. Hence

$$\psi(s) \leq \psi(\bar{s}) = \frac{a^{a/2}(2 - a)^{(2-a)/2}}{2C^{1-a/2}}, \qquad \text{for all } s \geq 0.$$

An elementary computation shows that

$$s^a \leq C^{a/2-1} s^2 + C^{a/2}, \qquad \text{for all } s \geq 0. \tag{4.5}$$

Arguing by contradiction, let us assume that problem (4.1) has a classical solution U. Consider the perturbed problem

$$\begin{cases} -\Delta u = p(x)g(u+\varepsilon) + A|\nabla u|^2 + B & \text{in } \Omega, \\ u > 0 & \text{in } \Omega, \\ u = 0 & \text{on } \partial\Omega, \end{cases} \qquad (4.6)$$

where $\varepsilon > 0$, and $A = \|q\|_\infty C^{a/2-1}$, $B = \|q\|_\infty C^{a/2}$. By virtue of (4.5) it follows that U is a sub-solution of (4.6). Set $v = e^{Au} - 1$. Then problem (4.6) becomes

$$\begin{cases} -\Delta v = Ap(x)(v+1)g\left(\dfrac{1}{A}\ln(v+1)+\varepsilon\right) + AB(v+1) & \text{in } \Omega, \\ v > 0 & \text{in } \Omega, \\ v = 0 & \text{on } \partial\Omega. \end{cases} \qquad (4.7)$$

We first remark that $V = e^{AU} - 1$ is a sub-solution of (4.7). On the other hand, since $AB < \lambda_1$, we conclude that there exists $w \in C^2(\overline{\Omega})$ such that

$$\begin{cases} -\Delta w = AB(w+1) & \text{in } \Omega, \\ w > 0 & \text{in } \Omega, \\ w = 0 & \text{on } \partial\Omega. \end{cases} \qquad (4.8)$$

Moreover, the maximum principle yields

$$c_1\varphi_1 \le w \le c_2\varphi_2 \quad \text{in } \Omega, \qquad (4.9)$$

for some positive constants c_1 and $c_2 > 0$. It is clear that w is a super-solution of (4.7). We claim that $V \le w$. To this aim, it suffices to prove that $U \le W$ in Ω, where $W = A^{-1}\ln(w+1)$ verifies

$$\begin{cases} -\Delta W = A|\nabla W|^2 + B & \text{in } \Omega, \\ W > 0 & \text{in } \Omega, \\ W = 0 & \text{on } \partial\Omega. \end{cases}$$

Assuming the contrary, we get that $\max_{x\in\overline{\Omega}}(U-W) > 0$ is achieved in some point $x_0 \in \Omega$. Then $\nabla(U-W)(x_0) = 0$ and

$$0 \le -\Delta(U-W)(x_0) = p(x_0)g(U(x_0)) + q(x_0)|\nabla U|^a(x_0) - A|\nabla W|^2(x_0) - B < 0,$$

which is a contradiction. Hence $U \le W$ in Ω, that is, $V \le w$ in Ω. By the sub- and super-solution method we deduce that there exists $v_\varepsilon \in C^2(\overline{\Omega})$ a solution of problem (4.7) such that

$$V \le v_\varepsilon \le w \quad \text{in } \Omega. \qquad (4.10)$$

Now we proceed to get our contradiction. Integrating in (4.7) and taking into account the fact that p is negative, we deduce

$$-\int_\Omega \Delta v_\varepsilon\, dx - A\int_\Omega p(x)g\left(\frac{1}{A}\ln(v_\varepsilon+1)+\varepsilon\right)dx \le AB\int_\Omega(w+1)\,dx.$$

Using monotonicity of g and the fact that $\ln(v_\varepsilon+1) \le v_\varepsilon$ in Ω, the above inequality yields

$$-\int_{\partial\Omega} \frac{\partial v_\varepsilon}{\partial n}\,ds - A\int_\Omega p(x)g\left(\frac{v_\varepsilon}{A}+\varepsilon\right)dx \le AB(\|w\|_\infty+1)|\Omega| < +\infty.$$

Since $\partial v_\varepsilon / \partial n \leq 0$ on $\partial\Omega$, the above relation implies

$$- \int_\Omega p(x) g\left(\frac{v_\varepsilon}{A} + \varepsilon\right) dx \leq M, \tag{4.11}$$

where $M = B(\|w\|_\infty + 1)|\Omega|$. Now, relations (4.10) and (4.11) imply

$$0 \leq - \int_\Omega p(x) g\left(\frac{w}{A} + \varepsilon\right) dx \leq M.$$

Therefore, for any compact subset $\omega \subset\subset \Omega$ we have

$$0 \leq - \int_\omega p(x) g\left(\frac{w}{A} + \varepsilon\right) dx \leq M.$$

Passing to the limit as $\varepsilon \searrow 0$ in the above inequality, it follows that

$$- \int_\omega p(x) g\left(\frac{w}{A}\right) dx \leq M, \qquad \text{for all } \omega \subset\subset \Omega.$$

This yields

$$- \int_\Omega p(x) g\left(\frac{w}{A}\right) dx \leq M. \tag{4.12}$$

On the other hand, the hypothesis $(g4)$ combined with (4.9) implies $g(w/A) \geq c_0 \varphi_1^{-\gamma}$ in Ω, for some $c_0 > 0$. The last inequality together with (4.4) and (4.12) produces $c \int_\Omega \varphi_1^{\beta-\gamma} dx \leq M$, where $\beta - \gamma < -1$. But, by a result of Lazer and McKenna (see [13]), $\int_\Omega \varphi_1^{-s} dx < +\infty$ if and only if $s < 1$. This contradiction shows that problem (4.1) has no classical solutions and the proof is now complete. $\qquad\square$

The situation changes radically in the case where p is positive in Ω, as established in the next result.

Theorem 4.2. *Assume that g satisfies $(g4)$ and the potential $p(x)$ is positive and fulfills (4.2). Then the following properties hold true.*

(i) *If $\beta \leq -2$, then problem (4.1) has no classical solutions.*
(ii) *If $\beta > -2$, then problem (4.1) has a unique solution u which, moreover, has the following properties:*
  (ii1) *there exist $M, m > 0$ such that*

$$m\, d(x)^{(2+\beta)/(1+\gamma)} \leq u(x) \leq M\, d(x)^{(2+\beta)/(1+\gamma)}, \qquad \textit{for all } x \in \Omega;$$

  (ii2) *if $\beta \geq \max\{0, \gamma - 3\}$, then u is in $H_0^1(\Omega)$;*
  (ii3) *if $2\beta \leq \gamma - 3$, then u does not belong to $H_0^1(\Omega)$.*

We refer to [11] for the proof of Theorem 4.2, as well as for a result concerning the entire solutions of problem (4.1).

Acknowledgment

The author is greatly indebted to Professor Haim Brezis, for his highest level guidance during the PhD and Habilitation theses at the Université Pierre et Marie Curie (Paris 6), as well as for suggesting to him several modern research subjects and directions of interest. *Bonne anniversaire, mon Professeur!*

References

[1] P. Bénilan, H. Brezis, and M. Crandall, *A semilinear equation in $L^1(\mathbb{R}^N)$*. Ann. Scuola Norm. Sup. Pisa **4** (1975), 523–555.

[2] J.T. Chayes, S.J. Osher, and J.V. Ralston, *On singular diffusion equations with applications to self-organized criticality.* Comm. Pure Appl. Math. **46** (1993), 1363–1377.

[3] F.-C. Cîrstea, M. Ghergu, and V. Rădulescu, *Combined effects of asymptotically linear and singular nonlinearities in bifurcation problems of Lane-Emden-Fowler type.* J. Math. Pures Appl. (Journal de Liouville) **84** (2005), 493–508.

[4] M. Coclite and G. Palmieri, *On a singular nonlinear Dirichlet problem.* Commun. Partial Diff. Equations **14** (1989), 1315–1327.

[5] M.G. Crandall, P.H. Rabinowitz, and L. Tartar, *On a Dirichlet problem with a singular nonlinearity.* Commun. Partial Diff. Equations **2** (1977), 193–222.

[6] P.G. de Gennes, *Wetting: statics and dynamics.* Review of Modern Physics **57** (1985), 827–863.

[7] M. Ghergu and V. Rădulescu, *Bifurcation and asymptotics for the Lane-Emden-Fowler equation.* C. R. Acad. Sci. Paris, Ser. I **337** (2003), 259–264.

[8] M. Ghergu and V. Rădulescu, *Sublinear singular elliptic problems with two parameters.* J. Differential Equations **195** (2003), 520–536.

[9] M. Ghergu and V. Rădulescu, *Bifurcation for a class of singular elliptic problems with quadratic convection term.* C. R. Acad. Sci. Paris, Ser. I **338** (2004), 831–836.

[10] M. Ghergu and V. Rădulescu, *Multiparameter bifurcation and asymptotics for the singular Lane-Emden-Fowler equation with a convection term.* Proc. Royal Soc. Edinburgh Sect. A (Mathematics) **135** (2005), 61–84.

[11] M. Ghergu and V. Rădulescu, *Singular elliptic problems with sublinear convection term and Kato potential in anisotropic media,* in preparation.

[12] L. Hörmander, *The Analysis of Linear Partial Differential Operators I.* Springer, Berlin, 1983.

[13] A.C. Lazer and P.J. McKenna, *On a singular nonlinear elliptic boundary value problem.* Proc. Amer. Math. Soc. **3** (1991), 720–730.

[14] P. Mironescu and V. Rădulescu, *The study of a bifurcation problem associated to an asymptotically linear function.* Nonlinear Anal., T.M.A. **26** (1996), 857–875.

[15] J. Ockendon, S. Howison, A. Lacey, and A. Movchan, *Applied Partial Differential Equations,* Oxford University Press, 2003.

[16] J. Shi and M. Yao, *On a singular nonlinear semilinear elliptic problem.* Proc. Roy. Soc. Edinburgh Sect. A **128** (1998), 1389–1401.

Vicenţiu Rădulescu
Department of Mathematics
University of Craiova
RO 200585 Craiova, Romania
URL: http://inf.ucv.ro/~radulescu
e-mail: radulescu@inf.ucv.ro

Progress in Nonlinear Differential Equations
and Their Applications, Vol. 63, 403–414
© 2005 Birkhäuser Verlag Basel/Switzerland

A Model for Hysteresis in Mechanics Using Local Minimizers of Young Measures

Marc Oliver Rieger

Abstract. A model for effects related to hysteresis in continuum mechanics is introduced. Its key idea is to consider local minimizers of Young measures describing the deformation gradient of a body. These Young measures are defined as limit of a quasi-static evolution by means of a gradient flow with respect to a special "regularized" Wasserstein metric. The model is described for the space-homogeneous case, and some one-dimensional examples show the occurrence of hysteresis and illustrate possible applications to fracture.

1. Introduction

Different phenomena in continuum mechanics, as elasticity, fracture and the formation of microstructures, are traditionally described by different models that are often very successful for specific problems, but may make it difficult to see common properties of these phenomena. In recent years, progress has been made in particular in describing hysteresis effects [20, 12, 15, 14] and fracture [8, 6] by variational methods.

In this article we introduce a new approach to hysteresis problems, based on the notion of Young measures. A *Young measure* (or *parameterized measure*) is a family of probability measures $(\nu_x)_{x\in\Omega}$ describing a "one-point statistic" for a sequence of oscillating functions f_j, i.e., ν_x describes (in a certain sense which can be made mathematically precise) the probability distribution of the values of the sequence f_j at x. A rigorous definition can be found, e.g., in [17, 16] and in the original work by L. C. Young [23].

A *gradient Young measure* is a Young measure associated with a sequence of gradients $f_j = \nabla u_j$. In the one-dimensional case, the classes of gradient Young measures and Young measures coincide. In this paper we will always deal with gradient Young measures, but usually just refer to them as *Young measures*.

Supported by the Center for Nonlinear Analysis under NSF Grant DMS-9803791.

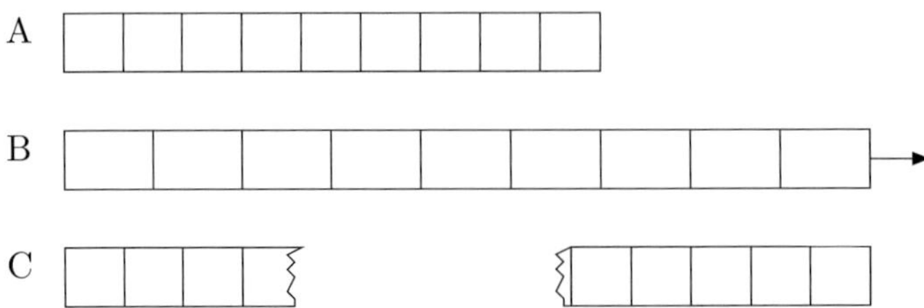

FIGURE 1. An elastic bar (A) can be deformed to some extend without breaking (B), although the broken configuration (C) would have a lower elastic energy. Hence it is not appropriate to consider this problem as *global* minimization of the elastic energy.

Young measures turned out to be a useful tool in the calculus of variations [21] and for describing microstructures in martensites [2] where assuming that the system is in a global energy minimum is a reasonable assumption, and hysteresis effects induced by the existence of local minima play a minor role. However, for other related problems as, e.g., fracture, this poses inherent difficulties: If we model, for instance, the energy density ϕ of an elastic bar as function of the deformation u_x, then we have to assume sublinear growth rate at infinity, e.g., $\phi(u_x) := \log(1 + |u_x|^2)$ or a Lennard-Jones potential, in order to obtain any fracture at all. But then the following problem arises: If we stretch the bar slightly, i.e., if we give $F = \int u_x \, dx$ a small positive value, then it is easy to find minimizing sequences with a limit with zero energy and deformation gradients concentrated in zero and infinity. This means that the elastic bar will break for arbitrarily small values of F, in contradiction with experiments (compare Fig. 1). This paradox was pointed out by Truskinovsky [22]. The problem arises because we are considering *global* energy minimization. Instead, we should look for *local* minimizers. As pointed out in [9], this might also be a useful strategy to solve the problems arising in dynamical elasticity problems where Young measure solutions (as obtained in [10, 18]) are often non-unique. But how should we introduce the notion of local minimizers in a natural way?

Recently, some progress had been made in the study of local minimizers of various variational problems in the class of *functions* [11, 7, 13, 3]. Although this approach has very interesting applications, in particular in micromagnetics, it does not use the concept of Young measures to describe possible microstructures. In particular in problems exhibiting different length scales it would be useful to introduce a notion of local minimizers for the *microstructure* itself, i.e., to define *local minimizers of Young measures*.

The fundamental idea which we propose in this article is to consider a quasistatic evolution of the Young measure following a natural gradient flow evolution. This enables us to find local minimizers that are more natural candidates for solutions of the system under study than the Young measure solutions obtained by global minimization.

In order to define a gradient flow reflecting the physical behavior of the system, it will be crucial to find an appropriate metric on the space of Young measures. The corresponding definitions will be given in Section 2. Although the concept is mathematically involved, it can be simplified in specific situations. This will be done in Section 3 where we apply the model to certain examples of hysteresis such as fracture. In particular, we will observe that Truskinovsky's paradox disappears in our model.

The goal of this article is to introduce the main ideas, thus we address chiefly motivations and basic concepts rather than technical questions and more involved applications. In particular, we do only give *formal* definitions at some instances. Many natural questions connected with our model are work in progress, further studies will be necessary to evaluate the usefulness of the proposed model for hysteresis effects in continuum mechanics. However, the examples illustrated in Section 3 seem to be very promising.

2. Evolution of Young measures

Our main idea is to describe the dynamics of a physical system via a gradient flow of Young measures with respect to a metric whose definition draws its motivation from the underlying physics. In this article we consider only problems which are homogenous in space.

The starting point for our model is to find an appropriate metric on the space of Young measures. It is natural to consider the well-studied Wasserstein metric, while considering Young measures as elements of the dual of Lipschitz continuous functions. More precisely, given two probability measures ν_1, ν_2 on the space X we set

$$d_W(\nu_1, \nu_2) := \frac{1}{2} \inf_\mu \left(\int_X \int_X |x - y| \, d\mu(x, y), \ \pi_1(\mu) = \nu_1, \ \pi_2(\mu) = \nu_2 \right), \qquad (1)$$

where $\pi_1(\mu)$ (resp. $\pi_2(\mu)$) are the projections of the measure μ on $X \times X$ on the first (resp. second) component, also known as marginals.

We will see later that only this type of Wasserstein metric (the L^1-Wasserstein metric) has the right scaling behavior needed for the gradient flow of our model.

However, it turns out that the topology induced by this Wasserstein metric is too weak. As illustrating one-dimensional example consider the two Young measures $\nu_0 := \delta_A$ and $\nu_1 := \delta_B$ for constants $A \neq B$. We can define a homotopy from ν_A to ν_B with respect to the Wasserstein metric by $\nu_\tau := (1 - \tau)\delta_A + \tau\delta_B$, and if the energy at B is less than the energy at A, then the energy of ν_τ will strictly decrease in τ, thus ν_0 cannot be a local minimum in the class of Young measures with respect to d_W. Hence the Wasserstein metric does not allow for the existence of local minima in these cases. A closer look at ν_τ shows that at $\tau = 0$ a new phase B forms instantaneously. In a certain sense the solution "tunnels" through a possible energy barrier between A and B, compare the similar situation of Fig. 4. In order to exclude such behavior, we "regularize" the Wasserstein metric and consider the limit where the regularizing term vanishes. In the following we denote by d_W the Wasserstein metric (1) and by d_H the Hausdorff metric.

Definition 2.1 (Regularized metrics). *Let $\varepsilon > 0$. We define the* regularized metric *$d_{\varepsilon,R}$ on the set of Young measures on $\mathbb{R}^{m \times n}$ with compact support by*

$$d_{\varepsilon,R}(\nu,\mu) \quad := \quad d_W(\nu,\mu) + \varepsilon \, d_H(\operatorname{supp} \nu, \operatorname{supp} \mu).$$

This definition takes into account that (if we assume continuous evolution of the Young measure) a sudden formation of a new phase (i.e., an instantaneous extension of the support of the Young measure) is forbidden. We can state the following lemma:

Lemma 2.2. *d_R is a metric.*

Proof. d_W is a metric on the set of probability measures on $\mathbb{R}^{m \times n}$ with compact support and d_H is symmetric and satisfies the triangle inequality, since it is a metric for subsets of $\mathbb{R}^{m \times n}$. $\qquad\square$

The space of Young measures equipped with the metric d_R is not necessarily complete: Take as an example the case $m = n = 1$ and $\varepsilon = 1$. The sequence $\{\nu_n\}$, where $\nu_n := \frac{1}{n+1}\delta_0 + \frac{n}{n+1}\delta_2$, is a Cauchy sequence, but does not converge: δ_2 is not the limit, since $d_R(\delta_2,\nu_n) > 1$ for all $n \in \mathbb{N}$. To fix these kinds of problems, we introduce a notion of convergence which differs slightly from the convergence induced by the metric. The physical reason for this is that, while the instantaneous formation of new phases should be prohibited (for $\varepsilon > 0$), the disappearance of phases should be allowed.

Definition 2.3. *We define $\mathcal{Y}_R$ as the space of probability measures on $\mathbb{R}^{N \times N}$ equipped with the metric d_R and the following notion of convergence:*

$$(\nu_n) \subset \mathcal{Y}_R, \ \nu_n \to \nu \quad :\Longleftrightarrow$$

(i) *(ν_n) is a Cauchy sequence with respect to d_R,*
(ii) *$\nu_n \to \nu$ with respect to d_W.*

Remark 2.4. *Definition 2.3 is independent of the choice of $\varepsilon > 0$.*

We have now established a metric and a convergence on the space of Young measures. The next step would be to define a differentiation on $\mathcal{Y}_R$ which can be used for defining a gradient flow. However, the structure of the metric d_R is mathematically difficult, therefore the following problem can (in the general case) only be understood in a formal sense:

Let $N \in \mathbb{N}$, $\gamma, \varepsilon > 0$ and let $\phi \colon \mathbb{R}^{N \times N} \to \mathbb{R}$ be a smooth function. Our aim is to study the initial value problem

$$\begin{cases} \gamma \nu_t(t) &= -D\langle \phi(t), \nu(t) \rangle, \\ \langle Id, \nu(t) \rangle &= F(t), \\ \nu(0) &= \nu_0. \end{cases} \tag{2}$$

Here we write $\langle \phi(t), \nu(t) \rangle := \int_{\mathbb{R}^{N \times N}} \phi(t, Y) d\nu(t)(Y)$. First let $\gamma, \varepsilon > 0$, later we will discuss the limit where $\varepsilon \to 0$ and $\gamma \to 0$ (quasi-static case). It will turn out that this model allows for the existence of local minimizers and for hysteresis effects. We will illustrate this in the following section by considering a special case where the formal definitions can be made rigorous. We remark that if we considered the problem analog to (2) with d_W instead of d_R as the underlying metric (i.e., with $\varepsilon = 0$), then we would get different solutions without local minimizers and hence without hysteresis effects.

Solutions of the general problem (2) might be obtained by approximating the Young measure $\nu(t)$ by a sum of finitely many Dirac masses. Approximations of a similar form have been used in [4, 5]. General results on the solvability of this approximating problem and the convergence of the approximating solutions to solutions of the full system (2) are work in progress.

3. One-dimensional systems

In this section we apply our ideas to a simplified situation which can be handled mathematically relatively easy. We consider only one-dimensional cases, i.e., $m = n = 1$ and we assume that the Young measure at every time t consists at most of two Dirac masses at $A(t)$ and $B(t)$, i.e.,

$$\nu(t) = \lambda(t)\delta_A(t) + (1 - \lambda(t))\delta_B(t).$$

Since $\nu(t)$ is positive we have $\lambda \in [0, 1]$. Moreover, since $\langle Id, \nu(t) \rangle = F(t)$ given, we have either $A(t) \le F(t) \le B(t)$ or $B(t) \le F(t) \le A(t)$. Without loss of generality we assume $A(t) \le F(t) \le B(t)$.

Under these assumptions on $\nu(t)$ (which can be motivated for the systems under investigation), the definition of the metric d_R gives immediately that λ, A and B are continuous functions of t as long as $\lambda \in (0, 1)$. If $\lambda(t) \to 0$ or $\lambda(t) \to 1$ as $t \to t_0$, then the definition of convergence yields that A (resp. B) may have a discontinuity at t_0 and $A(t_0) = B(t_0) = F(t_0)$. (This reflects the disappearing of a phase.)

In this simplified model we can derive evolution equations for A and B, valid whenever $\lambda(t) \in (0, 1)$. The variable λ is fixed by the condition $\langle Id, \nu(t) \rangle = F(t)$ which yields

$$\lambda(t) = \frac{F(t) - B(t)}{A(t) - B(t)}.$$

From now on, we often write A instead of $A(t)$ etc., and we use the abbreviation

$$\nu_{A,B} \quad = \quad \lambda\delta_A + (1 - \lambda)\delta_B. \tag{3}$$

We can rewrite (2) as a pair of two dynamical equations for A and B. We want to find the equation for A. To derive it, we first need to calculate the distance between $\nu_{A+h,B}$ and $\nu_{A,B}$ for a small $h > 0$. The regularizing part of $d_{\varepsilon,R}(\nu_{A+h,B}, \nu_{A,B})$ is easy and gives εh. To calculate the Wasserstein part we observe that the optimal probability measure μ in (1) can have a support only on the four points $(A+h, A)$, $(A + h, B)$, (B, A) and (B, B). Hence a short computation shows that

$$d_W(\nu_{A+h,B}, \nu_{A,B}) = \frac{B - F}{B - A}\, h.$$

Now we can calculate the gradient of the energy $\langle \phi, \nu \rangle$ to get

$$
\begin{aligned}
\frac{d}{dt} A(t) &= -\nabla \langle \phi(t), \nu_{A(t),B(t)} \rangle \\
&= -\lim_{h \to 0} \frac{\langle \phi(t), \nu_{A(t)+h,B(t)} \rangle - \langle \phi(t), \nu_{A(t),B(t)} \rangle}{d_R(\nu_{A(t)+h,B(t)}, \nu_{A(t),B(t)})} \\
&= -\lim_{h \to 0} \left[\left(\frac{B - A}{B - (A + h)} \phi(A + h) + \frac{F - (A + h)}{B - (A + h)} \phi(B) \right. \right. \\
&\qquad \left. \left. - \frac{B - F}{B - A} \phi(A) - \frac{F - A}{B - A} \phi(B) \right) \left(\frac{B - F}{B - A} + \varepsilon \right)^{-1} h^{-1} \right].
\end{aligned}
$$

We use the identity

$$
\left(\frac{B - F}{B - A} + \varepsilon \right)^{-1} = \frac{B - A}{B - F} - \varepsilon \frac{B - A}{(B - F)\left(\frac{B-F}{B-A} + \varepsilon \right)}
$$

and estimate

$$
\left| \varepsilon \frac{B - A}{(B - F)\left(\frac{B-F}{B-A} + \varepsilon \right)} \right| \leq \varepsilon \left| \frac{B - A}{B - F} \right|^2.
$$

Hence we get

$$
\begin{aligned}
\frac{d}{dt} A(t) &= -\lim_{h \to 0} \left[\left((\phi(A + h) - \phi(A)) + \frac{h\phi(A + h)}{B - (A + h)} \right. \right. \\
&\qquad \left. \left. + \frac{h(F - A)\phi(B) - h(B - A)\phi(B)}{(B - (A + h))(B - F)} \right) \frac{1}{h} \right] + O(\varepsilon) \\
&= -\phi'(A(t)) - \frac{\phi(A) - \phi(B)}{B - A} + O(\varepsilon).
\end{aligned}
$$

If we had chosen an L^p-Wasserstein metric with $p > 1$, the limit $h \to 0$ would be zero, since no mass transport between A and B would be allowed. Hence we had to choose $p = 1$ in the original definition.

Now we take the limit $\varepsilon \to 0$ and let the regularizing term in the metric d_R vanish to obtain

$$
\frac{d}{dt} A(t) = -\phi'(A(t)) - \frac{\phi(A) - \phi(B)}{B - A}. \tag{4}
$$

Similarly, we deduce that

$$
\frac{d}{dt} B(t) = -\phi'(B(t)) - \frac{\phi(A) - \phi(B)}{B - A}. \tag{5}
$$

Using standard ODE theory we get:

Lemma 3.1. *If $\phi \in C^1$ has quadratic growth, ϕ' has linear growth and ϕ is convex outside some interval $(-M, +M)$ and if F is smooth, then the system (4)–(5) admits a solution for as long as $A, B \in (0, 1)$.*

However, in general this solution does not satisfy the constraint $\lambda(t) \in (0, 1)$. If we define a solution according to the definition of convergence in $\mathcal{Y}_R$, we can solve this problem:

Definition 3.2. *Let $(A(t), B(t))_{t \geq 0}$ satisfy (4)–(5) whenever $\lambda(t)$ as given by (3) is contained in $(0, 1)$ and let $A(t) = B(t) = F(t)$ else. Suppose that $(A(t), B(t))$ is continuous for all t with $\lambda(t) \in (0, 1)$ and let $\lambda(t)$ be continuous for all $t \geq 0$. Moreover, assume that $A(0) = A_0$ and $B(0) = B_0$ are given. Then we call $(A(t), B(t))_{t \geq 0}$ a solution of the problem stated above.*

With this definition the following proposition is an immediate consequence of our derivation and Lemma 3.1.

Proposition 3.3. *If the assumptions of Lemma 3.1 hold and if $A_0 \leq F(0) \leq B_0$, then there exist a solution in the sense of Definition 3.2.*

Particularly interesting are stable solutions. We make the following observation for F constant and $\nu(0) = \delta_F$:

Proposition 3.4. *If ϕ is smooth and locally strictly convex at F, then the solution $\nu(t) = \delta_F$ is stable, i.e., there exists a neighborhood U around F such that every solution $\mu(t) = \lambda(t)\delta_{A(t)} + (1 - \lambda(t))\delta_{B(t)}$ with $\mu(0) \in U$ converges to $\nu(t)$. If ϕ is locally nonconvex at F, then the solution $\nu(t) = \delta_F$ cannot be stable.*

Proof. Let ϕ be locally strictly convex at F. Consider $\mu(0) = \lambda(0)\delta_{A(0)} + (1 - \lambda(0))\delta_{B(0)}$, where $|A(0) - F|, |B(0) - F| < \eta$. Let $\mu(t) = \lambda(t)\delta_{A(t)} + (1 - \lambda(t))\delta_{B(t)}$. Since

$$\phi(B) = \phi(A) + \phi'(A)(B - A) + \frac{1}{2}\phi''(A)(B - A)^2 + O(|B - A|^3),$$

we obtain from (4) that

$$\frac{d}{dt}A(t) = \frac{1}{2}\phi''(A)(B - A)^2 + O(\eta^3).$$

If $\eta > 0$ is sufficiently small, then we have $\phi''(A) \geq C > 0$ and $O(\eta^3) < C\eta^2$, hence

$$\frac{d}{dt}A(t) \geq \frac{C}{4}(B - A)^2,$$

and thus, by deducing the analogous formula for $B(t)$, we get $A(t), B(t) \to F$, so we have proved $\mu(t) \to \nu(t) = \delta_F$.

If, on the other hand, ϕ is locally nonconvex, then the same argument yields that $\mu(t)$ does not converge to $\nu(t) = \delta_F$. $\qquad\square$

We will apply (4)–(5) in the following sections to different potentials ϕ to describe physical phenomena associated with nonconvexity of the underlying energy density.

4. Applications

4.1. Microstructures and hysteresis

As a first example we study the formation of microstructures in an elastic bar with an energy density described by a two-well potential $\phi_1(X) := (1 - X^2)^2$. Since the

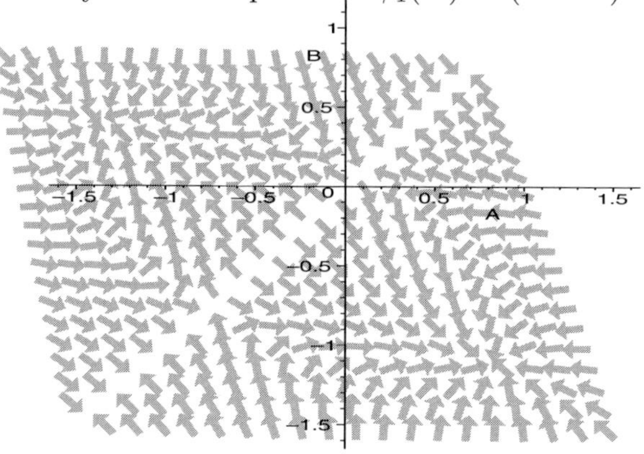

FIGURE 2. Phase diagram for $\phi_1(X) := (1 - X^2)^2$.

dynamical equations (4)–(5) do not depend on F, we can directly calcula gradient field, see Fig. 2. From this we see that $\nu(t) = \delta_F$ is only a stable s if $|F|$ is larger than some constant, in agreement with Proposition 3.4, sinc locally strictly convex for all F with $|F| > 1/\sqrt{3} =: F_{crit}$.

If we choose $F(t)$ non-constant on time, we can observe hysteresis The quasistatic solution $\nu(t)$ for $F(t) := 1 - \sin^2(\pi t)$ and $t \in [0, 1]$ is in Fig. 3. Hence the formation of microstructures takes place instantane but only after the deformation (measured with respect to the deformation uniform state $\nu = \delta_{+1}$, i.e., $F = 1$) exceeds a certain critical value, i.e. $F < F_{crit}$. On the other hand, the microstructure vanishes only after the process is reversed, i.e., when $F = 1$.

How can we explain this behavior microscopically? Our model tak account that to form a microstructure the crystal has to "evolve" throug mediate states (see Fig. 4). If they have a higher energy than the origin without microstructure, the system cannot overcome the energy barrier an in the local minimum, hence no microstructure forms. In our model exam is the case for $F > F_{crit}$.

Here, some comparison to observed hysteresis loops is in order: Fi formation of microstructures often shows very small hysteresis. Because classical Young measure models using global energy minimization can give description of such materials. However, our model captures additionally hy

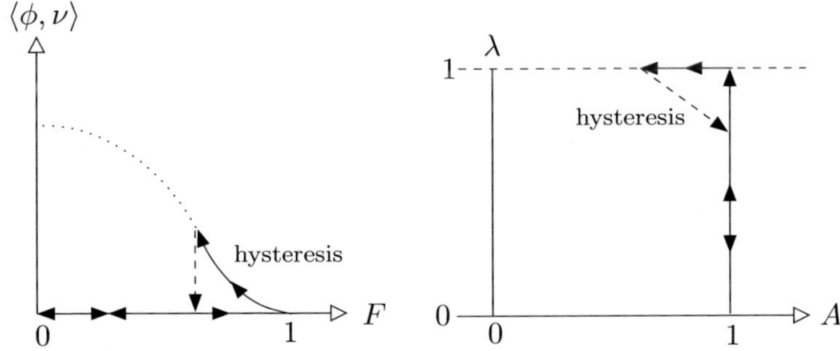

FIGURE 3. Quasistatic evolution with hysteresis loop, energy vs. deformation (left) and (A, λ)-plot (right) where $F(t) = \langle Id, \nu(t)\rangle$ and $\nu(t) = \lambda(t)\delta_{A(t)} + (1 - \lambda(t))\delta_{B(t)}$.

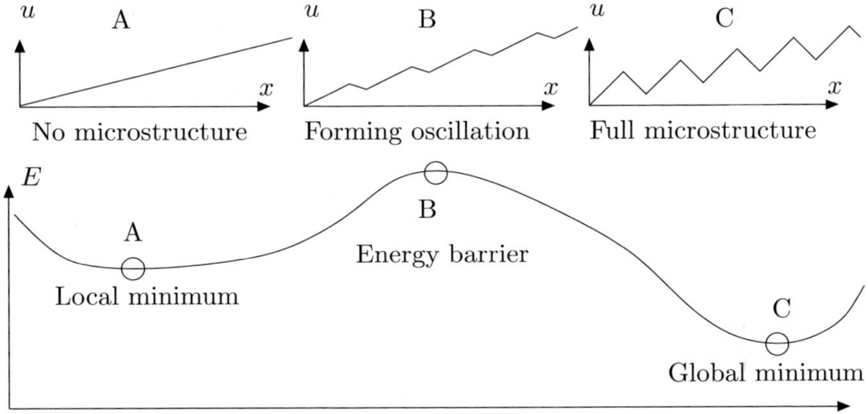

FIGURE 4. To form a microstructure with lower energy, a state with higher energy would have to be crossed.

effects. The amount of hysteresis is thereby solely defined by the form of the energy density ϕ. A steep form with a small convex region results in a small hysteresis effect, a large convex region results in a large hysteresis effect.

Second, the form of the hysteresis loops of Fig. 3 is incomplete when compared to experiments. This can be explained by one oversimplification in our model: We have restricted ourselves to space-homogeneous situations. However, the onset of microstructures in a real material will be local in space. An extension of our model to non-homogeneous situations could therefore reproduce hysteresis loops corresponding to the experiments.

4.2. Fracture

We consider an elastic bar and assume space-homogenous deformations. We can describe the deformation of the bar at a given time $t \geq 0$ by $F(t)$ where $F(t) = 0$ is the undeformed state. We model the elastic energy by a Perona-Malik type function $\phi(F) := \log(1 + |F|^2)$, and thus we are in the same situation as in the Truskinovsky paradox. We consider the case where the bar is slowly stretched with $F(t) = t$. The initial data is $\nu(0) = \delta_0$. It follows by Proposition 3.4 that the stable solution of our quasistatic limit problem (2) is $\nu(t) = \delta_{F(t)}$ as long as ϕ is locally strictly convex at $F(t)$. Hence the bar is not breaking in this range of deformations and its behavior is completely elastic. However beyond the critical point on which ϕ becomes locally nonconvex our solution becomes unstable. This corresponds to a fracture of the bar. To model this situation rigorously, we had to use the notion of Young measure varifolds [1] to capture the resulting infinite deformation gradients at the fracture, but even without using this notion we see that the Truskinovsky paradox does not occur in our model. This means that our model can describe the breaking of a bar, although it solely relies on the elastic energy without additional (phenomenological) "surface terms" for the energy contribution of a fracture.

An extension of this model to non-homogeneous situations is work in progress. First numerical experiments seem to show that in generic situations (i.e., when the problem is not set to be artificially uniform in space) the onset of fracture occurs in a single point when the stress exceeds a critical value [19]. It would be interesting to see whether also some forms of plasticity can be handled in this framework using a periodic energy density ϕ.

Acknowledgements

I am grateful to Kaushik Bhattacharya, Irene Fonseca, Robert Kohn, Massimiliano Morini, Stefan Müller, Luc Tartar and Johannes Zimmer for interesting discussions about the contents of this article and previous related works. Supported by the Center for Nonlinear Analysis under NSF Grant DMS-9803791

References

[1] J.J. ALIBERT AND G. BOUCHITTÉ, *Non-uniform integrability and generalized Young measures*, J. Convex Anal., 4 (1997), pp. 129–147.

[2] J.M. BALL AND R.D. JAMES, *Fine phase mixtures as minimizers of energy*, Arch. Rational Mech. Anal., 100 (1987), pp. 13–52.

[3] J.M. BALL, A. TAHERI, AND M. WINTER, *Local minimizers in micromagnetics and related problems*, Calc. Var. Partial Differential Equations, 14 (2002), pp. 1–27.

[4] C. CARSTENSEN, *Numerical analysis of microstructure*, in Theory and numerics of differential equations (Durham, 2000), Universitext, Springer, Berlin, 2001, pp. 59–126.

[5] C. CARSTENSEN AND M.O. RIEGER, *Young-measure approximations for elastodynamics with non-monotone stress-strain relations*, M2AN Math. Model. Numer. Anal., 38, 3 (2004), pp. 397–418.

[6] G. DAL MASO AND R. TOADER, *A model for the quasi-static growth of brittle fractures based on local minimization*, Math. Models Methods Appl. Sci., 12 (2002), pp. 1773–1799.

[7] A. DE SIMONE, *Hysteresis and imperfection sensitivity in small ferromagnetic particles*, Meccanica, 30 (1995), pp. 591–603. Microstructure and phase transitions in solids (Udine, 1994).

[8] G.A. FRANCFORT AND J.-J. MARIGO, *Revisiting brittle fracture as an energy minimization problem*, J. Mech. Phys. Solids, 46 (1998), pp. 1319–1342.

[9] K.-H. HOFFMANN AND T. ROUBÍČEK, *About the concept of measure-valued solutions to distributed parameter systems*, Math. Methods in the Applied Sciences, 18 (1995), pp. 671–685.

[10] D. KINDERLEHRER AND P. PEDREGAL, *Weak convergence of integrands and the Young measure representation*, SIAM J. Math. Anal., 23 (1992), pp. 1–19.

[11] R.V. KOHN AND P. STERNBERG, *Local minimisers and singular perturbations*, Proc. Roy. Soc. Edinburgh Sect. A, 111 (1989), pp. 69–84.

[12] P. KREJČÍ, *Evolution variational inequalities and multi-dimensional hysteresis operators*, in Nonlinear differential equations (Chvalatice, 1998), vol. 404 of Chapman & Hall/CRC Res. Notes Math., Chapman & Hall/CRC, Boca Raton, FL, 1999, pp. 47–110.

[13] J. KRISTENSEN AND A. TAHERI, *Partial regularity of strong local minimizers in the multi-dimensional calculus of variations*, Arch. Ration. Mech. Anal., 170, 1 (2003), pp. 63–89.

[14] A. MIELKE, *Analysis of energetic models for rate-independent materials*, in Proceedings of the International Congress of Mathematicians, Vol. III (Beijing, 2002), Beijing, 2002, Higher Ed. Press, pp. 817–828.

[15] A. MIELKE, F. THEIL, AND V.I. LEVITAS, *A variational formulation of rate-independent phase transformations using an extremum principle*, Arch. Ration. Mech. Anal., 162 (2002), pp. 137–177.

[16] S. MÜLLER, *Variational models for microstructure and phase transition*, in Calculus of Variations and Geometric Evolution Problems, S. Hildebrandt and M. Struwe, eds., no. 1713 in Lecture Notes in Mathematics, Springer-Verlag, Berlin etc., 1999, p. 211 ff.

[17] P. PEDREGAL, *Parametrized measures and variational principles*, Birkhäuser, 1997.

[18] M.O. RIEGER, *Young measure solutions for nonconvex elastodynamics*, SIAM Journal on Mathematical Analysis, 34 (2003), pp. 1380–1398.

[19] M.O. RIEGER AND J. ZIMMER, *Young measure flow as a model for damage.* in preparation.

[20] T. ROUBÍČEK, *Finite element approximation of a microstructure evolution*, Math. Methods Appl. Sci., 17 (1994), pp. 377–393.

[21] L. TARTAR, *Compensated compactness and applications to partial differential equations*, in Nonlinear analysis and mechanics: Heriot-Watt Symposium, Vol. IV, Pitman, Boston, Mass., 1979, pp. 136–212.

[22] L. TRUSKINOVSKY, *Fracture as a phase transformation*, in Contemporary research in mechanics and mathematics of materials, R.C. Batra and M.F. Beatty, eds., CIMNE, 1996, pp. 322–332.

[23] L.C. YOUNG, *Generalized curves and the existence of an attained absolute minimium in the calculus of variations*, Comptes Rendus de la Soc. des Sciences et de Lettres de Varsovie, classe III, 30 (1937), pp. 212–34.

Marc Oliver Rieger
Scuola Normale Superiore
Pisa, Italy
e-mail: `rieger@sns.it`

Progress in Nonlinear Differential Equations
and Their Applications, Vol. 63, 415–421
© 2005 Birkhäuser Verlag Basel/Switzerland

The Precise L^p-theory of Elliptic Equations in the Plane

Carlo Sbordone

To Haïm Brezis on his sixtieth birthday

1. Introduction

Fix a bounded domain $\Omega \subset \mathbb{R}^2$. For $K \geq 1$ we denote by $\mathcal{E}(K)$ the set of measurable matrix valued functions $A \in L^\infty(\Omega; \mathbb{R}^{2\times 2})$ such that

$$A(z) = A(z)^t \qquad \det A(z) = 1 \quad \text{a.e. } z \in \Omega, \tag{1.1}$$

satisfying the ellipticity bounds

$$\frac{|\xi|^2}{\sqrt{K}} \leq \langle A(z)\xi, \xi \rangle \leq \sqrt{K}|\xi|^2 \tag{1.2}$$

for all $\xi \in \mathbb{R}^2$ and a.e. $z \in \Omega$.

To any such a matrix we will associate three different linear differential operators:

the divergence form operator

$$\mathcal{L}[u] = \operatorname{div}\left(A^2(z)\nabla u\right) \qquad u \in W^{1,p}_{loc}, p > 1 \tag{1.3}$$

the non divergence form operator

$$\mathcal{M}[w] = Tr(A(z)D^2 w) \qquad w \in W^{2,p}_{loc}, p > 1 \tag{1.4}$$

and its formal adjoint

$$\mathcal{N}[v] = D^2(A(z)v) \qquad v \in L^p_{loc}(\Omega), p > 1. \tag{1.5}$$

For rather obvious reasons the natural domains of definition of these operators correspond to $p = 2$. However, we will discuss here optimal regularity results which require $2 \leq p < \frac{2K}{K-1}$.

The close connections between such equations and planar quasiconformal mappings will play a significant role in what follows. The heart of the matter is the estimate of Astala in his Area Distortion Theorem [A]. Another important tool is the sharp result on invertibility of complex Beltrami operators recently established in [AIS]. Let us begin with a well-known optimal regularity [A] [LN] [IS] for solutions to the divergence equation

$$\mathcal{L}[u] = 0.$$

Theorem 1.1. *Suppose A is a measurable symmetric matrix function satisfying (1.1) and (1.2). If $\frac{2K}{K+1} < q \leq 2 \leq p < \frac{2K}{K-1}$ and $u \in W^{1,q}_{\mathrm{loc}}(\Omega)$ is a solution in $\Omega \subset \mathbb{R}^2$ to the equation*

$$\mathcal{L}[u] = \mathrm{div}\,(A^2(x)\nabla u) = 0.$$

Then

$$u \in W^{1,p}_{\mathrm{loc}}(\Omega).$$

The range of exponents q and p is optimal.
Next we are concerned with interior regularity of solutions to the nondivergence equation

$$\mathcal{M}[w] = 0, \qquad \text{see (1.4)}.$$

Using Theorem 1.1 we shall demonstrate a new proof of the regularity of solutions to the nondivergence equation, first established in [AIM].

Theorem 1.2. *Let $A(z)$ verify (1.1) and (1.2). If $\frac{2K}{K+1} < q \leq 2 \leq p < \frac{2K}{K-1}$ and $w \in W^{2,q}_{\mathrm{loc}}(\Omega)$ is a solution to the equation*

$$\mathcal{M}[w] = 0$$

then

$$D^2 w \in L^p_{\mathrm{loc}}(\Omega).$$

The range of the exponents q and p is optimal, (see [AIM] and [AFS])
Concerning the adjoint operator we have the following improvement of a result from [MS].

Theorem 1.3. *Let $A(z)$ verify (1.1), (1.2) and $\frac{2K}{K+1} < q \leq 2 \leq p < \frac{2K}{K-1}$. If $v \in L^q_{loc}(\Omega)$ is a non-negative solution to the adjoint equation*

$$\mathcal{N}[v] = 0, \quad \text{see (1.5)}$$

then

$$v \in L^p_{\mathrm{loc}}(\Omega).$$

This result fails for $p = \frac{2K}{K-1}$.

2. The non divergence equation

As observed by D'Onofrio-Greco ([DG]) and D'Onofrio ([D]) there is a particular connection between the operators (1.3)–(1.4). This connection is emphasized by the following

Proposition 2.1. [DG] *Let* $w \in W_{loc}^{2,r}(\Omega)$, $r > 1$, *satisfy*

$$\begin{cases} \mathcal{M}[w] = 0 \\ AD^2 w = D^2 w A \quad \text{(commuting condition)}. \end{cases}$$

Then

$$\mathcal{L}[w_{x_1}] = \mathcal{L}[w_{x_2}] = 0.$$

This proposition also explains why we have chosen the square matrix A^2 for coefficient of the operator in divergence form. It is generally known [GT] that each of the partials w_{x_1} and w_{x_2} (w being a solution to a planar elliptic equation in non divergence form) solves its own elliptic equation divergence form. Here we would like to stress that, because of the commuting condition, w_{x_1} and w_{x_2} solve the same equation.

Remark 2.1. The commutation hypothesis $AD^2 w = D^2 w A$ is critical for Proposition 2.1. Our next proposition will tell us that this poses no restrictions of generality. We denote by $\mathbb{R}_{\text{sym}}^{2\times 2}$ the set of 2×2 symmetric matrices with real entries.

Proposition 2.2. [DG] *Given two matrices* A, $M \in \mathbb{R}_{sym}^{2\times 2}$, *there exists* $\tilde{A} \in \mathbb{R}_{\text{sym}}^{2\times 2}$ *that commutes with M and we still have*

$$Tr(\tilde{A}M) = Tr(AM).$$

Moreover, if $A \in \mathcal{E}(K)$, *then also* $\tilde{A} \in \mathcal{E}(K)$.

We are now in a position to prove Theorem 1.2.

Proof. Let $w \in W_{\text{loc}}^{2,q}(\Omega)$ solve the equation

$$Tr(A(z)D^2 w) = 0 \qquad \text{in } \Omega.$$

By Proposition 2.2 there exists $\tilde{A}(z) \in \mathcal{E}(K)$ such that

$$\begin{cases} Tr(\tilde{A}(z)D^2 w) = 0 \\ \tilde{A}(z)D^2 w = D^2 w \tilde{A}(z). \end{cases}$$

Applying Proposition 2.1 we deduce that

$$\text{div}\,(\tilde{A}^2(z)\nabla w_{x_1}) = 0$$

$$\text{div}\,(\tilde{A}^2(z)\nabla w_{x_2}) = 0.$$

By mean of Theorem 1.1 we then conclude with the desired regularity of w

$$w \in W_{loc}^{2,p}(\Omega). \qquad \square$$

3. The adjoint to non divergence equation

Questions connected with increasing the degree of integrability of a function satisfying a reverse Hölder inequality have numerous applications in the regularity theory of PDE's [G] [Gi]. In this section the reverse Hölder inequality have the same support B

$$\left(\fint_B v^p\right)^{\frac{1}{p}} \leq c \fint_B v \tag{3.1}$$

as those originally considered by F. Gehring in his celebrated paper [G] and are much stronger than the ones with ball B and double ball $2B$. In [MS] a sharp reverse Hölder inequality is proved for non negative solutions $v \in L^2_{loc}(\Omega)$ to the adjoint equation

$$\mathcal{N}[v] = D^2(A(z)v) = 0 \quad \text{in } \Omega$$

where A satisfies (1.1), (1.2), improving for the case $n = 2$ previous results of Fabes-Stroock [FS]. Specifically, the best integrability exponent of v in terms of the ellipticity constant K is determined as a consequence of inequality (3.1) to hold for $B = B(a,r) \subset B(a,2r) \subset \Omega$ and $2 \leq p < \frac{2K}{K-1}$.

Our aim here is to show that the assumption $v \in L^2_{loc}(\Omega)$ can be relaxed into

$$v \in L^q_{loc}(\Omega), \qquad \frac{2K}{K+1} < q \leq 2$$

and this is optimal too.

Our proof of Theorem 1.3 will rely on the following recent results of Astala-Iwaniec-Martin [AIM] on the interior regularity and maximum principle for solutions to the equation $\mathcal{M}[w] = h$, where $\mathcal{M}$ is defined in (1.4) with the assumptions (1.1) and (1.2).

Theorem 3.1. *[AIM] Let $h \in L^p(B_r)$ $B_r = B(0,r)$ for some $\frac{2K}{K+1} < p < \frac{2K}{K-1}$. Then the equation*

$$\mathcal{M}[w] = h$$

has a unique solution $w \in W^{2,p}$ vanishing on ∂B_r. Moreover we have the uniform estimate

$$\|D^2 w\|_{L^p(B_r)} \leq c\|h\|_{L^p(B_r)}.$$

Theorem 3.2. *[AIM] Suppose $\frac{2K}{K+1} < q \leq 2$, and $w \in W^{2,q}_{loc}(B_r)$ satisfies*

$$\begin{cases} \mathcal{M}[w] = h & a.e. \text{ in } B_r = B(0,r) \\ w = 0 & on \ \partial B_r. \end{cases}$$

Then, we have

$$\|w\|_{L^\infty(B_r)} \leq c(K,q)r^{2-\frac{2}{q}}\|h\|_{L^q(B_r)}. \tag{3.2}$$

The estimate no longer holds if $q \leq \frac{2K}{K+1}$.

We are now in a position to prove Theorem 1.3.

Proof. Assume $v \in L^{q_o}(\Omega)$ with $\frac{2K}{K+1} < q_o \leq 2$, v a non negative solution to the adjoint equation

$$\mathcal{N}[v] = 0,$$

that is:

$$\int_\Omega v\mathcal{M}[\varphi] = 0$$

for any $\varphi \in W^{2,q_o'}(\Omega)$ with compact support. Let $2 \leq p < \frac{2K}{K-1}$ and set $q = \frac{p}{(p-1)}$.

As in the proof of Theorem 3.1 in [MS] we make use of the dual formulation of the L^p-norm

$$\left(\int_{B_r} v^p\right)^{\frac{1}{p}} = \sup\left\{\int_{B_r} vh \ : \ h \geq 0, \, h \in C_0^1(B_r), \, \|h\|_{L^q(\mathbb{R}^2)} \leq 1\right\}. \tag{3.3}$$

Let us fix $h \in C_0^1(B_r)$, $\|h\|_{L^q} \leq 1$, $h \geq 0$, and solve the Dirichlet problem

$$\begin{cases} \mathcal{M}[w] = h & \text{in } B_{2r} \\ w = 0 & \text{on } \partial B_{2r} \end{cases}$$

for $w \in W^{2,2}(B_{2r})$. Next fix a non-negative function $\varphi_r \in C_0^1(B_{\frac{3}{2}r})$ such that $\varphi_r = 1$ on B_r and $|\frac{\partial^\alpha \varphi_r}{\partial x^\alpha}| \leq \frac{C\alpha}{r^{|\alpha|}}$.

Then, we have

$$\int_{B_r} vh \leq \int_{B_{2r}} v\mathcal{M}[w]\varphi_r$$

$$= -\int_{B_{2r}} vw\mathcal{M}[\varphi_r] - 2\int_{B_{2r}} v\langle A\nabla w, \nabla\varphi_r\rangle \tag{3.4}$$

$$\leq \frac{c}{r^2}\|w\|_{L^\infty(B_{2r})}\int_{B_{\frac{3}{2}r}} v + \frac{c}{r}\int_{B_{\frac{3}{2}r}} v|\nabla w|.$$

By (3.2) $\|w\|_{L^\infty(B_{2r})} \leq c(K,q)r^{\frac{2}{p}}$, hence (3.4) implies

$$\int_{B_r} vh \leq \frac{c}{r^2}r^{\frac{2}{p}}\int_{B_{\frac{3}{2}r}} v + \frac{c}{r}\left(\int_{B_{\frac{3}{2}r}} v\right)^{\frac{1}{2}}\left(\int_{B_{2r}} v|\nabla w|^2\right)^{\frac{1}{2}}. \tag{3.5}$$

We now estimate the last integral in the right-hand side, by using Caccioppoli inequality.

We have by (1.2)

$$\int_{B_{2r}} v|\nabla w|^2 \leq \sqrt{K}\int_{B_{2r}} v\langle A\nabla w, \nabla w\rangle = \sqrt{K}\int_{B_{2r}} v\left[\frac{\mathcal{M}[w^2]}{2} - wh\right].$$

By the previous Theorem 3.1 we deduce that $w \in W^{2,q_o'}(B_{2r}) \cap W_o^{1,q_o'}(B_{2r})$, and, also, by Sobolev embedding that $w^2 \in W^{2,q_o'}(B_{2r})$ where $q_o' = \frac{q_o}{(q_o-1)}$.

Since $w^2 = 0$ on ∂B_{2r}, and $\nabla(w^2) = 0$ on ∂B_{2r} we deduce

$$\int_{B_{2r}} v\mathcal{M}[w^2] = 0, \qquad \text{as} \quad \mathcal{N}[v] = 0.$$

Using again (3.2), yields

$$\int_{B_{2r}} v|\nabla w|^2 \leq 2\sqrt{K} \int_{B_{2r}} v|w|h$$

$$\leq 2\sqrt{K}\|w\|_{L^\infty(B_{2r})} \int_{B_r} vh \qquad (3.6)$$

$$\leq 2\sqrt{K}\,cr^{\frac{2}{p}} \int_{B_r} vh.$$

By (3.5) and (3.6) it follows that

$$\int_{B_r} vh \leq \frac{c}{r^{2(1-\frac{1}{p})}} \int_{B_{\frac{3}{2}r}} v + \frac{c}{r^{(1-\frac{1}{p})}} \left(\int_{B_{\frac{3}{2}r}} v\right)^{\frac{1}{2}} \left(\int_{B_r} vh\right)^{\frac{1}{2}}.$$

By the elementary inequality $\sqrt{a}\,\sqrt{b} \leq \frac{a}{2} + \frac{b}{2}$, we arrive at

$$\int_{B_r} vh \leq \frac{c}{r^{2(1-\frac{1}{p})}} \int_{B_{\frac{3}{2}r}} v + \frac{c}{r^{2(1-\frac{1}{p})}} \int_{B_{\frac{3}{2}r}} v + \frac{1}{2} \int_{B_r} vh.$$

Rearranging, it reduces to:

$$\int_{B_r} vh \leq \frac{c}{r^{2(1-\frac{1}{p})}} \int_{B_{\frac{3}{2}r}} v. \qquad (3.7)$$

Since h is arbitrary, by (3.3), (3.7) we obtain

$$\left(\fint_{B_r} v^p\right)^{\frac{1}{p}} \leq c\fint_{B_{\frac{3}{2}r}} v.$$

An application of Lemma 2.0 in [FS] concludes the proof. $\qquad\qquad\square$

Acknowledgments

The research has been supported by MIUR – PRIN 02 and GNAMPA – INdDAM.

References

[A]   K. Astala, *Area distortion of quasiconformal mappings*, Acta Math., 173, (1994), 37–60.

[AIM] K. Astala, T. Iwaniec and G. Martin, *Pucci's Conjecture and the Aleksandrov Inequality for Elliptic PDEs in the Plane* (2004) to appear.

[AIS] K. Astala, T. Iwaniec and E. Saksman, *Beltrami operators in the plane*, Duke Math. J. 107, 1, (2001), 27–56.

[AFS] K. Astala, D. Faraco and L. Székelyhidi, *Convex Integration and the L^p-theory of Elliptic Equations*, in preparation.

[B1]   P. Bauman, *Equivalence of Green's functions for diffusion operators in $\mathbb{R}^n$: a counterexample*, Proc. Amer. Math. Soc. 91(1) (1984), 64–68.

[D] L. D'Onofrio, *Relations between G-convergence of elliptic equations in the divergence and non-divergence form* in preparation.

[DG] L. D'Onofrio and L. Greco, *A counter-example in G-convergence of non-divergence elliptic operators*, Proc. Roy. Soc. Edin. 133A, (2003) 1299–1310.

[FS] E.B. Fabes and D.W. Stroock, *The L^p-integrability of Green's functions and fundamental solutions for elliptic and parabolic equations*, Duke Math. J. 51, 4 (1984), 997–1016.

[G] F. Gehring, *The L^p-integrability of the partial derivatives of a quasiconformal mapping*, Acta Math. 130 (1973), 265–277.

[Gi] M. Giaquinta, *Multiple Integrals in the Calculus of Variations and Nonlinear Elliptic System*, Ann. of Math. Studies 105 (1983).

[GT] D. Gilbarg and N.S. Trudinger, *Elliptic Partial Differential Equations of Second Order*, Springer 1998.

[IS] T. Iwaniec and C. Sbordone, *Quasiharmonic fields*, Ann. Inst. H. Poincaré, Analyse Nonlinéaire 18 (2001) 5, 519–572.

[LN] F. Leonetti and V. Nesi, *Quasiconformal solutions to certain first order systems and the proof of a conjecture of G.W. Milton*, J. Math. Pures Appl. (9) 76, 2 (1997), 109–124.

[MS] G. Moscariello and C. Sbordone, *Optimal L^p-properties of Green's functions for non divergence elliptic equations in two dimensions*, (to appear on Studia Math.).

Carlo Sbordone
Dipartimento di Matematica e Applicazioni "R. Caccioppoli"
Via Cintia
I-80126 Napoli, Italy
e-mail: `sbordone@unina.it`

Progress in Nonlinear Differential Equations
and Their Applications, Vol. 63, 423–431

Essential Spectrum and Noncontrollability of Membrane Shells

V. Valente

Abstract. The paper deals with the well-posedness and the controllability of vibrations for spherical membrane shells linearized around stable solutions of the nonlinear equilibrium problem. We characterize the minimizers of the non-convex energy functional and deduce the noncontrollability of the membrane shells from the existence of the essential spectrum for the tangent operators.

1. Introduction

The system of differential equations which describes the vibrations of linear elastic thin shells can be reported in the following general form

$$\mathbf{v}_{tt} + \mathbf{A}^m \mathbf{v} + h\,\mathbf{A}^f \mathbf{v} = 0 \qquad \text{in} \quad \Omega \times (0, T) \tag{1.1}$$

where Ω is the middle surface of the shell, T is a positive fixed time, $\mathbf{A}^m$ and $\mathbf{A}^f$ are the membrane and flexion operators respectively and h denotes the thinness parameter of the shell. The controllability of elastic shells has been intensively studied in the last years especially after the introducing of a constructive controllability method (HUM method) by J.L. Lions. In the pioneer paper [4] the exact controllability of hemispherical shells follows from the existence of an asymptotic gap for the eigenvalues of the operator $\mathbf{A}_h = \mathbf{A}^m + h\,\mathbf{A}^f$. The main theorem states that we have exact controllability for any fixed time

$$T \geq C/\sqrt{h}, \tag{1.2}$$

where C is a constant depending on the characteristic parameters of the material. In [5] the exact controllability of shallow spherical shells has been proved using multiplier techniques; also in this case we get to (1.2) under the further condition that the shallowness parameter is small enough. The shallowness condition can be removed, as we proved in [10], using Carleman estimates. Analogous situation appears in the controllability of shells of general shape [12]. For this reason we restrict our attention to the exact controllability of *spherical* shells for showing and understanding the phenomena which appear in the transition shell-membrane, that is when the thinness of the shell goes to zero. The time behavior (1.2) suggests to look at the controllability of the membrane problem. In a recent paper [8] an abstract theorem of noncontrollability for linear systems of mixed order, including spherical elastic membranes, has been proved. The non-controllability follows from the existence of the essential spectrum. Indeed under suitable hypothesis on the operator $\mathbf{A}^m$, by a reformulated Weyl characterization of the essential spectrum, it is possible to construct a sequence of initial data for getting to a contradiction of the sufficient and necessary condition of exact controllability. The noncontrollability is not the only phenomenon imputable to the existence of an essential spectrum. For example in [9] has been proved that, because the existence of the essential spectrum, heterogeneous membrane shells cannot be described in terms of homogenized systems.

Although we have that the linear membrane shell is not exactly controllable, one may think, following the return method proposed by Coron [2] for the controllability of incompressible fluids, that the controllability of the elastic membranes follows from the nonlinear terms. The main step of the return method is to prove the controllability of the system linearized around another nontrivial stable solution to the static problem. Hence the exact controllability problem for the nonlinear membranes has to be connected to the nonlinear equilibrium problem.

The analysis of the equilibria for nonlinear spherical caps has been carried out in ([6, 7, 13]) through a *almost dual* formulation which allows to define a more simple energy functional. The multiplicity of equilibria has been proved when the parameter of thinness is small enough. In particular the existence of a branch of stable solutions which tend to a limit unstressed solution (the everted shape) has been shown. More recently the problem of the eversion of thin caps has been recovered. Beside the trivial and the everted shape in [3] many infinitely nonregular stable solutions of the membrane problem have been predicted, in [1] a multiscale analysis by Γ-convergence of the energy functional allows to characterize the asymptotic solutions. These results suggests to linearize the membrane shell around the limit stable configurations and to look at the controllability problem for the derived system of differential equations. In this note we show that the linearized systems are not exactly controllable. The noncontrollability result is independent of the stable equilibrium solution around which we study the linearized system, since in any case the linear membrane operator has an nonempty essential spectrum.

2. The nonlinear problem

The axially symmetric vibrations of a thin spherical shell of unitary ray are described by the partial differential equations in the unknown $\mathbf{v} = (u, w)$, whose components represent the meridional and radial (oriented towards the center) displacements respectively,

$$\begin{aligned}
\ddot{u}\sin\theta &= (S\sin\theta)' - T\cos\theta - SV\sin\theta + h\mathcal{L}(V)\sin\theta \\
\ddot{w}\sin\theta &= (S+T)\sin\theta + (SV\sin\theta)' - h(\mathcal{L}(V)\sin\theta)'
\end{aligned} \tag{2.1}$$

where the dot stands for the partial derivative with respect to time and the prime stands for the partial derivative with respect to θ. The angle θ belongs to the set $(0, \theta_0)$ where θ_0 is the fixed opening angle of the shell which we assume to satisfy the condition $\theta_0 < \pi/2$, moreover we define

$$
\begin{aligned}
V &= u + w' \\
S &= (u' - w + \tfrac{1}{2}V^2) + \nu(u \cot\theta - w) \\
T &= \nu(u' - w + \tfrac{1}{2}V^2) + (u \cot\theta - w) \\
\mathcal{L}(V) &= V'' + V' \cot\theta - V(\nu + \cot^2\theta)
\end{aligned}
\tag{2.2}
$$

with $\nu \in [0, 1[$ a given material constant. The static formulation of (2.1) equipped with suitable homogeneous boundary conditions has been studied in [13, 6, 7] with particular regard to the appearance of multiple solutions and the characterization of everted shapes. The analysis for the eversion of thin shells has been carried out from theoretical and numerical point of view for shallow spherical shells and the existence of equilibrium configurations, interpretable as everted stressed configurations of the shell, has been proved for small values of the thinness parameter h.

The results obtained can be easily generalized to spherical shells without assuming shallowness hypothesis. Indeed the static equation associated to (2.1) can be rearranged (see [6, 13]) leading to the equivalent problem in the unknowns

$$
H = \frac{V}{\sin\theta}, \qquad G = \frac{S}{\cos\theta}
$$

which reads

$$
h\left(\frac{1}{\sin^3\theta} \frac{d}{d\theta}(\sin^3\theta \frac{dH}{d\theta})\right) - (1 + \nu)H = (1 + H\cos\theta)G
$$

$$
\frac{1}{\sin^3\theta} \frac{d}{d\theta}(\sin^3\theta \frac{dG}{d\theta}) - (1 - \nu)G = -\mu H(H\cos\theta + 2)
$$
$$\tag{2.3}$$

where the constant μ is equal to $(1 - \nu^2)/2$. We associate to the above system the boundary conditions

$$
H'(0) = G'(0) = 0, \quad G(\theta_0) = 0, \quad H'(\theta_0) + (1 + \nu)\cot\theta_0 H(\theta_0) = 0.
\tag{2.4}
$$

For the analysis of the problem we need the following functional spaces:

- $\mathcal{K}$ is the space of functions which are square integrable with respect to the weight $\sin^3\theta$ with scalar product and norm defined by

$$
((H_1, H_2)) = \int_0^{\theta_0} \sin^3\theta H_1(\theta) H_2(\theta) d\theta, \quad \|H\|^2 = \int_0^{\theta_0} \sin^3\theta H^2(\theta) d\theta
$$

- $\mathcal{K}_1$ is the space of functions which belong to $\mathcal{K}$ with their first derivatives. The space $\mathcal{K}_1$ is compactly imbedded to $\mathcal{K}$, the scalar product in $\mathcal{K}_1$ is defined by

$$
((H_1, H_2))_1 = \int_0^{\theta_0} \sin^3\theta H_1'(\theta) H_2'(\theta) d\theta + (1 + \nu)\sin^2\theta_0 \cos\theta_0 H_1(\theta_0) H_2(\theta_0)
$$

- $\mathcal{K}_0 = \{H \in \mathcal{K}_1 \;:\; H(\theta_0) = 0\}$ with scalar product denoted by $((H_1, H_2))_0$

Introduced the Green operator $G_0 \; : \; \mathcal{K} \to \mathcal{K}_0$, such that

$$G = -\mu\, G_0[H(H\cos\theta + 2)]$$

we see that the solutions of (2.3), (2.4) are the stationary points of the functional

$$J(H) = \frac{h}{2}\|H\|_1^2 + \frac{h}{2}(1+\nu)\|H\|^2 + \mu\frac{1}{4}\|G_0[H(2 + H\cos\theta)]\|_0^2. \qquad (2.5)$$

The functional $J(H)$ is lower semicontinuous in the weak topology of $\mathcal{K}_1$, moreover there is a positive constant h^* such that it is coercive an nonconvex for $h \le h^*$. The trivial solution $H \equiv 0$, the reference configuration of the shell, is the point of absolute minimum for the functional, moreover for h small enough, there are other local minimizers H_n corresponding to the thinness $h_n = 1/n$ which satisfy the property as $n \to \infty$

$$H_n \to -\frac{2}{\cos\theta} \quad \text{weakly in } \mathcal{K}_1$$

and can be denoted as everted solutions.

In terms of the previous variable V, besides the trivial stable solution $V = 0$ we can look at another stable solution of the limit problem $V = -2\tan\theta$ which is interpretable as the reflection with the respect to the horizontal plane of the middle surface of the membrane in its reference configuration. Moreover one can find infinite nonregular stable solutions (see [3]) defining

$$V = 0 \text{ if } \theta \in \Omega_i, \qquad V = -2\tan\theta, \text{ if } \theta \in [0,\theta_0]/\Omega_i \qquad (2.6)$$

where $\Omega_i = \bigcup_{k=1}^{i} \omega_k$ and $\{\omega_k\}$ is a sequence of subsets of $[0,\theta_0]$ such that $\omega_k \cap \omega_j = \varnothing$ for any $j \neq k$.

The asymptotic analysis of the functional (2.5) as $h \to 0^+$, in the hypothesis of shallowness, is studied in [1] using Γ-convergence techniques.

3. The linearized equations

With in mind the analysis of the previous section we are going to define the motion of linearized spherical membranes. Firstly we look at (2.1) and formally put $h = 0$. We introduce a function V_0 depending only on the variable θ and consider the linear system

$$\begin{aligned}
\ddot{u}\sin\theta &= (\mathcal{S}\sin\theta)' - \mathcal{T}\cos\theta - \mathcal{S}V_0\sin\theta \\
\ddot{w}\sin\theta &= (\mathcal{S}+\mathcal{T})\sin\theta + (\mathcal{S}V_0\sin\theta)'
\end{aligned} \qquad (3.1)$$

where

$$\begin{aligned}
\mathcal{S} &= (u' - w + VV_0) + \nu(u\cot\theta - w) \\
\mathcal{T} &= \nu(u' - w + VV_0) + (u\cot\theta - w) \\
V &= u + w'
\end{aligned} \qquad (3.2)$$

with the associated boundary conditions

$$u(0,t) = w'(0,t) = u(\theta_0,t) + V_0(\theta_0)w(\theta_0,t) = 0. \qquad (3.3)$$

We recall that $\theta_0 < \pi/2$ and we assume V_0 is a regular stable solution or a suitable regularization of a stable solution of the static problem described in the previous section such that

$$V_0 \in \mathcal{B}, \qquad \mathcal{B} = \{b \ : \ b(\theta),\, b'(\theta) \ \in L_\infty[0,\theta_0],\ b(0) = 0\} \tag{3.4}$$

Then we introduce the new variable $\mathbf{z} = (f,q)$ with $f = u + V_0\, w$ and $q = w - V_0\, u$, and look at the system compactly written as

$$\ddot{\mathbf{z}} - \mathbf{A}\,\mathbf{z} = 0 \tag{3.5}$$

with boundary conditions

$$f(0,t) = f(\theta_0, t) = 0 \tag{3.6}$$

Multiplying the first equation of (3.5) by $\tilde{f}(1+V_0^2)^{-1}\sin\theta$ and the second one by $\tilde{q}(1+V_0^2)^{-1}\sin\theta$, integrating by parts and taking into account the boundary conditions (3.6), we can get to the following variational setting

$$b(\ddot{\mathbf{z}},\tilde{\mathbf{z}}) \ + a(\mathbf{z},\tilde{\mathbf{z}}) = 0. \tag{3.7}$$

where

$$b(\mathbf{z},\tilde{\mathbf{z}}) = \int_0^{\theta_0} (f\,\tilde{f}\sin\theta + q\,\tilde{q}\sin\theta)(1+V_0^2)^{-1}d\theta \tag{3.8}$$

$$a(\mathbf{z},\tilde{\mathbf{z}}) = \int_0^{\theta_0} [e_1(\mathbf{z})e_1(\tilde{\mathbf{z}}) + e_2(\mathbf{z})e_2(\tilde{\mathbf{z}}) + \nu e_1(\mathbf{z})e_2(\tilde{\mathbf{z}}) + \nu e_2(\mathbf{z})e_1(\tilde{\mathbf{z}})]\sin\theta d\theta \tag{3.9}$$

and

$$e_1(\mathbf{z}) = f' - q - V_0'\frac{fV_0 + q}{1+V_0^2}, \qquad e_2(\mathbf{z}) = \frac{f\cot\theta - q}{1+V_0^2} - V_0\frac{f + q\cot\theta}{1+V_0^2}$$

Let $L^2(0,\theta_0;\sin\theta)$ the space of the square integrable functions with respect to the weight $\sin\theta$ with scalar product $(f,\tilde{f}) = \int_0^{\theta_0} f\tilde{f}\sin\theta d\theta$, we define the following Hilbert space

$$\mathcal{U} = \{f \ : \ f, f', f\cot\theta \in L^2(0,\theta_0;\sin\theta),\ f(0) = f(\theta_0) = 0\}$$

provided with the scalar product $(f,\tilde{f})_{\mathcal{U}} = \int_0^{\theta_0}(f'\tilde{f}' + f\tilde{f}\cot\theta)\sin\theta d\theta$, and the vector spaces

$$\mathcal{V} = \mathcal{U} \times L^2(0,\theta_0;\sin\theta), \qquad \mathcal{H} = L^2(0,\theta_0;\sin\theta) \times L^2(0,\theta_0;\sin\theta)$$

Consider the problem $(\mathcal{P})$:

let $T > 0$, $\mathbf{z}^0 \in \mathcal{V}$ and $\mathbf{z}^1 \in \mathcal{H}$ be given, find a function $\mathbf{z}(t)$ such that for any $\tilde{\mathbf{z}} \in \mathcal{V}$

$$b(\ddot{\mathbf{z}},\tilde{\mathbf{z}}) + a(\mathbf{z},\tilde{\mathbf{z}}) = 0, \quad \mathbf{z}(0) = \mathbf{z}^0,\ \dot{\mathbf{z}}(0) = \mathbf{z}^1.$$

Theorem 3.1 There exists a unique solution $\mathbf{z}$ to the problem $(\mathcal{P})$ such that $\mathbf{z} \in L^2(0,T;\mathcal{V})$, $\dot{\mathbf{z}} \in L^2(0,T;\mathcal{H})$

Proof. The proof of the theorem follows from Lemma 3.1, Lemma 3.2 (see [11]).

Lemma 3.1 The bilinear form $a(\mathbf{z},\tilde{\mathbf{z}})$ is continuous on $\mathcal{V} \times \mathcal{V}$.

Proof. The proof is trivial, indeed taking into account the conditions (3.4) one has that there is a positive constant c s.t.

$$a(\mathbf{z}, \tilde{\mathbf{z}}) \leq c \, \|\mathbf{z}\|_{\mathcal{V}} \, \|\tilde{\mathbf{z}}\|_{\mathcal{V}}.$$

Lemma 3.2 There exist two positive constants τ and C such that for any $\mathbf{z} \in \mathcal{V}$

$$\bar{a}(\mathbf{z}, \mathbf{z}) = a(\mathbf{z}, \mathbf{z}) + \tau \, b(\mathbf{z}, \mathbf{z}) \geq C\|\mathbf{z}\|_{\mathcal{V}}^2. \tag{3.10}$$

Proof. Arguing as in [4] we firstly observe that, since $|\nu| < 1$,

$$\bar{a}(\mathbf{z}, \mathbf{z}) \geq (1 - \nu) \int_0^{\theta_0} [(e_1(\mathbf{z}))^2 + (e_2(\mathbf{z}))^2] \sin\theta d\theta + \tau \, b(\mathbf{z}, \mathbf{z}).$$

We suppose now that the condition (3.10) does not hold true, then there is a sequence $\{\mathbf{z}_n\}$ such that

$$0 \leq \bar{a}(\mathbf{z}_n, \mathbf{z}_n) \leq \frac{1}{n}\|\mathbf{z}_n\|_{\mathcal{V}}^2 \qquad \forall \, n.$$

Let

$$\hat{f}_n = \frac{f_n}{\|\mathbf{z}_n\|_{\mathcal{V}}^2}, \quad \hat{q}_n = \frac{q_n}{\|\mathbf{z}_n\|_{\mathcal{V}}^2},$$

then $\hat{\mathbf{z}}_n$ verifies

$$\|\hat{\mathbf{z}}_n\|_{\mathcal{V}}^2 = 1, \quad \text{moreover} \quad \lim_{n \to \infty} \bar{a}(\hat{\mathbf{z}}_n, \hat{\mathbf{z}}_n) = 0. \tag{3.11}$$

That implies

$$\hat{q}_n \to 0, \qquad \hat{f}_n \to 0, \qquad e_1(\hat{\mathbf{z}}_n) = \hat{f}_n' - \hat{q}_n - V_0' \frac{\hat{f}_n V_0 + \hat{q}_n}{1 + V_0^2} \to 0,$$

$$e_2(\hat{\mathbf{z}}_n) = \frac{\hat{f}_n \cot\theta - \hat{q}_n}{1 + V_0^2} - V_0 \frac{\hat{f}_n + \hat{q}_n \cot\theta}{1 + V_0^2} \to 0,$$

in $L^2(0, \theta_0; \sin\theta)$. Hence, taking into account (3.4), also $\hat{f}_n' \to 0$, $\hat{f}_n \cot\theta \to 0$ in $L^2(0, \theta_0; \sin\theta)$ that is $\|\hat{\mathbf{z}}_n\|_{\mathcal{V}} \to 0$ and this is in contradiction with (3.11) $\quad\square$

3.1. Essential spectrum and noncontrollability

We have shifted our interest from the initial system (3.1)–(3.3) in the natural variables (u, w), representing the displacements of the shell, to the system (3.5), (3.6) in the auxiliary variables (f, q) because this new formulation allows to define the operator $\mathbf{A}$, a Douglis-Nirenberg matrix operator containing a principal part of order zero, and get an analogous situation like that studied in [8]. Indeed the system (3.5) explicitly reads

$$\ddot{f} = (1 + V_0^2)f'' - q'(1 + V_0' + V_0^2 + \nu(1 + V_0 \cot\theta)) + l.o.t.$$

$$\ddot{q} = f'(1 + V_0' + V_0^2 + \nu(1 + V_0 \cot\theta))$$

$$- q\frac{(1 + V_0 \cot\theta)^2 + 2\nu(1 + V_0 \cot\theta)(1 + V_0' + V_0^2) + (1 + V_0' + V_0^2)^2}{(1 + V_0^2)} + l.o.t.$$

We denote by $\sigma^0(\mathbf{A})(\theta, \xi)$ the matrix obtained from the principal part of the matrix $\mathbf{A}$ when we formally put $\frac{d}{d\theta} = i\xi$, it is easy to check that

$$\det \sigma^0(\mathbf{A} + \lambda\mathbf{I}) = 0 \qquad \Rightarrow \qquad \lambda = \frac{(1 - \nu^2)(1 + V_0 \cot \theta)^2}{1 + V_0{}^2}$$

that means the operator $\mathbf{A}$ has a nonempty essential spectrum $\sigma_{\mathrm{ess}}(\mathbf{A})$ given by

$$\sigma_{\mathrm{ess}}(\mathbf{A}) = \lambda(\theta) = \frac{(1 - \nu^2)(1 + V_0 \cot \theta)^2}{1 + V_0{}^2}. \tag{3.12}$$

Let $\bar\theta \in [0, \theta_0)$, we fix $\theta = \bar\theta$ and consider $\bar\lambda = \lambda(\bar\theta)$. Using the Weyl characterization of the essential spectrum we can find a sequence $\bar{\mathbf{z}}_n(\theta)$ such that (see [8])

$$a(\bar{\mathbf{z}}_n, \bar{\mathbf{z}}_n) \to \bar\lambda, \qquad \text{and} \quad b(\bar{\mathbf{z}}_n, \bar{\mathbf{z}}_n) = 1.$$

Moreover since $\bar{\mathbf{z}}_n$ has compact support in a subset $[0, \theta_0 - \delta]$ with δ a suitable positive number, we deduce that, for n great enough, at the boundary $\bar{\mathbf{z}}_n$ vanishes with all its derivatives. Now we can apply the noncontrollability result proved in [8] which we report for the reader in a general form. The problem of boundary exact controllability is the following:
(E.C.) *Given $T > 0$ and an initial state $\boldsymbol{\Phi}^0, \boldsymbol{\Phi}^1$ find the control function $\mathbf{g}$ such that the unique solution $\boldsymbol{\Phi}$ of*

$$\ddot{\boldsymbol{\Phi}} - \mathbf{A}\boldsymbol{\Phi} = 0 \qquad \text{in } \Omega \times (0, T)$$

$$\mathbf{B}\boldsymbol{\Phi} = \mathbf{g} \qquad \text{on } \Sigma = \Gamma \times (0, T) \ ,$$

$$\boldsymbol{\Phi}(0) = \boldsymbol{\Phi}^0 \ , \qquad \dot{\boldsymbol{\Phi}}(0) = \boldsymbol{\Phi}^1 \quad \text{in } \Omega$$

satisfies the following conditions:

$$\boldsymbol{\Phi}(T) = 0 \ , \qquad \dot{\boldsymbol{\Phi}}(T) = 0 \quad \text{in } \Omega \ .$$

Let us remark that $\Gamma \subseteq \partial\Omega$ and $\mathbf{B}$ defines a normal boundary conditions, i.e., there exists a complementary system $\mathbf{C}$ of boundary conditions such that $\{\mathbf{B}, \mathbf{C}\}$ are the reduced Cauchy data of $\mathbf{A}$

For all initial data of the associated homogeneous problem such that $\boldsymbol{\eta}^0 \in \mathcal{V}$ and $\boldsymbol{\eta}^1 \in \mathcal{H}$, we have the existence of a solution $\mathbf{g}$ *if and only if*

$$\left(\int_\Sigma (\mathbf{C}\boldsymbol{\eta})^2 \mathrm{d}\Sigma\right)^{1/2} \geq \mathrm{const.}\|\{\boldsymbol{\eta}^0, \boldsymbol{\eta}^1\}\|_{\mathcal{V} \times \mathcal{H}}. \tag{3.13}$$

As shown in [8] the existence of the essential spectrum violates the inequality (3.13).

In our case we can use the singular sequence $\bar{\mathbf{z}}_n$ to prove the noncontrollability result. Indeed the contradiction is easily shown if we take the following initial data of the homogeneous problem: $\boldsymbol{\eta}^0 = \bar{\mathbf{z}}_n$ and $\boldsymbol{\eta}^1 = \mathbf{0}$. We have $\mathbf{B}\bar{\mathbf{z}}_n = \bar{f}_n(\theta_0) = 0$ and the complementary boundary condition $\mathbf{C}\bar{\mathbf{z}}_n = e_1(\bar{\mathbf{z}}_n) + \nu e_2(\bar{\mathbf{z}}_n)|_{\theta=\theta_0} = 0$, while $a(\bar{\mathbf{z}}_n, \bar{\mathbf{z}}_n) \to \bar\lambda$ and hence $\|\bar{\mathbf{z}}_n\|_{\mathcal{V}} \to \bar c$ where $\bar c$ is a positive constant. We are in a situation which is in contradiction with the necessary and sufficient controllability condition (3.13).

Remark 3.1 We can weaken the hypothesis (3.4) and put, according to (2.6) and without loss of generality, $0 < \alpha < \theta_0$ and

$$V_0 = \begin{cases} -2\tan\theta, & \theta \in [0, \alpha) \\ 0, & \theta \in [\alpha, \theta_0] \end{cases} \tag{3.14}$$

then, for fixed positive ε with $\varepsilon < \min[\alpha, \theta_0 - \alpha]$ and for any function V_ε such that $V_\varepsilon \in \mathcal{B}$ and $V_\varepsilon \equiv V_0$ for $\theta \in [0, \alpha - \varepsilon] \cup [\alpha + \varepsilon, \theta_0]$, we can again carry out the analysis of this section and show hence the existence of the essential spectrum

$$\sigma_{\text{ess}}^\varepsilon = \lambda^\varepsilon(\theta) = \frac{(1 - \nu^2)(1 + V_\varepsilon \cot\theta)^2}{1 + V_\varepsilon^2} \tag{3.15}$$

For $\varepsilon \to 0$ the convergence of the solution to the motion equations of the linearized spherical membranes requires the introduction of suitable functional spaces and it will be the context of a forthcoming paper, on the contrary the convergence of the essential spectrum easily follows from (3.15).

The invariance of the controllability is imputable to the existence of a continuum of minimizers, as shown in [1] and [3], arising in the transition shell-membrane for passing from the reference configuration to the everted one.

Remark 3.2 The existence of the essential spectrum for all the tangent operators does not allow to apply the *return method* as in the case of the controllability of compressible fluids [2] and seems to exclude that the nonlinear terms can lead to the controllability of membrane shells.

Acknowledgement

I am gratefully indebted with G. Geymonat for useful discussions concerning the context of this paper.

The work has been supported in part by the European Community's Human Potential Programme under contract HPRN-CT-2002-00284 SMART-SYSTEMS.

References

[1] N. Ansini, A. Braides and V. Valente, *A multi-scale analysis by Γ-convergence of a shell-membrane transition*, (2004) preprint.

[2] J.M. Coron, *On the controllability of 2-D incompressible perfect fluids*, J. Math. Pures Appl., 75 (1996), pp. 155–188.

[3] G. Geymonat and A. Leger, *Nonlinear spherical caps and associated plate and membrane problems*, J. Elasticity , 57 (1999), pp. 171–200.

[4] G. Geymonat, P. Loreti and V. Valente, *Spectral problem for thin shells and exact controllability*, in Spectral Analysis of Complex Structures, Travaux en Cours n. 49, Hermann, Paris, 1995, pp. 35–57.

[5] G. Geymonat, P. Loreti and V. Valente, *Exact controllability of a shallow shell model*, International Series of Numerical Mathematics, Birkhäuser Verlag Basel, 107 (1992), pp. 85–97.

[6] G. Geymonat, M. Rosati and V. Valente, *Numerical analysis for eversion in elastic spherical caps equilibrium*, Computer Methods in Appl. Mech. Eng., 75 (1989), pp. 39–52.

[7] G. Geymonat, M. Rosati, V. Valente, The maximal thickness for everted equilibrium shapes of elastic spherical caps, *Calcolo* **27** (1990) 103–125.

[8] G. Geymonat and V. Valente, *A noncontrollability result for systems of mixed order*, SIAM J. Control Optim., 39 (2000), pp. 661–672.

[9] K. Hamdache and E. Sanchez-Palencia, *Homogénéisation d'un problème modèle en théorie des coques membranaires*, C.R. Acad. Sci., Serie I, 325 (1997), pp. 683–688.

[10] I. Lasiecka, R. Triggiani and V. Valente, *Uniform stabilization of spherical shells by boundary dissipation*, Ad. Diff. Eq., 1 (1996), pp. 635–674.

[11] J.L. Lions, Equations Differentielles Operationnelles et Problèmes aux Limites, Springer Verlag, 1961.

[12] B. Miara and V. Valente, *Exact controllability of a Koiter shell by a boundary action*, Journal of Elasticity, 52 (1999), pp. 267–287.

[13] P. Podio-Guidugli, M. Rosati, A. Schiaffino and V. Valente, *Equilibrium of an elastic spherical cap pulled at the rim*, SIAM J. Math. Anal., 20 (1989), pp. 643–663.

V. Valente
Istituto per le Applicazioni del Calcolo "M. Picone"
Viale del Policlinico 137
I-00161 Roma, Italy
e-mail: `valente@iac.rm.cnr.it`

Progress in Nonlinear Differential Equations
and Their Applications, Vol. 63, 433–451

The Porous Medium Equation.
New Contractivity Results

J.L. Vázquez

Dedicated to Prof. Haim Brezis on the occasion of his 60th birthday

Abstract. We review some lines of recent research in the theory of the porous medium equation. We then proceed to discuss the question of contractivity with respect to the Wasserstein metrics: we show contractivity in one space dimension in all distances d_p, $1 \leq p \leq \infty$, and show a negative result for the d_∞ metric in several dimensions. We end with a list of problems.

Mathematics Subject Classification (2000). 35K55, 35K65.

Keywords. Porous Medium Equation, contractivity, Wasserstein metrics.

1. Introduction

In the year 1976 I came under the influence of Haim Brezis during one of his first stays in Madrid. After years of formation in rather abstract mathematics, I was struck by his simple and clear approach to applied analysis, and I learnt that theory has to be tied to the detailed attention to your equation, its motivation and its estimates. This approach has influenced my doctoral thesis and every step of my research for almost 30 years.

Among the readings of those early years of my research in nonlinear PDEs, I specially remember the book *"Opérateurs maximaux monotones. . . "* [Br73] and the beautiful paper *"A semilinear equation in $L^1(R^N)$"* [BBC], recently appeared at the time. It led to my lasting interest in the solution of nonlinear elliptic equations of the form

$$A(u) + b(u) = f,$$

where A is an elliptic operator, possibly the Laplacian operator, $Au = -\Delta u$, possibly a more general object, like the p-Laplacian operator; b is a monotone function, or, more generally, a maximal monotone graph in the spirit of [Br73];

f is locally integrable function, mostly $f \in L^1(\Omega)$, where Ω is a bounded open set or $\mathbb{R}^N$, as in [BBC].

Brezis' book also led to a second lasting interest, the generation of semigroups by means of implicit time discretization, in the spirit of Crandall and Liggett's famous work [CL71], where the emphasis is led on the presence of contractive operators in Banach spaces. I have concentrated a large part of my research on that issue in the case of the so-called Porous Medium Equation (PME), $u_t = \Delta u^m$, or more generally, the Filtration Equation, $u_t = \Delta \Phi(u)$, where Φ is again a maximal monotone graph.

There was a third strand in those early years of my research, the study of equations with singularities and measures as data, that was in my case motivated by Brezis' work on the Thomas-Fermi problem, cf. [BB03], and the paper of Brezis and Véron [BrV80]. I contributed in my thesis the study of the equation $-\Delta u + e^u = \mu$, where μ is a measure that has a Dirac mass as a singular part, [V83c]; the results seem to have had an influence on Brezis-Marcus-Ponce [BMP]. But later on I drifted again to evolution models of the form $u_t = \Delta \Phi(u) \pm f(u)$; many of these models can be found in the book with V. Galaktionov [GV03]. I hope this will show sufficiently how much my later work has been influenced by the early period of my contact with Haim Brezis.

2. Progress in the PME

Turning now the attention to the PME, let me recall some salient advances in the theory of this equation that have happened in my surroundings, and prepare the results that will be proved in this paper:

(i) The study of the regularity of solutions of the PME and interfaces in one space dimension received a great impulse in the 1980's. Assuming that $x \in \mathbb{R}$ and the initial function $u(x,0)$ is nonnegative, continuous and compactly supported, and concentrating our attention to be specific on the right-hand interface, $x = s(t)$, it was proved that a finite waiting time t_* may appear, whose existence is characterized in [V83b]; the interface moves ($s'(t) > 0$) for $t > t_*$ [CF79]; the interface is C^∞ regular for $t > t_*$ [AV87]; lack of regularity is only possible at $t = t_*$ in the form of a corner point [ACV]. Even more, for compactly supported solutions, Angenent proves that the moving interfaces are analytic functions.

In this research an important role is played by the pressure defined as $p = cu^{m-1}$ with $c = m/(m-1)$. It satisfies the quadratic equation $p_t = (m-1)p\Delta p + |\nabla p|^2$. It is proved that p is a C^∞ function in the positivity set and up to the moving boundary, but not near the corner point.

(ii) Progress on the regularity of interfaces in several space dimensions has been much slower. After the proof of Hölder continuity of solutions and interfaces established in [CF80], the paper [CVW] showed that solutions of the N-dimensional PME with compactly supported initial data become regular for large times in the sense that there exists a time T such that for $t > T$ the power u^{m-1} (the pressure) of the solution becomes Lipschitz continuous and the free boundary is also a Lipschitz continuous hypersurface in space-time. This regularity has been improved to $C^{1,\alpha}$ by Caffarelli and Wolanski [CW90] and to C^∞ by Koch [K99].

(iii) The latter result was used by Lee and the author to give a very accurate description of the asymptotic behavior of the solutions as $t \to \infty$, [LV03], improving on the classical paper on asymptotic convergence by Friedman and Kamin [FK80]; see in this respect the survey paper [V03], which includes the fast diffusion case in the "good range" $m_c < m < 1$ with $m_c = (N-2)_+/N$. Substantial progress has been done in the asymptotics for $m \le m_c$ but there are still many open questions. Let us quote as a very recent contribution the work of Daskalopoulos and Hamilton [DH04] on logarithmic diffusion. A very important question in the study of asymptotics is establishing convergence rates: for the PME and Fast Diffusion equations, this has been done using the entropy dissipation method, where we must mention the work of Carrillo and Toscani [CT00] and Del Pino-Dolbeault [DP02] and many later ones. The author's contribution is in [CV03] where further references can be found.

(iv) The regularity situation turned out to be different for small times. Indeed, a new phenomenon was described that had a deep impact on the theory, namely the *focusing phenomenon*, first described by Aronson and Graveleau [AG93] (after the preliminary work [G72]). Simply stated, when the problem is posed in several dimensions and initial data $u_0(x)$ vanish in a ball $B_r(0)$ and are positive outside, the free boundary advances to fill the "hole" with increasing speed, so that in the last moment when the hole is filled the speed is infinite and the pressure ceases to be Lipschitz continuous. This lack of regularity has motivated a lot of research that we will review in Section 6; it is also the motivation for our new results in that section.

There have been many other important developments in the theory of the PME, whose description is out of the scope of these notes. I would like to mention some lines that have specially interested me: the theory of viscosity solutions of Caffarelli and the author [CV99], and the theory of singular solutions called *extended continuous solutions* for the fast diffusion range $m_c < m < 1$ developed with Chasseigne, [ChV02].

3. Contractivity and the Porous Medium Equation

In developing the theory of weak solutions and showing well-posedness and regularity for the standard parabolic equations, a key role has been played by different contractivity properties. Thus, the heat semigroup is contractive for all the Lebesgue norms, L^p, $1 \le p \le \infty$, when posed in the whole space $\mathbb{R}^N$ or in a bounded domain $\Omega \subset \mathbb{R}^N$ with either Dirichlet or Neumann zero boundary conditions.

3.1. The theory of the PME developed as a consequence of the results of [BBC] and [CL71] is based on the fact that the flow generated by the Porous Medium Equation

$$u_t = \Delta(u^m), \tag{3.1}$$

posed in $Q = \mathbb{R}^N \times (0, T)$, $0 < T \leq \infty$, with initial data

$$u(x, t) = u_0(x), \quad x \in \mathbb{R}^N, \tag{3.2}$$

generates a semigroup of contractions in the functional space $L^1(\mathbb{R}^N)$. This property is true for all exponents $m > 0$; it does not extend however to the range $m \leq 0$ because it can be shown that the solutions disappear through an initial layer [V92b] unless $N = 1$ and $m > -1$ or $N = 2$ and $m = 0$. There is a well-developed existence and uniqueness theory for the PME that says that we may take as u_0 any measurable and locally integrable function with some growth conditions at infinity if $m \geq 1$. A locally finite Radon measure μ with growth condition $\mu(B_R(0)) = O(R^\alpha)$, $\alpha = N + (2/(m-1))$ is also admissible. In the range $1 > m \geq (N-2)/N$ the locally integrable function can be replaced by a (locally finite) Radon measure. A unique weak solution is then obtained for the Cauchy problem and the maps $S_t = S_t(\Phi) : u_0 \mapsto u(t)$ generate a semigroup in a suitable function space (we write at times $u(t)$ instead of $u(x, t)$ without fear of confusion). Weak solutions are continuous functions.

The main qualitative difference between the ranges $m > 1$ and $m \leq 1$ is the property of finite propagation that holds only for $m > 1$. In that case, solutions with compactly supported data have the same property for all $t > 0$ and the separation between the sets $\{(x, t) : u(x, t) > 0\}$ and $\{(x, t) : u(x, t) = 0\}$ forms a free boundary whose location has to be determined.

Contrary to what happens to the Heat Equations, the PME is only contractive in L^1, though the flow is bounded in L^p for all $1 \leq p \leq \infty$, cf. [Ar86, Be76, V92].

3.2. Another contractive property occurs when the PME is posed in the framework of $H^{-1}(\Omega)$. Brezis [Br71] proved that when Φ is a maximal monotone graph, the operator $Au = -\Delta\Phi(u)$ is a subdifferential, hence maximal monotone, so that it generates a semigroup of contractions. This applies for instance when Φ is a power, $\Phi(u) = |u|^{m-1}u$, like in the PME case.

3.3. In the sequel we will study a further scenario, the contractivity properties of the PME with respect to Wasserstein distances for the Cauchy problem posed in $Q = \mathbb{R}^N \times (0, T)$ for some $T > 0$, possibly $T = \infty$. The space dimension is arbitrary, $N \geq 1$. We will consider only nonnegative data and solutions, $u_0, u \geq 0$, which is a standard assumption in diffusion theory. Otherwise, we are interested in putting minimal restrictions on the data u_0.

4. The Wasserstein metrics

Nowadays, the influence of probability and measure theory is strong in PDEs. One way it is felt is through the interest in the Wasserstein metrics, which arise naturally in the consideration of optimal transportation problems. Attention to the PME in this connection is associated with the famed paper by Otto [Ot01]. The Wasserstein metrics are defined on the set probability measures, $\mathcal{P}(\mathbb{R}^N)$ as follows: for any number $p > 0$ we define the Wasserstein distance d_p between two probability measures μ_1, μ_2 by the formula

$$(d_p(\mu_1, \mu_2))^p = \inf_{\pi \in \Pi} \int_{\mathbb{R}^N \times \mathbb{R}^N} |x - y|^p \, d\pi(x, y), \tag{4.3}$$

where $\Pi = \Pi(\mu_1, \mu_2)$ is the set of all transport plans that move the measure μ_1 into μ_2. Technically, this means that π is a probability measure on the product space $\mathbb{R}^N \times \mathbb{R}^N$ that has marginals μ_1 and μ_2. It can be proved that we may use transport functions $y = T(x)$ instead of transport plans (this is Monge's version of the transportation problem). In principle, for any two probability measures, the infimum may be infinite. But when $1 \leq p < \infty$, d_p defines a metric on the set $\mathcal{P}_p$ of probability measures with finite p-moments, $\int |x|^p d\mu < \infty$. A convenient reference for this topic is Villani's book [Vi03].

The metric d_∞ plays an important role in controlling the location of free boundaries, an important issue in porous medium flow. It has an independent definition as

$$d_\infty(\mu_1, \mu_2) = \inf_{\pi \in \Pi} d_{\pi,\infty}(\mu_1, \mu_2),$$
$$d_{\pi,\infty}(\mu_1, \mu_2) = \sup\{|x - y| : (x, y) \in \text{support}\,(\pi)\}.$$

In other words, $d_{\pi,\infty}(\mu_1, \mu_2)$ is the maximal distance incurred by the transport plan π, i.e., the supremum of the distances $|x - y|$ such that $\pi(A) > 0$ on all small neighborhoods A of (x, y). We call this metric the *maximal transport distance* between μ_1 and μ_2; d_∞ can be also defined as the limit of the metrics d_p as $p \to \infty$. See in this respect [Mc].

For the applications in more general diffusion theories and also as an essential step of our proofs below, we need to extend the scope of these metrics to deal with data μ_1 and μ_2 are nonnegative Radon measures, on the condition that they should have the same total mass

$$\mu_1(\mathbb{R}^N) = \mu_2(\mathbb{R}^N), \tag{4.4}$$

positive but not necessarily unity.

We want to discuss the contractivity of the porous medium flow posed in the whole space with respect to the Wasserstein metrics. We will present two different results: on the one hand, our original ideas for the proof of the contractivity of the flow of equation (3.1) posed in $\mathbb{R}$, for all norms W_p with $1 < p \leq \infty$. On the other hand, as a brandnew contribution, the proof of the failure of the contractivity in d_∞ in all dimensions $N \geq 2$. The result extends to d_p for sufficiently large p.

5. Contractivity in One Space dimension

The contractivity of the PME flow for the Wasserstein metrics is well established in one space dimension. As we will explain, the d_∞ contractivity of the PME flow in $\mathbb{R}$ is a consequence of the *Shifting Comparison* results proved by the author in 1983, see [V83]. However, the important optimal transportation connection has been observed only recently, when this aspect came to the forefront. The d_∞ contractivity in the framework was rediscovered in [CT03], and the contractivity in all d_p's in $\mathbb{R}$ is proved by [CGT]; the proof uses the special characterization of the Wasserstein metrics in 1 dimension together with the contractivity properties of the p-Laplacian equation, that appears associated to the inverse distribution function. We recall next the whole one-dimensional proof, and we use the opportunity formulate the results for general data.

5.1. Shifting Comparison

We review the results of our paper [V83], where we described the technique of Shifting comparison, that we formulate here in full generality.

Given two initial distributions μ_1 and μ_2, nonnegative Radon measures, not necessarily with the same total mass, we solve the 1-D Porous Medium Equation to produce solutions $u_1(x,t)$, $u_2(x,t)$, which are continuous and bounded functions. We introduce the distribution functions

$$U_1(x,t) = \int_{-\infty}^{x} u_1(s,t)\,dx, \quad U_2(x,t) = \int_{-\infty}^{x} u_1(s,t)\,dx. \tag{5.1}$$

In the same way, we define at $t = 0$

$$U_1(x,0) = \mu_1((-\infty, x]), \quad U_2(x,0) = \mu_2((-\infty, x]).$$

Proposition 1. *Under the above assumptions, the relation*

$$U_1(x,0) \leq U_2(x,0) \quad \forall x \tag{5.2}$$

implies a similar ordering for all $t > 0$, i.e.,

$$U_1(x,t) \leq U_2(x,t) \quad \forall x. \tag{5.3}$$

In the case where μ_1 and μ_2 are absolutely continuous measures, given by integrable densities $d\mu_i = u_{0i}dx$, it is proved in Lemma 2.1 of [V83] and is called Comparison by Shifting, For the general case of measures, we only need to pass to the limit on a sequence of approximations using the continuous dependence of the solutions on the initial data, [BC81], [Pi83]. But it maybe better to remark that the integrals $U(x,t)$ satisfy the equation $U_t = (U_x^m)_x$. This is an equation with gradient-dependent diffusivity, usually written in the standard form

$$U_t = (|U_x|^{m-1}U_x)_x, \tag{5.4}$$

but here $U_x = u \geq 0$, so there is no difference. It is usually called a p-Laplacian equation; we will call it here m-Laplacian equation, since the letter p is already used for the index of the metric). In this setting, Shifting Comparison is just the standard maximum principle for Equation (5.4), and it is known that it holds for all bounded data U_{0i} satisfying (5.2), on the condition that the solutions conserve the relation at infinity, i.e., $U_1(\infty, t) \leq U_2(\infty, t)$, which means that

$$\int u_1(s,t)ds \leq \int u_2(s,t)\,ds,$$

a relation that follows from conservation of mass once it holds for $t = 0$ by assumption.

5.1. Proof of d_∞ contractivity

Let us examine now the relation of this comparison with transport techniques. To being with, we may say that Shifting Comparison is a transport technique, since given any two compactly supported measures with $\mu_1(\mathbb{R}) \leq \mu_2(\mathbb{R})$, we may always obtain a relation of the type (5.2) after shifting the second mass distribution to the left in a rigid way (since the equation is invariant under translations). The result says that a comparison of distribution functions holds then between the shifted u_1 and u_2 for all times $t > 0$. To be precise, let

$$c(\mu_1, \mu_2) = \min\{c \in \mathbb{R}, \ U_1(x, 0) \leq U_2(x + c, 0) \quad \forall x\}$$

be the minimum translation to the left that μ_2 has to undergo for its distribution function to be larger or equal than that of μ_1. Assume that such a distance is finite. Since the equation is invariant under translations, Proposition 1 implies that for all $t > 0$

$$U_1(x, t) \leq U_2(x + c(\mu_1, \mu_2), t) \ \forall x.$$

Reversing the roles of μ_1 and μ_2 we obtain the minimal translation distance in the other direction, $c(\mu_2, \mu_1)$, that may be finite only if $\mu_1(\mathbb{R}) \geq \mu_2(\mathbb{R})$.

Assuming that the measures have *the same total mass*, $\mu_1(\mathbb{R}) = \mu_2(\mathbb{R})$, it is now an easy matter to check that the maximal transport distance between μ_1 and μ_2 is given by

$$d_\infty(\mu_1, \mu_2) = \max\{c(\mu_1, \mu_2), c(\mu_2, \mu_1)\}.$$

and is finite if both $c(\mu_1, \mu_2)$ and $c(\mu_1, \mu_2)$ are. This may serve as a definition of $d_\infty(\mu_1, \mu_2)$, even if μ_1 and μ_2 do not have compact support.

Theorem 2. *Let μ_1 and μ_2 be finite nonnegative Radon measures on the line and assume that $\mu_1(\mathbb{R}) = \mu_2(\mathbb{R})$ and $d_\infty(\mu_1, \mu_2)$ is finite. Let $u_i(x, t)$ the continuous weak solution of the PME with initial data μ_i. Then for every $t_2 > t_1 > 0$*

$$d_\infty(u_1(\cdot, t_2), u_2(\cdot, t_2)) \leq d_\infty(u_1(\cdot, t_1), u_2(\cdot, t_1)) \leq d_\infty(\mu_1, \mu_2). \qquad (5.5)$$

Proof. Since we have shown that the quantities $c(\mu_1, \mu_2)$ and $c(\mu_1, \mu_2)$ are monotonically non-increasing in time, so does the metric d_∞. When the data are not compactly supported we may apply an approximation process, so that the properties of the left and right displacements will be conserved. $\qquad \square$

Remark. Note that we are not assuming that the initial measures have compact support. Actually, by taking limits we may extend the result to any two locally finite nonnegative measures, with finite or infinite total mass, satisfying the conditions for existence.

Extension. The same result applies to the solutions of the Filtration Equation $u_t = \Phi(u)_{xx}$ on the condition that it generates a semigroup that conserves mass. This happens for instance when Φ is continuous, nondecreasing with $\Phi(0) = 0$ and $\Phi(\infty) = \infty$.

5.2. The cases $1 \le p < \infty$

The analysis of this case has been performed by J.A. Carrillo in [C04] and the key observation goes as follows. For positive data the solutions are positive everywhere and we can invert the distribution functions $z = U_i(x,t)$ for fixed t and obtain an inverse distribution $x = U_i^{-1}(z)$. In terms of this function the expression for the distance d_p at time t is just the expression

$$\int_0^M |x_1 - x_2|^p \, dz,$$

where $x_i(z,t) = U_i^{-1}(z,t)$. Now, it is well known that the "inverse" of the m-Laplacian equation $U_t = (U_x^m)_x$ is another equation of the same type, precisely

$$x_t = (-(x_z)^{-m})_z = x_z^{-m-1} x_{zz}. \tag{5.6}$$

(See a detailed discussion of this issue in [V03b] where it is related to the Bäcklund transform.) The diffusivity, $D = x_z^{-m-1}$, is now singular at $x_z = 0$. The proof of monotonicity of the d_p distance is just reduced to proving that monotone solutions of this equation are well defined an have the property of accretivity in all L^p spaces, $1 \le p \le \infty$. This property of p-Laplacian equations is well known, cf. [Be76].

We refer to [C04] for complete details of the calculation of contractivity of the d_p metrics for the porous medium equation. Asymptotic convergences with decay rates are obtained.

Remarks.

1) As in the case $p = \infty$, we can obtain one-directional versions of the distance that are also monotone in time. They are given by the integrals

$$\int_0^M |(x_1 - x_2)_+|^p \, dz.$$

A more general result holds.

Theorem 3. *Let $u_i(x,t)$, $i = 1,2$, be continuous and nonnegative weak solutions of the PME with initial data μ_i, where μ_1 and μ_2 are finite nonnegative Radon measures on the line with $\mu_1(\mathbb{R}) = \mu_2(\mathbb{R})$. Let F be a convex, continuous and nonnegative function. Then, the quantity*

$$J_t^F(u_1, u_2) := \int_0^M F(x_1(z,t) - x_2(z,t)) \, dz \tag{5.7}$$

is monotone non-increasing in time.

2) We point our that other uses of the inverse m-Laplacian equation (5.6) and its connection with the so-called Bäcklund transform can be found in [V03b]. In [BV04] it is used in a problem of image processing.

3) A generalization of the results of this section to filtration equations of the form $u_t = \Phi(u)_{xx}$ is immediate if Φ satisfies assumptions as mentioned above, cf. [BC81].

5.3. Interface location control

The control of the d_∞ metric, i.e., the maximal transport distance, has an immediate consequence on the control of the location of the free boundaries when $m > 1$, since obviously the distance of two mass distributions is not less than the distance between their supports,

$$
\begin{aligned}
d_\infty(\mu_1, \mu_2) &\geq \sup\{d(x, K_2), x \in K_1\}, \\
d_\infty(\mu_1, \mu_2) &\geq \sup\{d(x, K_1), x \in K_2\},
\end{aligned}
\tag{5.8}
$$

where K_i is the support of μ_i. In this way we can derive an interesting asymptotic consequence: if u is a solution with initial data supported in the interval $[-a, a]$ and having mass $M = \int u_0\, dx$, we may use shifting comparison, i.e., d_∞ contraction, with respect to the Barenblatt solution $U(x, t; M)$ of the same mass to localize the free boundaries of u at time t with an error or at most $2a$ of the explicit free boundaries of U, see the very precise result in [V83]. An analogous result is maybe true, but remains unproven in several dimensions.

6. The contractivity question in several space dimensions

In view of the results of the preceding section, the question is posed whether the PME flow is also contractive with respect to the Wasserstein d_p distances when the space dimension $N \geq 2$.

There are some positive results. Thus, Carrillo, McCann and Villani [CMV] have recently proved the d_2-contractivity in all space dimensions, developing ideas of Otto's seminal paper on gradient flows [Ot01], see also Agueh's [Ag]. McCann has given a proof that the result is true for all distances d_p, $1 \leq p \leq \infty$ for the heat equation $u_t = \Delta u$, which is the limit case $m = 1$ of the PME, see [C04].

In view of these facts, the author was convinced of the positive result for all p's and tried hard to prove it during the summer of 2004. Sadly, the result is false, at least for d_∞ and d_p with large p.

Theorem 4. *The PME flow is not d_∞- contractive for any dimension $N \geq 2$ and any exponent $m > 1$. It is not even exponentially increasing. The same is true for d_p if $p > p(m, N)$ for some finite $p(m, N) > N$ that we explicitly estimate.*

By exponentially increasing we mean that there exist constants C and ω (not depending on the data) such that for any two solutions $u_1(t), u_2(t) \geq 0$ with the same mass and finite p-moments, we have

$$
d_p(u_1(t), u_2(t)) \leq C\, d_p(u_1(0), u_2(0))e^{\omega t}.
\tag{6.1}
$$

The proof relies on the construction of a counterexample that shows that the d_∞ may grow in time, even blow up in a sense in finite time, cf. Lemmas 7 and 8. The phenomenon that underlies our proof is the famous *focusing phenomenon* mentioned in Section 2. We give the full details in six steps. The first three review the needed facts about focusing solutions.

I. The focusing solutions revisited

The paper [AG93] considers self-similar solutions of the PME that we can write in the form

$$U(x,t) = (T-t)^\alpha F(x(T-t)^{-\beta}), \tag{6.2}$$

with the compatibility condition $(m-1)\alpha = 2\beta - 1$. It assumes that the profiles $F(\eta)$ $(\eta = x(T-t)^{-\beta})$ are radially symmetric, $F(\eta) = F(r)$, $r = |\eta|$, and examines the behavior of the possible solutions of the ODE that must be satisfied by the profiles $F(r)$:

$$(F^m)'' + \frac{N-1}{r}(F^m)' + \alpha F - \beta r F' = 0,$$

The analysis in a suitable phase plane allows the authors to show that there exists a precise value of the parameter β, let us call it β_* (it depends on m and N), such that a corresponding profile F can be found with the following properties:

(i) F is continuous, nonnegative and radially symmetric: $F = F(r)$, $r = |\eta|$;

(ii) F vanishes for $0 < r < a$ and is C^∞ and strictly increasing for $r > a$; $U(x,t)$ given by (6.2) is a weak solution of the PME, and it is even a classical solution in the positivity set, i.e., for $|x| > a(T-t)^{\beta_*}$.

Actually, paper [AG93] performs all computations in terms of the pressure variable $p = mu^{m-1}/(m-1)$, which has a self-similar formula

$$P(x,t) = (T-t)^{2\beta-1}G(x(T-t)^{-\beta}), \tag{6.3}$$

with $G = (m/(m-1))F^{m-1}$. The following limit behavior is also established:

(iii) There exists $c > 0$ such that $G(r)r^{-\varepsilon} \to c$ as $r \to \infty$ if $\varepsilon = (2\beta_* - 1)/\beta_*$.

As a consequence of this property and formula (6.3), the limit profile of the focusing solution is known:

$$\lim_{t\to T} P(x,t) = c|x|^\varepsilon,$$

We call these profiles found by Aronson and Graveleau the *AG profiles*. We remark that for all $N \geq 1$ a one-parameter family of focusing solutions is obtained; they can be normalized by fixing $a = 1$, or to any other positive value. We indicate the family when needed with the notation $G(\eta; a)$.

The main fact proved in [AG93] about these special solutions is the estimate on the value of the exponent β_* and the regularity of G.

Proposition 5. *For $N = 1$ we have $\beta_* = 1$ and $P(x,t)$ is Lipschitz continuous. On the contrary, for $N \geq 2$ it turns out that $1/2 < \beta_*(m,N) < 1$ and $P(x,t)$ is only locally Hölder continuous for some Hölder exponent ε less than 1.*

Let us mention that for $N = 1$ it is well known that a solution with these characteristics corresponds to $\beta_* = 1$, and the solution is in fact the travelling wave, which in terms of the pressure variable says

$$P(x,t) = \frac{m}{m-1}U(x,t)^{m-1} = c(x - c(T-t))_+$$

with a free parameter $c = a > 0$. For $N \geq 2$ the exponent β_* does not come from a priori physical or dimensional considerations and is called an *anomalous exponent*; in Zel'dovich's words we have a *self-similarity of the second kind*, a topic that is beautifully explored in Barenblatt's book [BV96].

It follows from the proposition that $\varepsilon \in (0,1)$ for $N \geq 2$, hence P is not Lipschitz continuous near $x = 0$, $t = T$. It is further proved in [AGV] that $\beta_*(m, N) \to 1/2$ if $m \to \infty$, while it tends to 1 as $m \to 1$, always for $N \geq 2$. The monotonicity of β_* as a function of m has been subsequently proved in [ABH]. We will explain next why the value of β_* and the regularity of $G \sim F^{m-1}$ matter to us.

II. Propagation and hole filling

Let us examine some of the remarkable consequences of this result when seen from the point of view of mass transport. It is well known that the PME can be viewed as a mass conservation law for a density u transported with speed V in the usual form

$$u_t + \nabla \cdot (u V) = 0. \tag{6.4}$$

In order for u to satisfy the porous medium equation, the *particle speed* must be defined as $V(x,t) = -m u^{m-1} \nabla u = -\nabla p$, which is known to be a form of the famous Darcy law of flow propagation, cf. [Ar86] or [V92].

The PME has finite speed of propagation, a fact that has a clear interpretation when we apply the equation to model groundwater infiltration as in [Bo03], or gas flow in porous media, as in [Lei45], [Mu37]. But, contrary to a popular misconception, that does not mean that the pointwise speed V of the flow has to be finite everywhere. The boundedness of the particle speed is true in one space dimension but not necessarily in two or more. Let see how this happens in our example. Its free boundary (in other words, the front that separates the empty region from the wet region when we use groundwater infiltration imagery), is given by the surface Γ with equation

$$|x| = a(T - t)^{\beta_*}. \tag{6.5}$$

The advance speed of this surface in time is given by the formula

$$V_f(t) = \beta_* a(T - t)^{\beta_* - 1}. \tag{6.6}$$

Note that:

(i) the speed V_f can be calculated both geometrically, as the value of the normal front speed, and also dynamically, as the limit value of the internal particle speed $V(x,t)$ as $(x,t) \to \Gamma$. Internal means defined in the wet region, where $V > 0$ and the solution is C^∞; there, V is given by Darcy's law. This version of Darcy's law is rigorously proved in the pointwise sense for the focusing solutions;

(ii) the front advances towards the origin and it reaches it precisely at $t = T$;

(iii) we come now to a key point in our argument: if $\beta_* < 1$ the speed V_f tends to infinity as $t \to T$. We conclude that the focusing solutions have a diverging front speed as they approach the focusing time;

(iv) on the contrary, the speed is finite in $Q = \mathbb{R}^N \times (0, T)$ away from a neighborhood of $(x = 0, t = T)$.

III. Asymptotic convergence

The properties of the focusing solutions and the corresponding exponents have been studied by a number of authors in the radially symmetric case, like [AV95] and [AA95]. Moreover, the results have been extended to nonradial solutions in [AA01], [AABL], and they have been studied for other equations, like in [AGV] and [AA03].

The contribution of paper [AA95] deals with the problem of deciding how generic is the focusing behaviour described by the AG solutions. The answer turns out to be positive for solutions of the PME with radially symmetric initial data.

Proposition 6. *Let $u_0(x)$ be a nonnegative, radially symmetric, continuous and compactly supported initial function, which is positive for $r_1 < |x| < r_2$ and zero otherwise. Let $u(x, t)$ the corresponding solution of the PME. Then there exist $T > 0$ and $a > 0$ such that, as $t \to T$ (with $t < T$), $u(r, t)$ tends to the self-similar solution (6.2) with parameter a in the following sense:*

(i) *if $p(r, t)$ is the pressure of the solution, then for each fixed $\eta = x(T - t)^{-\beta^*} \in [0, \infty)$,*

$$\lim_{t \to T} p(\eta(T - t)^{\beta_*}, t)\,(T - t)^{-2\beta_* + 1} = G(\eta; a). \tag{6.7}$$

(ii) *The inner interface converges: if $R(t) = \sup\{|x| : p(x, t) = 0\}$ is the radius of the hole of p at time t, then*

$$\lim_{t \to T} R(t)/(T - t)^{\beta_*} = a. \tag{6.8}$$

(We have changed the notations and statement form of [AA95] for convenience.)

IV. Blow-up of the distance ratio

We proceed now with the proof of non-contractivity for the d_∞ distance. We take a solution $u_1(x, t)$ with data as in Proposition 6. We assume further that the pressure $p_1(x, 0)$ is positive and smooth in the annulus $r_1 < |x| < r_2$, and zero otherwise, with nonzero and finite radial derivative at the endpoints, $p_{1,r}(r_1, 0), p_{1,r}(r_2, 0) \neq 0$; this condition is imposed to ensure moving free boundaries, hence regular, from the start. Let T be the extinction time of u_1 and let the mass $M = \int u_1(x, t)\,dx$ be finite and constant. Let finally $|x| = r_1(t)$ be the internal free boundary, that converges to zero as $t \to T$. We take a small time increment $h > 0$ and consider as second solution

$$u_2(x, t) = u_1(x, t + h),$$

which is defined for $0 \leq t \leq T - h$. We now consider in $I_h = [0, T - h)$ the following distance between the two mass distributions

$$d_h(t) = d_\infty(u_1(t), u_2(t)), \tag{6.9}$$

which is defined for $0 \leq t < T - h$.

Lemma 7. *As $h \to 0$ we have*

$$\sup_{t \in I_h} \frac{d_h(t)}{d_h(0)} \geq C\, h^{\beta - 1}. \tag{6.10}$$

for a constant C depending only on $u_1(0)$.

Proof. (i) We first examine the behavior of $d_h(0)$ for small $h > 0$. In view of the finite speed of the initial function and the interpretation of the PME as a mass transport equation, we may transport the mass distribution $u_1(0)$ into the mass distribution $u_2(0) = u_1(h)$ using the trajectories $X(x,t)$, defined by

$$\frac{dX}{dt} = V(X,t), \quad X(y,0) = x$$

for all x such that $r_1 < |x| < r_2$. This mass transport approach is known as the Lagrangian formulation of the PME and is perfectly described in several references, like [GMS], [SV96], [Sh01]; cf. the monograph [MPS] for the topic of Eulerian versus Lagrangian systems of coordinates in Continuum Mechanics.

Since the solution u_1 is regular and has finite velocity for a small times by known local regularity results, it follows that there is a constant C such that the cost of transportation along this plan is bounded by Ch, where C is an upper bound for the speed $|V|$ in $0 \leq t \leq h$. If we take the infimum among all admissible plans, this quantity may only go down, hence,

$$d_h(0) \leq Ch.$$

(ii) The second part of the proof follows easily from the focusing geometry: any transportation map from the continuous distribution $u_1(t)$ to $u_2(t)$ has to transport all elements of mass in the support of $u_1(t)$, which lie outside the ball of radius $r_1(t)$, into all the elements of the support of $u_2(t)$, which are spread in the complement of the ball of radius $r_2(t) = r_1(t + h) < r_1(t)$. By virtue of the asymptotic behavior of Proposition 6, we have

$$d_h(t) \geq r_1(t) - r_2(t) \sim a((T - t)^\beta - (T - t - h)^\beta)$$

which behaves like ch^β as $t \to T - h$. $\qquad\square$

This result is enough to show that the PME flow cannot be contractive in the d_∞ distance (taking h small). Since T is finite and determined by $u_1(r,0)$, it even proves the part about exponential growth in d_∞-distance contained in Theorem 4.

V. Scaling argument

In order to show the extension of the non-contraction result, we eliminate the possible objection that the result has been obtained only for small initial distances $d(t)$. We now define the scaled functions

$$u_1^h(x,t) = \frac{1}{h^q} u_1(hx, ht), \quad u_2^h(x,t) = \frac{1}{h^q} u_2(hx, ht),$$

with $q = 1/(m-1)$. We have $u_2^h(x,t) = u_1^h(x, t+1)$. We also have scaled blow-up times $T_{1,h} = T/h$, $T_{2,h} = (T/h) - 1$, as well as scaled hole radius

$$r_1^h(t) = \frac{1}{h} r_1(ht).$$

Besides,

$$d_\infty(u_1^h(t), u_2^h(t)) = \frac{1}{h} d_\infty(u_1(ht), u_2(ht)).$$

It follows that the sequence $d_\infty(u_1^h(0), u_2^h(0))$ is bounded, while for $h \to 0$

$$d_\infty(u_1^h(T_{2,h}), u_2^h(T_{2,h})) \to \infty.$$

VI. The case $p < \infty$

We repeat the proof with a slight variation. Using the distance

$$d_{p,h}(t) = d_p(u_1(t), u_2(t)),$$

we obtain the following result.

Lemma 8. *As $h \to 0$ we have*

$$\sup_{t \in I_h} \frac{d_{p,h}(t)}{d_{p,h}(0)} \geq C\, h^{\mu-1}. \tag{6.11}$$

for a constant C depending only on $u_1(0)$; $\mu = \mu(m, N, p) > 0$ is given below as an explicit function of β_*.

Proof. Following the same outline, one of the estimates is immediate. Indeed, we have the standard relation between distances for $p = \infty$ and $p < \infty$ when defined in bounded sets: $d_p(f_1, f_2) \leq C_1 d_\infty(f_1, f_2)$ where C_1 depends on the length of the support of $f_1(x)$ and $f_2(x)$. Therefore,

$$d_{p,h}(0) \leq C_1 d_h(0) \leq C_2 h.$$

The other estimate concerns the behavior near $t = T - h$. We have to transport all the mass of $u_1(x, T-h)$, lying in $|x| \geq r_1(T-h)$, into the profile $u_2(x, T_h) = u_1(x, T) = c|x|^\varepsilon$. Now, the part of the distribution $u_2(T-h)$ contained in the ball of radius $r_1(T-h)/2$ must have travelled are at least a distance $d(x) \geq r_1(T-h)/2$. Taking into account the value of $r_1(t)$ and setting $\rho = (a/2)h^{\beta_*}$, the cost of the transportation at $t = T - h$ can be computed as

$$(d_{p,h}(T-h))^p = \int d(x)^p u_2(x, T-h)dx \geq C \int_{|x| \leq \rho} (\rho/2)^p |x|^{\varepsilon/(m-1)}\, dx = C\rho^\gamma$$

with $\gamma = p + N + (\varepsilon/(m-1))$; here, $d(x)$ is the transportation length for the particle that ends up at x. It follows that

$$d_{p,h}(T-h) \geq C h^{\gamma\beta_*/p} \tag{6.12}$$

for all small $h > 0$. The result follows with $\mu = \gamma\beta_*/p$. C depends only on m and N and the initial data. $\qquad\square$

End of proof of the theorem. We easily check that $\mu(m, N, p) \to \beta_* < 1$ as $p \to \infty$ for fixed $m > 1$ and $N \geq 2$; this means that there exists a finite $p(m, N)$ such that for all $p > p(m, N)$ the last part of the statement of Theorem 4 holds. In fact, $p(m, N)$ is given by

$$p(m, N) = \frac{1}{1 - \beta_*}(N\beta_* + \frac{2\beta_* - 1}{m - 1}), \tag{6.13}$$

which comes from putting $\mu < 1$ for $p > p(m, N)$. Since $\beta_* > 1/2$, it follows that $p(m, N) > N$. We also know that β_* decreases with m and tends to $1/2$ as $m \to \infty$; it follows that $p(m, N)$ decreases with m and $p(\infty, N) = N$. On the contrary, $p(1, N) = +\infty$. $\qquad\square$

7. Open problems and comments

A number of related problems are naturally posed after the preceding exposition. Let me state four that I consider rather immediate, and about which I would very much like to receive answers.

Problem 1. Determine the exact range of p's for which the PME flow is contractive when posed in the several-dimensional space, $x \in \mathbb{R}^N$.

 We already have answers for $p = 2$ (yes), and for $p > p(m, N)$ (no).

Problem 2. Decide whether the exponential growth (6.1) is true under some extra assumptions. The set of data to which this estimate applies must be specified and should not be too small. C may depend on some norm of the data.

Problem 3. Do the preceding results depend on whether the solutions have some kind of focusing? State a theorem that excludes such geometry and proves contractivity, or at least exponential growth.

Problem 4. Study similar problems for Fast Diffusion, $m < 1$, where there is no focusing. Extend to the more general Filtration Equation,

$$u_t = \Delta\Phi(u), \tag{7.1}$$

where Φ is a monotone nondecreasing function satisfying certain growth conditions.

Extension. The application of these ideas to the p-Laplacian is interesting and will be done elsewhere.

Acknowledgment

Author partially supported by MCYT Project BMF2002-04572-C02-02 (Spain) and EU Programme TMR FMRX-CT98-0201. I thank J.A. Carrillo for information, comments and suggestions.

References

[Ag]   M. AGUEH. Existence of solutions to degenerate parabolic equations via the Monge-Kantorovich theory, *Georgia Institute of Technology, Preprint,* 2002; to appear in Adv. Differential Equations.

[An88]   S. ANGENENT. Large-time asymptotics of the porous media equation, in *Nonl. Diff. Equat. and Their Equil. States I,* (Berkeley, CA, 1986), W.-M. Ni, L.A. Peletier and J. Serrin eds., MSRI Publ. **12**, Springer Verlag, Berlin, 1988.

[AA95]   S.B. ANGENENT, D.G. ARONSON. The focusing problem for the radially symmetric porous medium equation, *Comm. Partial Differential Equations* **20** (1995), 1217–1240.

[AA96]   S.B. ANGENENT, D.G. ARONSON. Self-similarity in the post-focussing regime in porous medium flows. *European J. Appl. Math.* **7** (1996), no. 3, 277–285.

[AA01]   S.B. ANGENENT, D.G. ARONSON. Non-axial self-similar hole filling for the porous medium equation. *J. Amer. Math. Soc.* **14** (2001), no. 4, 737–782.

[AA03]   S.B. ANGENENT, D.G. ARONSON. The focusing problem for the Eikonal Equation. *Journal of Evolution Equations* **3** (2003), no. 1, 137–151.

[AABL]   S.B. ANGENENT, D.G. ARONSON, S.I. BETELÚ, J. LOWENGRUB. Focusing of an elongated hole in porous medium flow. *Physica D,* **151** (2001), 228–252.

[Ar86]   D.G. ARONSON. The Porous Medium Equation, in *Nonlinear Diffusion Problems,* Lecture Notes in Math. **1224**, A. Fasano and M. Primicerio eds., Springer-Verlag New York, 1986, pp. 12–46.

[ABH]   D.G. ARONSON, J.B. VAN DEN BERG, J. HULSHOF. Parametric dependence of exponents and eigenvalues in focussing porous media flows, *European J. Appl. Math.* **14** (2003), no. 4, 485–512.

[ACV]   D.G. ARONSON, L.A. CAFFARELLI, J.L. VÁZQUEZ. Interfaces with a corner point in one-dimensional porous medium flow. *Comm. Pure Appl. Math.* **38** (1985), no. 4, 375–404.

[AG93]   D.G. ARONSON, J.A. GRAVELEAU. self-similar solution to the focusing problem for the porous medium equation. *European J. Appl. Math.* **4** (1993), no. 1, 65–81.

[AGV]   D.G. ARONSON, O. GIL, J.L. VÁZQUEZ. Limit behaviour of focusing solutions to nonlinear diffusions. *Comm. Partial Differential Equations* **23** (1998), no. 1-2, 307–332.

[AV87]   D.G. ARONSON, J.L. VAZQUEZ. Eventual C^∞-regularity and concavity for flows in one-dimensional porous media. *Arch. Rational Mech. Anal.* **99** (1987), no. 4, 329–348.

[AV95]   D.G. ARONSON, J.L. VÁZQUEZ. Anomalous exponents in Nonlinear Diffusion. *Journal Nonlinear Science* **5,** 1 (1995), 29–56.

[BV96]   G. I. BARENBLATT. *Scaling, Self-Similarity, and Intermediate Asymptotics,* Cambridge Univ. Press, Cambridge, 1996. Updated version of *Similarity, Self-Similarity, and Intermediate Asymptotics,* Consultants Bureau, New York, 1979.

[BV04]   G.I. BARENBLATT AND J.L. VÁZQUEZ. Nonlinear diffusion and image contour enhancement. *Interfaces and Free Boundaries,* **6** (2004), 31–54.

[Be76]   PH. BÉNILAN. Opèrateurs accrétifs et semigroupes dans les espaces L^p $(1 \le p \le \infty)$, *France-Japan Seminar,* Tokyo, 1976.

[BB03]   PH. BÉNILAN, H. BREZIS. Nonlinear problems related to the Thomas-Fermi equation. Dedicated to Philippe Bénilan. *J. Evol. Equ.* **3** (2003), no. 4, 673–770.

[BBC]   PH. BÉNILAN, H. BREZIS, M.G. CRANDALL. A semilinear equation in $L^1(R^N)$, *Ann. Scuola Norm. Sup. Pisa* Cl. Sci. (4) **2** (1975), 523–555.

[BC81]   PH. BÉNILAN AND M.G. CRANDALL. The continuous dependence on φ of Solutions of $u_t - \Delta\varphi(u) = 0$, *Indiana Univ. Math. J.* **30** (1981), 161–177

[Bo03]   J. BOUSSINESQ. Recherches théoriques sur l'écoulement des nappes d'eau infiltrés dans le sol et sur le débit de sources. *Comptes Rendus Acad. Sci. / J. Math. Pures Appl.* **10** (1903/04), pp. 5–78.

[Br71]   H. BREZIS. Monotonicity methods in Hilbert spaces and some applications to nonlinear partial differential equations. *Contributions to nonlinear functional analysis* (Proc. Sympos., Math. Res. Center, Univ. Wisconsin, Madison, Wis., 1971), pp. 101–156. *Academic Press*, New York, 1971.

[Br73]   H. BREZIS. *"Opérateurs maximaux monotones et semi-groupes de contractions dans les espaces de Hilbert"*, North-Holland, 1973.

[BMP]   H. BREZIS, M. MARCUS, A.C. PONCE. A new concept of reduced measure for nonlinear elliptic equations. *C. R. Math. Acad. Sci. Paris* **339** (2004), no. 3, 169–174.

[BrV80]   H. BREZIS, L. VÉRON. Removable singularities for some nonlinear elliptic equations. *Arch. Rational Mech. Anal.* **75** (1980/81), no. 1, 1–6.

[CF79]   L.A. CAFFARELLI, A. FRIEDMAN. Regularity of the free boundary for the one-dimensional flow of gas in a porous medium, *Amer. Jour. Math.* **101** (1979), 1193-1218.

[CF80]   L.A. CAFFARELLI, A. FRIEDMAN. Regularity of the free boundary of a gas flow in an n-dimensional porous medium, *Indiana Univ. Math. J.* **29** (1980), 361–391.

[CV99]   L.A. CAFFARELLI, J.L. VÁZQUEZ. Viscosity solutions for the porous medium equation, *Proc. Symposia in Pure Mathematics* volume **65**, in honor of Profs. P. Lax and L. Nirenberg, M. Giaquinta et al. eds, 1999, 13–26.

[CVW]   L.A. CAFFARELLI, J.L. VÁZQUEZ, N.I. WOLANSKI. Lipschitz-continuity of solutions and interfaces of the N-dimensional porous medium equation, *Indiana Univ. Math. J.* **36** (1987), 373–401.

[CW90]   L.A. CAFFARELLI, N.I. WOLANSKI. $C^{1,\alpha}$ regularity of the free boundary for the N-dimensional porous media equation, *Comm. Pure Appl. Math.* ,**43** (1990), 885–902.

[CGT]   J.A. CARRILLO, M.P. GUALDANI, G. TOSCANI. Finite speed of propagation in porous media by mass transportation methods. *C. R. Acad. Sci. Paris Ser. I* **338** (2004), 815–818.

[C04]   J.A. CARRILLO. EDPs de difusión y transporte óptimo de masa, *Bol. Soc. Mat. Apl.* **28** (2004), 129–154 [in Spanish].

[CMV]   J. A. CARRILLO, R. McCANN, C. VILLANI. Contractions in the 2-Wasserstein length space and thermalization of granular media. *preprint HYKE2004-036*, www.hyke.org.

[CT00]   J. A. CARRILLO, G. TOSCANI. Asymptotic L^1-decay of solutions of the porous medium equation to self-similarity, *Indiana Univ. Math. J.* **49** (2000), no. 1, 113–142.

[CT03]   J. A. CARRILLO, G. TOSCANI. Wasserstein metric and large-time asymptotics of nonlinear diffusion equations. *Preprint HYKE2003-067*, www.hyke.org. To appear in Proceedings of the conference in honor of S. Rionero 2003.

[CV03]   J.A. CARRILLO, J.L. VÁZQUEZ. Fine asymptotics for fast diffusion equations. *Comm. Partial Differential Equations* **28** (2003), no. 5-6, 1023–1056.

[ChV02] E. CHASSEIGNE, J.L. VÁZQUEZ. Theory of extended solutions for fast diffusion equations in optimal classes of data. Radiation from singularities. *Arch. Ration. Mech. Anal.* **164** (2002), no. 2, 133–187.

[CL71]   M.G. CRANDALL, T.M. LIGGETT. Generation of semi-groups of nonlinear transformations on general Banach spaces. *Amer. J. Math.* **93** (1971) 265–298.

[DH04]   P. DASKALOPOULOS, R. HAMILTON. Geometric estimates for the logarithmic fast diffusion equation. *Comm. Anal. Geom.* **12** (2004), no. 1-2, 143–164.

[DP02]   M. DEL PINO, J. DOLBEAULT. Best constants for Gagliardo-Nirenberg inequalities and applications to nonlinear diffusions. *J. Math. Pures Appl.* (9), **81** (2002), no. 9, 847–875.

[FK80]   A. FRIEDMAN, S. KAMIN. The asymptotic behavior of gas in an N-dimensional porous medium. *Trans. Amer. Math. Soc.* **262** (1980), 551–563.

[GV03]   V.A. GALAKTIONOV, J.L. VÁZQUEZ. *"A Stability Technique for Evolution Partial Differential Equations. A Dynamical Systems Approach"*. PNLDE 56 (Progress in Non-Linear Differential Equations and Their Applications), Birkhäuser Verlag, 2003. 377 pages. ISBN 0-8176-4146-7, English, 391 pages.

[G72]   J. GRAVELEAU. Quelques solutions auto-semblables pour l'équation de la chaleur non-linéaire, *Rapport interne C.E.A.*, 1972 [in French].

[GMS]   M.E. GURTIN, R.C. MCCAMY, E. SOCOLOVSKI. A coordinate transformation for the porous media equation that renders the free boundary stationary, *Quart. Appl. Math.* **42** (1984), no. 3, 345–357.

[K99]   H. KOCH. Non-Euclidean singular integrals and the porous medium equation, University of Heidelberg, *Habilitation Thesis,* 1999, http://www.iwr.uniheidelberg.de/groups/amj/koch.html

[LV03]   K.A. LEE, J.L. VÁZQUEZ. Geometrical properties of solutions of the porous medium equation for large times. *Indiana Univ. Math. J.* **52** (2003), no. 4, 991–1016.

[Lei45] L.S. LEIBENZON. General problem of the movement of a compressible fluid in a porous medium, *Izv. Akad. Nauk SSSR*, Geography and Geophysics **9** (1945), 7–10 [in Russian].

[Mc]   R. MCCANN. Stable rotating binary stars and fluid in a tube. *Preprint,* www.math.toronto.edu/~mccann/.

[MPS]   A.M. MEIRMANOV, V.V. PUKHNACHOV, S.I. SHMAREV. *"Evolution equations and Lagrangian coordinates"*. de Gruyter Expositions in Mathematics, 24. Walter de Gruyter & Co., Berlin, 1997.

[Mu37]   M. MUSKAT. *The Flow of Homogeneous Fluids Through Porous Media*, McGraw-Hill, New York, 1937.

[Ot01]   F. OTTO. The geometry of dissipative evolution equations. The porous medium equation. *Comm. Partial Diff. Eqns.* **26**, 1-2 (2001), 101–174.

[Pi83]   M. Pierre. Uniqueness of the solutions of $u_t - \Delta\phi(u) = 0$ with initial datum a measure, *Nonlinear Anal. T. M. A.* **6** (1982), 175–187.

[Sh01]   S.I. Shmarev. Lagrangian coordinates in free boundary problems for multidimensional parabolic equations. *Elliptic and parabolic problems* (Rolduc/Gaeta, 2001), 274–282, World Sci. Publishing, River Edge, NJ, 2002.

[SV96]   S.I. Shmarev, J.L. Vázquez. The regularity of solutions of reaction-diffusion equations via Lagrangian coordinates. *NoDEA Nonlinear Differential Equations Appl.* **3** (1996), no. 4, 465–497.

[V83]   J.L. Vázquez. Asymptotic behaviour and propagation properties of the one-dimensional flow of gas in a porous medium, *Trans. Amer. Math. Soc.* **277** (1983), 507–527 (announced in International Congress on Free Boundary Problems, Theory and Applications, Montecatini, Italy, 1981).

[V83b]   J.L. Vázquez. Waiting times. The interfaces of one-dimensional flows in porous media, *Trans. Amer. Math. Soc.* **277** (1983), 507–527.

[V83c]   J.L. Vázquez. On a semilinear equation in R^2 involving bounded measures. *Proc. Roy. Soc. Edinburgh Sect. A* **95** (1983), no. 3-4, 181–202.

[V92]   J.L. Vázquez. An Introduction to the Mathematical Theory of the Porous Medium Equation, in *Shape Optimization and Free Boundaries*, M. C. Delfour ed., Math. and Phys. Sciences, Series C, Kluwer Acad. Publ., Dordrecht-Boston-Leiden, 1992. Pages 347–389.

[V92b]   J.L. Vázquez. Nonexistence of solutions for nonlinear heat equations of fast diffusion type, *J. Math. Pures Appl.* **71** (1992), pp. 503–526.

[V03]   J.L. Vázquez. Asymptotic behaviour for the porous medium equation posed in the whole space. Dedicated to Philippe Bénilan. *J. Evol. Equ.* **3** (2003), no. 1, 67–118.

[V03b]   J.L. Vázquez. Darcy's law and the theory of shrinking solutions of fast diffusion equations. *SIAM J. Math. Anal.* **35** (2003), no. 4, 1005–1028

[Vi03]   C. Villani. *"Topics in Optimal Transportation"*, American Mathematical Society, Providence, Rh. I., 2003.

J.L. Vázquez
Departamento de Matemáticas
Universidad Autónoma de Madrid
E-28046 Madrid, Spain
e-mail: juanluis.vazquez@uam.es

Progress in Nonlinear Differential Equations
and Their Applications, Vol. 63, 453–464

Large Solutions of Elliptic Equations with Strong Absorption

Laurent Véron

Dedicated to H. Brezis on the occasion of his 60th birthday

Abstract. We present some general results dealing with existence and uniqueness of solutions of $-\Delta u + g(x, u) = 0$ in a domain $\Omega \subset \mathbb{R}^N$, which satisfy $\lim_{\mathrm{dist}(x,\partial\Omega)\to 0} u(x) = \infty$, where g is a continuous nonnegative function. We emphasize the links between the regularity of the boundary and the existence of such solutions. The cases $g(x, r) = \rho^\alpha(x) r_+^q$ and $g(x, r) = \rho^\alpha(x) e^{br}$ are thoroughly investigated.

Mathematics Subject Classification (2000). 35J60, 35J65.

Keywords. Boundary blow-up, Keller-Osserman condition, singular solutions.

Introduction

Let Ω be a domain in $\mathbb{R}^N$ and $g \in C(\Omega \times \mathbb{R}; \mathbb{R}_+)$. A solution u of

$$-\Delta u + g(x, u) = 0 \quad \text{in } \Omega \tag{0.1}$$

is called a large solution if it holds

$$\lim_{\substack{\rho(x)\to 0 \\ x \in K \cap \Omega}} u(x) = \infty \tag{0.2}$$

for any compact $K \subset \bar{\Omega}$, where $\rho(x) = \mathrm{dist}(x, \partial\Omega)$. The first study of such solutions has been performed by Bieberbach [6] and Rademacher [17] in the special case $g(u) = e^u$ and $N = 2$ or 3. Later on, Loewner and Nirenberg [10] proved the existence and uniqueness of a positive function u satisfying (0.2) and

$$-\frac{4(N-1)}{N-2}\Delta u + u_+^{(N+2)/(N-2)} = 0 \tag{0.3}$$

in a smooth bounded domain Ω. Let ds be the Euclidean metric in $\mathbb{R}^N$; a consequence of this result is that the metric $d\sigma = u^{2/(n-2)} ds$ is complete, has scalar

454 L. Véron

curvature -1 and is invariant under the Möbius transformations (see also [1]). In 1990 Bandle and Marcus proved in [2] the uniqueness of a large solution to

$$-\Delta u + u_+^q = 0 \tag{0.4}$$

when $q > 1$ and $\partial\Omega$ is smooth and compact. They also give precise expansion of u in [3] and [4]. Independently, existence and uniqueness results of large solutions are obtained by Véron [20] for large solutions of general second order elliptic equations

$$-Lu + u_+^q = 0 \tag{0.5}$$

in the same type of domains. The first striking observation concerning (0.1) is due to Keller [8] and Osserman [16]. Their result is the following. *Assume there is a nondecreasing function h such that $g(x,r) \geq h(r)$ for all $(x,r) \in \Omega \times \mathbb{R}$ and*

$$\int_a^\infty \left(\int_0^t h(s)ds\right)^{-1/2} dt < \infty \quad \forall a > 0. \tag{0.6}$$

Then there exists a nonincreasing function η defined on $\mathbb{R}_+$, depending only on h and N, and satisfying

$$\begin{aligned} \lim_{r\to 0} \eta(r) &= \infty \\ \lim_{r\to\infty} \eta(r) &= 0, \end{aligned} \tag{0.7}$$

such that any solution u of (0.1) in Ω verifies

$$u(x) \leq \eta(\rho(x)) \quad \forall x \in \Omega. \tag{0.8}$$

Condition (0.6) is called the *Keller-Osserman condition*. Under the additional assumption

$$g(x,r) \geq g(x,r') \quad \forall x \in \Omega \quad \forall r \geq r', \tag{0.9}$$

it is easy to derive the existence of a maximal solution $\bar{u}$ to (0.1) in Ω provided there exists at least one solution of the same equation. However, the maximal solution may not be a large solution, and even if it is the case, uniqueness may not hold.

Keller-Osserman condition is too restrictive for a function which depends truly on the x variable. It can be weakened in a natural way by introducing the *local Keller-Osserman condition relative to Ω*. This condition is as follows. For any compact subset $K \in \Omega$, there exists a nondecreasing function h_K defined on $\mathbb{R}_+$, with positive value, such that

$$g(x,r) \geq h_K(r) \quad \forall(x,r) \in K \times \mathbb{R}, \tag{0.10}$$

and $h = h_K$ satisfies (0.6). We first prove

Theorem 1. *Let Ω be a domain in $\mathbb{R}^N$ and $g \in C(\Omega \times \mathbb{R}; \mathbb{R}_+)$ satisfy the local Keller-Osserman condition. Assume also (0.9) holds. Then either (0.1) has no solution in Ω, or there exists a maximal solution.*

The easiest method to prove that the maximal solution is actually a large solution is to associate to each point of the boundary a solution which admits this point as its unique singular point. We say that g satisfies *the local weak singularity assumption* in Ω, if for any compact subset $K \subset \bar{\Omega}$, there exist $\gamma = \gamma_K > 0$ and $\eta = \eta_K > 0$ such that, for any $a \in K$,

$$\limsup_{x \to a} \left(\Gamma(x-a)\right)^{-1} \int_{\Omega \cap B_\eta(a)} \Gamma(x-y))g(y, \gamma \Gamma(y-a))dy < \gamma, \tag{0.11}$$

uniformly in K, where

$$\Gamma(y) = \begin{cases} (N(N-2)\alpha_N)^{-1}|y|^{2-N} & \text{if } N \geq 3 \\ (2\pi)^{-1}\ln(1/|x|) & \text{if } N = 2, \end{cases} \tag{0.12}$$

and α_N is the volume of the unit ball in $\mathbb{R}^N$. Notice that (0.11) implies that $g(., \gamma\Gamma(.-a)) \in L^1(\Omega \cap B_\eta(a))$.

Theorem 2. *Let Ω be a domain in $\mathbb{R}^N$ and $g \in C(\Omega \times \mathbb{R}; \mathbb{R}_+)$ satisfy the local Keller-Osserman condition and the local weak singularity assumption on $\partial\Omega$. Assume also (0.9) holds and that there exists of solution of (0.1) which is bounded below in the neighborhood of $\partial\Omega$. Then the maximal solution $\bar{u}$ is a large solution.*

Uniqueness of large solutions is established in assuming convexity, besides the monotonicity assumption (0.9). The following technical result is at the core of our approach.

Theorem 3. *Let Ω be a bounded domain in $\mathbb{R}^N$ and $g \in C(\Omega \times \mathbb{R}; \mathbb{R}_+)$ satisfy*

$$r \mapsto g(x, r) \text{ be nondecreasing and convex } \forall x \in \Omega. \tag{0.13}$$

Assume Equation (0.1) admits a bounded subsolution Φ and a maximal solution $\bar{u}$ in Ω and there exist two constants $m = m(\Omega, g) > 1$ and $\delta = \delta(\Omega, g) > 0$ such that

$$0 \leq \bar{u}(x) \leq mu(x), \qquad \forall x \in \Omega \quad s.t. \ \rho(x) \leq \delta \tag{0.14}$$

for any large solution. Then there exists at most one large solution.

We apply our results to equations of the following types

$$-\Delta u + \rho^\alpha(x)u_+^q = 0 \qquad \alpha > -2, \ q > 1, \tag{0.15}$$

and

$$-\Delta u + \rho^\alpha(x)e^{bu} = 0 \qquad \alpha > -2, \ b > 0. \tag{0.16}$$

1. Proof of the main results

Proof of Theorem 1. We assume that there exists a solution V of (0.1). Let $\{\Omega_n\}$ be a sequence of smooth bounded domains such that

$$\Omega_n \subset \bar{\Omega}_n \subset \Omega_{n+1} \quad \forall n \in \mathbb{N}, \ \bigcup_{n \in \mathbb{N}} \Omega_n = \Omega.$$

Because g satisfies the local Keller-Osserman assumption, for each $\bar{\Omega}_n$ there exists a nondecreasing positive function $h_{\bar{\Omega}_n}$ satisfying (0.6) and such that $g(x, r) \geq h_{\bar{\Omega}_n}(r)$ for every $(x, r) \in \Omega \times \mathbb{R}$. Thus, for any $n \in \mathbb{N}$, there exists a maximal solution u_n

to (0.1) in Ω_n. Furthermore, since $\partial\Omega_n$ is smooth and g is continuous in $\bar\Omega \times \mathbb{R}$, u_n is constructed as the increasing limit, as $k \to \infty$, of the solutions $v = u_{n,k}$ of

$$\begin{cases} -\Delta v + g(x,v) = 0 & \text{in } \Omega_n \\ v = k & \text{on } \partial\Omega_n. \end{cases} \tag{1.1}$$

Moreover $u_n \geq u_p|_{\Omega_n} \geq V|_{\Omega_n}$ for $p > n$, because $g(x,.)$ is monotone. Thus $u = \lim_{n\to\infty} u_n$ is a solution of (0.1). Finally, any continuous solution of the same equation is dominated by u_n in Ω_n, for n large enough, which means that u is the maximal solution.

Remark. Even if Ω is bounded, the existence of at least one solution of (0.1) may not hold if $g(x,r)$ blows up too strongly when $\rho(x)$ tends to 0. For example, it is proved in [18] that there exists no solution of

$$-\Delta u + \rho^{-2}(x)e^{2u} = 0$$

in the unit disk in the plane. However the technique herein is of a local nature and the method extends easily to show that, for any $a > 0$, inequality

$$-\Delta u + \rho^{-2}(x)e^{au} \leq 0,$$

admits no solution in any smooth bounded domain Ω. In the same way, it it is established in the same reference that inequality

$$-\Delta u + \rho^{-2}(x)u^q \leq 0$$

$(q > 1)$ admits no positive solution in the unit ball B_1 in $\mathbb{R}^N$. The proof extends easily to any bounded smooth domain.

Proof of Theorem 2. Step 1: Construction of fundamental solutions. We denote by $\tilde g$ the extension of g by 0 outside Ω. If $s > 0$, let $G_s(y)$ be the solution of

$$\begin{cases} -\Delta v = \delta_0 & \text{in } B_s \\ v = 0 & \text{on } \partial B_s. \end{cases} \tag{1.2}$$

Then $0 \leq G_s(y) \leq \Gamma(y)$, for any s if $N \geq 3$ and for $s \leq 1$ if $N = 2$. For $k > 0$ we set $\tilde g_k(x,r) = \inf\{k, \tilde g(x,r)\}$. If $0 < \eta \leq 1$, $\gamma > 0$ and $a \in \partial\Omega$, we define an operator $T : L^1(B_\eta(a)) \mapsto L^1(B_\eta(a))$ by

$$v = T(w) \iff \begin{cases} -\Delta v = \gamma\delta_0 - \tilde g_k(x,w) & \text{in } B_\eta(a) \\ v = 0 & \text{on } \partial B_\eta(a). \end{cases} \tag{1.3}$$

The operator T is compact, with bounded image, thus there exists $v_k \in L^1(B_\eta(a))$ such that $T(v_k) = v_k$. Equivalently

$$\begin{cases} -\Delta v_k + \tilde g_k(x,v_k) = \gamma\delta_0 & \text{in } B_\eta(a) \\ v_k = 0 & \text{on } \partial B_\eta(a), \end{cases} \tag{1.4}$$

and v_k is unique since $\tilde g_k(x,.)$ is monotone. Moreover

$$v_k(x) \leq \gamma G_\eta(x - a) \leq \gamma\Gamma(x - a),$$

which implies

$$0 \leq \tilde g_k(x, v_k(x)) \leq \tilde g(x, \gamma G_\gamma(x - a)) \leq \tilde g(x, \gamma\Gamma(x - a)).$$

If $\ell > k$, $\tilde{g}_k(x, r) \leq \tilde{g}_\ell(x, r)$, thus $v_k \geq v_\ell$. Finally, if η and γ are chosen such that $g(., \gamma\Gamma(. - a)) \in L^1(B_\eta(a))$ (which is locally uniformly possible by assumption (0.11)), the Lebesgue theorem asserts that v_k converges to some $v \in L^1(B_{r_a}(a))$. Clearly $\tilde{g}(x, v) \in L^1(B_{r_a}(a))$ and v is solution of

$$\begin{cases} -\Delta v + \tilde{g}(x, v) = \gamma\delta_0 & \text{in } B_{r_a}(a) \\ v = 0 & \text{on } \partial B_{r_a}(a). \end{cases} \tag{1.5}$$

Step 2: Comparison. Assume V is a solution of (0.1) bounded below in a neighborhood of $\partial\Omega$. Then the sequence of large solutions $\{u_n\}$ introduced in the proof of Theorem 1 is bounded below by V in Ω_n. Let K be a compact subset of $\bar{\Omega}$, with corresponding parameters $\eta = \eta_K$ and $\gamma = \gamma_K$ for property (0.11) to hold. There exists $n_K \in \mathbb{N}$ such that $n \geq n_K$ implies $\text{dist}(a, \partial\Omega_n) < \eta$, for any $a \in K \cap \partial\Omega$. Let v_a be the solution v of (1.5). If $m = m_K$ is the infimum of V on $K \cap \Omega$, $v_a - m_-$ is a subsolution of (1.5). Thus $u_n \geq v_a - m_-$ in $\Omega_n \cap B_\eta(a)$, and $\lim_{n\to\infty} u_n = \bar{u} \geq v_a - m_-$ in $\Omega \cap B_\eta(a)$. Since

$$v_a(x) \geq \gamma G_\eta(x) - \int_{\Omega \cap B_\eta(a)} \Gamma(x - y))g(y, \gamma\Gamma(y - a))dy, \tag{1.6}$$

and

$$\lim_{x \to a} \Gamma^{-1}(x - a)G_\eta(x) = 1,$$

it follows by (0.11),

$$\bar{u}(x) \geq \epsilon\Gamma(x - a) \quad \forall x \in B_\gamma(a), \tag{1.7}$$

for some $\epsilon = \epsilon_K > 0$, uniformly with respect to $a \in K$.

Step: End of the proof. Let $K \subset \bar{\Omega}$ be compact, and $\eta = \eta_K$ and $\gamma = \gamma_K$ are defined as in Step 2. If $x \in K \cap \Omega$ satisfies $\rho(x) \leq \gamma$, let $a \in \partial\Omega$ such that $\rho(x) = |x - a|$. Clearly (1.7) implies that $\bar{u}$ satisfies (0.2). $\qquad\square$

The proof of the next result is an adaptation of the proof of [13, Th. 02], which, itself, is derived from a method introduced in [12].

Proof of Theorem 3. Step 1: Construction of a minimal large solution. For the sake of completeness, we shall recall it. Let $\Omega_\sigma = \{x \in \Omega : \rho(x) \leq \sigma\}$. Since $\partial\Omega$ is compact, there exists $\sigma_0 \in (0, \delta)$ such that $\bar{u}(x) \geq 1$ for any $x \in \Omega_{\sigma_0}$. Thus any large solution u satisfies

$$u(x) \geq m^{-1} \quad \forall x \in \Omega_{\sigma_0}.$$

Let Z be the solution of

$$\begin{cases} -\Delta Z + g(x, Z) = 0 & \text{in } \Omega'_{\sigma_0} = \Omega \setminus \Omega_{\sigma_0} \\ Z = 0 & \text{on } \partial\Omega'_{\sigma_0}, \end{cases} \tag{1.8}$$

then any large solution is bounded below in Ω by $m = \inf\{m_K^{-1}, \min Z\}$. For any $x \in \Omega$, we set

$$\underline{u}(x) =: \inf\{u(x) : u \text{ large solution }\}.$$

There exists a countable dense set $\mathbf{Q} \subset \Omega$ and a sequence of large solutions $\{u_n\}$ such that $u_n(x)$ decreases to $\underline{u}(x)$ for any $x \in \mathbf{Q}$. Since the sets of functions $\{u_n\}$ and $\{g(., u_n\}$ are locally bounded in Ω, it can be also assumed that $\{u_n\}$ converges

458 L. Véron

to $\underline{u}$ in the $C^1_{loc}(\Omega)$ topology. Thus $\underline{u}$ is a solution of (0.1) in Ω and $\underline{u}(x) \geq \bar{u}(x)/m$ in Ω_{σ_0}, by density. Therefore $\underline{u}$ is the minimal large solution.

Step 2: End of the proof. Suppose the maximal solution $\bar{u}$ is different from the minimal large solution. Then $\bar{u} > \underline{u}$ by the strong maximum principle. If we set

$$w = \underline{u} - \frac{1}{2m}(\bar{u} - \underline{u}),$$

there holds

$$-\Delta w + g(x, w) = g\left(x, (1 + \frac{1}{2m})\underline{u} - \frac{\bar{u}}{2m}\right) - (1 + \frac{1}{2m})g(x, \underline{u}) + \frac{1}{2m}g(x, \bar{u}).$$

Clearly

$$\frac{2m}{1 + 2m}\left[(1 + \frac{1}{2m})\underline{u} - \frac{\bar{u}}{2m}\right] + \frac{1}{1 + 2m}\bar{u} = \underline{u},$$

then

$$\frac{2m}{1 + 2m}g\left(x, (1 + \frac{1}{2m})\underline{u} - \frac{\bar{u}}{2m}\right) + \frac{1}{1 + 2m}g(x, \bar{u}) \geq g(x, \underline{u}),$$

by the convexity assumption. Therefore w is a supersolution in Ω, larger than $(m + 1)u/2m$ in Ω_σ. Let Φ be a bounded subsolution of (0.1). For any $\theta \in [0, 1]$ the function $w_\theta = \theta\underline{u} + (1 - \theta)\Phi$ satisfies

$$-\Delta w_\theta + g(x, w_\theta) = g(x, \theta\underline{u} + (1 - \theta)\Phi - \theta g(x, \underline{u}) - (1 - \theta)g(x, \Phi)$$
$$+ (1 - \theta)(g(x, \Phi) - \Delta\Phi) \leq 0.$$

If we fix $0 < \theta < (m + 1)/2m$, the function $x \mapsto (w_\theta - w)_+$ has compact support in Ω. By monotonicity and the comparison principle, the inequality $w_\theta < w$ holds in Ω. Using a classical result on the existence of solution to nonlinear elliptic equations in presence of two ordered sub and super-solutions (see [13, Prop. 2.1]), there exists a solution u_1 of (0.1) such that $w_\theta \leq u_1 \leq w$ in Ω. Therefore u_1 is a large solution smaller than $\underline{u}$, contradiction. $\square$

Remark. The assumption of existence of a bounded subsolution Φ can be replaced by the following: *For any $M \in \mathbb{R}$ there exists $\theta_M > 0$ such that*

$$g(x, \theta r) \leq \theta g(x, r) \quad \forall \theta \in (0, \theta_M], \ \forall x \in \Omega, \ \forall r \geq M. \tag{1.9}$$

If this assumption is satisfied, the function $x \mapsto \theta\underline{u}(x)$ is a subsolution of (0.1), which blows up on $\partial\Omega$, and is dominated by w provided $\theta < (m + 1)/2m$. The remaining of the proof is unchanged.

2. Applications

Theorem 4. *Let $\Omega \subset \mathbb{R}^N$ be a domain with a compact boundary such that $\partial\Omega = \partial\bar{\Omega}^c$. Then for any $\alpha \geq 0$ and $1 < q < (N + \alpha)/(N - 2)$ if $N > 2$, or any $q > 1$ if $N = 2$, there exists a unique large solution to*

$$-\Delta u + \rho^\alpha(x)u_+^q = 0 \quad in \ \Omega. \tag{2.1}$$

Proof. When $\alpha = 0$ the result is established in [13]. It is clear that the function $(x, r) \mapsto \rho^\alpha(x) r_+^q$ satisfies the local Keller-Osserman condition with $h_K(r) = \text{dist}^\alpha(K, \partial\Omega) r_+^q$, for any compact subset $K \subset \Omega$. Since the zero function is a solution of (2.1), there exists a maximal solution in Ω. Let $a \in \partial\Omega$, $s \in (0,1)$ and H be the solution of

$$\begin{cases} -\Delta H = \rho^\alpha(x)\Gamma^q(x-a) & \text{in } B_s(a) \\ \quad H = 0 & \text{on } \partial B_s(a). \end{cases} \tag{2.2}$$

Since $\alpha \geq 0$, $H \leq \tilde{H}$ where

$$\begin{cases} -\Delta\tilde{H} = |x-a|^\alpha \, \Gamma^q(x-a) & \text{in } B_s(a) \\ \quad \tilde{H} = 0 & \text{on } \partial B_s(a). \end{cases} \tag{2.3}$$

Because $\tilde{H} \in L^1(B_s(a))$ is radial with respect to a, there holds

$$\lim_{x \to a} \Gamma^{-1}(x-a)\tilde{H}(x) = 0.$$

Furthermore $\tilde{H}(x) \leq H^*(x-a) \leq C\tilde{H}(x)$ for some $C > 0$, under the condition $|x-a| \leq s/2$, where H^* is defined by

$$H^*(z) = \int_{B_s(0)} \Gamma(z-y) \, |y|^\alpha \, \Gamma^q(y) dy.$$

Therefore (0.11) holds and the maximal solution is a large solution. At end, the Keller-Osserman condition implies that any solution u of (2.1) satisfies

$$u(x) \leq C\rho^{-(2+\alpha)/(q-1)}(x) \quad \forall x \in \Omega, \tag{2.4}$$

where $C = C(N, q, \alpha) > 0$. If

$$\ell_{q,N,\alpha} = \left(\left(\frac{2+\alpha}{q-1} \right) \left(\frac{2q+\alpha}{q-1} - N \right) \right)^{1/(q-1)},$$

the function v_S defined by

$$v_S(x) = \ell_{q,N,\alpha} |x|^{-(2+\alpha)/(q-1)}, \tag{2.5}$$

is a singular solution of

$$-\Delta v + |x|^\alpha v^q = 0$$

in $\mathbb{R}^N \setminus \{0\}$. Therefore, for any $a \in \bar{\Omega}^c$, the restriction to Ω of $x \mapsto v_{S,a}(x) = v_S(x-a)$; is a locally bounded subsolution of (2.1) in Ω. Since $\partial\Omega = \partial\bar{\Omega}^c$, for any $a \in \partial\Omega$, there exists a sequence $\{a_n\} \subset \bar{\Omega}^c$ converging to a. If u is a large solution of (2.1) in Ω, there holds

$$u(x) \geq v_{S,a_n}(x) \implies u(x) \geq v_{S,a}(x) \quad \forall x \in \Omega.$$

If we choose in particular $a \in \partial\Omega$ to be a point such that $|x-a| = \rho(x)$, we derive

$$\ell_{q,N,\alpha}\rho^{-(2+\alpha)/(q-1)}(x) \leq u(x) \leq C_{q,N,\alpha}\rho^{-(2+\alpha)/(q-1)}(x). \tag{2.6}$$

Thus (0.14) holds in whole Ω and uniqueness follows. $\qquad\square$

If $\alpha < 0$, the situation is much more complicated and we have only a partial answer to the existence and uniqueness problem.

460   L. Véron

Theorem 5. *Let $\Omega \subset \mathbb{R}^N$ ($N \geq 2$) be a bounded domain such that $\partial\Omega = \partial\bar{\Omega}^c$, $\alpha \in (-2, 0)$ and $q > 1$.*

  (i) *If $\partial\Omega$ is Lipschitz continuous and $q \in (1, (N + \alpha)/(N - 2))$ ($q > 1$ if $N = 2$), the maximal solution of (2.1) is a large solution.*

  (ii) *If $\partial\Omega$ is C^2 and $q > 1$, the maximal solution is the unique large solution.*

Proof. Step 1: The maximal solution is a large solution. For proving that (0.11) holds, we can assume $a = 0$. There exists a Lipschitz continuous diffeomorphism which transforms $\Omega \cap B_s(0)$ into $\mathbb{R}_+^N \cap B_1(0) = \{y = (y_1, y') : y_1 > 0, y_1^2 + |y'|^2 \leq 1\}$, and

$$\int_{\Omega \cap B_s(0)} \rho^\alpha(y)\Gamma^q(y)dy \leq C \int_0^1 \int_{|y'| \leq 1} y_1^\alpha \Gamma^q(y)dy' \, dy_1. \tag{2.7}$$

We assume $N \geq 3$, the case $N = 2$ being treated similarly. Then

$$\int_0^1 \int_{|y'| \leq 1} y_1^\alpha \, |y|^{q(2-N)} \, dy' \, dy_1$$

$$= \int_0^1 y_1^{\alpha+1-(q-1)(N-2)} \int_{|z'| \leq 1/y_1} (1 + |z|'^2)^{q(2-N)/2} dz' \, dy_1$$

$$= |S^{N-2}| \int_0^1 y_1^{\alpha+1-(q-1)(N-2)} \int_0^{1/y_1} (1 + |r|^2)^{q(2-N)/2} r^{N-2} dr \, dy_1$$

$$= \frac{|S^{N-2}|}{\alpha + N - q(N-2)} \int_0^\infty (1 + |r|^2)^{q(2-N)/2} r^{N-2} dr.$$

Thus $x \mapsto \rho^\alpha(x)\Gamma^q(x - a) \in L^1(B_s(a))$, for any $a \in \partial\Omega$ and $0 < s < 1$. For $\epsilon > 0$, we write

$$|x - a|^{N-2} \int_{\Omega \cap B_s(a)} |x - y|^{2-N} \rho^\alpha(y)\Gamma^q(y)dy = I_\epsilon + J_\epsilon,$$

where

$$I_\epsilon = |x - a|^{N-2} \int_{\Omega \cap B_\epsilon(a)} |x - y|^{2-N} \rho^\alpha(y)\Gamma^q(y)dy,$$

and

$$J_\epsilon = |x - a|^{N-2} \int_{\Omega \cap (B_s(a) \setminus B_\epsilon(a))} |x - y|^{2-N} \rho^\alpha(y)\Gamma^q(y)dy.$$

By the previous change of coordinates, I_ϵ and J_ϵ are smaller respectively than $\tilde{I}_\epsilon$ and $\tilde{J}_\epsilon$ (up to a fixed multiplicative positive constant), where

$$\tilde{I}_\epsilon = |x|^{N-2} \int_{\substack{|y| \leq \epsilon \\ y_1 > 0}} |x - y|^{2-N} y_1^\alpha \, |y|^{q(2-N)} \, dy$$

and

$$\tilde{J}_\epsilon = |x|^{N-2} \int_{\substack{\epsilon < |y| \leq 1 \\ y_1 > 0}} |x - y|^{2-N} y_1^\alpha \, |y|^{q(2-N)} \, dy.$$

Clearly, if $|x| < \epsilon$,

$$\tilde{J}_\epsilon \le |x|^{N-2} (\epsilon - |x|)^{2-N} \int_{|y|\le 1} y_1^\alpha \, |y|^{q(2-N)} \, dy = C\, |x|^{N-2} (\epsilon - |x|)^{2-N}. \qquad (2.8)$$

For $\tilde{I}_\epsilon$, we write $x = (x_1, x')$ with $x_1 > 0$ and evaluate the differential of

$$x \mapsto \mathcal{I}_\epsilon(x) = \tilde{I}_\epsilon = \int_{\substack{|y|\le \epsilon \\ y_1 > 0}} |x - y|^{2-N} \, y_1^\alpha \, |y|^{q(2-N)} \, dy,$$

under the constraint $|x| = r$. Then

$$D\mathcal{I}_\epsilon(x) = (2-N) \int_{\substack{|y|\le \epsilon \\ y_1 > 0}} |x - y|^{-N} \, y_1^\alpha \, |y|^{q(2-N)} \, (x-y) dy = \lambda x.$$

for some $\lambda \in \mathbb{R}$. Therefore

$$\int_{\substack{|y|\le \epsilon \\ y_1 > 0}} |x - y|^{-N} \, y_1^\alpha \, |y|^{q(2-N)} \, y dy = \mu x,$$

for some $\mu \in \mathbb{R}$. In particular

$$\int_{\substack{|y|\le \epsilon \\ y_1 > 0}} |x - y|^{-N} \, y_1^\alpha \, |y|^{q(2-N)} \, y' dy = \mu x'.$$

But the value of the above integral is invariant under any orthogonal transformation in the y'-space and so is the vector $\mu x'$. This implies $\mu x' = 0$. If $\mu = 0$, we would have

$$\int_{\substack{|y|\le \epsilon \\ y_1 > 0}} |x - y|^{-N} \, y_1^\alpha \, |y|^{q(2-N)} \, y_1 dy = 0,$$

a contradiction. Consequently $x' = 0$ and the maximal value of $\tilde{I}_\epsilon$ for fixed $|x| = r$ is achieved for $x' = 0$. Furthermore this value is smaller than

$$\tilde{I}_{\epsilon,0} = r^{N-2} \int_0^\epsilon \int_{|y'|<\epsilon} \left((r - y_1)^2 + |y'|^2 \right)^{(2-N)/2} y_1^\alpha \, |y|^{q(2-N)} \, dy$$

$$= r^{N+\alpha+q(2-N)} \int_0^{\epsilon/r} \int_{|y'|<\epsilon/r} \left((1 - y_1)^2 + |y'|^2 \right)^{(2-N)/2} y_1^\alpha \, |y|^{q(2-N)} \, dy.$$

Finally

$$\lim_{x\to 0} \tilde{I}_\epsilon = 0,$$

and (0.11) is satisfied.

Step 2: The smooth case. In [18, Th. 7.1-(ii)] the result is established with $\Omega = B_1$ and $q = (N+2)/(N-2)$. However, with minor modifications and by using techniques of localization used in [20], it is not difficult to verify that any large solution u of (2.1) in a bounded C^2 domain Ω endows the following boundary behavior,

$$\lim_{\rho(x)\to 0} \rho^{(2+\alpha)/(q-1)}(x)u(x) = \left(\frac{(2+\alpha)(\alpha+q-1)}{(q-1)^2} \right)^{1/(q-1)}. \qquad (2.9)$$

Since any couple of large solutions (u_1, u_2) satisfies

$$\lim_{\rho(x)\to 0} \frac{u_1(x)}{u_2(x)} = 1, \tag{2.10}$$

uniqueness follows by convexity. $\qquad\square$

Remark. We believe that regularity assumption in the case $\alpha < 0$, which appears very far from the case $\alpha \geq 0$, is too restrictive. It would be interesting to search if the results of Theorem 4 are valid in any bounded domain Ω, in all the cases $1 < q < (N+\alpha)/(N-2)$ and $\alpha > -2$.

Theorem 6. *Let $\Omega \subset \mathbb{R}^N$ be a bounded domain such that $\partial\Omega = \partial\bar\Omega^c$. Then for any $\alpha > -2$ and $b > 0$ there exists a unique large solution to*

$$-\Delta u + \rho^\alpha(x)e^{bu} = 0 \quad \text{in } \Omega, \tag{2.11}$$

(i) *either if $N = 2$ and $\alpha \geq 0$,*

(ii) *or if $N \geq 3$ and $\partial\Omega$ is C^2.*

Proof. When $N = 2$ the result is already known if $\partial\Omega$ is C^2 or if $\alpha = 0$ (see respectively [18]) and [13]). By using the method of [19, Lemma 1.6], it is easy to establish that, for any $x \in \Omega$ and $0 < R < \rho(x) \leq 1$, the following inequality holds

$$u(x) \leq \frac{1}{b} \ln\left(\frac{4N}{b(\rho(x) - R)^\alpha)R^2}\right). \tag{2.12}$$

The optimal choice $R = (1 + \alpha/2)\rho(x)$ leads to

$$u(x) \leq \frac{2+\alpha}{b} \ln\left(\frac{c}{\rho(x)}\right) \quad \forall x \in \Omega, \tag{2.13}$$

with $c = 16N\alpha^\alpha/(2+\alpha)b$.

Step 1: The case $N = 2$. As in Theorem 5, the verification of (0.11) is reduced to prove that, for some $\gamma > 0$,

$$\int_0^1 \int_{|y'|} y_1^\alpha \exp((2\pi)^{-1}b\gamma \ln(1/|y|))dy = \int_0^1 \int_{|y'|} y_1^\alpha |y|^{-b\gamma/2\pi} < \infty, \tag{2.14}$$

and

$$\lim_{x\to a} (\ln(1/|x-a|))^{-1} \int_0^1 \int_{|y'|} \ln(1/|x-y|)y_1^\alpha |y|^{-b\gamma/2\pi} dy = 0. \tag{2.15}$$

The two estimates hold clearly by Hölder's inequality for $0 < \gamma \leq \gamma_0$ small enough. This fact implies that the maximal solution is a large solution. Let $R > 0$ such that $\bar\Omega \subset B_R$. As in the proof of Theorem 4, for any $a \in \bar\Omega^c$ and $\text{dist}(a\Omega) < 1$, any large solution of (2.1) in Ω is bounded below by $v = v_{\gamma,a}$, the solution of

$$\begin{cases} -\Delta v + |x-a|^\alpha e^{bv} = \gamma\delta_a & \text{in } B_R(a) \\ v = 0 & \text{on } \partial B_R(a). \end{cases} \tag{2.16}$$

This solution exists for $0 < \gamma \leq \gamma_0$. Furthermore

$$\frac{\gamma}{2\pi} \ln \left(\frac{R}{|x - a|} \right) - w \leq v_{\gamma, a} \leq \frac{\gamma}{2\pi} \ln \left(\frac{R}{|x - a|} \right),$$

where

$$\begin{cases} -\Delta w = |x - a|^\alpha \left(\dfrac{R}{|x - a|} \right)^{\gamma b / 2\pi} & \text{in } B_R(a) \\ w = 0 \quad \text{on } \partial B_R(a). \end{cases} \tag{2.17}$$

Choosing $\gamma \leq \gamma_1$ such that the right-hand side of (2.17) belongs to $L^2(B_R(a))$, implies that w remains bounded. Let $x \in \Omega$ and $a \in \partial\Omega$ such that $|x - a| = \rho(x)$. Introducing a sequence $\{a_n\} \subset \bar{\Omega}^c$ converging to a, we derive that

$$\frac{\gamma}{2\pi} \ln \left(\frac{R}{\rho(x)} \right) - w \leq u(x) \leq \frac{2 + \alpha}{b} \ln \left(\frac{c}{\rho(x)} \right), \tag{2.18}$$

holds for any large solution. Therefore (0.14) is verified and uniqueness follows.

Step 2: The smooth case. The boundary expansion performed in [18, Th. 7.2-(ii)] in the two-dimensional case with $\Omega = B_1$ can be easily extended to a bounded C^2 N-dimensional domain. It follows that any large solution u satisfies

$$u(x) = \frac{2 + \alpha}{b} \ln \left(\frac{1}{\rho(x)} \right) + O(1) \quad \text{as } \rho(x) \to 0. \tag{2.19}$$

This implies that (2.10) holds and uniqueness follows. $\qquad\square$

In [11] and [13], the uniqueness of a large solution (may be without existence) to equations (2.1) and (2.11) in the case $\alpha = 0$, is established when Ω is a bounded domain the boundary of which is locally a continuous graph. In [18, Th. 7.1–7.2] uniqueness is obtained for the same equations when $\alpha \geq 0$ and $\Omega = B_1$. In this last reference, the proof is based upon a scaling technique, and it is not difficult to see this method works straightforwardly if B_1 is replaced by domain Ω which admits a point x_0 such that $t \mapsto \rho(x_0 + t(x - x_0))$ is nonincrecreasing for any $x \in \Omega$ and $t > 1$. Actually the scaling method of [18] and the local translation method of [11] and [13] rely on the same principle and the following result can be obtained without severe difficulties.

Theorem 7. *Let $\Omega \subset \mathbb{R}^N$ be a bounded domain. If $\partial\Omega$ is locally a continuous graph and $\alpha \geq 0$, Equations (2.1) and (2.11) admit at most one large solution.*

References

[1] C. Bandle and H. Leutwiler, *On a quasilinear elliptic equation and a Riemannian metric invariant under Möbius transformations.* Aequationes Math. **42** (1991), 166–181.

[2] C. Bandle and M. Marcus, *Large solutions of semilinear elliptic equations with "singular" coefficients.* in Optimization and Nonlinear Analysis, Pitman Res. Notes Math. Ser., **244**, Longman Sci. Tech., Harlow, (1992), 25–38.

[3] C. Bandle and M. Marcus, *Large solutions of semilinear elliptic equations: existence, uniqueness and asymptotic behavior.* J. Anal. Math. **58** (1992), 9–24.

[4] C. Bandle and M. Marcus, *Asymptotic behavior of solutions and their derivative for semilinear elliptic problems with blow-up on the boundary.* Ann. I.H.P., Analyse Non Linéaire **12** (1995), 155–171.

[5] Ph. Bénilan and H. Brezis, *Nonlinear problems related to the Thomas-Fermi equation.* J. Evol. Equ. **3** (2004), 673–770.

[6] L. Bieberbach, $\Delta u = e^u$ *und die automorphen Funktionen.* Math. Annalen **77** (1916), 173–212.

[7] P. Frank and R. von Mises, *Die Differential- und Integralgleichungen der Mechanik und Physik* **I**, Second Edit., Rosenberg, New York (1943).

[8] J.B. Keller, *On solutions of* $\Delta u = f(u)$. Comm. Pure Appl. Math. **10** (1957), 503–510.

[9] D. Labutin, *A Wiener regularity for large solutions of nonlinear equations.* Ark. Mat. **41** (2003), 307–339.

[10] C. Loewner and L. Nirenberg, *Partial differential equations invariant under conformal or projective transformations.* Contributions to Analysis, L. Ahlfors et al., eds. (1972), 245–272.

[11] M. Marcus and L. Véron, *Uniqueness and asymptotic behaviour of solutions with boundary blow-up for a class of nonlinear elliptic equations.* Ann. Inst. H. Poincaré **14** (1997), 237–274.

[12] M. Marcus and L. Véron, *The boundary trace of positive solutions of semilinear elliptic equations: the subcritical case.* Arch. Rat. Mech. Anal. **144** (1998), 201–231.

[13] M. Marcus and L. Véron, *Existence and uniqueness results for large solutions of general nonlinear elliptic equations.* J. Evol. Equ. **3** (2004), 637–652.

[14] M. Marcus and L. Véron, *Maximal solutions of semilinear elliptic equations with locally integrable forcing term,* Isr. J. Math., to appear.

[15] M. Marcus and L. Véron, *Boundary trace of positive solutions of nonlinear elliptic inequalities.* Ann. Sc. Norm. Sup. Pisa (5) Vol. **III** (2004), 481–531.

[16] R. Osserman, *On the inequality* $\Delta u \geq f(u)$. Pacific J. Math. **7** (1957), 1641–1647.

[17] H. Rademacher, *Einige besondere Probleme partieller Differentialgleichungen.* See [7, pp. 838–845].

[18] Ratto A., Rigoli M. and Véron L., *Scalar curvature and conformal deformation of hyperbolic space.* J. Funct. Anal. **121** (1994), 15–77.

[19] J.L. Vazquez and L. Véron, *Singularities of elliptic equations with an exponential nonlinearity.* Math. Ann. **269** (1984), 119–135.

[20] L. Véron, *Semilinear elliptic equations with uniform blow-up on the boundary.* J. Analyse Math. **59** (1992), 231–250.

Laurent Véron
Laboratoire de Mathématiques et Physique Théorique
CNRS UMR 6083, Faculté des Sciences, Parc de Grandmont
F-37200 Tours, France
e-mail: `veronl@univ-tours.fr`

Progress in Nonlinear Differential Equations
and Their Applications, Vol. 63, 465–470
© 2005 Birkhäuser Verlag Basel/Switzerland

Relaxation in Presence of Pointwise Gradient Constraints

Elvira Zappale

Let U be a set, and $G\colon U \to [-\infty, +\infty]$. At a general level, the relaxation process for G consists in the introduction on U of a suitable topology, say τ, and in the determination of the sequential τ-lower semicontinuous envelope of G, say $\overline{G}$, defined as the greatest sequentially τ-lower semicontinuous functional on U less than or equal to G. This is done in order to approach the minimization problem of G on U since, if suitable growth conditions on G hold, it turns out that $\overline{G}$ has minima on U, and

$$\min_U \overline{G} = \inf_U G.$$

Let Ω be a smooth bounded open subset of $\mathbb{R}^N$ and $g : \Omega \times \mathbb{R}^N \to [0, +\infty[$ be a Carathéodory function with a growth of order p, then the relaxation process for the integral functional

$$G\colon u \in W^{1,p}(\Omega) \mapsto \int_\Omega g(x, \nabla u)dx$$

has been developed by choosing τ as the weak-$W^{1,p}(\Omega)$ topology if $p \in [1, +\infty[$ or the weak*-$W^{1,\infty}(\Omega)$ one if $p = +\infty$.

With this choice, a by now classical result says that

$$\overline{G}(u) = \int_\Omega g^{**}(x, \nabla u)dx \text{ for every } u \in W^{1,p}(\Omega),$$

where g^{**} is the Carathéodory function defined for every $x \in \Omega$ as the convex envelope of $g(x, \cdot)$.

The growth conditions and the continuity properties of g ensure that smooth functions are sufficient to determine the relaxation process and the same formula as above holds if one replaces G with the functional

$$G_{C^1}\colon u \in W^{1,p}(\Omega) \mapsto \begin{cases} \int_\Omega g(x, \nabla u)dx & \text{if } u \in W^{1,p}(\Omega) \cap C^1(\Omega) \\ +\infty & \text{if } u \in W^{1,p}(\Omega) \setminus C^1(\Omega), \end{cases}$$

thus obtaining $\overline{G} = \overline{G_{C^1}}$.

In the 'standard' results the gradients of the admissible functions are allowed to lie in the whole of $\mathbb{R}^N$, but several situations in the applications, say in elastic-plastic torsion theory, in nonlinear elastomer modelling, and in optimal control problems for learning processes, lead to relaxation problems where the admissible configurations are subjected to pointwise constraints on the gradient.

The relaxation process in presence of pointwise gradient constraints can be approached by studying $\overline{G}$ when

$$G \colon u \in W^{1,p}(\Omega) \mapsto \int_\Omega \{g(x, \nabla u) + I_{C(x)}(\nabla u)\}dx,$$

where

$$C : x \in \Omega \to C(x) \in \mathcal{P}(\mathbb{R}^N) \tag{1}$$

is a multifunction whose values describe the constraint, and $\mathcal{P}(\mathbb{R}^N)$ denotes the set of all the subsets of $\mathbb{R}^N$.

Clearly $G(u)$ is finite if $u \in W^{1,p}(\Omega)$ and

$$\nabla u(x) \in C(x) \text{ for a.e. } x \in \Omega, \tag{2}$$

therefore such functions play a role in the definition of $\overline{G}$. On the other hand it is clear that general results as those described above are not available.

We want to provide a first approach to the study of relaxation processes for G in presence of pointwise gradient constraints, when C has a concrete dependence on x.

Take a Carathéodory function g as above, a multivalued constraint function C such that

$$C(x) \text{ is convex for a.e. } x \in \Omega \tag{3}$$

and

$$\{x \in \mathbb{R}^N : |z| \le r\} \subseteq C(x) \subseteq \{z \in \mathbb{R}^N : |z| \le R\} \tag{4}$$

for a.e. $x \in \Omega$ and some $0 < r \le R$. Then

$$\overline{G}(u) = \inf \left\{ \begin{array}{l} \liminf_{h \to +\infty} \int_\Omega g(x, \nabla u_h)dx : \{u_h\} \subseteq W^{1,\infty}(\Omega), \\ \nabla u_h(x) \in C(x) \text{ for a.e. } x \in \Omega, \\ u_h \to u \text{ in } L^\infty(\Omega) \end{array} \right\} \begin{array}{l} \text{for every} \\ u \in W^{1,\infty}(\Omega). \end{array} \tag{5}$$

A first remark that naturally arises is that, if no smoothness assumptions on C are assumed, then, due to the presence of constraint conditions, there can be zones in $\{(x, z) \in \Omega \times \mathbb{R}^N : z \in C(x)\}$ that can be accessed only by graphs of gradients of functions in $W^{1,\infty}(\Omega)$, and not by graphs of continuous gradients in Ω. This suggests that, a priori, in this case one cannot expect that an equality as $\overline{G} = \overline{G}_{C^1(\Omega)}$ holds, and that separate results must be established. Moreover, it also proposes the problem of the definition of the zones in $\{(x, z) \in \Omega \times \mathbb{R}^N : z \in C(x)\}$ that can be accessed by graphs of continuous gradients in Ω.

In [DAMZ] these zones are identified through the 'essential lower limit' of $C(x)$, namely

$$\widetilde{C}: x \in \Omega \to \operatorname*{ess\,lim\,inf}_{y \to x} C(y) = \bigcup_{I \in \mathcal{N}_\Omega(x)} \bigcup_{\substack{Z \subseteq I \\ \mathcal{L}^N(Z)=0}} \bigcap_{y \in I \setminus Z} C(y), \tag{6}$$

and the following relaxation results are proven.

Theorem 1. *Let $p = +\infty$ and Ω a bounded open set with Lipschitz boundary in $\mathbb{R}^N$. Assume that $g : \Omega \times \mathbb{R}^N \to [0, +\infty[$ is a Carathéodory function, and satisfies the following growth condition*

$$\forall r \geq 0, \exists a_r \in L^1(\Omega) \text{ such that } g(x,z) \leq a_r(x)$$

$$\text{for a.e. } x \in \Omega \text{ and every } z \in \mathbb{R}^N \text{ with } |z| \leq r.$$

Let C be a multifunction as in (1), verifying (3), (4) and such that

$$\{(x,z) \in \Omega \times \mathbb{R}^N : z \in C(x)\} \in \mathcal{L}(\Omega) \times \mathcal{B}(\mathbb{R}^N). \tag{7}$$

Then there exists $\overline{g} : \Omega \times \mathbb{R}^N \to [0, +\infty], \mathcal{L}(\Omega) \times \mathcal{B}(\mathbb{R}^N)$-measurable, with $\overline{g}(x, \cdot)$ convex and lower semicontinuous for a.e. $x \in \Omega$, such that

$$\overline{G}(u) = \int_\Omega \overline{g}(x, \nabla u)dx \text{ for every } u \in PC^1(\Omega).$$

$(PC^1(\Omega)$ denotes the set of piecewise C^1 functions on Ω.) Furthermore a more shrinking result can be given for $\overline{G}_{PC^1(\Omega)}$, where the last functional is defined as it follows

$$\overline{G}_{PC^1(\Omega)}(u) = \inf \left\{ \liminf_{h \to +\infty} \int_\Omega g(x, \nabla u_h)dx : \{u_h\} \subseteq PC^1(\Omega), \ \nabla u_h(x) \in C(x) \right\}$$

$$\text{for every } u \in W^{1,\infty}(\Omega).$$

Theorem 2. *Under the same set of assumptions of Theorem 1, the following representation holds*

$$\overline{G}_{PC^1(\Omega)}(u) = \int_\Omega (g(x, \cdot) + I_{\widetilde{C}(x)})^{**}(\nabla u)dx,$$

for every $u \in PC^1(\Omega)$.

Remark 3. *If, from one hand, the representation result for $\overline{G}$ does not specialize the function $\overline{g}$, on the other, by means of several examples, one can show that*

$$(g + I_C)^{**} \neq \overline{g} \neq (g + I_{\widetilde{C}})^{**}.$$

The previous results can be specialized in the case when the multifunction C is a sphere of varying radius.

Theorem 4. *Let Ω and g be as in Theorem 1 and assume that $\varphi : \Omega \to [0, +\infty[$ be the $\mathcal{L}(\Omega)$-measurable and such that*

$$r \leq \varphi(x) \text{ for a.e. } x \in \Omega$$

for some $r \leq R \in]0, +\infty[$. *Then*

$$\overline{G}_{PC^1(\Omega)}(u) = \int_{\Omega} (g(x, \cdot) + I_{\overline{B_{\mathrm{ess\,lim\,inf}_{y \to x}\, \varphi(y)}(0)}})^{**}(\nabla u)dx \qquad (8)$$

for every $u \in PC^1(\Omega)$.

By $\mathrm{ess\,lim\,inf}_{y \to x}\, \varphi(y)$ we mean the essential lower limit of the function φ, namely

$$\sup_{I \in \mathcal{N}_\Omega(x)} \ \sup_{Z \subseteq I, \mathcal{L}^N(Z)=0} \ \inf_{y \in Z \setminus I} \varphi(y) \ \text{for every } y \in \Omega. \qquad (9)$$

Clearly Theorem 1, 2 and 4 provide just a first approach to deal with relaxation with pointwise constraints and with this setting we are only able to give formulas on $PC^1(\Omega)$ and not on the set of Lipschitz functions.

In order to obtain more detailed representation formulas than that in Theorem 2 one could ask more regularity on the constraint set C.

Indeed, the following result can be obtained as a corollary of Theorem 5.3 in [DAZ].

Theorem 5. *Under the same set of assumptions of Theorems 1 and 2 and by requiring that the multifunction C verifies the following*

$$\mathrm{int}(C(x)) \subseteq \overline{\mathrm{ess\,lim\,inf}_{y \to x} C(y)} \ \text{for a.e. } x \in \Omega \qquad (10)$$

it results that

$$\overline{G}_{PC^1(\Omega)}(u) = \overline{G}(u) = \int_{\Omega} (g(x, \cdot) + I_{C(x)})^{**}(\nabla u)dx$$

for every $u \in PC^1(\Omega)$.

On the other hand one can read the previous representation formulas in the light of *inner semicontinuity* (cf. [RW] for a detailed analysis of this and related notions). To this end before stating the next result, recall that if $C : \Omega \to \mathcal{P}(\mathbb{R}^N)$ is a multifunction, the *inner limit* of C at x is defined by

$$\liminf_{y \to x} C(y) = \{z \in \mathbb{R}^N : \forall \{x_n\} \subseteq \Omega, x_n \to x, \exists z_n \to z, z_n \in C(x_n) \forall n \in \mathbb{N}\} \quad (11)$$

and a multifunction C is said to be inner semicontinuous at x if

$$C(x) \subseteq \liminf_{y \to x} C(y). \qquad (12)$$

Corollary 6. *Under the same set of assumptions of Theorems 1 and 2 and by requiring that the multifunction C is inner semicontinuous for a.e. $x \in \Omega$, the following representation results can be written*

$$\overline{G}_{PC^1(\Omega)}(u) = \overline{G}(u) = \int_{\Omega} (g(x, \cdot) + I_{C(x)})^{**}(\nabla u)dx$$

for every $u \in PC^1(\Omega)$.

Proof. The result can be easily obtained as a corollary of Theorem 5.6 in [DAZ].
$\square$

Remark 7. *It is worthwhile to observe that Corollary 6 can be seen also as a consequence of Theorem 5 since inner semicontinuity is a stronger condition than (10) (cf. Section 3 in [DAZ]).*

Assumption (10) or (12) still do not allow us to provide representation results on the whole space of Lipschitz functions.

Representation on Lipschitz functions can be achieved by assuming continuity type conditions on the dependence of C on the x variable, to this extent the 'essential upper limit' has been introduced

$$\operatorname*{ess\,lim\,sup}_{y \to x} C(y) = \bigcap_{I \in \mathcal{N}_\Omega(x)} \bigcap_{\substack{Z \subseteq I \\ \mathcal{L}^N(Z)=0}} \bigcup_{y \in I \setminus Z} C(y).$$

The following theorem holds (Corollary 5.8 in [DAZ]).

Theorem 8. *Under the same assumptions on g and C of Theorem 5, and the following hypothesis*

$$\operatorname{int}(\operatorname*{ess\,lim\,sup}_{y \to x} C(y)) \subseteq \overline{\operatorname*{ess\,lim\,inf}_{y \to x} C(y)} \ for \ a.e. \ x \in \Omega \tag{13}$$

the formula below has been proved

$$\overline{G}(u) = \overline{G_{PC^1(\Omega)}}(u) = \int_\Omega (g(x, \cdot) + I_{C(x)})^{**}(\nabla u) dx \tag{14}$$

for every Lipschitz function u.

Remark 9. *From one hand we observe that the crucial assumption in order to prove Theorem 8, can be read as a continuity assumption on the constraint set C, because*

$$\operatorname{int}(\operatorname*{ess\,lim\,sup}_{y \to x} C(y)) = \operatorname{int}(\operatorname*{ess\,lim\,inf}_{y \to x} C(y)) = \operatorname{int}(C(x))$$

for a.e. $x \in \Omega$, (see Remark 5.9 in [DAZ]), from another we have to say that continuity conditions on C can be given through the 'classical' notions of inner and outer limits of multifunctions. As for Corollary 6 one could prove formula (14) replacing (13) by the more shrinking hypotheses

$$C(x) \ is \ continuous \ for \ a.e. \ x \in \Omega$$

one we recall that a multifunction C is said to be continuous if

$$\operatorname*{lim\,sup}_{y \to x} C(y) \subseteq C(x) \subseteq \operatorname*{lim\,inf}_{y \to x} C(y),$$

where, following [RW] the outer limit of C at x is given by

$$\operatorname*{lim\,sup}_{y \to x} C(y) := \{z \in \mathbb{R}^N : \exists \{x_n\} \subseteq \Omega, x_n \to x \ and \ z_n \to z, z_n \in C(x_n) \forall n \in \mathbb{N}\}.$$

It has to be emphasized that Theorem 8 in [DAZ] also inhibits the non-identity phenomena among $\overline{g}$, $(g + I_C)^{**}$ and $(g + I_{\widetilde{C}})^{**}$, thus extending the by now classical formula $\int_\Omega (g + I_C)^{**}(x, \nabla u) dx$ obtained in the case of constant multifunctions C .

As a more readable case one can consider the case when the multifunction $C(x)$ is a sphere of the type $B_{\varphi(x)}(0)$ with continuous radius. In this case the following formula can be proved

$$\overline{G}(u) = \overline{G_{PC^1(\Omega)}}(u) = \int_{\Omega} (g(x,\cdot) + I_{\overline{B_{\varphi(x)}(0)}})^{**}(\nabla u)dx$$

for every Lipschitz function u.

Still, it remains the open problem of extending the formulae above on the whole of $W^{1,\infty}(\Omega)$ only under measurable pointwise gradient constraints.

Acknowledgments. The content of these notes summarizes more general results contained in [DAMZ] and [DAZ].

References

[Br]   H. BREZIS, Multiplicateur de Lagrange en Torsion Elasto-Plastique, *Arch. Rational Mech. Anal.*, **49**, (1973), 32–40.

[B]   G. BUTTAZZO, Semicontinuity, relaxation and integral representation in the calculus of variations, Pitman Research Notes in Mathematics Series, 207. Harlow: Longman Scientific & Technical; New York: John Wiley & Sons. (1989)

[CDA]   L. CARBONE, R. DE ARCANGELIS, Unbounded functionals in the calculus of variations. Representations, relaxation, and homogenization, Chapman & Hall/CRC Research Notes in Mathematics, **125** Boca Raton, Chapman & Hall/CRC. xiii, (2002)

[CV]   C. CASTAING, M. VALADIER, "Convex Analysis and Measurable Multifunctions", Lecture Notes in Math. **580**, Springer, Berlin (1977).

[DM]   G. DAL MASO, "An Introduction to Γ-Convergence", Progr. Nonlinear Differential Equations Appl. **8**, Birkhäuser, Boston (1993).

[DAMZ]   R. DE ARCANGELIS, S. MONSURRO', E. ZAPPALE, On the Relaxation and the Lavrentieff Phenomenon for Variational Integrals with Pointwise Measurable Gradient Constraints, Calc. Var. Partial Diff. Eq. **21**, n. 4 (2004), 357–400.

[DAZ]   R. DE ARCANGELIS, E. ZAPPALE, On the Relaxation of Variational Integrals with Pointwise Continuous- Type Gradient Constraints, Applied Mathematics and Optimization (2005), electronic.

[ET]   I. EKELAND, R. TEMAM, "Convex Analysis and Variational Problems", Stud. Math. Appl. **1**, North-Holland, Amsterdam (1976).

[RW]   R.T. ROCKAFELLAR, R.J-B. WETS, "Variational Analysis", Grundlehren Math. Wiss. **317**, Springer, Berlin (1998).

Elvira Zappale
DIIMA
Università degli Studi di Salerno
via Ponte Don Melillo
I-84084 Fisciano (SA), Italia
e-mail: `zappale@diima.unisa.it`